369 0246922

MicroRNAs in Cancer

MicroRNAs in Cancer

Editors

César López-Camarillo
Oncogenomics and Cancer Proteomics Laboratory
Autonomous University of Mexico City
Mexico City
México

Laurence A. Marchat
Multidisciplinary Program in Biomedicine
National School of Medicine and Homeopathy-IPN
Mexico City
Mexico

CRC Press
Taylor & Francis Group
Boca Raton   London   New York

CRC Press is an imprint of the
Taylor & Francis Group, an **informa** business

A SCIENCE PUBLISHERS BOOK

CRC Press
Taylor & Francis Group
6000 Broken Sound Parkway NW, Suite 300
Boca Raton, FL 33487-2742

© 2013 Copyright reserved
CRC Press is an imprint of Taylor & Francis Group, an Informa business

Cover photograph reproduced by kind courtesy of Dr. Lizeth A. Fuentes Mera, Molecular Biology Laboratory, General Hospital Manuel Gea Gonzalez, Mexico.

Library of Congress Cataloging-in-Publication Data

MicroRNAs in cancer / editors, César López-Camarillo,
          Laurence A. Marchat.
     p. ; cm.
  Includes bibliographical references and index.
  ISBN 978-1-4665-7676-6 (hardcover : alk. paper)
  I. López-Camarillo, César. II. Marchat, Laurence A.
  [DNLM: 1. MicroRNAs--physiology. 2. Genes, Tumor Suppressor.
          3. MicroRNAs--therapeutic use. 4. Neoplasms-
          -genetics. 5. Neoplasms--therapy. QU 58.7]

  572.88--dc23
                                                    2012041992

Visit the Taylor & Francis Web site at
http://www.taylorandfrancis.com

CRC Press Web site at
http://www.crcpress.com

Science Publishers Web site at
http://www.scipub.net

PREFACE

Cancer is a complex group of diseases characterized by the presence of cells with uncontrolled growth and high proliferation capability. It is the leading cause of death in most countries, affecting millions of people worldwide and representing a global health problem. Cancer has huge economic, social and psychological implications. Even till now cancer remains a major clinical challenge due to its frequently poor prognosis and limited treatment options in many cases. Thus there is an urgent need for the development of new molecular methods that will allow an early diagnosis and better treatment.

The recent development of "omics" technologies, such as proteomics and transcriptomics has opened new research areas for scientists working in cancer. Notably, the discovery of the relevance of microRNAs in the mechanisms leading to tumorigenesis has allowed a better understanding of the molecular events involved in carcinogenesis. MicroRNAs are a class of small non-coding single-stranded RNAs that play important roles in most biological processes by regulating gene expression through transcriptional repression and degradation of protein-coding messenger RNAs. MicroRNA biology is a cutting-edge topic in the basic and biomedical research fields. Dysregulation of microRNA expression levels emerges as an important mechanism that triggers the loss or gain of functions in cancer cells, acting as tumor suppressors and oncogenes.

This book is an invaluable collection of the recent findings and reviews of leading experts about the current understanding of roles of microRNAs in the development, progression, invasion and metastasis of the most common cancers. It also reviews their potential for applications as novel molecular markers for early diagnosis and prognosis, as well as for the identification of therapeutic targets. In addition, it provides discussions about their potential use in translational medicine. Topics include an integrated overview of microRNAs functions in a variety of neoplasias including breast, prostate, colorectal, cervix, gastric and lung cancer as well as melanoma, and medulloblastoma. In addition, this volume includes a comprehensive coverage of key high-throughput technologies that have been developed for clinically oriented microRNAs profiling. Perspectives of quantitative sciences including bioinformatics, system biology and mathematical modeling are also discussed. This book constitutes an essential reference for the graduates students, post-docs, researchers and for scientific and medical community by providing the up-to-date discoveries of an international team of renowned experts in the field about microRNAs and their important roles in cancer translational research.

CONTENTS

CHAPTER 1

BREAST CANCER MicroRNAs: SIGNALING NETWORKS AND CLINICAL APPLICATIONS

Gabriel Eades,[a] Yuan Yao[b] and Qun Zhou[c,]*

ABSTRACT

The abundant, non-coding regulatory molecules, microRNAs, are heavily implicated in the molecular pathobiology of breast cancer. In this review, we will discuss how microRNA networks contribute to key pathways and processes in breast cancer including hormone and growth factor signaling as well as drug resistance. We will also describe the connection between microRNAs, epigenetic pathways, and gene expression in breast cancer. Finally, we will explore recent work evaluating microRNAs as clinical tools and will discuss the potential for microRNAs to serve as prognostic biomarkers and therapeutic targets in breast cancer.

1. INTRODUCTION

Breast cancer is a heterogeneous disease consisting of several histological types and multiple molecular subtypes driven by genetic and epigenetic

Department of Biochemistry and Molecular Biology, University of Maryland School of Medicine, 108 N. Greene St. Baltimore, MD 21201, 410-706-1615.
[a]Email: geade001@umaryland.edu
[b]Email: yyao@som.umaryland.edu
[c]Email: qzhou@som.umaryland.edu
*Corresponding author

changes in gene expression (Buerger et al. 1999, Lakhani 1999, Simpson et al. 2005, Sotiriou and Pusztai 2009). Breast cancer is one of several hormone-related cancers. Estrogen, an important risk factor for breast cancer, is associated with initiation and progression of disease (Yager and Davidson 2006). Despite being highly curable with early detection and intervention, breast cancer is currently the second leading cause of cancer deaths among women (Siegel et al. 2011). The majority of breast cancer mortality results from drug resistance, disease progression, and development of metastases (Steeg 2006). As such, there is significant interest in identifying new molecular biomarkers and therapeutic targets that might prove to be clinically valuable for treatment or monitoring of disease. A newly discovered class of regulatory molecules, microRNAs (miRs), has been implicated in several diseases including cancer (Calin and Croce 2006a). Advances in understanding the molecular pathways underlying breast cancer have found that miRs play important roles in the development and progression of disease.

MiRs are known to regulate cellular renewal, differentiation and apoptosis (Bartel 2004). As such, it is unsurprising that they play important roles in tumorigenesis. Expression profiling has discovered dysregulation of miRs in tumor tissues when compared to normal tissue for nearly every type of cancer (Calin and Croce 2006b). MiR-encoding genes are subject to the same genomic hits that disrupt protein-coding genes. Cancer-related aberrant miR levels have been demonstrated to result from translocations (Sonoki et al. 2005), defects in processing, copy number variation (Zhang et al. 2006), and epigenetic silencing (Lujambio et al. 2007). Roughly 50% of microRNAs were shown to be located in cancer-associated regions of the genome (Calin et al. 2004). In addition, some miRs are contained within introns of protein coding oncogenes or tumor suppressor genes and are subject to the over-expression or silencing of their host genes (Westholm and Lai 2011). Furthermore, mutations in regulatory elements of miR encoding genes have been identified that disrupt expression levels of miRs (Wu et al. 2008). Although not related to miRs proper, mutations of target sites within mRNAs or over-expression of RNA binding proteins that may compete for access near miR target sites, all could lead to dysregulation of miR-regulated pathways in cancer. Currently, the miRBase database (release 18) predicts greater than 1500 human microRNAs (Kozomara and Griffiths-Jones 2011). Various estimates predict that anywhere from 30–60% of all protein coding genes may be regulated by miRs (Rajewsky 2006, Friedman et al. 2009). Adding to the difficulty of understanding these complex signaling networks is the fact that miRs can target multiple mRNAs and multiple miRs can target the same mRNA. Consequently, the outcome of dysregulation of an individual miR is made more complicated by the cell-specific expression of redundant and/or overlapping miRs as well as the presence or absence of various target mRNAs.

Here we will present the recent advances and new discoveries that are adding to our growing understanding of the roles miRs play in breast cancer biology. We will examine the unique miR profiles of breast tumors compared to normal breast tissue and study in detail the role of miRs in regulating estrogen signaling. Next, we will explore the epigenetic regulation of miRs and examine the reverse relationship, showing how miRs regulate epigenetic machinery and contribute to altered epigenetic activity in breast cancer. Finally, we will discuss the future prospects of miRs as biomarkers for early detection or monitoring of disease as well as the potential for miRs and miR regulated pathways to be therapeutic targets and will discuss any progress being made in animal models or clinical trials.

2. ALTERED microRNA EXPRESSION IN BREAST CANCER

Iorio and colleagues reported one of the first global profiles of miR expression in breast cancer. Using microarrays containing probes for 245 miRs, this group conducted miR profiling of 76 breast tumors and 10 samples of normal breast tissue. Using ANOVA analysis to compare miR expression of breast tumors and normal tissue controls, Iorio et al. identified 29 miRs with significantly altered expression in breast tumors. Among the most significantly dysregulated miRs were miR-145 and miR-125b, which were down-regulated, and miR-21 and miR-155, which were over-expressed in breast tumors (Iorio et al. 2005). Although many of these miRs are frequently dysregulated in breast cancer, it is now known that different breast tumors often possess quite unique miR expression profiles.

It is well established that breast cancer is not a uniform disease and is instead found in many histological types and molecular subtypes which may arise in different cells of origin and undergo malignant progression through acquiring different genetic lesions (Buerger et al. 1999, Lakhani 1999, Simpson et al. 2005, Sotiriou and Pusztai 2009). Sempere et al. made an important discovery by utilizing *in situ* hybridization (ISH) to examine miR dysregulation in breast tumors. They discovered that miRs demonstrate unique patterns of expression in different epithelial subpopulations of normal breast. Specifically, they found that miR-145 and miR-205 expression was restricted to myoepithelial populations, whereas miR-21, miR-141, let-7a, and miR-214 were restricted to luminal epithelial cell populations. Furthermore, they found that miR-145, miR-205, and let-7a expression was decreased or lost in matching tumor tissues whereas miR-21 was frequently over-expressed (Sempere et al. 2007). These findings suggest the possibility that different types of breast cancer may involve unique dysregulation of miRs.

2.1 MicroRNA Profiles of Different Molecular Subtypes of Breast Cancer

Distinct mRNA profiles characterize the different molecular subtypes of breast cancer (Sotiriou and Pusztai 2009). Blenkiron et al. were among the first to compare miR expression profiles of these different molecular subtypes. Using mRNA expression profiles, they first classified 51 breast tumors according to previously established molecular categories: luminal A (n=15), luminal B (n=9), basal-like (n=16), HER2+ (n=5), and normal-like (n=6). Next, they preformed miR expression profiling and found that the miRs were differentially expressed between different molecular subtypes. They observed altered expression of 38 miRs across some or all of the subtypes. Furthermore, when comparing luminal A (n=15) and basal-like tumors (n=16), the two largest tumor subtypes in their study, they were able to correctly classify tumors based solely on miR expression (Blenkiron et al. 2007). In a related study, Mackiewicz et al. performed miR expression profiling of cell models of luminal (MCF-7), HER2+ (SK-BR-3), and basal (MDA-MB-231; triple-negative) breast cancers, using the non-tumorigenic mammary epithelial cell line MCF-10A as a control for comparison. They observed altered expression of 21 miRs, including miR-200c and miR-221/222, in triple-negative/basal (MDA-MB-231) cells compared to control cells (MCF-10A). Furthermore, they found that 19 of these 21 miRs showed inverse expression between triple-negative cells (MDA-MB-231) and luminal or HER2+ cells (MCF-7 or SK-BR-3), suggesting that these 19 miRs might serve as a unique molecular profile of triple-negative breast cancer (Mackiewicz et al. 2011). Other studies examining miR profiles in triple-negative primary tumors have implicated the involvement of several of these same miRs (e.g. miR-221/222) (Radojicic et al. 2011). In another study, Foekens et al. examined altered miR expression in ERα+ lymph node negative (LNN) breast tumors. They found that unsupervised Pearson correlation grouped tumor samples into three clusters based on miR expression profiles. One of these clusters significantly overlapped with luminal B classification, offering further support that miRs are differentially expressed among the different subtypes of breast cancer while also suggesting that miR profiles might add information to enrich our current molecular classifications of breast cancer (Foekens et al. 2008).

Genetically engineered mouse (GEM) models have been used to examine the contributions of specific oncogenes and tumor suppressors to breast cancer development. GEM models have been found to generate tumors that recapitulate key features of human breast cancer (Vargo-Gogola and Rosen 2007, Borowsky 2011); some even generate tumors that closely resemble specific molecular subtypes of human breast cancer (Shoushtari et al. 2007). Zhu et al. conducted miR profiling of tumors generated from several GEM models of breast cancer: MMTV-driven -H-Ras (luminal B), -HER2, -c-Myc (luminal B), -PymT (luminal B), and -Wnt1 (basal-like) transgenics as well as C3/SV40 T/t-antigen transgenic mice, p53$^{fl/fl}$MMTV-cre transplant model, and BRCA1$^{fl/fl}$p53$^{+/-}$MMTV-cre mice (basal-like). They performed miR expression profiling on tumors from each of the different mouse models and conducted unsupervised hierarchical cluster analysis on the resulting miR expression profiles. They observed clustering of different tumor profiles based on tumor lineage; GEM models with basal phenotypes showed similar miR profiles and GEM models with luminal phenotypes showed similar miR profiles (Zhu et al. 2011a).

2.2 MicroRNA Profiles of Different Histological Types of Breast Cancer

MiRs are also differentially expressed among various histological types of breast cancer. Hannafon et al. examined miR expression in preinvasive Ductal Carcinoma *in situ* (DCIS) (n=8) and matched normal tissues (n=8). They identified altered expression of 29 miRs between DCIS and normal tissues including miR-125b, miR-182, and miR-183. Out of the 29 miRs, 15 were previously known to be dysregulated in invasive breast cancers. This suggests that DCIS may possess unique and common alterations in miR expression when compared to invasive carcinomas (Hannafon et al. 2011). There have also been comparisons of miR expression in unilateral and bilateral breast cancers. Iyevleva et al. examined expression of a panel of nine miRs (all previously associated with breast cancer) in bilateral (n=80) and unilateral cases of breast cancer (n=40). They found significant over-expression of miR-21, miR-10b, and miR-31 in bilateral breast cancers compared to unilateral breast cancers. They observed over-expression of these three miRs in cases of both synchronous and metachronous bilateral breast cancers (Iyevleva et al. 2011). Finally, altered miR expression has also been examined in lobular breast cancers. Giricz et al. identified altered expression of 85 miRs in lobular neoplasias compared to normal tissue. Among these, three miRs (miR-375, miR-182, and miR-183) were associated with lobular neoplasia progression and were up-regulated in invasive lobular carcinomas (ILC) compared to lobular carcinoma *in situ* (LCIS) (Giricz et al. 2012).

2.3 Altered microRNA Expression in the Tumor Microenvironment

In addition to cancer cells, tumors contain endothelial cells, immune cells, and activated stromal cells that together constitute the tumor microenvironment that plays important roles in promoting tumor growth (Whiteside 2008). Sempere et al. sought to investigate alterations in miR expression in breast tumors by using fluorescence based ISH and immunohistochemistry (IHC) to examine miR expression and protein markers within breast tumor tissues (Sempere et al. 2010). Among the miRs they examined by ISH were miR-155 and miR-21, both frequently over-expressed in breast tumors (Barbarotto et al. 2008). Intriguingly, Sempere et al. found that high levels of miR-155 in tumors were restricted to a subpopulation of infiltrating immune cells. Likewise, they found that over-expression of miR-21 was not found in cancer cells but in tumor-associated fibroblasts that expressed vimentin and smooth muscle actin protein (Sempere et al. 2010). This was later confirmed by another study probing miR-21 expression in invasive breast tumors. Again, high levels of miR-21 in breast tumors were observed not in cancer cells but in cancer-associated fibroblasts (Rask et al. 2011). These findings may explain variations in reported miR profiles that are performed by different approaches, some directly analyzing RNA in tumor tissue and others extracting RNA from isolated cancer cells. It is likely that miRs may contribute to tumorigenesis not only through altered expression in tumor cells but also through regulation of the tumor microenvironment.

3. MicroRNAs AND ESTROGEN SIGNALING PATHWAYS

Although estrogen is important in the normal growth, development, and homeostasis of mammary glands it is equally recognized that estrogen is involved in the development of breast cancer (Stingl 2011). In premenopausal women, estradiol (E2) is produced in the ovaries in granulosa cells, whereas in postmenopausal women, estradiol is produced by conversion of testosterone by aromatase enzymes in adipose tissues (Johnston and Dowsett 2003). Our primary understanding of estrogen signaling pathways comes from studying the involvement of Estrogen Receptor α (ERα) in controlling gene expression in response to E2. ERα is a hormone receptor, a subfamily of nuclear receptors. ERα protein is composed of two transactivation domains, a DNA binding domain, and a ligand-binding domain. The current understanding of ERα signaling involves E2 binding to ERα, subsequent dimerization of receptors, recruitment to and binding to Estrogen Response Element (ERE) sites on DNA as well as the subsequent recruitment of co-activator or co-repressor proteins and transcriptional machinery (all of this occurring cyclically) (Heldring et al. 2007).

Estrogen contributes to breast tumorigenesis by providing pro-growth signaling through estrogen receptor α (ERα) and via the damaging effects of genotoxic estrogen metabolites (Yager and Davidson 2006). ERα is a critical prognostic marker for breast cancer that is important in selecting therapy as well as predicting aggressiveness of disease. Nearly two-thirds of breast tumors are ERα positive where ERα serves as a direct therapeutic target for hormonal therapies (Jones and Buzdar 2004). Recently, miRs have been found to play a critical role in regulating estrogen signaling in breast cancer providing further evidence for the contribution of miRs to fundamental aspects of breast cancer biology.

3.1 MicroRNAs Regulate E2 Signaling

As previously mentioned, miRs target multiple mRNAs and a single mRNA may be subject to regulation by numerous miRs in cell-specific contexts. Several groups have examined the possibility that miRs may be important regulators of E2 signaling in breast cancer cells. As a result, there are reports of multiple miRs regulating E2 signaling via several independent mechanisms.

3.1.1 Direct targeting of the ERα mRNA 3'UTR

Among the first miRs reported to target ERα mRNA 3'UTR is miR-206. miR-206 is frequently dysregulated in breast cancers and has been found to inversely correlate with ERα expression with ERα negative tumors showing high levels of miR-206. Adams et al. reported that miR-206 targeted two sites in the ERα mRNA 3'UTR and found that ectopic over-expression of miR-206 in ERα+ MCF-7 cells reduced ERα mRNA and protein levels (Adams et al. 2007). miR-206 over-expression also resulted in growth arrest of MCF-7 cells by abrogating estrogen-induced growth signaling (Kondo et al. 2008). Next, miR-221/222 has also been shown to regulate ERα by targeting two sites in ERα mRNA 3'UTR. Like miR-206, miR-221/222 levels inversely correlate with ERα expression in breast cancer cells. Zhao et al. found that knockdown of miR-221/222 in MDA-MB-468 (ERα negative) cells partially restored ERα expression. Remarkably, miR-221/222 knockdown in MDA-MB-468 cells imparted Tamoxifen (a Selective Estrogen Receptor Modulator, SERM) sensitivity in this previously resistant (ERα negative) cell line. Likewise, miR-221/222 over-expression in MCF-7 and T47D cells (ERα+ Tamoxifen sensitive) decreased their sensitivity to Tamoxifen (Zhao et al. 2008). Although miR-221/222 negatively regulate ERα, it was demonstrated that miR-221/222 over-expression promoted proliferation of breast cancer cells, thereby suggesting that miR-221/222 may be involved in the growth and survival of ERα negative breast tumors (Di Leva et al. 2010). Yet another miR, miR-22, was also found to regulate ERα by directly targeting a conserved site in ERα mRNA 3'UTR (Pandey and Picard 2009). Expression profiling revealed that miR-22 levels inversely correlate with

ERα expression in breast cancer cells and patient samples (Xiong et al. 2010). Over-expression of miR-22 was found to reduce ERα protein levels in breast cancer cells (Pandey and Picard 2009). Interestingly, this study was not able to confirm earlier reports regarding miR-206 or miR-221/222 targeting, finding that neither reduced activity of a luciferase reporter containing ERα mRNA 3'UTR in their attempts (Pandey and Picard 2009). Later, Xiong et al. confirmed that miR-22 directly targeted ERα when testing miR expression libraries for their ability to regulate a luciferase reporter containing the 4.7kb ERα mRNA 3'UTR. Furthermore, they found that miR-22 over-expression suppressed growth of ERα+ breast cancer cells (Xiong et al. 2010).

There are recent reports of other miRs that target ERα mRNA 3'UTR. First, Leivonen et al. identified 21 miRs that down-regulate expression of ERα protein by using a protein lysate microarray to interrogate the impact of a miR expression library on ERα levels in breast cancer cells. Among the 21 candidate miRs identified, follow up studies found that miR-18a, miR-18b, miR-193b, and the previously mentioned miR-206 could directly target a luciferase reporter of ERα mRNA 3'UTR. Further characterization found that miR-18 expression was higher in ERα negative breast tumors compared to ERα+ tumors (Leivonen et al. 2009). Next, it was also reported that the let-7 family of tumor suppressor miRs could regulate ERα expression by directly targeting ERα mRNA 3'UTR. let-7a/b/i expression was found to inversely correlate with ERα expression and over-expression of let-7a/b/i was found to down-regulate ERα protein expression in breast cancer cells (Zhao et al. 2011).

3.1.2 Coding region targeting of ERα

MiRs can also negatively regulate mRNA through targeting amino acid coding regions, although this mechanism was found to be less efficient than targeting in the 3'UTR (Rigoutsos 2009). This discovery increased the potential for any transcript to be subject to regulation via miR targeting. MiR-145, which is frequently down-regulated in breast cancers, was found to target a site in the coding region of ERα mRNA. Ectopic over-expression of miR-145 resulted in decreased ERα protein levels in breast cancer cells (Spizzo et al. 2009).

3.1.3 Indirect regulation of ERα expression

MiRs can also indirectly regulate ERα levels. In one such study miR-27a was found to regulate ERα expression by targeting a negative regulator of ERα. Li et al. found that miR-27a directly targeted ZBTB10, a protein that represses Sp1, Sp3, and Sp4 expression. In MCF-7 breast cancer cells, basal expression of ERα was found to be dependent on Sp1 expression. Knockdown of miR-27a was found to inhibit activity of an estrogen response element (ERE) containing luciferase reporter (Li et al. 2010). Next, De Souza et al. found that miR-375 also indirectly regulates ERα expression. miR-375 was found to be over-expressed in ERα+ breast cancer cell lines via epigenetic mechanisms. Inhibiting miR-375 resulted in decreased ERα expression and decreased activity of a ERE containing luciferase reporter. Furthermore, knockdown of miR-375 inhibited proliferation of breast cancer cells. Exploring the mechanism for these changes, they found that miR-375 directly targeted Ras dexamethasone-induced 1 (RASD1) mRNA 3'UTR. They found that RASD1 over-expression inhibited MCF-7 breast cancer cell growth by inhibiting ERα expression (de Souza et al. 2010).

3.1.4 Targeting co-regulatory proteins

In addition to regulating ERα expression, miRs have been found to regulate other proteins important in controlling estrogen regulated gene expression. Nuclear receptor regulation of gene expression typically involves the recruitment of co-activators or co-repressors to regulatory sites of genes in conjunction with ligand bound receptor (Heldring et al. 2007). In addition to directly targeting ERα, miR-206 has also been shown to target the important ERα co-activator proteins steroid receptor coactivator-1 and 3 (SRC-1 & SRC-3), histone acetyltransferases that are known to be involved in ERα regulation of many estrogen target genes. As such, over-expression of miR-206 reportedly repressed estrogen signaling in breast cancer cells even upon ERα rescue (exogenous ERα lacking a 3'UTR) (Adams et al. 2009). Several reports indicate that SRC-3/Amplified in Breast Cancer 1 (AIB1), is also subject to regulation by other miRs in breast cancer. First, miR-17-5p was reported to directly target AIB1 mRNA in breast cancer cells, which express low levels of miR-17-5p (Hossain et al. 2006). Finally, miR-20a over-expression (bioinformatically predicted to target the AIB1 mRNA 3'UTR) was found to reduce AIB1 protein levels in breast cancer cells (Yu et al. 2008).

3.1.5 Targeting ERβ

Most of what is known about estrogen signaling involves ERα activation and the regulation of downstream pro-growth and proliferative gene expression. Less is known concerning function of ERβ, a suspected tumor suppressor that is shown to be anti-proliferative *in vitro* and *in vivo* (Thomas and Gustafsson 2011). A recent study found that miR-92, a suspected oncogenic miR in breast cancer, directly targeted ERβ1 in ERα+ MCF-7 breast cancer cells (Al-Nakhle et al. 2010). ERβ1 levels were inversely correlated with miR-92 expression in breast cancer cell lines and patient tissues. Furthermore, it was reported that E2 treatment resulted in increased expression of miR-92, linking E2 signaling with ERβ1 repression in breast cancer cells.

3.2 ERα Modulates microRNA Expression Revealing microRNA-mediated Feedback Loops

In a study of estrogen-induced mammary carcinogenesis in female ACI rats, expression profiling revealed alteration in the levels of 34 miRs after six weeks treatment with high levels of E2. It was found that dysregulation of miR expression was an early event in tumorigenesis occurring by six weeks of E2 treatment, whereas preneoplastic lesions became evident only after 12 weeks of treatment (Kovalchuk et al. 2007).

Studies in human breast cancer cell lines have also found that estrogen signaling alters miR expression. Recently, it was discovered that miR-221/222 and ERα (a miR-221/222 target) are involved in a negative regulatory loop in which ERα directly suppresses miR-221/222 transcription. Di Leva et al. produced compelling evidence that ERα directly suppresses miR-221/222 transcription. First, in ERα+ MCF-7 cells, knockdown of ERα resulted in increased expression of miR-221/222. Next, treatment of ERα negative MDA-MB-231 breast cancer cells with 5-Aza-2′deoxycytidine (5Aza-2′dC) demethylating agent led to the re-expression of ERα, and subsequently resulted in reduced expression of miR-221/222. Finally, using chromatin immunoprecipitation (ChIP) it was found that ERα binds directly to an ERE in the miR-221/222 promoter, which is then accompanied by recruitment of transcriptional co-repressors including nuclear receptor co-repressor (NCoR) and silencing mediator of retinoid and thyroid hormone receptors (SMRT) (Di Leva et al. 2010). Estrogen signaling also reportedly regulates expression of miR-206, which targets ERα mRNA as well as SRC-1/3 co-regulatory proteins in breast cancer cells. E2 treatment of MCF-7 breast cancer cells resulted in decreased miR-206 expression, whereas treatment with an ERβ selective agonist increased miR-206 expression (Adams et al. 2007). It has also been reported that E2 treatment of MCF-7 cells suppresses expression of miR-34b (Lee et al. 2011). Lee et al. found that miR-34b directly targets cyclin D1, an E2-induced gene. They found that over-expression of miR-34b inhibited ERα+ tumor growth *in vivo* (Lee et al. 2011).

3.2.1 Genome wide studies of E2-regulated microRNAs

In addition to exploring E2 regulation of individual miRs, some studies have examined the effect of E2 treatment on the global expression of miRs in breast cancer cells. Bhat-Nakshatri et al. found that E2 treatment of the MCF-7 cells (containing a bicistronic control vector) resulted in altered expression of 28 miRs (Bhat-Nakshatri et al. 2009). Among the 21 miRs found to be E2-inducible genes were miR-17-5p and the let-7 family, both previously implicated in the regulation of E2 signaling proteins ERα and AIB1 (Zhao et al. 2011, Hossain et al. 2006). Among the seven miRs found to be E2-repressible genes was miR-27a (Bhat-Nakshatri et al. 2009), previously found to indirectly regulate ERα (Li et al. 2010). Next, Maillot et al. conducted a similar study that produced quite different results. They found that E2 treatment of MCF-7 cells resulted in global down-regulation of miRs (Maillot et al. 2009). In this study, forced expression of several of the down-regulated miRs resulted in inhibition of E2-induced growth effects. The varying results of these two studies may be explained by cell line variations. Maillot et al. examined parental MCF-7 cells (Maillot et al. 2009) and Bhat-Nakshatri et al. examined MCF-7 cells containing a bicistronic control vector (Bhat-Nakshatri et al. 2009).

Xu et al. employed a unique approach to identify miRs regulated by ERα (Xu et al. 2011). Using a bioinformatics approach they analyzed published ChIP-chip data to examine predicted ERα binding sites in regulatory regions of miR genes. 59 miRs were identified that contained at least one ERα binding site in their regulatory regions. They examined a small selection of miRs (eight candidate miRs) from this group using qRT-PCR. All of the miRs they examined showed expression changes following E2 treatment in MCF-7 cells. Among these miRs was miR-145, previously found to target ERα (Spizzo et al. 2009). In this study, miR-145 was found to be down-regulated following E2 treatment (Xu et al. 2011).

3.2.2 ERα regulation of microRNA biogenesis

A study by Castellano et al. revealed added complexities regarding E2 regulation of miR expression. They found that E2 treatment of MCF-7 cells resulted in increased expression of polycistronic pri-miR-17-92 and its paralogue pri-mir-106a-363, while not resulting in an immediate increase in mature transcripts of the encoded miRs (miR-17/18a/19a/19b/19b-1/20a/20b/92-1) (Castellano et al. 2009). Notably, several of these mature transcripts, including miR-18a/19b/20b, target ERα mRNA 3′UTR and some, miR-20a/20b/17(5p), target co-activator AIB1 mRNA 3′UTR (Yu et al. 2008, Castellano et al. 2009, Chang et al. 2011). Castellano et al. discovered that processing of pri-miR-17-92 to pre-miR stem loops was not delayed or rate-limiting, however, processing by Dicer was delayed by at least 12 h. They were only able to detect increased expression of mature miRs from this cluster between 24 and 72 h after E2 treatment, suggesting that this provides negative feedback for attenuating E2 signaling (Castellano et al. 2009). Several other studies have also suggested that E2 signaling might impact miR biogenesis. Yamagata et al. found that E2-bound ERα could associate with Drosha complex and inhibit the processing of some pri-miRs into pre-miRs (Yamagata et al. 2009). Additionally, Cochrane et al. found that miR-221/222, which are negatively regulated by ERα, directly target Dicer1 mRNA 3'UTR, further linking ERα activity and miR biogenesis (Cochrane et al. 2010).

Cumulatively, these studies demonstrate that in addition to contributing to E2-induced growth and proliferation, many of the miRs regulated by E2 may be involved in feedback mechanisms responsible for attenuating or fine-tuning estrogen signaling (Fig. 1).

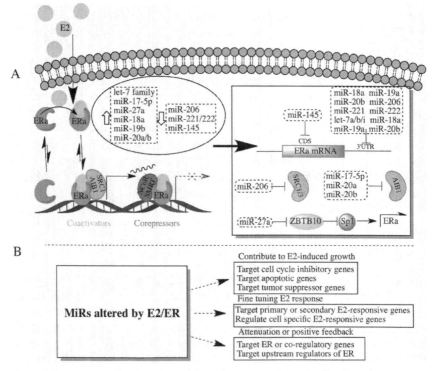

Figure 1. Estrogen-regulated miRs participate in autoregulatory feedback mechanisms. (A) E2 signaling modulates expression of numerous miRs that can directly or indirectly regulate ERα or co-regulatory proteins. E2 binding to ERα leads to ERα dimerization and localization to gene regulatory regions on DNA, followed by recruitment or dissociation of co-regulatory proteins. In response to E2, many miRs are up- or down-regulated that are involved in regulatory feedback loops through targeting ERα or SRC-1 and AIB1 co-activators. (B) In response to E2 signaling, miRs may regulate pathways that promote growth, fine tune the cellular response to E2, or provide positive or negative feedback.

Color image of this figure appears in the color plate section at the end of the book.

3.2.3 Xenoestrogen regulation of microRNA expression

Numerous studies have provided evidence for a link between exposure to xenoestrogens or synthetic estrogens and the development of breast cancer. Diethylstilbestrol (DES) exposure of pregnant women has been shown to increase risk to both mother and daughter for the development of breast cancer (Hoover et al. 2011). Hsu et al. found that DES treatment of normal breast progenitor cells (isolated and grown in mammosphere culture) resulted in altered expression of 82 miRs, including the down-regulation of miR-9-3. They found miR-9-3 was down-regulated by DES treatment via epigenetic mechanisms including altered histone methylation (H3K27me3 and H3K9me2) and DNA methylation (Hsu et al. 2009).

4. MicroRNAs AND GROWTH FACTOR SIGNALING PATHWAYS

Growth factor signaling through the ErbB family of receptor tyrosine kinases (RTK) plays important roles in mammary gland development as well as in the development of breast cancer (Stern 2003). ErbB receptors provide pro-survival signaling through downstream signaling cascades including the MAPK and PI3K/Akt pathways (Linggi and Carpenter 2006). The ErbB2/HER2 RTK is found over-expressed in 25–30% of breast tumors where HER2 over-expression is associated with aggressive disease, distant metastasis, and poor clinical outcome (Moasser 2007). The HER2 protein is a target for several drugs used for treating certain molecular types of breast cancer that are characterized by HER2 amplification (Arteaga et al. 2012). Like HER2, ErbB1/EGFR has also been implicated in breast cancer, specifically in difficult to treat triple-negative tumors, where EGFR may serve as a therapeutic target (Foley et al. 2010). Finally, it has been found that ErbB3/HER3 also plays an important role in breast cancer by forming heterodimers with ErbB2 and activating PI3K/Akt survival signaling in breast cancer cells (Holbro et al. 2003). Recently, a role for miRs in regulating the ErbB family in breast cancer has been recognized.

4.1 MicroRNAs Regulate the ErbB Family

Webster et al. identified EGFR mRNA as a direct target for miR-7 in breast cancer cells. Over-expression of miR-7 reduced EGFR protein levels and inhibited activation of downstream Akt signaling. They found that miR-7 further regulated EGFR signaling by also targeting downstream signaling protein Raf1 (Webster et al. 2009). MiRs also regulate other ErbB family members in breast cancer, including ErbB2 and ErbB3. miR-125a and miR-

125b are both down-regulated in HER2 over-expressing breast cancers (HER2+) (Mattie et al. 2006), and both directly target the 3'UTRs of HER2 and HER3 mRNAs (Scott et al. 2007). Over-expression of miR-125a/b was found to down-regulate HER2 and HER3 protein levels and decrease phosphorylation of the downstream signaling proteins Erk1/2 and Akt (Scott et al. 2007). miR-205 is also down-regulated in HER2+ breast cancers and was similarly found to target HER3 mRNA (Iorio et al. 2009).

4.2 Growth Factor Signaling Modulates microRNA Expression

Avrahaim et al. found that 23 miRs were rapidly down-regulated upon EGF stimulation of MCF-10A non-tumorigenic mammary epithelial cells. This group of miRs was also found down-regulated in breast cancers driven by EGFR or HER2 signaling (Avraham et al. 2010). Similarly, several reports indicate that HER2 signaling regulates the expression of miRs. Huang et al. discovered that miR-21 is up-regulated in HER2+ breast cancers where it targets programmed cell death 4 mRNA (PDCD4), a suppressor of metastasis. Furthermore, ectopic over-expression of HER2 was found to up-regulate miR-21 expression in non-HER2-amplified breast cancer cells (Huang et al. 2009). Next, Adachi et al. found that ectopic over-expression of HER2 in MCF-10A mammary epithelial cells resulted in down-regulation of miR-205, whereas knockdown of HER2 in HER2+ breast cancer cells resulted in up-regulation of miR-205 (Adachi et al. 2011). Intriguingly, a study using next-generation sequencing analysis of miR expression identified a new candidate miR, miR-4278, encoded within an intron of the ErbB2 gene that is over-expressed along with HER2 protein in HER2+ breast cancers (Persson et al. 2011).

5. MicroRNAs AND THE BREAST CANCER EPIGENOME

There has been extensive research into the contributions of epigenetics to the development of cancer (Esteller 2008, Baylin and Jones 2011). It is well established that breast tumors possess an altered epigenetic landscape where many tumor suppressor genes are often silenced via epigenetic inactivation (Baylin 2005). Multiple epigenetic mechanisms are involved in altering breast cancer gene expression including DNA methylation, histone methylation, and histone deacetylation (Jovanovic et al. 2010). Several drugs have been developed to target epigenetic machinery, including histone deacetylase inhibitors (HDACi) and DNA methyltransferase inhibitors (DNMTi), in an attempt to restore expression of silenced tumor suppressors and kill tumor cells (Baylin and Jones 2011, Tsai and Baylin 2011). There are numerous ongoing studies investigating the potential for epigenetic therapies to treat breast cancer.

5.1 Epigenetic Regulation of microRNAs

In breast cancer, miR expression is often dysregulated through epigenetic mechanisms. This is unsurprising, given that studies of normal breast tissues have revealed a major role for epigenetic regulation of miRs in controlling differentiation and cell fate. Vrba et al. examined epigenetic regulation of miRs in mammary epithelial cells (HMECs) and mammary fibroblasts (HMFs) (Vrba et al. 2011). They found that 10% of miRs expressed by either cell type were subject to unique, cell type-specific epigenetic regulation, mostly involving DNA methylation or trimethylation of histone 3 lysine 27 (H3K27me3) (Vrba et al. 2011).

5.1.1 DNA methylation and microRNA dysregulation

Among the miRs epigenetically regulated in breast cancer are the miR-200 family members: miR-200b/miR-200a/miR-429 and miR-200c/miR-141. Promoter hypermethylation and histone deacetylation were found to be associated with loss of miR-200 family expression in invasive breast cancer cells (Neves et al. 2010, Eades et al. 2011). It is interesting to note that the miR-200 family is also regulated via DNA methylation and histone modifications in normal breast tissue (Vrba et al. 2011). In addition to the miR-200 family, promoter hypermethylation has been found to be associated with frequent down-regulation of several prospective tumor-suppressor miRs in breast cancer including: miR-335 (suppressor of metastasis (Png et al. 2011)), miR-195/497 (targets Raf1 mRNA (Li et al. 2011)), miR-125b (targets proto-oncogene ETS1 (Zhang et al. 2011a)), miR-34a (targets E2F3 and BCL2 (Lodygin et al. 2008)), and miR-34b/c (targets Notch4 (Yu et al. 2012)). Genome-wide approaches have revealed new insights into the epigenetic regulation of miR expression in breast cancer. Lehmann et al. conducted a genome-wide study of DNA methylation in breast cancer cell lines and control tissues using combined bisulphite restriction analysis (COBRA) and bisulphite sequencing. Their results revealed that regulatory regions of miR-9-1, miR-124a3, miR-148, miR-152, and miR-663 were hypermethylated in breast cancer cells when compared to normal tissues (Lehmann et al. 2008). Epigenetic therapies have also been useful tools in examining global epigenetic regulation of miRs in breast cancer. Treatment of several breast cancer cell lines with 5Aza-2'dC (a DNMT inhibitor) resulted in significant alterations in miR expression (Radpour et al. 2011).

5.1.2 Histone methylation and microRNA dysregulation

Aberrant histone methylation is also frequently involved in alterations of miR expression in breast cancer. As previously mentioned, miR-375 is highly expressed in ERα+ breast cancer cells and targets RASD1, a negative regulator of ERα expression. Multiple epigenetic mechanisms were found to contribute to miR-375 over-expression in ERα+ breast cancer cells including promoter DNA hypomethylation and loss of the repressive chromatin mark histone 3 lysine 9 dimethylation (H3K9me2) at the miR-375 locus (de Souza et al. 2010). Aberrant histone methylation was also shown to regulate miR-9-3 expression in mammary epithelial progenitor cells treated with the xenoestrogen DES. Repressive chromatin marks H3K27me3 and H3K9me2 were found at the miR-9-3 locus in mammary epithelial progenitors after DES treatment (Hsu et al. 2009).

5.1.3 Histone deacetylation and microRNA dysregulation

Chang et al. discovered a novel role for tumor suppressor BRCA1 in breast cancer (Chang et al. 2011). They found that the R1699 variant of BRCA1 negatively regulates the oncomiR miR-155 in breast cancer by binding to a region in the miR-155 promoter and recruiting HDAC2, resulting in deacetylation of H2A and H3 histones (Chang et al. 2011). They found that BRCA1-deficient breast cancer cell lines (HCC1937 and MDA-MB-436) showed 50 to 100-fold higher expression of miR-155 compared to BRCA1 expressing breast cancer cell lines (MDA-MB-231 and MCF-7) (Chang et al. 2011), suggesting a role for epigenetic dysregulation of miR-155 in hereditary breast cancers.

5.2 MicroRNA Regulation of Epigenetic Machinery

MiRs have been found to regulate several critical epigenetic enzymes known to be involved in tumorigenesis. In lung cancer and acute myeloid leukemia, tumor suppressor miR-29b has been reported to directly target DNMT3a and DNMT3b and also indirectly target DNMT1 (Fabbri et al. 2007, Garzon et al. 2009). In prostate cancer it was reported that miR-449a directly targets HDAC1 mRNA (Noonan et al. 2009). Recently, several reports have identified miR regulation of epigenetic machinery in breast cancer, linking altered miR expression to altered epigenetic regulation and altered epigenetic landscapes of breast cancer.

5.2.1 MicroRNA regulation of polycomb group complexes

Polycomb group (PcG) genes are involved in transcriptional repression of gene expression through post-translational histone modifications carried out through two protein complexes. Polycomb repressor complex 2 (PRC2) di/tri-methylates H3K27 via EZH1/EZH2 histone methyltransferases and PRC1 recognizes H3K27me2/3 modifications and subsequently monoubiquitylates histone 2A lysine 119 (H2AK119ub) via RING1A and RING1B ubiquitin ligases (Richly et al. 2011). Although much is still unclear concerning the mechanism of PcG suppression of gene expression, it is well established that this is a critical regulator of gene expression in embryonic stem cells that is suspected to contribute to many cancers where members of PRC1/PRC2 have been found up-regulated (e.g. Bmi-1 and EZH2 are frequently over-expressed in breast cancers (Collett et al. 2006, Bachmann et al. 2006, Guo et al. 2011)).

Members of the miR-200 family of tumor suppressors have been reported to target multiple important epigenetic regulators. Shimono et al. reported that miR-200c directly targeted Bmi-1 mRNA 3'UTR (a member of PRC1) in breast cancer cells. Over-expression of miR-200c or knockdown of Bmi-1 was found to suppress self-renewal of breast CSCs (Shimono et al. 2009). Next, Iliopoulos et al. discovered that miR-200b directly targeted Suz12, a member of PRC2. Suz12 knockdown or miR-200b over-expression was found to reduce growth of breast CSCs (Iliopoulos et al. 2010). Finally, in transformed mammary epithelial cells, miR-200a has been found to target the 3'UTR of SIRT1, a class III histone deacetylase, member of PRC4, and a suspected oncogene in breast cancer (Eades et al. 2011). By targeting BMI1, Suz12, and SIRT1, the miR-200 family is a critical regulator of PcG complexes in breast cancer.

Other miRs have also been reported to target PcG genes in breast cancer. Derfoul et al. reported that miR-214 expression is inversely correlated with expression of EZH2 histone methyltransferase in breast cancer cells. miR-214 was found to directly target EZH2 mRNA 3'UTR and over-expression of miR-214 was shown to down-regulate EZH2 protein levels and reduce breast cancer cell growth (Derfoul et al. 2011). Similarly, Zhang et al. reported that miR-26a also targets the EZH2 mRNA 3'UTR. They found that miR-26a over-expression decreased EZH2 protein levels and inhibited breast tumor growth *in vivo* (Zhang et al. 2011b).

6. MicroRNAs AND BREAST CANCER DRUG RESISTANCE

Around 30% of patients with early stage breast cancer will have recurrent disease (Gonzalez-Angulo et al. 2007). For patients with locally invasive disease or metastatic disease, adjuvant systemic therapy is usually given in addition to surgery or radiation. Systemic therapy consists mainly of chemotherapy and for some patients, hormonal therapy or monoclonal antibodies. Although an estimated 90% of patients with primary breast cancer and 50% of patients with metastatic breast cancer generally show some initial response to chemotherapy, drug resistance either intrinsic or acquired is estimated to occur in as many as 50% of these patients which can often lead to therapeutic failure, and death (O'Driscoll and Clynes 2006). Gaining a better understanding of the molecular basis of drug resistance will identify new molecular targets and pathways that can be manipulated to prevent or overcome resistance and more effectively treat breast cancer. As is the case for several cancers, miRs have been found to play critical roles in breast cancer drug resistance (Table 1).

6.1 Mechanisms of Drug Resistance to Cytotoxic Chemotherapy

6.1.1 Defective apoptosis

In a study of Paclitaxel (Taxol) resistance in breast cancer, Zhou et al. identified up-regulation of miR-125b, miR-221/222, and miR-923 in several Taxol resistant breast cancer cell lines when compared to their parental (drug-sensitive) cell lines. Follow up studies revealed that miR-125b directly targeted Bak1 mRNA, a pro-apoptotic member of the Bcl-2 gene family. Knocking down miR-125b or rescuing Bak1 expression restored Taxol sensitivity in resistant cells (Zhou et al. 2010). By inhibiting the pro-apoptotic Bak1 protein, miR-125b may block Paclitaxel-induced apoptosis, leading to drug resistance.

6.1.2 Altered expression of tubulin isotopes

Expression of class III β-tubulin (TUBB3) has been associated with resistance to microtubule targeting drugs (Tommasi et al. 2007, Stengel et al. 2009). Cochrane et al. discovered that TUBB3 was a direct target of miR-200c (Cochrane et al. 2009), which is often silenced in invasive breast cancers (Neves et al. 2010). Furthermore, they found that miR-200c restoration resulted in increased sensitivity of breast cancer cells to Taxol (Cochrane et al. 2009). Loss of miR-200c expression may promote resistance to microtubule targeting agents through increasing TUBB3 expression.

Table 1. MicroRNAs Associated with Breast Cancer Drug Resistance.

miR	Known target	Drug resistance	(References)
⇑ miR-125b	Bak1	Paclitaxel	(Zhou et al. 2010)
⇓ miR-125b	HER2	Letrozole, Anastrozole, and Exemestane	(Masri et al. 2010)
⇑ miR-21	PTEN	Doxorubicin	(Wang et al. 2011)
⇓ miR-200c	TUBB3, MDR1	Paclitaxel, Doxorubicin	(Chen et al. 2008, Cochrane et al. 2009, Chen et al. 2011)
⇓ miR-451	MDR1, 14-3-3ζ	Doxorubicin, Tamoxifen	(Kovalchuk et al. 2008, Bergamaschi 2012)
⇓ miR-345	MDR1	Cisplatin	(Pogribny et al. 2010)
⇓ miR-7	MDR1	Cisplatin	(Pogribny et al. 2010)
⇑ miR-221/222	ERα, p27^{kip1}	Tamoxifen	(Manavalan et al. 2011)
⇑ miR-181	TIMP3	Tamoxifen	(Miller et al. 2008)
⇑ miR-128	TGF-βR1	Letrozole	(Masri et al. 2010)
⇓ miR-15a/16	Bcl-2	Tamoxifen	(Cittelly et al. 2010)

*Arrows indicate up- or down-regulation of miR expression in resistant breast cancers.

6.1.3 Loss of tumor suppressor PTEN

Previously, studies have shown that loss of PTEN function can lead to Doxorubicin resistance (Steelman et al. 2008). Wang et al. found that miR-21 up-regulation was associated with Doxorubicin resistance in breast cancer cells. Comparisons of gene expression in parental MCF-7 cells and Doxorubicin resistant cells (MCF-7/ADR) revealed up-regulation of miR-21 and down-regulation of tumor suppressor PTEN (a known miR-21 target (Meng et al. 2007)). Restoration of PTEN or knockdown of miR-21 returned Doxorubicin sensitivity to MCF-7/ADR cells (Wang et al. 2011).

6.1.4 MDR1/P-Glycoprotein

Kovalchuk et al. identified altered expression of 137 miRs as well as down-regulation of Dicer and Argonaute 2 in Doxorubicin resistant breast cancer cells. They found that miR-451, which was silenced in Doxorubicin resistant cells, directly targeted multidrug resistance 1 (MDR1) mRNA, which codes for an ATP-binding transporter that contributes to multidrug resistance by functioning as a cellular efflux pump for drugs and toxins (Kovalchuk et al. 2008). Similarly, Chen et al. examined miR expression in Doxorubicin resistant breast cancer cells and discovered dramatic down-regulation of miR-200c. Exogenous over-expression of miR-200c restored Doxorubicin sensitivity and also down-regulated MDR1/PgP expression. Furthermore, they found that miR-200c was down-regulated in tumors from breast cancer patients who were non-responders to neoadjuvant chemotherapy when compared to tumors of patients who showed a positive response (Chen et al. 2011).

Studies have also examined the role of miRs in acquired resistance of breast cancer cells to Cisplatin. Expression profiling of MCF-7 parental cells and a Cisplatin-resistant variant revealed altered expression of 103 miRs. The most significantly altered miRs included miR-146a, miR-10a, miR-221/222, miR-345, and miR-200b/c. MiR-345 and miR-7, both down-regulated in Cisplatin resistant cells, were found to directly target MDR1 mRNA (Pogribny et al. 2010).

6.2 Mechanisms of Drug Resistance to Endocrine Therapy

Drugs targeting ERα or the production of estrogens are often given to women with early stage breast cancers, to prevent recurrence, as well as patients with ERα+ metastatic disease (Burstein and Griggs 2010, Osborne and Schiff 2011). For patients with ERα+ metastatic disease, hormonal therapy is among the most effective therapies, however, due to intrinsic and acquired resistance, only 30% of patients show tumor regression and 20% of patients show stable disease (Osborne and Schiff 2011). Multiple studies have implicated miRs in intrinsic and acquired resistance to hormone therapies (Table 1 & Fig. 2).

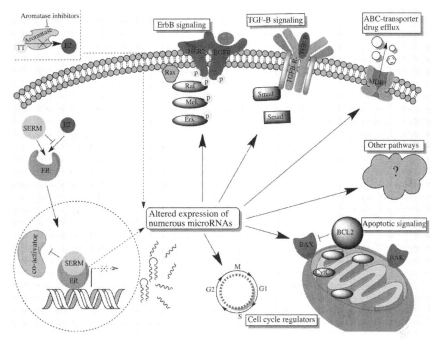

Figure 2. Dysregulation of miRs is associated with endocrine resistance. Selective estrogen receptor modulators (SERM) such as Tamoxifen can block E2/ERα interactions whereas aromatase inhibitors can block the conversion of testosterone (TT) to E2 in adipose tissue. Treatment with these drugs leads to altered miR expression, which may result in hormone resistance. In hormone resistant breast cancers, miRs were found to regulate several pathways known to be associated with drug resistance including growth factor/HER2 pathways, TGF-β growth inhibitory pathways, cell cycle inhibitory pathways, and intrinsic apoptosis pathways, among others.

Color image of this figure appears in the color plate section at the end of the book.

6.2.1 Loss of ERα

Several reports implicate miRs in Tamoxifen resistance (a selective estrogen receptor modulator; SERM). Manavalan et al. compared miR expression of Tamoxifen sensitive MCF-7 cells and Tamoxifen resistant LY2 breast cancer cells (LY2 were derived from serial passage of MCF-7 in the presence of a Raloxifene precursor). They identified altered expression of 97 miRs in LY2 cells. Among the dysregulated miRs, miR-181 and miR-222 were found to be over-expressed in Tamoxifen resistant LY2 cells (Manavalan et al. 2011).

As previously discussed, miR-221/222 are known to target ERα mRNA 3'UTR (Zhao et al. 2008). Manavalan et al. found that miR-221/222 over-expression imparted Tamoxifen resistance in MCF-7 cells whereas miR-221/222 knockdown sensitized MDA-MB-468 (Tamoxifen resistant) cells to Tamoxifen (Manavalan et al. 2011). It is well established that hormone-resistant tumors may lose ERα expression (Osborne and Schiff 2011), as such miR-221/222 over-expression may be involved in the transition of tumors away from estrogen dependence.

6.2.2 Cell-cycle inhibitors

Miller et al. examined miR expression in MCF-7 parental cells and a Tamoxifen resistant MCF-7 variant. They found altered expression of 15 miRs in Tamoxifen resistant breast cancer cells. Among the miRs altered in resistant cells, miR-181b and miR-221/222 were again found to be up-regulated. Furthermore, they found that miR-221/222 were over-expressed in HER2+ breast tumors compared to HER2 negative tumors. In addition to ERα they identified a new target of miR-221/222, the cell-cycle inhibitor p27^{kip1}, which was down-regulated in Tamoxifen resistant cells. Restoration of p27^{kip1} was found to return Tamoxifen sensitivity to resistant breast cancer cells (Miller et al. 2008).

6.2.3 Matrix metalloprotease inhibitors

Recently, Lu et al. reported that miR-181 and miR-221/222 share a common target, a metalloprotease inhibitor TIMP3 mRNA. TIMP3 was found to be down-regulated in Tamoxifen resistant breast cancer cells and TIMP3 restoration was found to inhibit growth of Tamoxifen resistant breast cancer cells. Furthermore, TIMP3 knockdown in MCF-7 parental cells reduced their sensitivity to Tamoxifen. It was demonstrated that treatment with either anti-miR-222 or anti-miR-181b, in combination with Tamoxifen, suppressed growth of Tamoxifen-resistant xenograft tumors in nude mice (Lu et al. 2011).

6.2.4 Defective apoptosis

Bergamaschi et al. have found that 14-3-3ζ protein is up-regulated in breast cancer cells following Tamoxifen treatment. 14-3-3 isoforms, including 14-3-3ζ, are known to sequester and inhibit the pro-apoptotic gene BAD (Masters et al. 2001, Hekman et al. 2006). Bergamaschi et al. found that 14-

3-3ζ is up-regulated via down-regulation of miR-451, which was shown to directly target 14-3-3ζ mRNA (Bergamaschi and Katzenellenbogen 2012). Previously, high levels of 14-3-3ζ in breast tumors were found to correlate with Tamoxifen treatment failure and decreased disease-free survival (Frasor et al. 2006).

6.2.5 Loss of TGF-β growth inhibition

Several studies have shown that miRs are involved in resistance to other endocrine therapies besides Tamoxifen. First, Masri et al. found that miR-128a was up-regulated in Letrozole resistant breast cancer cells (resistant MCF-7 variant stably expressing aromatase). MiR-128a was found to target TGF-βR1 and reduce sensitivity of Letrozole resistant breast cancer cells to TGF-β growth inhibitory signaling (Masri et al. 2010). Loss of TGF-β growth inhibitory effects has been previously associated with antiestrogen resistance (Buck and Knabbe 2006). Next, Rao et al. found that Fulvestrant treatment resulted in miR-221/222 up-regulation. They found that miR-221/222 were over-expressed in Fulvestrant resistant MCF-7 cells (MCF-7-F) where knockdown of miR-221/222 inhibited proliferation. Next, they examined global gene expression of MCF-7-F cells following miR-221/222 knockdown and found an enrichment of target genes within the TGF-β pathway. Furthermore, they found that miR-221/222 knockdown in MCF-7 cells increased sensitivity to TGF-β1-induced growth inhibition (Rao et al. 2011).

6.2.6 HER2 signaling

HER2 provides survival signals even in the presence of therapies inhibiting E2/ERα and HER2 signaling has been associated with breast cancer endocrine resistance (Osborne and Schiff 2011). In particular, HER2 signaling has been associated with resistance to aromatase inhibitors (AI) (Shin et al. 2006, Sabnis et al. 2009). Several studies suggest that miRs are involved in HER2-associated endocrine resistance. First, miR-125b (known to target HER2 (Mattie et al. 2006)) was found to be down-regulated in AI-resistant cell lines (Masri et al. 2010). Next, Cittelly et al. found that ectopic over-expression of a HER2 isoform, HERΔ16, in MCF-7 cells resulted in Tamoxifen resistance of subsequent MCF-7 xenografts (Cittelly et al. 2010). They observed up-regulation of the anti-apoptotic gene BCL-2 in MCF-7/HERΔ16 cells. Furthermore they found that HERΔ16 expression resulted in down-regulation of miR-15a and miR-16, both previously shown to target BCL-2 mRNA 3'UTR (Cimmino et al. 2005).

6.3 Other Roles for microRNAs in Breast Cancer Drug Resistance

Trastuzumab is a targeted therapy used to treat HER2+ breast cancers. Gong et al. found that miR-21 is up-regulated in multiple HER2+ breast cancer cell lines that acquired resistance through long-term exposure to Trastuzumab antibody. MiR-21 was previously shown to inhibit the tumor suppressor PTEN (Meng et al. 2007). Knocking down miR-21 was found to restore Trastuzumab sensitivity to these cell lines *in vitro* and *in vivo* (Gong et al. 2011).

One suspected culprit behind breast cancer multidrug resistance is breast cancer stem cells (CSCs) (Al-Ejeh et al. 2011). Multiple reports have found roles for miRs in regulating drug resistance of breast CSCs. First, Zhu et al. identified that miR-128 was reduced in breast tumor initiating cells (BT-IC) isolated from patient tissues and breast cancer cell lines. Furthermore, low miR-128 expression in patient tumor tissue was found to correlate with chemoresistance and poor survival. They found that over-expression of miR-128 sensitized breast CSCs to doxorubicin-induced apoptosis. The polycomb gene Bmi-1 and the multidrug transporter ABCC5 were found to be direct targets of miR-128 (Zhu et al. 2011b). Next, Iliopoulous et al. found that miR-200b down-regulation was critical to breast CSC growth and self-renewal. They found that breast cancer xenografts treated with Doxorubicin in combination with synthetic miR-200b prevented tumor relapse that arose in Doxorubicin-only treatments (Iliopoulos et al. 2010).

7. CLINICAL RELEVANCE OF MicroRNAs IN BREAST CANCER

Despite improvements in detection and therapeutic management, breast cancer remains the second leading cause of cancer deaths in women (Siegel et al. 2011). As such there still exists a clear need in the clinic for high quality biomarkers that will aid in patient evaluation and therapeutic selection. Many recent reports suggest that miRs may prove clinically valuable as prognostic biomarkers for breast cancer and may reveal new mechanisms for therapeutic intervention (Table 2).

7.1 Tumor microRNA Profiles as Prognostic Biomarkers

Defects in miR processing, including altered expression of Drosha or Dicer enzymes, is generally thought to contribute to miR dysregulation for some breast cancers. Dedes et al. examined the prognostic value of Drosha and Dicer expression in 245 breast cancer patients receiving

Table 2. MicroRNAs as Biomarkers for Breast Cancer Prognosis.

miR	Tumor	Prognosis	(References)
miR-210	ER+/ ER-/Triple -	shorter TDM, lower DRFS	(Foekens et al. 2008, Camps et al. 2008, Rothé et al. 2011, Buffa et al. 2011)
miR-7	ER+	shorter TDM	(Foekens et al. 2008)
miR-128a	ER+	shorter TDM, lower DRFS	(Foekens et al. 2008, Buffa et al. 2011)
miR-516-3p	ER+	shorter TDM	(Foekens et al. 2008)
miR-767/769-3p	ER+	lower DRFS	(Buffa et al. 2011)
miR-135a	ER+	higher DRFS	(Buffa et al. 2011)
miR-27b	ER-	lower DRFS	(Buffa et al. 2011)
miR-144	ER-	lower DRFS	(Buffa et al. 2011)
miR-342/140/30c	ER-	lower DRFS	(Buffa et al. 2011)
mir-21	serum		(Wang et al. 2010, Wu et al. 2011)
miR-155	serum		(Wang et al. 2010, Roth et al. 2010)
miR-195	whole blood		(Heneghan et al. 2010a, Heneghan et al. 2010b)

TDM: time to distant metastasis; DRFS: distant relapse-free survival

adjuvant chemotherapy. They identified Drosha down-regulation in 18% of the samples and Dicer down-regulation in 46% of the samples. Drosha down-regulation was found to associate with several prognostic indicators including high tumor grade, high Ki-67 expression, Bcl2 negative, HER2+, and TOPO2A amplification, whereas Dicer down-regulation was associated with ERα negative, PR negative, Bcl2 negative, high grade, high Ki-67, and triple-negative phenotypes (Dedes et al. 2011). Disruption in biogenesis and possible global down-regulation of miRs (except for miRs that undergo non-canonical processing (Miyoshi et al. 2010)) appears to correlate with several aggressive types of breast cancer i.e. HER2+ and triple-negative. Several studies have identified specific miR profiles that may possess prognostic value for breast cancer.

Foekens et al. performed an analysis of miR expression profiles of ERα+ (n=185) and ERα negative (n=114) tumors to explore associations of miR signatures and breast cancer aggressiveness. They found that in ERα+ breast tumors, four miRs (miR-7, miR-128a, miR-210, and miR-516-3p) were differentially expressed in patients with short *vs.* long time to distant metastasis (TDM). Additionally they found that miR-210 expression was associated with early relapse in ERα negative tumors. Furthermore, miR-210 levels were also associated with poor outcome in triple-negative (ERα- PR- HER2-) breast tumors (n=69) (Foekens et al. 2008). Next, Rothé et al. examined miR expression profiles of 56 breast tumors to probe for any associations of miR expression and clinical outcome. They found altered expression of 19 miRs associated with ERα status and 20 miRs associated with tumor grade. Expression levels of miR-148a and miR-210 were found to be significantly associated with relapse-free survival rates. Examination of a second cohort of 89 ERα+ patients that underwent five years Tamoxifen treatment revealed that high levels of miR-210 were significantly associated with poor clinical outcome in a multivariate analysis adjusted for both age and tumor size (Rothé et al. 2011). Similarly, Camps et al. examined miR-210 expression levels in 219 breast tumors and also found a significant inverse correlation with disease-free survival and overall survival in a Cox multivariate analysis (Camps et al. 2008).

Recently, Buffa et al. examined miR expression in 207 well annotated breast tumors with 10 year follow ups to search for independent prognostic indicators of distant relapse-free survival (DRFS). Using Cox regression analysis to examine miR expression and clinical covariates they found significant association of miR expression and DRFS independent of patient age, tumor size, tumor grade, ERα status, chemotherapy or Tamoxifen treatment status, HER2 signaling, hypoxia, stem cell, invasion, immune response, or apoptosis. In particular, high expression of miR-767-3p, miR-128a, and miR-769-3p were associated with poor prognosis in ERα+ cancers and high expression of miR-135a was associated with good prognosis. In ERα negative cancers, high expression of miR-27b, miR-144, and miR-210 was associated with poor prognosis, whereas high expression of miR-342, miR-140, and miR-30c were associated with good prognosis (Buffa et al. 2011).

Altered expression of miRs can also be detected in preneoplastic lesions. Sempere et al. examined miR expression in breast tumors by *in situ* hybridization. They identified down-regulation of miR-145 in samples of atypical hyperplasia and ductal carcinoma *in situ* (DCIS) (Sempere et al. 2007). It is possible that miR profiling of early stage breast cancers might someday predict disease recurrence and help inform clinical decisions.

7.2 Circulating microRNAs, Tools for Breast Cancer Detection?

MicroRNAs (miRs) have been detected in peripheral circulation of breast cancer patients that distinguish tumor-bearing individuals from healthy individuals (Heneghan et al. 2010b). These miRs are demonstrably stable and may represent surrogates of tumor miR levels (Chen et al. 2008). Already, tumor miR profiles are known to be of prognostic value for clinicians in evaluation of metastasis of unknown primary origin and in classification of tumor subtypes (Rosenfeld et al. 2008, Lebanony et al. 2009, Bishop et al. 2010). Recent work evaluates the prognostic capacity of circulating miRs, a method that many hypothesize could be a useful tool in early cancer detection.

7.2.1 Circulatory microRNA profiles

Several studies have examined circulatory levels of candidate miRs, known to be dysregulated in breast tumors. Examining six candidate miRs, Wang et al. found over-expression of miR-21, miR-106a, and miR-155 and down-regulation of miR-126, miR-199a, and mir-335 in sera of breast cancer patients (n=68) compared to healthy controls (n=40). They discovered that serum levels of miR-21, miR-106a, and miR-155 increased with tumor grade. Furthermore, they found a significant correlation between expression of these six miRs in tumor tissue and serum ($R^2 = 0.853$), indicating that serum miRs might serve as surrogate markers for tumor miR expression (Wang et al. 2010). Similarly, Heneghan et al. examined a panel of seven candidate miRs (miR-10b, miR-21, miR-145, miR-155, miR-195, miR-16, and let-7a) in the whole blood of breast cancer patients (n=83) and healthy controls (n=44). They found significantly higher levels of miR-195 and let-7a in blood from breast cancer patients compared to healthy controls. Furthermore, circulatory levels of miR-195a and let-7a were significantly lower in post-operative blood samples following surgery (Heneghan et al. 2010b). Finally, Roth et al. examined the expression of four miRs (miR-10b, miR-34a, miR-141, and miR-155) in the serum of 89 breast cancer patients and 29 healthy controls. miR-155 was found to be over-expressed in serum from patients with primary breast tumors (M0) compared to healthy controls. In serum from patients with metastatic disease (M1), miR-34a and miR-10b were found to be over-expressed when compared to M0 patients' and control patients' sera (Roth et al. 2010). Utilizing a different approach, Wu et al. employed next generation sequencing technology to create a profile of candidate miRs (altered by 5-fold in breast tumor tissues) that were then used to screen patient and control serum. They found that miR-29a and miR-21 were significantly up-regulated in serum of breast cancer patients (n=20) compared to healthy controls (n=20) (Wu et al. 2011).

The only currently reported genome-wide examination of circulatory miRs in breast cancer was conducted by Zhao et al. They performed microarray profiling of miR expression in plasma of breast cancer patients (n=20) and healthy controls (n=20). They found 26 miRs to be differentially expressed in breast cancer patient plasma, thus greatly expanding the number of potential circulatory miR biomarkers (Zhao et al. 2010).

7.2.2 Early detection

Recently, Heneghan et al. examined seven miRs (miR-10b, miR-21, miR-145, miR-155, miR-195, miR-16, and let-7a) in whole blood from 163 cancer patients (including breast, prostate, colon, melanoma, and renal cancers) and 63 aged matched healthy controls. They found that miR-195 was over-expressed (on average 25-fold) in serum from breast cancer patients (n=83) but not in healthy controls (n=63) or other cancer patients (n=80). This shows that miR-195 may be a biomarker specific to breast cancers. Furthermore, they found that circulatory miR-195 levels were significantly higher in patients with noninvasive *in situ* breast cancers when compared to healthy controls (Heneghan et al. 2010a). This indicates that miR-195 may serve as a blood-based biomarker for early detection of breast cancer.

7.2.3 Potential problems

Several issues need to be addressed to clarify the value of circulatory miR profiling. First, proper controls need to be established for normalizing small RNA levels in plasma, serum, or whole blood samples. Several normalization controls often used in circulatory miR studies are miR-16 and U6 snRNA. Appaiah et al. found that U6 was over-expressed in serum of breast cancer patients (n=39) compared to age-matched controls (n=40) and they found that miR-16 showed notable variability between samples, raising questions as to the suitability of either as proper controls (Appaiah et al. 2011). It is also important to expand our knowledge on the basic biology behind circulatory miRs in normal and disease states. Several studies have offered support that circulatory miR profiles may serve as surrogate biomarkers for tumor miR profiles. However, Pigati et al. found evidence for selective export of miRs in normal mammary epithelial cells and breast cancer cells. They found that some miRs e.g. miR-451 and mir-1246, were selectively exported from breast cancer cells showing a higher expression extracellularly. In breast cancer cells, greater than 90% of miR-451 was secreted, whereas in normal mammary epithelial cells, miR-451 was preferentially retained (Pigati et al. 2010).

7.3 Genetic Variation and Breast Cancer Susceptibility

The clinical value of miRs extends beyond the prognostic value of utilizing tumor miR or circulatory miR profiles to evaluate disease. It has been found that miRs are also involved in germline variations that contribute to breast cancer susceptibility.

7.3.1 SNPs in microRNA genes

Duan et al. identified a single-nucleotide polymorphism (SNP) in miR-125a that blocked processing of pri-miR-125a to stem-loop pre-miR-125a (Duan et al. 2007). As previously mentioned, miR-125a has been found to directly target both HER2 and HER3 mRNAs (Scott et al. 2007). In a follow up study, Li et al. examined the frequency of this miR-125a SNP in breast cancer patients. They examined normal breast tissue DNA from 72 breast cancer patients and found that six of them (8.3%) were heterozygous for this minor allele (the same was true for their tumor DNA). Examination of

869 control patients failed to detect any other instances of this allele. Based on this, they concluded that this SNP in pri-miR-125a was likely a germline mutation that elevated breast cancer risk (Li et al. 2009).

7.3.2 SNPs in microRNA binding elements in target mRNAs

Nicoloso et al. examined several SNPs known to be associated with increased breast cancer risk and tested whether any of these SNPs might alter miR/mRNA interactions. They identified SNPs in transcribed regions of TGF-βR1 and XRCC1 that affect targeting sites of miR-187 and miR-138 respectively. With luciferase assays, they demonstrated that SNPs in these alleles completely blocked interaction with miR-187 and miR-138. Furthermore, they discovered that a SNP in the coding region of BRCA1 that is associated with breast cancer risk interferes with BRCA1 regulation by miR-638 (Nicoloso et al. 2010). Similarly, in a study examining miR regulation of the ryanodine receptor gene 3 (RYR3), a calcium channel, Zhang et al. found that a SNP in the RYR3 mRNA 3'UTR interfered with miR-367 targeting. Furthermore, they found that this SNP was associated with increased breast cancer risk and poor overall survival (Zhang et al. 2011c).

7.3.3 SNPs in microRNA processing enzymes

Germline variations in miR processing proteins have also been shown to contribute to breast cancer risk. Leaderer et al. found that rs11544382, an Exportin-5 (XPO5) missense SNP variant, was associated with increased breast cancer risk. Bioinformatic analysis predicted that rs11544382 would impact protein function (Leaderer et al. 2011). XPO5 is responsible for shuttling pre-miRs to the cytoplasm and is critically important in miR biogenesis (Kim 2005). This study raises the possibility that SNPs in other miR processing enzymes may also contribute to increased breast cancer risk.

7.4 MicroRNAs as Therapeutic Tools

Along with the discovery that miRs function as oncogenes and tumor suppressor genes came interest in developing therapeutic approaches to modulate miR expression and modify miR regulated pathways. With the aim of developing miR replacement therapies or miR inhibition therapies, multiple methods have been described for manipulating miR levels *in vitro* including synthetic precursor miRs, pri-miR constructs, 2'O-methyl/locked nucleic acid (2'OMe/LNA) oligonucleotide antagonists (antagomiRs), 'target-protecting' oligonucleotides, and tandem miR response element containing 'sponge' constructs. There has also been extensive investigation into gene therapy approaches for delivering miR-based therapy, including both non-viral (liposomes and nanoparticle based vectors) and viral delivery methods (adenovirus, lentivirus, and adeno-associated virus) (Wang and Wu 2009).

7.4.1 Animal models

Several studies have examined potential therapeutic applications of miR-based therapies in mouse models of breast cancer. First, Kim et al. investigated the effectiveness of miR-145 replacement therapy for treating breast cancer. miR-145 is a bona fide tumor suppressor miR down-regulated in many breast cancers that regulates genes important to proliferation, invasion, and stem cell phenotype (Iorio et al. 2005, Sachdeva et al. 2009, Xu et al. 2009, Sachdeva and Mo 2010). Kim et al. found that injection of adenoviral-delivered miR-145 (Ad-miR-145) into established orthotopic breast tumor xenografts (MDA-MB-231 cells) inhibited tumor growth in nude mice. They also found that Ad-miR-145 in combination with 5-fluorouracil (5-FU) was more effective at inhibiting tumor growth than either treatment individually (Kim et al. 2011). Next, the miR-200 family of tumor suppressor miRs have also been investigated as candidates for miR replacement therapies. Shimono et al. found that lentiviral delivery of miR-200c reduced growth of breast tumor xenografts in NOD/SCID mice (Shimono et al. 2009). Similarly, Iliopoulous et al. reported that pretreatment with miR-200b mimic reduced tumor growth of transformed mammary epithelial cell xenografts in nude mice. They found that combination treatment of Doxorubicin and miR-200b mimic prevented tumor relapse that was observed with Doxorubicin treatment alone (Iliopoulos et al. 2010). Next, Ma et al. examined the therapeutic potential of miR-10b inhibition for suppressing breast cancer metastasis. MiR-10b is a suspected metastamiR (metastasis regulating miR) that is associated with invasion and metastasis in breast cancer (Ma et al. 2007). Ma et al. examined the effect of miR-10b inhibition on a syngenic xenograft model of breast cancer involving orthotopically implanting 4T1 cells in mammary fat pads of immunocompetent BALB/c mice, resulting in large primary tumors and lung metastasis within four weeks. They found that tail vein injection of miR-10b antagomiR, beginning on day two, did not block primary tumor formation but strongly suppressed lung metastasis. They also examined the efficacy of applying a retroviral miR-10b sponge inhibitor prior to inoculation and found a 90% reduction in lung metastasis with no effect on primary tumor size. Moreover, they performed toxicity studies of mice treated with miR-10b antagomiR and found treatment to be well tolerated by the mice, concluding that miR-10b inhibition therapy might be an effective anti-metastasis therapy (Ma et al. 2010).

7.4.2 Clinical trials

There are not yet any clinical trials of miR-based therapies for treating breast cancer. Nonetheless, there are currently many clinical trials conducting miR profiling for several cancers in search of better biomarkers. In one of these studies, investigators will perform miR profiling of breast cancer patients before and after treatment, collecting tissue and blood samples for identification of biomarkers and potential novel therapeutic targets (City of Hope Medical Center 2012). Despite the lack of a study in cancer patients, there is however, an ongoing Phase II study using an LNA-based inhibitor of miR-122 (a liver specific miR) to treat Hepatitis C (Santaris Pharma A/S 2012). Before additional studies of miR-based therapy can begin, more work in pre-clinical animal models is needed to determine the ideal targets, the best delivery methods, and toxicity. Nevertheless, it is likely only a matter of time before the first miR-based cancer therapies reach early clinical trials.

8. CONCLUSIONS

It is evident that miRs play critical roles in the development and progression of breast cancers. One of the most heavily investigated components of breast cancer is the involvement of the hormone estrogen in breast tumorigenesis. Several miRs are dysregulated in breast cancer that directly modulate estrogen signaling through targeting ERα or through targeting co-regulatory molecules. In addition, estrogen signaling in breast cancer cells alters the expression of numerous miRs either by direct or indirect regulation via ERα. It is likely that miRs may contribute to feedback pathways that attenuate or fine-tune estrogen signaling via regulating ERα or downstream primary and secondary target genes. Dysregulation of miRs in mammary epithelium may lead to loss of inhibitory feedback on estrogen signaling leading to a proliferative phenotype and the development of preneoplastic lesions.

It is suspected that tyrosine kinase receptors including EGFR and HER2 may play important roles in the development of drug resistance to endocrine therapies by promoting growth and proliferation in the absence of ERα activity (Giuliano et al. 2011). Some miRs are dysregulated in ERα negative tumors that directly target EGFR, HER2, or downstream members of MAPK and PI3K/AKT signaling cascades. Dysregulation of miRs that regulate estrogen signaling and growth factor signaling may promote the development of endocrine resistance. Further investigation into the roles miRs in regulating estrogen signaling, ERα negative tumor growth, and therapeutic resistance may provide a better understanding of the malignant progression of breast cancer and identify important pathways and potential therapeutic targets for overcoming endocrine resistance.

In normal mammary epithelium, miRs are important regulators of cell fate and differentiation. Epigenetic regulation is often involved in controlling cell-type specific expression of miRs in normal breast tissue. In breast cancer, miRs are both subject to aberrant epigenetic regulation and involved in the altered regulation of epigenetic machinery including HDACs and HMTs. Currently, epigenetic therapies are being explored as potential breast cancer therapeutics. Since it is clear that treatment with epigenetic therapies results in restoration of many tumor suppressor miRs, miR modulation may serve as a potential mechanism by which epigenetic therapies kill tumor cells.

Many studies have examined miR profiles of breast tumors hoping to identify better biomarkers that may serve as prognostic indicators and help inform therapeutic selection. Expression profiling has revealed that different molecular subtypes of breast cancer express unique profiles of miRs. Yet it is equally clear that miR profiling results in clustering of some breast tumors in groups that do not overlap with molecular subtypes generated through mRNA profiling, meaning that miR profiles may add information to improve current molecular classifications of breast cancer. Recent reports have identified many miRs that provide valuable prognostic information for breast cancers such as predicting time to disease progression, risk of distant metastasis, and overall survival.

Concerning the use of circulatory miRs as biomarkers, a noteworthy goal would be the detection of early stage malignancy and if possible, premalignancy in the form of preneoplastic lesions. The possibility of curative resection upon the early identification of many of these lesions, places enormous emphasis on developing sensitive and specific biomarkers that are testable in a non-invasive manner. Circulatory miRs in plasma/ serum certainly fit these criteria, however, it has yet to be shown how clinic-ready these kinds of diagnostic tools are. Screening for cancer in healthy populations is a very complex issue given the great costs and sheer number of individuals needing to be screened in order to save lives. However, work will continue to evaluate circulatory miRs for early cancer detection and miR profiling may someday augment existing tests in high-risk cancer populations where cost-benefit analysis are more likely to support testing. But as mentioned above, miR profiling should find its way into the clinic as a prognostic tool much sooner.

Currently, many avenues of miR-based therapies are being examined *in vitro* and *in vivo*. Ideas ranging from tumor suppressor miR replacement therapy to oncomiR inhibition therapies are currently being explored. Current research into gene therapy is addressing how best to deliver nucleic acid to cells via viral or non-viral means. *In vivo* delivery of synthetic precursor miRs or antagomiR inhibitors has proven effective in suppressing tumor growth, and in combinatorial therapies, has proven capable of preventing or delaying development of drug resistance. Better understanding of the effectiveness of these therapies as well as their toxicity should come from ongoing studies in preclinical models. MiR-based prognostic tools are already making their way into the clinic where miR-based therapeutics may soon augment existing therapies and provide more options for tailoring therapies to individual patients' breast tumors.

9. SUMMARY POINTS

- MiRs are dysregulated in breast cancers compared to normal breast tissues and different molecular subtypes of breast cancer possess unique miR profiles.
- MiRs are important regulators of estrogen signaling in breast cancer where they are frequently found to function in regulatory feedback pathways.
- Aberrant epigenetic regulation of miRs is common in breast cancer where miRs also regulate critical epigenetic machinery.
- MiRs have been shown to contribute mechanistically to breast cancer resistance to both cytotoxic chemotherapy and hormone therapies.
- MiRs have the potential to serve as novel biomarkers and therapeutic targets for breast cancer.

ACKNOWLEDGEMENTS

The authors would like to acknowledge support from NCI T32 (G.E.), and from the Maryland Stem Cell Fund (Q.Z.), American Cancer Society (Q.Z.), FAMRI (Q.Z.), and NCI R01 CA163820 (Q.Z.).

ABBREVIATIONS

miRs	:	MicroRNAs
ERα	:	Estrogen Receptor alpha positive breast cancer
HER2	:	Human EGF (Epidermal Growth Factor) Receptor 2 over-expressing breast cancer
GEM	:	Genetically engineered mouse
MMTV	:	Mouse Mammary Tumor Virus
DCIS	:	Ductal Carcinoma *in situ*
ILC	:	Invasive Lobular Carcinoma

LCIS	:	Lobular Carcinoma *In Situ*
ISH	:	*In Situ* Hybridization
IHC	:	Immunohistochemistry
E2	:	Estrogen
3'UTR	:	3' untranslated region
CDS	:	Coding sequence
ERE	:	Estrogen Response Element
ChIP	:	Chromatin Immunoprecipitation
qRT-PCR	:	Quantitative Reverse Transcription Polymerase Chain Reaction
H3K27me2/3	:	Histone 3 Lysine 27 di/tri-methylation
HDAC	:	Histone deacetylase
DNMT	:	DNA Methyltransferase
HMT	:	Histone Methyltransferase
5'Aza-2'dC	:	5'Aza-2'-deoxycytidine
PcG	:	Polycomb Group
CSC	:	Cancer Stem Cell
SNP	:	Single Nucleotide Polymorphism
LNA	:	Locked nucleic acid

REFERENCES

Adachi, R., S. Horiuchi, Y. Sakurazawa, T. Hasegawa, K. Sato, and T. Sakamaki. 2011. ErbB2 down-regulates microRNA-205 in breast cancer. Biochem Biophys Res Commun. 411(4): 804–808.

Adams, B.D., H. Furneaux, and B.A. White. 2007. The micro-ribonucleic acid (miRNA) miR-206 targets the human estrogen receptor-α (ERα) and represses ERα messenger RNA and protein expression in breast cancer cell lines. Mol Endocrinol. 21(5): 1132–1147.

Adams, B.D., D.M. Cowee, and B.A. White. 2009. The role of miR-206 in the epidermal growth factor (EGF) induced repression of estrogen receptor-α (ERα) signaling and a luminal phenotype in MCF-7 breast cancer cells. Mol Endocrinol. 23(8): 1215–1230.

Al-Ejeh, F., C.E. Smart, B.J. Morrison, G. Chenevix-Trench, J.A. López, S.R. Lakhani et al. 2011. Breast cancer stem cells: Treatment resistance and therapeutic opportunities. Carcinogenesis. 32(5): 650–658.

Al-Nakhle, H., P.A. Burns, M. Cummings, A.M. Hanby, T.A. Hughes, S. Satheesha et al. 2010. Estrogen receptor β1 expression is regulated by miR-92 in breast cancer. Cancer Res. 70(11): 4778–4784.

Appaiah, H., C. Goswami, L. Mina, S. Badve, G. Sledge, Y. Liu et al. 2011. Persistent upregulation of U6:SNORD44 small RNA ratio in the serum of breast cancer patients. Breast Cancer Res. 13(5): R86.

Arteaga, C.L., M.X. Sliwkowski, C.K. Osborne, E.A. Perez, F. Puglisi, and L. Gianni. 2012. Treatment of HER2-positive breast cancer: Current status and future perspectives. Nat Rev Clin Oncol. 9(1): 16–32.

Avraham, R., A. Sas-Chen, O. Manor, I. Steinfeld, R. Shalgi, G. Tarcic et al. 2010. EGF decreases the abundance of MicroRNAs that restrain oncogenic transcription factors. Sci Signal. 3(124): ra43.

Bachmann, I.M., O.J. Halvorsen, K. Collett, I.M. Stefansson, O. Straume, S.A. Haukaas et al. 2006. EZH2 expression is associated with high proliferation rate and aggressive tumor subgroups in cutaneous melanoma and cancers of the endometrium, prostate, and breast. J Clin Oncol. 24(2): 268–273.

Barbarotto, E., T.D. Schmittgen, and G.A. Calin. 2008. MicroRNAs and cancer: Profile, profile, profile. Int J Cancer. 122(5): 969–977.

Bartel, D.P. 2004. MicroRNAs: Genomics, biogenesis, mechanism, and function. Cell. 116(2): 281–297.

Baylin, S.B. 2005. DNA methylation and gene silencing in cancer. Nat Clin Prac Oncol. 2: S4.

Baylin, S.B. and P.A. Jones. 2011. A decade of exploring the cancer epigenome—biological and translational implications. Nat Rev Cancer. 11(10): 726–734.

Bergamaschi, A. and B.S. Katzenellenbogen. 2012. Tamoxifen downregulation of miR-451 increases 14-3-3[zeta] and promotes breast cancer cell survival and endocrine resistance. Oncogene. 31(1): 39–47.

Bhat-Nakshatri, P., G. Wang, N.R. Collins, M.J. Thomson, T.R. Geistlinger, J.S. Carroll et al. 2009. Estradiol-regulated microRNAs control estradiol response in breast cancer cells. Nucleic Acids Res. 37(14): 4850–4861.

Bishop, J.A., H. Benjamin, H. Cholakh, A. Chajut, D.P. Clark, and W.H. Westra. 2010. Accurate classification of Non-Small cell lung carcinoma using a novel MicroRNA-based approach. Clin Cancer Res. 16(2): 610–619.

Blenkiron, C., L. Goldstein, N. Thorne, I. Spiteri, S. Chin, M. Dunning et al. 2007. MicroRNA expression profiling of human breast cancer identifies new markers of tumor subtype. Genome Biol. 8(10): R214.

Borowsky, A.D. 2011. Choosing a mouse model: Experimental biology in Context—The utility and limitations of mouse models of breast cancer. Cold Spring Harb Perspect Biol. 3(9): a009670.

Buck, M.B. and C. Knabbe. 2006. TGF-beta signaling in breast cancer. Ann N Y Acad Sci. 1089(1): 119–126.

Buerger, H., F. Otterbach, R. Simon, K. Schäfer, C. Poremba, R. Diallo et al. 1999. Different genetic pathways in the evolution of invasive breast cancer are associated with distinct morphological subtypes. J Pathol. 189(4): 521–526.

Buffa, F.M., C. Camps, L. Winchester, C.E. Snell, H.E. Gee, H. Sheldon et al. 2011. microRNA-associated progression pathways and potential therapeutic targets identified by integrated mRNA and microRNA expression profiling in breast cancer. Cancer Res. 71(17): 5635–5645.

Burstein, H.J. and J.J. Griggs. 2010. Adjuvant hormonal therapy for early-stage breast cancer. Surg Oncol Clin N Am. 19(3): 639–647.

Calin, G.A., C. Sevignani, C.D. Dumitru, T. Hyslop, E. Noch, S. Yendamuri et al. 2004. Human microRNA genes are frequently located at fragile sites and genomic regions involved in cancers. Proc Natl Acad Sci USA. 101(9): 2999–3004.

Calin, G.A. and C.M. Croce. 2006a. MicroRNA-cancer connection: The beginning of a new tale. Cancer Res. 66(15): 7390–7394.

Calin, G.A. and C.M. Croce. 2006b. MicroRNA signatures in human cancers. Nat Rev Cancer. 6(11): 857–866.

Camps, C., F.M. Buffa, S. Colella, J. Moore, C. Sotiriou, H. Sheldon et al. 2008. Hsa-miR-210 is induced by hypoxia and is an independent prognostic factor in breast cancer. Clin Cancer Res. 14(5): 1340–1348.

Castellano, L., G. Giamas, J. Jacob, R.C. Coombes, W. Lucchesi, P. Thiruchelvam et al. 2009. The estrogen receptor-α-induced microRNA signature regulates itself and its transcriptional response. Proc Natl Acad Sci USA. 106(37): 15732–15737.

Chang, S., R. Wang, K. Akagi, K. Kim, B.K. Martin, L. Cavallone et al. 2011. Tumor suppressor BRCA1 epigenetically controls oncogenic microRNA-155. Nat Med. 17(10): 1275–1282.

Chen, X., Y. Ba, L. Ma, X. Cai, Y. Yin, K. Wang et al. 2008. Characterization of microRNAs in serum: A novel class of biomarkers for diagnosis of cancer and other diseases. Cell Res. 18(10): 997–1006.

Chen, J., W. Tian, H. Cai, H. He, and Y. Deng. 2011. Down-regulation of microRNA-200c is associated with drug resistance in human breast cancer. Med Oncol. (in press).

Cimmino, A., G.A. Calin, M. Fabbri, M.V. Iorio, M. Ferracin, M. Shimizu et al. 2005. miR-15 and miR-16 induce apoptosis by targeting BCL2. Proc Natl Acad Sci USA. 102(39): 13944–13949.

Cittelly, D.M., P.M. Das, V.A. Salvo, J.P. Fonseca, M.E. Burow, and F.E. Jones. 2010. Oncogenic HER2Δ16 suppresses miR-15a/16 and deregulates BCL-2 to promote endocrine resistance of breast tumors. Carcinogenesis. 31(12): 2049–2057.

City of Hope Medical Center. MIRNA profiling of breast cancer in patients undergoing neoadjuvant or adjuvant treatment for locally advanced & inflammatory breast cancer. In: ClinicalTrials.gov [Internet]. Bethesda (MD): National Library of Medicine (US). 2000-[cited 2012 Jan 27]. Available from http://clinicaltrials.gov/ct2/show/NCT01231386 NLM Identifier: NCT01231386.

Cochrane, D.R., N.S. Spoelstra, E.N. Howe, S.K. Nordeen, and J.K. Richer. 2009. MicroRNA-200c mitigates invasiveness and restores sensitivity to microtubule-targeting chemotherapeutic agents. Mol Cancer Ther. 8(5): 1055–1066.

Cochrane, D.R., D.M. Cittelly, E.N. Howe, N.S. Spoelstra, E.L. McKinsey, K. LaPara et al. 2010. MicroRNAs link estrogen receptor alpha status and dicer levels in breast cancer. Horm Cancer. 1(6): 306–319.

Collett, K., G.E. Eide, J. Arnes, I.M. Stefansson, J. Eide, A. Braaten et al. 2006. Expression of enhancer of zeste homologue 2 is significantly associated with increased tumor cell proliferation and is a marker of aggressive breast cancer. Clin Cancer Res. 12(4): 1168–1174.

de Souza, R.S., A. Breiling, N. Gupta, M. Malekpour, M. Youns, R. Omranipour et al. 2010. Epigenetically deregulated microRNA-375 is involved in a positive feedback loop with estrogen receptor α in breast cancer cells. Cancer Res. 70(22): 9175–9184.

Dedes, K.J., R. Natrajan, M.B. Lambros, F.C. Geyer, M. Lopez-Garcia, K. Savage et al. 2011. Down-regulation of the miRNA master regulators drosha and dicer is associated with specific subgroups of breast cancer. Eur J Cancer. 47(1): 138–150.

Derfoul, A., A.H. Juan, M.J. Difilippantonio, N. Palanisamy, T. Ried, and V. Sartorelli. 2011. Decreased microRNA-214 levels in breast cancer cells coincides with increased cell proliferation, invasion and accumulation of the polycomb Ezh2 methyltransferase. Carcinogenesis. 32(11): 1607–1614.

Di Leva, G., P. Gasparini, C. Piovan, A. Ngankeu, M. Garofalo, C. Taccioli et al. 2010. MicroRNA cluster 221-222 and estrogen receptor α interactions in breast cancer. J Natl Cancer Inst. 102(10): 706–721.

Duan, R., C. Pak, and P. Jin. 2007. Single nucleotide polymorphism associated with mature miR-125a alters the processing of pri-miRNA. Hum Mol Genet. 16(9): 1124–1131.

Eades, G., Y. Yao, M. Yang, Y. Zhang, S. Chumsri, and Q. Zhou. 2011. miR-200a regulates SIRT1 expression and epithelial to mesenchymal transition (EMT)-like transformation in mammary epithelial cells. J Biol Chem. 286(29): 25992–26002.

Esteller, M. 2008. Epigenetics in cancer. N Engl J Med. 358(11): 1148–1159.

Fabbri, M., R. Garzon, A. Cimmino, Z. Liu, N. Zanesi, E. Callegari et al. 2007. MicroRNA-29 family reverts aberrant methylation in lung cancer by targeting DNA methyltransferases 3A and 3B. Proc Natl Acad Sci U S A. 104(40): 15805–15810.

Foekens, J.A., A.M. Sieuwerts, M. Smid, M.P. Look, V. de Weerd, A.W.M. Boersma et al. 2008. Four miRNAs associated with aggressiveness of lymph node-negative, estrogen receptor-positive human breast cancer. Proc Natl Acad Sci USA. 105(35): 13021–13026.

Foley, J., N.K. Nickerson, S. Nam, K.T. Allen, J.L. Gilmore, K.P. Nephew et al. 2010. EGFR signaling in breast cancer: Bad to the bone. Semin Cell Dev Biol. 21(9): 951–960.

Frasor, J., E.C. Chang, B. Komm, C. Lin, V.B. Vega, E.T. Liu et al. 2006. Gene expression preferentially regulated by tamoxifen in breast cancer cells and correlations with clinical outcome. Cancer Res. 66(14): 7334–7340.

Friedman, R.C., K.K. Farh, C.B. Burge, and D.P. Bartel. 2009. Most mammalian mRNAs are conserved targets of microRNAs. Genome Res. 19(1): 92–105.

Garzon, R., S. Liu, M. Fabbri, Z. Liu, C.E.A. Heaphy, E. Callegari et al. 2009. MicroRNA-29b induces global DNA hypomethylation and tumor suppressor gene reexpression in acute myeloid leukemia by targeting directly DNMT3A and 3B and indirectly DNMT1. Blood. 113(25): 6411–6418.

Giricz, O., P.A. Reynolds, A. Ramnauth, C. Liu, T. Wang, L. Stead et al. 2012. Hsa-miR-375 is differentially expressed during breast lobular neoplasia and promotes loss of mammary acinar polarity. J Pathol. 226(1): 108–119.

Giuliano, M., R. Schifp, C.K. Osborne, and M.V. Trivedi. 2011. Biological mechanisms and clinical implications of endocrine resistance in breast cancer. Breast. 20: S42–S49.

Gong, C., Y. Yao, Y. Wang, B. Liu, W. Wu, J. Chen et al. 2011. Up-regulation of miR-21 mediates resistance to trastuzumab therapy for breast cancer. J Biol Chem. 286(21): 19127–19137.

Gonzalez-Angulo, A.M., F. Morales-Vasquez, and G.N. Hortobagyi. 2007. Overview of resistance to systemic therapy in patients with breast cancer. Adv Exp Med Biol. 608: 1–22.

Guo, B., Y. Feng, R. Zhang, L. Xu, M. Li, H. Kung et al. 2011. Bmi-1 promotes invasion and metastasis, and its elevated expression is correlated with an advanced stage of breast cancer. Mol Cancer. 10(1): 10.

Hannafon, B., P. Sebastiani, I.M. de, J. Lu, and C. Rosenberg. 2011. Expression of microRNA and their gene targets are dysregulated in preinvasive breast cancer. Breast Cancer Res. 13(2): R24.

Hekman, M., S. Albert, A. Galmiche, U.E.E. Rennefahrt, J. Fueller, A. Fischer et al. 2006. Reversible membrane interaction of BAD requires two C-terminal lipid binding domains in conjunction with 14-3-3 protein binding. J Biol Chem. 281(25): 17321–17336.

Heldring, N., A. Pike, S. Andersson, J. Matthews, G. Cheng, J. Hartman et al. 2007. Estrogen receptors: How do they signal and what are their targets. Physiol Rev. 87(3): 905–931.

Heneghan, H.M., N. Miller, R. Kelly, J. Newell, and M.J. Kerin. 2010a. Systemic miRNA-195 differentiates breast cancer from other malignancies and is a potential biomarker for detecting noninvasive and early stage disease. Oncologist. 15(7): 673–682.

Heneghan, H.M., N. Miller, A.J. Lowery, K.J. Sweeney, J. Newell, and M.J. Kerin. 2010b. Circulating microRNAs as novel minimally invasive biomarkers for breast cancer. Ann Surg. 251(3): 499–505.

Holbro, T., R.R. Beerli, F. Maurer, M. Koziczak, C.F. Barbas, and N.E. Hynes. 2003. The ErbB2/ ErbB3 heterodimer functions as an oncogenic unit: ErbB2 requires ErbB3 to drive breast tumor cell proliferation. Proc Natl Acad Sci U S A. 100(15): 8933–8938.

Hoover, R.N., M. Hyer, R.M. Pfeiffer, E. Adam, B. Bond, A.L. Cheville et al. 2011. Adverse health outcomes in women exposed in utero to diethylstilbestrol. N Engl J Med. 365(14): 1304–1314.

Hossain, A., M.T. Kuo, and G.F. Saunders. 2006. Mir-17-5p regulates breast cancer cell proliferation by inhibiting translation of AIB1 mRNA. Mol Cell Biol. 26(21): 8191–8201.

Hsu, P., D.E. Deatherage, B.A.T. Rodriguez, S. Liyanarachchi, Y. Weng, T. Zuo et al. 2009. Xenoestrogen-induced epigenetic repression of microRNA-9-3 in breast epithelial cells. Cancer Res. 69(14): 5936–5945.

Huang, T., F. Wu, G.B. Loeb, R. Hsu, A. Heidersbach, A. Brincat et al. 2009. Up-regulation of miR-21 by HER2/neu signaling promotes cell invasion. J Biol Chem. 284(27): 18515–18524.

Iliopoulos, D., M. Lindahl-Allen, C. Polytarchou, H.A. Hirsch, P.N. Tsichlis, and K. Struhl. 2010. Loss of miR-200 inhibition of Suz12 leads to polycomb-mediated repression required for the formation and maintenance of cancer stem cells. Molecular Cell. 39(5): 761–772.

Iorio, M.V., M. Ferracin, C. Liu, A. Veronese, R. Spizzo, S. Sabbioni et al. 2005. MicroRNA gene expression deregulation in human breast cancer. Cancer Res. 65(16): 7065–7070.

Iorio, M.V., P. Casalini, C. Piovan, G. Di Leva, A. Merlo, T. Triulzi et al. 2009. microRNA-205 regulates HER3 in human breast cancer. Cancer Res. 69(6): 2195–2200.

Iyevleva, A., E. Kuligina, N. Mitiushkina, A. Togo, Y. Miki, and E. Imyanitov. 2011. High level of miR-21, miR-10b, and miR-31 expression in bilateral vs. unilateral breast carcinomas. Breast Cancer Res Treat. 131(3): 1049–1059.

Johnston, S.R.D. and M. Dowsett. 2003. Aromatase inhibitors for breast cancer: Lessons from the laboratory. Nat Rev Cancer. 3(11): 821–831.

Jones, K.L. and A.U. Buzdar. 2004. A review of adjuvant hormonal therapy in breast cancer. Endocr Relat Cancer. 11(3): 391–406.

Jovanovic, J., J.A. Ronneberg, J. Tost, and V. Kristensen. 2010. The epigenetics of breast cancer. Mol Oncol. 4(3): 242–254.

Kim, V.N. 2005. MicroRNA biogenesis: Coordinated cropping and dicing. Nat Rev Mol Cell Biol. 6(5): 376–385.

Kim, S., J. Oh, J. Shin, K. Lee, K.W. Sung, S.J. Nam et al. 2011. Development of microRNA-145 for therapeutic application in breast cancer. J Controlled Release. 155(3): 427–434.

Kondo, N., T. Toyama, H. Sugiura, Y. Fujii, and H. Yamashita. 2008. miR-206 expression is down-regulated in estrogen receptor α Positive human breast cancer. Cancer Res. 68(13): 5004–5008.

Kovalchuk, O., V.P. Tryndyak, B. Montgomery, A. Boyko, K. Kutanzi, F. Zemp et al. 2007. Estrogen-induced rat breast carcinogenesis is characterized by alterations in DNA methylation, histone modifications, and aberrant microRNA expression. Cell Cycle. 6(16): 2010–2018.

Kovalchuk, O., J. Filkowski, J. Meservy, Y. Ilnytskyy, V.P. Tryndyak, V.F. Chekhun et al. 2008. Involvement of microRNA-451 in resistance of the MCF-7 breast cancer cells to chemotherapeutic drug doxorubicin. Mol Cancer Ther. 7(7): 2152–2159.

Kozomara, A. and S. Griffiths-Jones. 2011. miRBase: Integrating microRNA annotation and deep-sequencing data. Nucleic Acids Res. 39: D152–D157.

Lakhani, S.R. 1999. The transition from hyperplasia to invasive carcinoma of the breast. J Pathol. 187(3): 272–278.

Leaderer, D., A.E. Hoffman, T. Zheng, A. Fu, J. Weidhaas, T. Paranjape et al. 2011. Genetic and epigenetic association studies suggest a role of microRNA biogenesis gene exportin-5 (XPO5) in breast tumorigenesis. Int J Mol Epidemiol Genet. 2(1): 9–18.

Lebanony, D., H. Benjamin, S. Gilad, M. Ezagouri, A. Dov, K. Ashkenazi et al. 2009. Diagnostic assay based on hsa-miR-205 expression distinguishes squamous from nonsquamous non small-cell lung carcinoma. J Clin Oncol. 27(12): 2030–2037.

Lee, Y., J. Lee, C. Ho, Q. Hong, S. Yu, C. Tzeng et al. 2011. miRNA-34b as a tumor suppressor in estrogen-dependent growth of breast cancer cells. Breast Cancer Res. 13(6): R116.

Lehmann, U., B. Hasemeier, M. Christgen, M. Müller, D. Römermann, F. Länger et al. 2008. Epigenetic inactivation of microRNA gene hsa-mir-9-1 in human breast cancer. J Pathol. 214(1): 17–24.

Leivonen, S., R. Makela, P. Ostling, P. Kohonen, S. Haapa-Paananen, K. Kleivi et al. 2009. Protein lysate microarray analysis to identify microRNAs regulating estrogen receptor signaling in breast cancer cell lines. Oncogene. 28(44): 3926–3936.

Li, W., R. Duan, F. Kooy, S.L. Sherman, W. Zhou, and P. Jin. 2009. Germline mutation of microRNA-125a is associated with breast cancer. J Med Genet. 46(5): 358–360.

Li, X., S. Mertens-Talcott, S. Zhang, K. Kim, J. Ball, and S. Safe. 2010. MicroRNA-27a indirectly regulates estrogen receptor α expression and hormone responsiveness in MCF-7 breast cancer cells. Endocrinology. 151(6): 2462–2473.

Li, D., Y. Zhao, C. Liu, X. Chen, Y. Qi, Y. Jiang et al. 2011. Analysis of MiR-195 and MiR-497 expression, regulation and role in breast cancer. Clin Cancer Res. 17(7): 1722–1730.

Linggi, B. and G. Carpenter. 2006. ErbB receptors: New insights on mechanisms and biology. Trends Cell Biol. 16(12): 649–656.

Lodygin, D., V. Tarasov, A. Epanchintsev, C. Berking, T. Knyazeva, H. Körner et al. 2008. Inactivation of miR-34a by aberrant CpG methylation in multiple types of cancer. Cell Cycle. 7(16): 2591–2600.

Lu, Y., S. Roy, G. Nuovo, B. Ramaswamy, T. Miller, C. Shapiro et al. 2011. Anti-microRNA-222 (anti-miR-222) and -181B suppress growth of tamoxifen-resistant xenografts in mouse by targeting TIMP3 protein and modulating mitogenic signal. J Biol Chem. 286(49): 42292–42302.

Lujambio, A., S. Ropero, E. Ballestar, M.F. Fraga, C. Cerrato, F. Setién et al. 2007. Genetic unmasking of an epigenetically silenced microRNA in human cancer cells. Cancer Res. 67(4): 1424–1429.

Ma, L., J. Teruya-Feldstein, and R.A. Weinberg. 2007. Tumour invasion and metastasis initiated by microRNA-10b in breast cancer. Nature. 449(7163): 682–688.

Ma, L., F. Reinhardt, E. Pan, J. Soutschek, B. Bhat, E.G. Marcusson et al. 2010. Therapeutic silencing of miR-10b inhibits metastasis in a mouse mammary tumor model. Nat Biotech. 28(4): 341–347.

Mackiewicz, M., K. Huppi, J. Pitt, T. Dorsey, S. Ambs, and N. Caplen. 2011. Identification of the receptor tyrosine kinase AXL in breast cancer as a target for the human miR-34a microRNA. Breast Cancer Res Treat. 130(2): 663–679.

Maillot, G., M. Lacroix-Triki, S. Pierredon, L. Gratadou, S. Schmidt, V. Bénès et al. 2009. Widespread estrogen-dependent repression of microRNAs involved in breast tumor cell growth. Cancer Res. 69(21): 8332–8340.

Manavalan, T.T., Y. Teng, S.N. Appana, S. Datta, T.S. Kalbfleisch, Y. Li et al. 2011. Differential expression of microRNA expression in tamoxifen-sensitive MCF-7 versus tamoxifen-resistant LY2 human breast cancer cells. Cancer Letters. 313(1): 26–43.

Masri, S., Z. Liu, S. Phung, E. Wang, Y. Yuan, and S. Chen. 2010. The role of microRNA-128a in regulating TGFbeta signaling in letrozole-resistant breast cancer cells. Breast Cancer Res Treat. 124(1): 89–99.

Masters, S.C., H. Yang, S.R. Datta, M.E. Greenberg, and H. Fu. 2001. 14-3-3 inhibits bad-induced cell death through interaction with serine-136. Mol Pharmacol. 60(6): 1325–1331.

Mattie, M., C. Benz, J. Bowers, K. Sensinger, L. Wong, G. Scott et al. 2006. Optimized high-throughput microRNA expression profiling provides novel biomarker assessment of clinical prostate and breast cancer biopsies. Mol Cancer. 5(1): 24.

Meng, F., R. Henson, H. Wehbe-Janek, K. Ghoshal, S.T. Jacob, and T. Patel. 2007. MicroRNA-21 regulates expression of the PTEN tumor suppressor gene in human hepatocellular cancer. Gastroenterology. 133(2): 647–658.

Miller, T.E., K. Ghoshal, B. Ramaswamy, S. Roy, J. Datta, C.L. Shapiro et al. 2008. MicroRNA-221/222 confers tamoxifen resistance in breast cancer by targeting p27Kip1. J Biol Chem. 283(44): 29897–29903.

Miyoshi, K., T. Miyoshi, and H. Siomi. 2010. Many ways to generate microRNA-like small RNAs: Non-canonical pathways for microRNA production. Mol Genet Genomics. 284(2): 95–103.

Moasser, M.M. 2007. The oncogene HER2: Its signaling and transforming functions and its role in human cancer pathogenesis. Oncogene. 26(45): 6469–6487.

Neves, R., C. Scheel, S. Weinhold, E. Honisch, K. Iwaniuk, H. Trompeter et al. 2010. Role of DNA methylation in miR-200c/141 cluster silencing in invasive breast cancer cells. BMC Res Notes. 3(1): 219.

Nicoloso, M.S., H. Sun, R. Spizzo, H. Kim, P. Wickramasinghe, M. Shimizu et al. 2010. Single-nucleotide polymorphisms inside MicroRNA target sites influence tumor susceptibility. Cancer Res. 70(7): 2789–2798.

Noonan, E.J., R.F. Place, D. Pookot, S. Basak, J.M. Whitson, H. Hirata et al. 2009. miR-449a targets HDAC-1 and induces growth arrest in prostate cancer. Oncogene. 28(14): 1714–1724.

O'Driscoll, L. and M. Clynes. 2006. Biomarkers and multiple drug resistance in breast cancer. Curr Cancer Drug Targets. 6(5): 365–384.

Osborne, C.K. and R. Schiff. 2011. Mechanisms of endocrine resistance in breast cancer. Annu Rev Med. 62(1): 233–247.

Pandey, D.P. and D. Picard. 2009. miR-22 inhibits estrogen signaling by directly targeting the estrogen receptor α mRNA. Mol Cell Biol. 29(13): 3783–3790.

Persson, H., A. Kvist, N. Rego, J. Staaf, J. Vallon-Christersson, L. Luts et al. 2011. Identification of new MicroRNAs in paired normal and tumor breast tissue suggests a dual role for the ERBB2/Her2 gene. Cancer Res. 71(1): 78–86.

Pigati, L., S.C.S. Yaddanapudi, R. Iyengar, D. Kim, S.A. Hearn, D. Danforth et al. 2010. Selective release of MicroRNA species from normal and malignant mammary epithelial cells. PLoS ONE. 5(10): e13515.

Png, K.J., M. Yoshida, X.H. Zhang, W. Shu, H. Lee, A. Rimner et al. 2011. MicroRNA-335 inhibits tumor reinitiation and is silenced through genetic and epigenetic mechanisms in human breast cancer. Genes Dev. 25(3): 226–231.

Pogribny, I.P., J.N. Filkowski, V.P. Tryndyak, A. Golubov, S.I. Shpyleva, and O. Kovalchuk. 2010. Alterations of microRNAs and their targets are associated with acquired resistance of MCF-7 breast cancer cells to cisplatin. Int J Cancer. 127(8): 1785–1794.

Radojicic, J., A. Zaravinos, T. Vrekoussis, M. Kafousi, D.A. Spandidos, and E.N. Stathopoulos. 2011. MicroRNA expression analysis in triple-negative (ER, PR and Her2/neu) breast cancer. Cell Cycle. 10(3): 507.

Radpour, R., Z. Barekati, C. Kohler, M.M. Schumacher, T. Grussenmeyer, P. Jenoe et al. 2011. Integrated epigenetics of human breast cancer: Synoptic investigation of targeted genes, MicroRNAs and proteins upon demethylation treatment. PLoS ONE. 6(11): e27355.

Rajewsky, N. 2006. microRNA target predictions in animals. Nat Genet. 38: S8–S13.

Rao, X., G. Di Leva, M. Li, F. Fang, C. Devlin, C. Hartman-Frey et al. 2011. MicroRNA-221/222 confers breast cancer fulvestrant resistance by regulating multiple signaling pathways. Oncogene. 30(9): 1082–1097.

Rask, L., E. Balslev, S. Jorgensen, J. Eriksen, H. Flyger, S. Moller et al. 2011. High expression of miR-21 in tumor stroma correlates with increased cancer cell proliferation in human breast cancer. APMIS. 119(10): 663–673.

Richly, H., L. Aloia, and L. Di Croce. 2011. Roles of the polycomb group proteins in stem cells and cancer. Cell Death and Dis. 2: e204.

Rigoutsos, I. 2009. New tricks for animal MicroRNAs: Targeting of amino acid coding regions at conserved and nonconserved sites. Cancer Res. 69(8): 3245–3248.

Rosenfeld, N., R. Aharonov, E. Meiri, S. Rosenwald, Y. Spector, M. Zepeniuk et al. 2008. MicroRNAs accurately identify cancer tissue origin. Nat Biotech. 26(4): 462–469.

Roth, C., B. Rack, V. Muller, W. Janni, K. Pantel, and H. Schwarzenbach. 2010. Circulating microRNAs as blood-based markers for patients with primary and metastatic breast cancer. Breast Cancer Res. 12(6): R90.

Rothé, F., M. Ignatiadis, C. Chaboteaux, B. Haibe-Kains, N. Kheddoumi, S. Majjaj et al. 2011. Global MicroRNA expression profiling identifies MiR-210 associated with tumor proliferation, invasion and poor clinical outcome in breast cancer. PLoS ONE. 6(6): e20980.

Sabnis, G., A. Schayowitz, O. Goloubeva, L. Macedo, and A. Brodie. 2009. Trastuzumab reverses letrozole resistance and amplifies the sensitivity of breast cancer cells to estrogen. Cancer Res. 69(4): 1416–1428.

Sachdeva, M., S. Zhu, F. Wu, H. Wu, V. Walia, S. Kumar et al. 2009. p53 represses c-myc through induction of the tumor suppressor miR-145. Proc Natl Acad Sci U S A. 106(9): 3207–3212.

Sachdeva, M. and Y. Mo. 2010. MicroRNA-145 suppresses cell invasion and metastasis by directly targeting mucin 1. Cancer Res. 70(1): 378–387.

Santaris Pharma A/S. Multiple ascending dose study of miravirsen in treatment-naïve chronic hepatitis C subjects. In: ClinicalTrials.gov [Internet]. Bethesda (MD): National Library of Medicine (US). 2000- [cited 2012 Jan 27]. Available from http://clinicaltrials.gov/ct2/show/NCT01200420 NLM Identifier: NCT01200420.

Scott, G.K., A. Goga, D. Bhaumik, C.E. Berger, C.S. Sullivan, and C.C. Benz. 2007. Coordinate suppression of ERBB2 and ERBB3 by enforced expression of micro-RNA miR-125a or miR-125b. J Biol Chem. 282(2): 1479–1486.

Sempere, L.F., M. Christensen, A. Silahtaroglu, M. Bak, C.V. Heath, G. Schwartz et al. 2007. Altered MicroRNA expression confined to specific epithelial cell subpopulations in breast cancer. Cancer Res. 67(24): 11612–11620.

Sempere, L.F., M. Preis, T. Yezefski, H. Ouyang, A.A. Suriawinata, A. Silahtaroglu et al. 2010. Fluorescence-based codetection with protein markers reveals distinct cellular compartments for altered MicroRNA expression in solid tumors. Clin Cancer Res. 16(16): 4246–4255.

Shimono, Y., M. Zabala, R.W. Cho, N. Lobo, P. Dalerba, D. Qian et al. 2009. Downregulation of miRNA-200c links breast cancer stem cells with normal stem cells. Cell. 138(3): 592–603.

Shin, I., T. Miller, and C.L. Arteaga. 2006. ErbB receptor signaling and therapeutic resistance to aromatase inhibitors. Clin Cancer Res. 12(3): 1008s–1012s.

Shoushtari, A.N., A.M. Michalowska, and J.E. Green. 2007. Comparing genetically engineered mouse mammary cancer models with human breast cancer by expression profiling. Breast Dis. 28(1): 39–51.

Siegel, R., E. Ward, O. Brawley, and A. Jemal. 2011. Cancer statistics, 2011. CA Cancer J Clin. 61(4): 212–236.

Simpson, P.T., J. Reis-Filho, T. Gale, and S.R. Lakhani. 2005. Molecular evolution of breast cancer. J Pathol. 205(2): 248–254.

Sonoki, T., E. Iwanaga, H. Mitsuya, and N. Asou. 2005. Insertion of microRNA-125b-1, a human homologue of lin-4, into a rearranged immunoglobulin heavy chain gene locus in a patient with precursor B-cell acute lymphoblastic leukemia. Leukemia. 19(11): 2009–2010.

Sotiriou, C. and L. Pusztai. 2009. Gene-expression signatures in breast cancer. N Engl J Med. 360(8): 790–800.

Spizzo, R., M.S. Nicoloso, L. Lupini, Y. Lu, J. Fogarty, S. Rossi et al. 2009. miR-145 participates with TP53 in a death-promoting regulatory loop and targets estrogen receptor-[alpha] in human breast cancer cells. Cell Death Differ. 17(2): 246–254.

Steeg, P.S. 2006. Tumor metastasis: Mechanistic insights and clinical challenges. Nat Med. 12(8): 895–904.

Steelman, L.S., P.M. Navolanic, M.L. Sokolosky, J.R. Taylor, B.D. Lehmann, W.H. Chappell et al. 2008. Suppression of PTEN function increases breast cancer chemotherapeutic drug resistance while conferring sensitivity to mTOR inhibitors. Oncogene. 27(29): 4086–4095.

Stengel, C., S.P. Newman, M.P. Leese, B.V.L. Potter, M.J. Reed, and A. Purohit. 2009. Class III [beta]-tubulin expression and in vitro resistance to microtubule targeting agents. Br J Cancer. 102(2): 316–324.

Stern, D.F. 2003. ErbBs in mammary development. Exp Cell Res. 284(1): 89–98.

Stingl, J. 2011. Estrogen and progesterone in normal mammary gland development and in cancer. Horm Cancer. 2(2): 85–90.

Thomas, C. and J. Gustafsson. 2011. The different roles of ER subtypes in cancer biology and therapy. Nat Rev Cancer. 11(8): 597–608.

Tommasi, S., A. Mangia, R. Lacalamita, A. Bellizzi, V. Fedele, A. Chiriatti et al. 2007. Cytoskeleton and paclitaxel sensitivity in breast cancer: The role of beta-tubulins. Int J Cancer. 120(10): 2078–2085.

Tsai, H. and S.B. Baylin. 2011. Cancer epigenetics: Linking basic biology to clinical medicine. Cell Res. 21(3): 502–517.

Vargo-Gogola, T. and J.M. Rosen. 2007. Modelling breast cancer: One size does not fit all. Nat Rev Cancer. 7(9): 659–672.

Vrba, L., J.C. Garbe, M.R. Stampfer, and B.W. Futscher. 2011. Epigenetic regulation of normal human mammary cell type-specific miRNAs. Genome Res. 21(12): 2026–2037.

Wang, V. and W. Wu. 2009. MicroRNA-based therapeutics for cancer. BioDrugs. 23(1): 15–23.

Wang, F., Z. Zheng, J. Guo, and X. Ding. 2010. Correlation and quantitation of microRNA aberrant expression in tissues and sera from patients with breast tumor. Gynecol Oncol. 119(3): 586–593.

Wang, Z., B. Lu, H. Wang, Z. Cheng, and Y. Yin. 2011. MicroRNA-21 modulates chemosensitivity of breast cancer cells toÂ Doxorubicin by targeting PTEN. Arch Med Res. 42(4): 281–290.

Webster, R.J., K.M. Giles, K.J. Price, P.M. Zhang, J.S. Mattick, and P.J. Leedman. 2009. Regulation of epidermal growth factor receptor signaling in human cancer cells by MicroRNA-7. J Biol Chem. 284(9): 5731–5741.

Westholm, J.O. and E.C. Lai. 2011. Mirtrons: microRNA biogenesis via splicing. Biochimie. 93(11): 1897–1904.

Whiteside, T.L. 2008. The tumor microenvironment and its role in promoting tumor growth. Oncogene. 27(45): 5904–5912.

Wu, M., N. Jolicoeur, Z. Li, L. Zhang, Y. Fortin, D. L'Abbe et al. 2008. Genetic variations of microRNAs in human cancer and their effects on the expression of miRNAs. Carcinogenesis. 29(9): 1710–1716.

Wu, Q., Z. Lu, H. Li, J. Lu, L. Guo, and Q. Ge. 2011. Next-generation sequencing of MicroRNAs for breast cancer detection. J Biomed Biotechnol. (in press).

Xiong, J., D. Yu, N. Wei, H. Fu, T. Cai, Y. Huang et al. 2010. An estrogen receptor α suppressor, microRNA-22, is downregulated in estrogen receptor α-positive human breast cancer cell lines and clinical samples. FEBS J. 277(7): 1684–1694.

Xu, N., T. Papagiannakopoulos, G. Pan, J.A. Thomson, and K.S. Kosik. 2009. MicroRNA-145 regulates OCT4, SOX2, and KLF4 and represses pluripotency in human embryonic stem cells. Cell. 137(4): 647–658.

Xu, J., X. Zhou, and C. Wong. 2011. Genome-wide identification of estrogen receptor alpha regulated miRNAs using transcription factor binding data. *In:* Bioinformatics - Trends and Methodologies. InTech. 559–574.

Yager, J.D. and N.E. Davidson. 2006. Estrogen carcinogenesis in breast cancer. N Engl J Med. 354(3): 270–282.

Yamagata, K., S. Fujiyama, S. Ito, T. Ueda, T. Murata, M. Naitou et al. 2009. Maturation of MicroRNA is hormonally regulated by a nuclear receptor. Mol Cell. 36(2): 340–347.

Yu, Z., C. Wang, M. Wang, Z. Li, M.C. Casimiro, M. Liu et al. 2008. A cyclin D1/microRNA 17/20 regulatory feedback loop in control of breast cancer cell proliferation. J Cell Biol. 182(3): 509–517.

Yu, F., Y. Jiao, Y. Zhu, Y. Wang, J. Zhu, X. Cui et al. 2012. MicroRNA 34c gene down-regulation via DNA methylation promotes self-renewal and epithelial-mesenchymal transition in breast tumor-initiating cells. J Biol Chem. 287(1): 465–473.

Zhang, L., J. Huang, N. Yang, J. Greshock, M.S. Megraw, A. Giannakakis et al. 2006. microRNAs exhibit high frequency genomic alterations in human cancer. Proc Natl Acad Sci USA. 103(24): 9136–9141.

Zhang, Y., L. Yan, Q. Wu, Z. Du, J. Chen, D. Liao et al. 2011a. miR-125b is methylated and functions as a tumor suppressor by regulating the ETS1 proto-oncogene in human invasive breast cancer. Cancer Res. 71(10): 3552–3562.

Zhang, B., X. Liu, J. He, C. Zhou, M. Guo, M. He et al. 2011b. Pathologically decreased miR-26a antagonizes apoptosis and facilitates carcinogenesis by targeting MTDH and EZH2 in breast cancer. Carcinogenesis. 32(1): 2–9.

Zhang, L., Y. Liu, F. Song, H. Zheng, L. Hu, H. Lu et al. 2011c. Functional SNP in the microRNA-367 binding site in the 3'UTR of the calcium channel ryanodine receptor gene 3 (RYR3) affects breast cancer risk and calcification. Proc Natl Acad Sci USA. 108(33): 13653–13658.

Zhao, J., J. Lin, H. Yang, W. Kong, L. He, X. Ma et al. 2008. MicroRNA-221/222 negatively regulates estrogen Receptor α and is associated with tamoxifen resistance in breast cancer. J Biol Chem. 283(45): 31079–31086.

Zhao, H., J. Shen, L. Medico, D. Wang, C.B. Ambrosone, and S. Liu. 2010. A pilot study of circulating miRNAs as potential biomarkers of early stage breast cancer. PLoS ONE. 5(10): e13735.

Zhao, Y., C. Deng, J. Wang, J. Xiao, Z. Gatalica, R. Recker et al. 2011. Let-7 family miRNAs regulate estrogen receptor alpha signaling in estrogen receptor positive breast cancer. Breast Cancer Res Treat. 127(1): 69–80.

Zhou, M., Z. Liu, Y. Zhao, Y. Ding, H. Liu, Y. Xi et al. 2010. MicroRNA-125b confers the resistance of breast cancer cells to paclitaxel through suppression of pro-apoptotic bcl-2 antagonist killer 1 (Bak1) expression. J Biol Chem. 285(28): 21496–21507.

Zhu, M., M. Yi, C.H. Kim, C. Deng, Y. Li, D. Medina et al. 2011a. Integrated miRNA and mRNA expression profiling of mouse mammary tumor models identifies miRNA signatures associated with mammary tumor lineage. Genome Biol. 12(8): R77.

Zhu, Y., F. Yu, Y. Jiao, J. Feng, W. Tang, H. Yao et al. 2011b. Reduced miR-128 in breast Tumor-Initiating cells induces chemotherapeutic resistance via bmi-1 and ABCC5. Clin Cancer Res. 17(22): 7105–7115.

CHAPTER 2

MicroRNAs IN BREAST CANCER DEVELOPMENT

Mei Yang,[1,2,a,*] Zhe Xu,[3] Yuxin Zhang[1,b] and Chenfang Zhang[1,c]

ABSTRACT

MicroRNAs (miRNAs) play an important role in several major events during life, such as embryonic development, cellular differentiation, proliferation, apoptosis, and metabolism. The mis-regulation of miRNAs is linked to the development of cancer. In breast cancer, the expression of miRNAs is frequently deregulated, and specific miRNAs have been identified as oncogenes or tumor suppressors that modulate all of the stages of breast cancer progression. The unique properties of miRNAs suggest that they may be useful for the early clinical diagnosis, classification, and grading of breast cancer. Additionally, miRNAs may aid in determining prognosis and patient response to treatment. A

[1]Department of General Surgery, General Hospital of Guangzhou Military Command of PLA, Guangzhou, China.
[a]Email: bayberry513@yahoo.com
[b]Email: zhangyuxin-53825@163.com
[c]Email: zcf.gz@163.com
[2]Department of Biochemistry, University of Hong Kong, Pokfulam, Hong Kong SAR, China.
[3]Department of Ophthalmology, General Hospital of Guangzhou Military Command of PLA, Guangzhou, China.
Email: darkeye119@yahoo.com
*Corresponding author

comparative analysis of miRNA expression between physiologic and pathophysiologic states and a clarification of the role of miRNAs in breast cancer development are essential for determining the mechanism underlying miRNA-mediated cancer development and progression. Such information may uncover new objectives and methodologies for the treatment of breast cancer. Recently, miRNA-related breast cancer studies have shifted from the examination of single molecules to the exploration of entire pathways or even entire networks. Studies have also shifted from examining the single cell phenotype to examining tissues as a whole throughout the multiple stages of breast cancer development. Moreover, extracellular miRNAs have been identified and are being thoroughly studied. In the tumor microenvironment, we found that miRNAs secreted by macrophages can promote breast cancer cell invasion. This finding reveals a new mechanism by which exogenous miRNAs can regulate the progression of breast cancer. In this chapter, we will discuss the emerging role of miRNAs in breast cancer with respect to the following: (1) miRNAs as oncogenes and tumor suppressors in breast cancer; (2) miRNAs in breast cancer metastasis, including the function of miRNAs in breast cancer stem cells; and (3) exogenous and secretory miRNAs. Specific examples from each category will be described.

1. INTRODUCTION

Breast cancer is the most common malignant tumor and the second leading cause of cancer death in women (Kong et al. 2010). Recently, because of advances in both basic and clinical research, there have been substantial advances in life-saving treatments for breast cancer, bringing new hope and excitement to breast cancer patients. Additionally, because of the discovery of oncogenes and tumor suppressors, breast cancer treatment has become more individualized. Among the key discoveries, microRNAs (miRNAs) hold a very important place. miRNAs are small, non-coding RNAs that are, on average, 22 nucleotides in length. miRNAs function as post-transcriptional regulators that bind to complementary sequences on target mRNAs, resulting in translational repression or target degradation and gene silencing (Bartel 2004, Bartel 2009). In the human genome, over 1000 miRNAs have been identified, and approximately 60% of mammalian genes are targeted by at least one miRNA (Lewis et al. 2005). Studies have demonstrated that miRNA expression profiles are highly correlated with mammary gland development. Because human miRNA genes are frequently located in fragile genomic regions (Calin et al. 2004), miRNAs are sensitive to alterations that occur prior to, or during, breast cancer development. Indeed, miRNAs have emerged as important factors that are involved in almost all stages of breast cancer, including cancer initiation and metastasis

(Iorio et al. 2005, Silveri et al. 2006). Moreover, the miRNA profiles of healthy mammalian epithelial cells are distinct from those of breast cancer cells (Bockmeyer et al. 2011). And miRNA detection studies in frozen and formalin-fixed tissues and plasma/serum have identified miRNAs that are involved in breast cancer initiation (oncomiRs or tumor suppressors), progression and metastasis (metastamirs). Additionally, miRNA studies have quickly progressed from the molecular basis of cancer to clinical applications (Ferracin et al. 2010).

2. ONCOMIRS IN BREAST CANCER

An oncomiR is a miRNA that is involved in cancer. Typically, miRNAs are properly transcribed and processed, and they subsequently bind to their target mRNAs, resulting in normal cellular growth, proliferation, differentiation or apoptosis. In contrast, oncogenic miRNAs may be amplified or overexpressed as a result of gene amplification, increased promoter activation, or increased efficiency in miRNA processing. These miRNAs typically target tumor suppressor genes. Therefore, their overexpression results in the increased degradation of tumor suppressor mRNAs and subsequent tumor formation (Esquela-Kerscher and Slack 2006). Evidence has shown that the expression of several miRNAs correlates with breast cancer initiation and progression. The identification of additional oncomiRs might prove useful in the diagnosis and treatment of breast cancer.

2.1 miR-155

miR-155 is processed from an exon of a non-coding RNA that is transcribed from the B-cell Integration Cluster (BIC) located on chromosome 21 (Sethupathy et al. 2007). The pre-miRNA stem loop structure and mature miR-155 single strand sequence are shown in Fig. 1. miR-155 is involved in multiple biological processes, including hematopoiesis, inflammation and immunity. The deregulation of miR-155 is associated with cancer (Faraoni et al. 2009), and miR-155 overexpression has been verified in a number of neoplastic diseases, suggesting its involvement in carcinogenesis (Esquela-Kerscher and Slack 2006). By comparing breast cancer and normal breast tissues, Marilena V. Iorio (Iorio et al. 2005) and Zhu (Zhu et al. 2010) found that miR-155 was consistently up-regulated in breast cancer. And miR-155 expression is also associated with advanced TNM (tumor-lymph node-metastasis) clinical stage, tumor proliferation, estrogen and progesterone receptor status, and poor prognosis. In a mouse model, the transgenic over-expression of miR-155 resulted in the development of cancer (Tili et al. 2009). In mechanistic studies, Ovcharenko found that miR-155 could

miR-155: stem loop structure

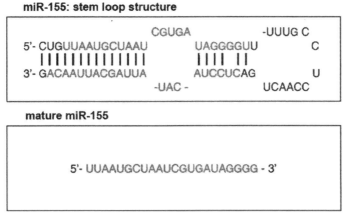

mature miR-155

5'- UUAAUGCUAAUCGUGAUAGGGG - 3'

Figure 1. The pre-miR-155 stem loop structure and the mature miR-155 single strand sequence.

suppress caspase-3 activity and apoptosis in breast cancer cells (Ovcharenko et al. 2007). Shuai Jiang revealed that miR-155 functions as an oncomiR in breast cancer by targeting the tumor suppressor gene socs1 (suppressor of cytokine signaling 1) (Jiang et al. 2010). In experiments performed by William Kong, the ectopic expression of miR-155 induced cell survival and chemo-resistance to multiple agents through the repression of FOXO3a (Kong et al. 2010). Collectively, these studies suggest that miR-155 can be considered a potential biomarker for breast cancer diagnosis and a therapeutic target for breast cancer treatment.

2.2 miR-21

miR-21 was one of the first mammalian miRNAs to be identified. The gene encoding this miRNA is located on the plus strand of chromosome 17q23.2 (55273409–55273480) within a coding gene (TMEM49, also called vacuole membrane protein). To date, many in-depth functional studies have suggested that miR-21 displays oncogenic activities. Differences in the expression level of miR-21 between matched normal breast tissue and breast cancer samples have established miR-21 as an oncomiR (Si et al. 2007). The potential effects of miR-21 are reflected in many aspects of breast cancer. For example, Rask's studies demonstrated that miR-21 expression correlates with lymph node status in patients. And miR-21 expression was predominantly observed in lymph node-positive breast cancer samples, indicating that miR-21 expression correlates with a high risk for breast cancer metastasis (Rask et al. 2011). Additional evidence suggests that miR-21 is involved in the resistance of HER2-positive breast cancer to trastuzumab treatment (Gong et al. 2011). Interestingly, miR-21 is overexpressed in pregnancy-associated breast cancer. This overexpression correlates with clinico-pathological features, including the loss of hormone receptor expression and lymph node positivity (Walter et al. 2011). Moreover, bilateral and unilateral breast cancers exhibit distinct miRNA expression patterns, and miR-21 expression is significantly higher in bilateral breast cancer (Iyevleva et al. 2011). The clinical analyses mentioned above support the hypothesis that miR-21 is involved in breast cancer progression and suggest that miR-21 may be a marker for poor prognosis. *In vitro* and *in vivo* studies have also been conducted to verify the functions of miR-21. When miR-21 expression was suppressed, breast cancer cell growth was reduced. In a xenograft carcinoma mouse model, miR-21 down-regulation inhibited breast tumor growth (Si et al. 2007). Much effort has been put into uncovering miR-21 up-regulators and downstream targets. Recently, studies indicated that RNA helicase DDX5 (also known as p68) regulates miR-21 expression. Indeed, a knockdown of DDX5 resulted in the down-regulation of miR-21 and the up-regulation of PDCD4, a known miR-21 target (Wang et al. 2011). Lan et al. demonstrated that TCF4 (transcription factor 7 like 2) directly participates in the transcriptional activation of miR-21 by binding to the miR-21 promoter region (Lan et al. 2011). In estrogen receptor α positive breast cancer, which shows increased miR-21 expression in comparison to estrogen receptor α negative breast cancer, estradiol (E2) could repress miR-21 expression and increase its target gene expression (Wickramasinghe et al. 2009). With regard to the downstream effects of miR-21, the tumor suppressor genes TPM1 (tropomyosin 1) (Zhu et al. 2007), PTEN, BCL2, PDCD4 (programmed cell death protein 4) (Frankel and Lu 2008, Gong and Walter 2011) and RHOB (Connolly et al. 2010) have been identified as miR-21 targets. These targets mediate the biological effects of miR-21 in breast cancer cells, further supporting the hypothesis that miR-21 is an oncogenic miRNA. In short, because of its important role in breast cancer progression, further studies on miR-21 may contribute to improved clinical stratification according to growth rates, thereby facilitating the development of more tailored treatments for breast cancer patients.

Additional miRNAs, such as miR-27a (Scott et al. 2006, Yu et al. 2010), have been identified as oncomiRs in breast cancer. Studies on these oncomiRs suggest that miRNAs may be important targets for future cancer therapeutics.

3. BREAST CANCER SUPPRESSOR miRNAs

Expression analyses of miRNAs in normal tissue and tumor samples demonstrate that tumor tissues exhibit a unique miRNA signature (Hammond 2007). And among the altered miRNAs, many have strongly reduced expression. Because such down-regulated miRNAs have been identified as suppressors of neoplasms, they are referred to as tumor suppressors (Babashah and Soleimani 2011, Li et al. 2011). Here, we discuss recent findings on several well-known miRNAs that inhibit breast tumor formation and progression.

3.1 let-7 Family

The let-7 (lethal-7) gene was first discovered in the nematode as a key developmental regulator. Using the BLAST (basic local alignment search tool) research tool, let-7 was later identified as the first known human miRNA (Pasquinelli and Reinhart 2000). The human let-7 family consists of 13 members located on 9 different chromosomes, and the mature form of each let-7 family member is highly conserved across species (Boyerinas et al. 2010). The aberrant expression of let-7 has been associated with poor prognosis in several types of cancer (Takamizawa et al. 2004, Yang et al. 2008, Ueda et al. 2010), and let-7 is widely viewed as a tumor suppressor miRNA. In Yu Fengyan's studies (Yu et al. 2007), the forced expression of let-7 reduced the malignancy and self-renewal capacity of breast tumor initiated cells (BT-ICs), which may be the main cell type responsible for breast cancer formation and metastasis. let-7 expression was shown to be markedly reduced in BT-ICs, and its expression increased with BT-ICs differentiation. The overexpression of let-7 in BT-ICs resulted in reduced cell proliferation and mammosphere formation *in vitro* as well as decreased tumor formation and metastasis *in vivo*. H-RAS and HMGA2 are the known targets of let-7. Silencing H-RAS in a BT-IC enriched cell line resulted in a reduction in the capacity for cell self-renewal. Silencing HMGA2 enhanced cell differentiation.

Another member of the let-7 family, let-7g (Qian et al. 2011), was reported to be specifically involved in breast cancer development. The diminished expression of let-7g has been significantly associated with breast cancer lymph node metastasis and poor patient survival. let-7g expression was observed to be progressively reduced from normal breast epithelium to ductal carcinoma *in situ* to invasive ductal carcinoma, indicating a role in both breast cancer initiation and progression. Moreover, in breast cancer cells, the depletion of let-7g promoted cell migration, invasion and distant metastasis. These studies verified known targets of let-7g and revealed that abrogation of let-7g expression promotes breast cancer cell migration and invasion through targets FN1 (fibronectin 1) and GAB2 (Grb2-associated binding protein 2), and consequent activation of the p44/42 MAPK (mitogen-activated protein kinase) pathway. Additionally, let-7g reduction results in the increased expression of MMP2/MMP9. Other let-7 family members, including let-7a, let-7b, let-7c, let-7d, let-7e, let-7f, let-7i, and miR-98 have also been studied in different types of cancer. Among them, the reduced expression of let-7a and let-7c (Iorio et al. 2005, Sempere et al. 2007, Qian et al. 2011) was observed in breast cancer; however, further studies are required to verify the functions of these miRNAs and to link their expression with breast cancer patient survival.

As let-7 behaves as tumor suppressor, the restoration of let-7 expression may be a useful therapeutic option in breast cancers in which its expression has been lost.

3.2 miR-200 Family

The miR-200 family consists of miR-200a, miR-200b, miR-200c, miR-141 and miR-429. This family exists as two clusters in the genome. miR-200a, miR-200b and miR-429 are located on chromosome 1, while miR-200c and miR-141 are on chromosome 12 (Korpal et al. 2008). Growing evidence suggests that the miR-200 family is involved in breast cancer metastasis, and the tumor suppressor role of this family has been well studied. The miR-200 family is believed to inhibit epithelial-mesenchymal transition (EMT), a process that is thought to initiate metastasis by enhancing tumor cell motility. Philip A. Gregory (Gregory et al. 2008, Gregory et al. 2011), Manav Korpal (Korpal et al. 2008) and Gabriel Eades (Eades et al. 2011) used a well-established model in which TGF-β was used to induce EMT in cell lines. They found that the miR-200 family members were selectively down-regulated in those cells that had undergone EMT. Forced expression of the miR-200 family could sufficiently prevent TGF-β-induced EMT. Consistent with these findings, a loss of miR-200 expression was observed in breast cancer cell lines with a mesenchymal phenotype as well as in metastatic breast cancer specimens (Gregory et al. 2008). In malignant myoepithelioma, miR-200c and miR-429 were also expressed at a significantly lower level compared with luminal A, luminal B, and basal-like breast cancers (Bockmeyer et al. 2011). Because of the significant role of the miR-200 family in EMT and cancer metastasis, studies on the mechanisms that regulate miR-200 expression and the targets of miR-200 family members are required. Genetic and/or epigenetic mechanisms have been demonstrated to regulate the expression levels of the miR-200 family members. Under one genetic mechanism, miR-200 targets ZEB1/SIP1 and, in turn, ZEB1/SIP1 suppresses miR-200. This mechanism creates a negative feedback loop that regulates the EMT process (Bracken et al. 2008). The down-regulation of Dicer, a key component of the miRNA processing machinery, also results in the reduced expression of miR-200 (Martello et al. 2010). In addition to genetic mechanisms, hypermethylation of the promoter CpG islands of miR-200 also results in reduced miR-200 expression, as was verified by Lukas Vrba's group (Vrba et al. 2010). The miR-200 family can also regulate the expression of ZEB1/ZEB2, which are transcriptional repressors of E-cadherin (Gregory et al. 2008, Korpal et al. 2008). The loss of the miR-200 family results in the loss of the epithelial markers E-cadherin and ZO-1 and the induction of the mesenchymal markers fibronectin and N-cadherin. Breast cancer metastasis is the main cause of death in women, and epithelial-mesenchymal transition is the initiating step in metastasis. Therefore, restoration of miR-200 family could be an effective way to prevent breast cancer metastasis.

Other miRNAs, such as miR-206 (O'Day and Lal 2010), miR-96, miR-205 (Wu H et al. 2009), the miR-34 family (Yu et al. 2010, Lee et al. 2011), and the miR 17/20 cluster (Yu et al. 2010) also behave as tumor suppressors. Collectively, these findings suggest a new therapeutic strategy for breast cancer, whereby the restoration of miRNA production may block cancer cell growth or induce differentiation.

4. BREAST CANCER METASTAMIRS

Breast cancer metastasis, which results from the dissemination of primary tumor cells to distant sites, is the major cause of cancer-related mortality. Considering the heterogeneity of breast cancer and its resistance to traditional treatment (Bertos and Park 2011), understanding the metastatic process and improving early detection and monitoring methods are particularly important. Recently, research has shown that tumor metastasis may be promoted by the enhanced expression of pro-metastatic miRNAs and/or the down-regulation of anti-metastatic miRNAs. miRNAs have been confirmed to be key regulators of breast cancer metastasis (Negrini and Calin 2008, Kasinski and Slack 2011). Therefore, determining which molecules regulate miRNA expression, targets of those miRNAs, as well as the mechanism underlying this regulation, is vital. Indeed, such studies may provide a biological basis for the targeting of miRNAs in breast cancer treatment. The functional importance of several miRNAs involved in breast tumor progression and metastasis are discussed below.

4.1 miR-10b

The miR-10b gene is located on chromosome 2q31.1. This miRNA is causally associated with breast cancer metastasis. Robert A. Weinberg's group demonstrated that miR-10b was highly expressed in metastatic breast cancer cells and that the overexpression of miR-10b in non-metastatic breast tumors initiated robust invasion and metastasis (Ma et al. 2007); however, when Marilena V. Iorio et al. compared the miRNA expression patterns of normal breast tissue and primary breast tumors, they found that miR-10b was expressed at lower levels in primary breast tumors (Iorio et al. 2005). In this regard, Robert A. Weinberg's group further measured the expression of miR-10b in metastasis-free and metastasis-positive patient breast tumor samples. Compared with normal breast tissues, although the expression level of miR-10b was lower in metastasis-free breast tumor samples, more than half of the metastasis-positive breast tumor samples had higher expression levels of miR-10b. This evidence fully demonstrated a specific function for miR-10b in breast cancer metastasis and allowed for it to be classified as a metastamir.

How does the miR-10b metastamir function to promote breast cancer cell metastasis? Lilly Y. W. Bourguignon showed that the transcription factor Twist could bind to the promoter site of miR-10b and up-regulate miR-10b expression. As a result, the expression level of the tumor suppressor protein (HOXD10) was reduced, and breast tumor cell invasion increased (Bourguignon et al. 2010). These studies revealed that miR-10b mediates pathways in breast cancer migration, invasion and metastasis.

4.2 miR-520c and miR-373

miR-520c and miR-373 are other examples of metastasis-promoting miRNAs in breast cancer. In 2008, Qihong Huang and his colleagues thoroughly examined the roles of miR-520c and miR-373 in a breast cancer model and analyzed their expression in clinical samples. They found that miR-520c and miR-373 stimulated breast cancer cell migration and invasion *in vitro* and *in vivo*. These metastasis-promoting effects were achieved via the suppression of CD44. Moreover, in clinical breast cancer metastasis samples, the expression level of miR-373 was significantly increased (Huang et al. 2008). The role of the miR-520c and miR-373 metastamirs in the promotion of cancer cell invasion was confirmed not only in breast cancer but also in fibrosarcoma (Liu and Wilson 2012) and prostate cancer (Yang et al. 2009). In fibrosarcoma cells, miR-373 and miR-520c expression led to the decreased expression of mTOR and SIRT1, and this decreased expression subsequently resulted in an increase in the expression of MMP9, which can enhance cell migration and invasion. In prostate cancer cells, miR-373 and miR-520c were shown to enhance cell migration and invasion by suppressing CD44 expression.

4.3 miR-221/miR-222

Recently, miR-221/miR-222 was identified as basal-like subtype-specific miRNAs in breast cancer and was shown to promote EMT in breast cancer cells. miR221/222 was reported to act downstream of the oncogenic RAS-RAFMEK pathway and to mediate metastasis through increased migration and invasion (Shah and Calin 2011, Stinson et al. 2011). Additionally, an elevated expression of miR-221/miR-222 was observed in estrogen-α negative breast cancer cells. Here, an interaction between miR-221/miR-222 and estrogen-α resulted in tamoxifen resistance, suggesting that cells harboring both estrogen-α and miR-221/miR-222 represent an aggressive breast cancer cell type. Therefore, depletion of miR-221/miR-222 could be an alternative way to prevent breast cancer cell metastasis and/or make cancer cells sensitive to tamoxifen treatment.

Additionally, miR-155, which has been identified as an oncomiR in breast cancer, was shown by William Kong to regulate the EMT process. miR-155 was reported to be a target of the TGF-β/Smad4 pathway, and a knockdown of miR-155 was reported to reduce the occurrence of TGF-β-induced EMT (Kong et al. 2008).

4.4 miRNAs in Breast Cancer Stem Cells

When discussing breast cancer metastasis, it is important to consider a cancer cell subtype that has recently emerged as a key player in this process. Recently, in breast cancer research, a subset of cancer cells that possess stem cell-like properties has been discovered. These cells, which are referred to as cancer stem cells (CSCs) or tumor-initiating cells (TICs), play crucial roles in tumor initiation, metastasis and the resistance of tumors to anticancer therapies (Korkaya et al. 2011, O'Brien et al. 2011, Sanguinetti et al. 2011). Isolated CSCs have been defined as CD44(+)/CD24(-/low)/lineage(-) cells (Ponti et al. 2006). Interestingly, CSCs have been demonstrated to possess a distinct miRNA expression profile, and some miRNAs have been shown to be deregulated in CSCs (Liu and Tang 2011). The expression of miR-181 has been shown to be increased in the breast CSC population, and the overexpression of this miRNA was shown to induce CSC formation (Wang et al. 2011). On the other hand, the reduced expression of miR-128 in breast CSCs has also been shown to induce chemodrug resistance (Zhu et al. 2011). Likely, let-7 (Yu et al. 2007) and miR-34c (Yu et al. 2012) were also found to be decreased in CSCs, and the overexpression of these miRNAs resulted in the reduced ability of CSCs to initiate breast tumor formation. Furthermore, distant metastasis was also suppressed. The significant role of CSCs in tumor formation and metastasis as well as studies showing the importance of CSC miRNAs in cancer cell transformation, epithelial-mesenchymal transition, and chemotherapeutic resistance indicate that miRNAs act as major regulators of the biological behavior of breast cancer CSCs (Ma and Guan 2011). Therefore, targeting stemness-related miRNAs could result in a reduction of the breast cancer stem cell population, thereby changing the features of breast cancer. Thus, targeting stemness-related miRNAs may prevent relapse and provide new hope for cancer prevention.

In contrast to miRNAs that are overexpressed during the metastatic process, the expression of a subset of miRNAs is lost during breast cancer progression. Such miRNAs are identified as breast cancer metastasis suppressors. The expression of miR-335 (Tavazoie et al. 2008) was reported to be lost in breast tumors from relapsed patients, and this loss of expression was associated with poor metastasis-free survival. When miR-335 expression was restored, the invasion of metastatic breast cancer cells was inhibited. These metastasis-suppressing effects were mediated by miR-335 targeting of the SRY-box containing transcription factor SOX4 and the extracellular matrix component tenascin C.

Similarly, miR-340 has been shown to inhibit breast cancer cell migration and invasion, and a loss of miR-340 expression has been associated with lymph node metastasis and a shorter overall survival (Wu et al. 2011). In breast cancers that have metastasized to the brain, the expression of miR-1258 was found to be significantly reduced, and the forced overexpression of miR-1258 was found to suppress breast cancer metastasis to the brain (Zhang et al. 2011).

In short, these studies substantiate the hypothesis that specific miRNAs can modulate the progression of breast cancer, resulting in either more aggressive or less metastatic cancer phenotypes. And those data provide potential targets for preventing breast cancer metastasis.

5. EXOGENOUS AND SECRETORY miRNAs

As miRNAs are promising potential biomarkers for breast cancer, there is an urgent need to screen for the presence of oncomiRs/tumor suppressor miRNAs in healthy individuals and breast cancer patients. Researchers have begun to investigate whether the expression of a select panel of miRNAs within breast tumor tissues and matching serum samples correlates with disease (Wang et al. 2010). To date, however, the detection of most miRNAs is still limited to tissue specimens, although clinical serum specimens have become more accessible. Because of recent advances in miRNA detection, miRNAs present in fluid have been described. miRNAs in blood have also been discovered (Mitchell et al. 2008). Therefore, apart from the endogenously expressed miRNAs in cancer cells, the exogenous miRNAs that are secreted by cancer cells themselves and/or non-cancerous cells in the tumor microenvironment have been intensely studied. A protective coat (microvesicles, also termed as exosomes) that surrounds secreted miRNAs has also been discovered. This coat protects the miRNAs from the ubiquitously expressed RNases (Lehmann et al. 2008, Yang et al. 2011). To communicate with one another and to drive oncogenesis, cancer cells release miRNAs. Keiichi Ohshima's group discovered that gastric cancer cells release members of the let-7 family, via exosomes, into the extracellular environment (Ohshima et al. 2010). In addition to the cancer cells themselves, cells of the tumor microenvironment can also release regulatory miRNAs. Kosaka et al. reported that normal epithelial cells can inhibit tumor cell growth via exosomal tumor suppressor miRNAs (Kosaka et al. 2011). In our studies, we have shown that tumor associated macrophages (TAMs) in the microenvironment, which are closely correlated with poor breast cancer prognosis and are believed to enhance mammary tumor progression and metastasis (Aharinejad et al. 2004) (Levano et al. 2011), can release miRNAs that regulate breast cancer cell invasion (Yang et al. 2011).

Because miRNA expression profiles are different among cell types, we initially screened for macrophage-specific miRNAs. Interestingly, when we mimicked the tumor microenvironment by co-culturing TAMs and breast cancer cells in the absence of direct contact, the expression of a macrophage-specific miRNA (miR-223) was increased in the cancer cells (Yang et al. 2011). To verify the shuttling of the miRNA from macrophages to breast cancer cells, we used a control miRNA, lin-4, which is expressed in neither macrophages nor breast cancer cells. A luciferase reporter construct carrying the lin-4 complementary sequence was transfected into breast cancer cells. Simultaneously, lin-4 mimics were transfected into alternatively-activated macrophages. Then, breast cancer cells and alternatively-activated macrophages were co-cultured in the absence of direct contact. After co-culture, we observed decreased luciferase activity in the breast cancer cells, which indicates an increase in lin-4 expression occurred in the cancer cells (Yang et al. 2011) and lin-4 transportation from macrophages to cancer cells should exist. Next, we used a fluorescently labeled miRNA (miR-223) to directly view miRNA transfer. In this transfer assay, macrophages loaded with cy3-miR-223 were co-cultured with breast cancer cells (for details please refer to Fig. 2). And in order to distinguish fluorescent miRNAs between macrophages and cancer cells, after co-culture, we stained the fluorescent cancer cells using the macrophage marker CD68 to show that the fluorescent miRNAs were indeed in cancer cells (Fig. 2). In a functional assay, we verified that miR-223 can promote breast cancer cell invasion in a manner similar as alternatively activated macrophages. In the following, we separated the exosomes (Fig. 3) that had been released by the macrophages. Those exosomes could be uptaken by breast cancer cells and promote cancer cell invasion. Our findings demonstrated the effects of exogenous miRNAs on breast cancer cells and paved the way for further investigations on the interactions between breast cancer cells and the tumor microenvironment.

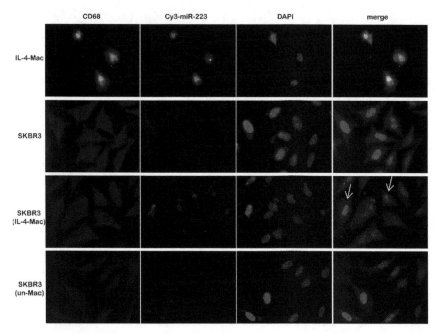

Figure 2. Shuttling of fluorescently-labeled miRNAs from macrophages to breast cancer cells. SKBR3 cells were cultured alone or co-cultured with IL-4-activated (IL-4-Mac) or unactivated macrophages (un-Mac) that were pre-transfected with Cy3-miR-223. Both macrophages and SKBR3 cells were then stained using the macrophage marker CD68 followed by an Alexa Fluor 488-conjugated secondary antibody. DAPI was used to visualize nuclei. Fluorescence was observed by fluorescence microscopy. Arrows indicate the Cy3 signal in SKBR3 cells, and images are shown at 1,000X magnification.

Color image of this figure appears in the color plate section at the end of the book.

In addition, a growing list of reports have identified miRNAs in the blood or fluid of patients (Pu et al. 2010, Zhao et al. 2011). Such miRNAs may be used as diagnostic or prognostic biomarkers or for dynamic observations.

6. CONCLUSIONS

miRNAs are intricately involved in breast cancer development (Negrini and Calin 2008). In the context of breast cancer, a balance exists between the repression of miRNA targets and the regulation of miRNA expression. And a second balance, between gene suppression and gene activation, is controlled by miRNAs (Negrini and Calin 2008). Alteration of any component in this system could disrupt the balance, resulting in a variety of cancer phenotypes. Today, both the main targets and the regulators of several miRNAs involved in breast cancer development and progression

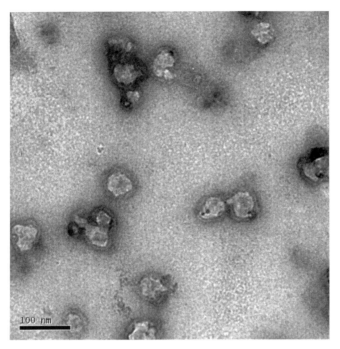

Figure 3. Exosomes secreted from IL4-activated macrophages were collected and observed by electron microscopy.

have been uncovered. Multiple oncogenic miRNAs, tumor suppressor miRNAs and metastamirs have also been fully characterized. Furthermore, researchers have identified and classified multiple exogenous miRNAs, which will result in improved noninvasive methodologies for the screening and monitoring of breast cancer. These remarkable discoveries and advances in the miRNA field have forced breast cancer investigators to consider the contributions of nonconventional, noncoding RNA genes to the development of cancer. Additionally, the translation of bench work to bedside applications has become apparent. Currently, miRNA-expression signatures are used to classify cancers and to identify miRNAs markers that may have prognostic potential (Lu et al. 2005, Esquela-Kerscher, Slack and Volinia 2006). With all of the efforts and advances made, miRNA-based cancer therapies, involving synthetic anti-miRNA oligonucleotides and synthetic miRNA mimics, may possibly be tailored to specific cancer types and individual patients; however, hurdles still remain. The potential for miRNA toxicity, caused by off-target gene silencing, must be considered, as individual miRNAs can simultaneously regulate several genes and pathways. The efficiency of the delivery of miRNA-based drugs must also be considered. Although many vectors have been developed to target specific organ and cell types, the adequate delivery of miRNAs to live animals, let alone humans, is still challenging. Lastly, multiple factors capable of modulating miRNA function exist in the tumor microenvironment. Such complexity requires that miRNA-based treatment strategies require a more complete understanding of all present molecules and methods of regulation. Therefore, to make progress in miRNA research and applications, uncovering the molecular basis of miRNA alteration and regulation and developing new technologies to aid in miRNA identification and delivery will be arduous, but necessary, tasks.

7. SUMMARY POINTS

- Breast cancer is the most common malignant tumor in women, and metastasis is the main cause of relapse and death.
- microRNAs (miRNAs) are intricately involved in the progression of breast cancer, and their potential as promising targets for breast cancer therapy has been demonstrated.
- Several oncogenic miRNAs (oncomiRs) and tumor suppressor miRNAs have been screened for and characterized in breast cancer.
- miRNAs that promote breast cancer metastasis (metastamirs) target many genes that are related to cancer cell dissemination, including those genes involved in EMT (epithelial-mesenchymal transition). In the population of breast cancer stem cells that may be responsible for tumor formation and metastasis, miRNAs function as key regulators.
- Exogenous miRNAs could target breast cancer cells and function to promote breast cancer development.

ABBREVIATIONS

CSCs	:	cancer stem cells
EMT	:	epithelial-mesenchymal transition
FN1	:	fibronectin 1
GAB2	:	Grb2-associated binding protein 2
MAPK	:	mitogen-activated protein kinase
TICs	:	tumor-initiating cells
TNM	:	T=primary tumor; N=regional lymph nodes; M=distant metastasis

REFERENCES

Aharinejad, S., P. Paulus, M. Sioud, M. Hofmann, K. Zins, R. Schafer, E.R. Stanley, and D. Abraham. 2004. Colony-stimulating factor-1 blockade by antisense oligonucleotides and small interfering RNAs suppresses growth of human mammary tumor xenografts in mice. Cancer Research. 64: 5378–84.

Babashah, S. and M. Soleimani. 2011. The oncogenic and tumour suppressive roles of microRNAs in cancer and apoptosis. European Journal of Cancer. 47: 1127–37.

Bartel, D.P. 2004. MicroRNAs: genomics, biogenesis, mechanism, and function. Cell. 116: 281–97.

Bartel, D.P. 2009. MicroRNAs: target recognition and regulatory functions. Cell. 136: 215–33.

Bertos, N.R. and M. Park. 2011. Breast cancer—one term, many entities? The Journal of Clinical Investigation. 121: 3789–96.

Bockmeyer, C.L., M. Christgen, M. Muller, S. Fischer, P. Ahrens, F. Langer, H. Kreipe, and U. Lehmann. 2011. MicroRNA profiles of healthy basal and luminal mammary epithelial cells are distinct and reflected in different breast cancer subtypes. Breast Cancer Research and Treatment. 130: 735–45.

Bourguignon, L.Y., G. Wong, C. Earle, K. Krueger, and C.C. Spevak. 2010. Hyaluronan-CD44 interaction promotes c-Src-mediated twist signaling, microRNA-10b expression, and RhoA/RhoC up-regulation, leading to Rho-kinase-associated cytoskeleton activation and breast tumor cell invasion. The Journal of Biological Chemistry. 285: 36721–35.

Boyerinas, B., S.M. Park, A. Hau, A.E. Murmann, and M.E. Peter. 2010. The role of let-7 in cell differentiation and cancer. Endocrine-related Cancer. 17: F19–36.

Bracken, C.P., P.A. Gregory, N. Kolesnikoff, A.G. Bert, J. Wang, M.F. Shannon, and G.J. Goodall. 2008. A double-negative feedback loop between ZEB1-SIP1 and the microRNA-200 family regulates epithelial-mesenchymal transition. Cancer Research. 68: 7846–54.

Calin, G.A., C. Sevignani, C.D. Dumitru, T. Hyslop, E. Noch, S. Yendamuri, M. Shimizu, S. Rattan, F. Bullrich, M. Negrini, and C.M. Croce. 2004. Human microRNA genes are frequently located at fragile sites and genomic regions involved in cancers. Proceedings of the National Academy of Sciences of the United States of America. 101: 2999–3004.

Connolly, E.C., K. Van Doorslaer, L.E. Rogler, and C.E. Rogler. 2010. Overexpression of miR-21 promotes an *in vitro* metastatic phenotype by targeting the tumor suppressor RHOB. Molecular cancer research. 8: 691–700.

Eades, G., Y. Yao, M. Yang, Y. Zhang, S. Chumsri, and Q. Zhou. 2011. miR-200a regulates SIRT1 expression and epithelial to mesenchymal transition (EMT)-like transformation in mammary epithelial cells. The Journal of Biological Chemistry. 286: 25992–6002.

Esquela-Kerscher, A. and F.J. Slack. 2006. Oncomirs-microRNAs with a role in cancer. Nature reviews Cancer. 6: 259–69.

Faraoni, I., F.R. Antonetti, J. Cardone, and E. Bonmassar. 2009. miR-155 gene: a typical multifunctional microRNA. Biochimica et Biophysica Acta. 1792: 497–505.

Ferracin, M., A. Veronese, and M. Negrini. 2010. Micromarkers: miRNAs in cancer diagnosis and prognosis. Expert Review of Molecular Diagnostics. 10: 297–308.

Frankel, L.B., N.R. Christoffersen, A. Jacobsen, M. Lindow, A. Krogh, and A.H. Lund. 2008. Programmed cell death 4 (PDCD4) is an important functional target of the microRNA miR-21 in breast cancer cells. The Journal of Biological Chemistry. 283: 1026–33.

Gong, C., Y. Yao, Y. Wang, B. Liu, W. Wu, J. Chen, F. Su, H. Yao, and E. Song. 2011. Up-regulation of miR-21 mediates resistance to trastuzumab therapy for breast cancer. The Journal of Biological Chemistry. 286: 19127–37.

Gregory, P.A., A.G. Bert, E.L. Paterson, S.C. Barry, A. Tsykin, G. Farshid, M.A. Vadas, Y. Khew-Goodall, and G.J. Goodall. 2008. The miR-200 family and miR-205 regulate epithelial to mesenchymal transition by targeting ZEB1 and SIP1. Nature Cell Biology. 10: 593–601.

Gregory, P.A., C.P. Bracken, E. Smith, A.G. Bert, J.A. Wright, S. Roslan, M. Morris, L. Wyatt, G. Farshid, Y.Y. Lim, G.J. Lindeman, M.F. Shannon, P.A. Drew, Y. Khew-Goodall, and

G.J. Goodall. 2011. An autocrine TGF-beta/ZEB/miR-200 signaling network regulates establishment and maintenance of epithelial-mesenchymal transition. Molecular Biology of the Cell. 22: 1686–98.

Hammond, S.M. 2007. MicroRNAs as tumor suppressors. Nature Genetics. 39: 582–3.

Huang, Q., K. Gumireddy, M. Schrier, C. le Sage, R. Nagel, S. Nair, D.A. Egan, A. Li, G. Huang, A.J. Klein-Szanto, P.A. Gimotty, D. Katsaros, G. Coukos, L. Zhang, E. Pure, and R. Agami. 2008. The microRNAs miR-373 and miR-520c promote tumour invasion and metastasis. Nature Cell Biology. 10: 202–10.

Iorio, M.V., M. Ferracin, C.G. Liu, A. Veronese, R. Spizzo, S. Sabbioni, E. Magri, M. Pedriali, M. Fabbri, M. Campiglio, S. Menard, J.P. Palazzo, A. Rosenberg, P. Musiani, S. Volinia, I. Nenci, G.A. Calin, P. Querzoli, M. Negrini, and C.M. Croce. 2005. MicroRNA gene expression deregulation in human breast cancer. Cancer Research. 65: 7065–70.

Iyevleva, A.G., E.S. Kuligina, N.V. Mitiushkina, A.V. Togo, Y. Miki, and E.N. Imyanitov. 2011. High level of miR-21, miR-10b, and miR-31 expression in bilateral vs. unilateral breast carcinomas. Breast Cancer Research and Treatment.131(3): 1049–1059.

Jiang, S., H.W. Zhang, M.H. Lu, X.H. He, Y. Li, H. Gu, M.F. Liu, and E.D. Wang. 2010. MicroRNA-155 functions as an OncomiR in breast cancer by targeting the suppressor of cytokine signaling 1 gene. Cancer Research. 70: 3119–27.

Kasinski, A.L. and F.J. Slack. 2011. Epigenetics and genetics. MicroRNAs en route to the clinic: progress in validating and targeting microRNAs for cancer therapy. Nature Reviews Cancer. 11: 849–64.

Kong, W., H. Yang, L. He, J.J. Zhao, D. Coppola, W.S. Dalton, and J.Q. Cheng. 2008. MicroRNA-155 is regulated by the transforming growth factor beta/Smad pathway and contributes to epithelial cell plasticity by targeting RhoA. Molecular and Cellular Biology. 28: 6773–84.

Kong, W., L. He, M. Coppola, J. Guo, N.N. Esposito, D. Coppola, and J.Q. Cheng. 2010. MicroRNA-155 regulates cell survival, growth, and chemosensitivity by targeting FOXO3a in breast cancer. The Journal of Biological Chemistry. 285: 17869–79.

Korkaya, H., S. Liu, and M.S. Wicha. 2011. Breast cancer stem cells, cytokine networks, and the tumor microenvironment. The Journal of Clinical Investigation. 121: 3804–9.

Korpal, M., E.S. Lee, G. Hu, and Y. Kang. 2008. The miR-200 family inhibits epithelial-mesenchymal transition and cancer cell migration by direct targeting of E-cadherin transcriptional repressors ZEB1 and ZEB2. The Journal of Biological Chemistry. 283: 14910–4.

Kosaka, N., H. Iguchi, Y. Yoshioka, K. Hagiwara, F. Takeshita, and T. Ochiya. 2011. Competitive interactions of cancer cells and normal cells via secretory microRNAs. The Journal of Biological Chemistry. 287(2): 1397–1405.

Lan, F., X. Yue, L. Han, Z. Shi, Y. Yang, P. Pu, Z. Yao, and C. Kang. 2011. Genome-wide identification of TCF7L2/TCF4 target miRNAs reveals a role. International Journal of Oncology. 40(2): 519–526.

Lee, Y.M., J.Y. Lee, C.C. Ho, Q.S. Hong, S.L. Yu, C.R. Tzeng, P.C. Yang, and H.W. Chen. 2011. MicroRNA 34b as a tumor suppressor in estrogen-dependent growth of breast cancer cells. Breast Cancer Research. 13(6): R116.

Lehmann, B.D., M.S. Paine, A.M. Brooks, J.A. McCubrey, R.H. Renegar, R. Wang, and D.M. Terrian. 2008. Senescence-associated exosome release from human prostate cancer cells. Cancer Research. 68: 7864–71.

Levano, K.S., E.H. Jung, and P.A. Kenny. 2011. Breast cancer subtypes express distinct receptor repertoires for tumor-associated macrophage derived cytokines. Biochemical and Biophysical Research Communications. 411: 107–10.

Lewis, B.P., C.B. Burge, and D.P. Bartel. 2005. Conserved seed pairing, often flanked by adenosines, indicates that thousands of human genes are microRNA targets. Cell. 120: 15–20.

Li, W., D.G. Lebrun, and M. Li. 2011. The expression and functions of microRNAs in pancreatic adenocarcinoma and hepatocellular carcinoma. Chinese Journal of Cancer. 30: 540–50.

Liu, C. and D.G. Tang. 2011. MicroRNA regulation of cancer stem cells. Cancer Research. 71: 5950–4.

Liu, P. and M.J. Wilson. 2012. miR-520c and miR-373 upregulate MMP9 expression by targeting mTOR and SIRT1, and activate the Ras/Raf/MEK/Erk signaling pathway and NF-kappaB factor in human fibrosarcoma cells. Journal of Cellular Physiology. 227: 867–76.

Lu, J., G. Getz, E.A. Miska, E. Alvarez-Saavedra, J. Lamb, D. Peck, A. Sweet-Cordero, B.L. Ebert, R.H. Mak, A.A. Ferrando, J.R. Downing, T. Jacks, H.R. Horvitz, and T.R. Golub. 2005. MicroRNA expression profiles classify human cancers. Nature. 435: 834–8.

Lu, Z., M. Liu, V. Stribinskis, C.M. Klinge, K.S. Ramos, N.H. Colburn, and Y. Li. 2008. MicroRNA-21 promotes cell transformation by targeting the programmed cell death 4 gene. Oncogene. 27: 4373–9.

Ma, L., J. Teruya-Feldstein, and R.A. Weinberg. 2007. Tumour invasion and metastasis initiated by microRNA-10b in breast cancer. Nature. 449: 682–8.

Ma, S. and X.Y. Guan. 2011. MiRegulators in cancer stem cells of solid tumors. Cell Cycle. 10: 571–2.

Martello, G., A. Rosato, F. Ferrari, A. Manfrin, M. Cordenonsi, S. Dupont, E. Enzo, V. Guzzardo, M. Rondina, T. Spruce, A.R. Parenti, M.G. Daidone, S. Bicciato, and S. Piccolo. 2010. A MicroRNA targeting dicer for metastasis control. Cell. 141: 1195–207.

Mitchell, P.S., R.K. Parkin, E.M. Kroh, B.R. Fritz, S.K. Wyman, E.L. Pogosova-Agadjanyan, A. Peterson, J. Noteboom, K.C. O'Briant, A. Allen, D.W. Lin, N. Urban, C.W. Drescher, B.S. Knudsen, D.L. Stirewalt, R. Gentleman, R.L. Vessella, P.S. Nelson, D.B. Martin, and M. Tewari. 2008. Circulating microRNAs as stable blood-based markers for cancer detection. Proceedings of the National Academy of Sciences of the United States of America. 105: 10513–8.

Negrini, M. and G.A. Calin. 2008. Breast cancer metastasis: a microRNA story. Breast Cancer Research. 10(2): 203.

O'Brien, C.S., G. Farnie, S.J. Howell, and R.B. Clarke. 2011. Breast cancer stem cells and their role in resistance to endocrine therapy. Hormones and Cancer. 2: 91–103.

O'Day, E. and A. Lal. 2010. MicroRNAs and their target gene networks in breast cancer. Breast Cancer Research. 12(2): 201.

Ohshima, K., K. Inoue, A. Fujiwara, K. Hatakeyama, K. Kanto, Y. Watanabe, K. Muramatsu, Y. Fukuda, S. Ogura, K. Yamaguchi, and T. Mochizuki. 2010. let-7 microRNA family is selectively secreted into the extracellular environment via exosomes in a metastatic gastric cancer cell line. PloS one. 5: e13247.

Ovcharenko, D., K. Kelnar, C. Johnson, N. Leng, and D. Brown. 2007. Genome-scale microRNA and small interfering RNA screens identify small RNA modulators of TRAIL-induced apoptosis pathway. Cancer Research. 67: 10782–8.

Pasquinelli, A.E., B.J. Reinhart, F. Slack, M.Q. Martindale, M.I. Kuroda, B. Maller, D.C. Hayward, E.E. Ball, B. Degnan, P. Muller, J. Spring, A. Srinivasan, M. Fishman, J. Finnerty, J. Corbo, M. Levine, P. Leahy, E. Davidson, and G. Ruvkun. 2000. Conservation of the sequence and temporal expression of let-7 heterochronic regulatory RNA. Nature. 408: 86–9.

Ponti, D., N. Zaffaroni, C. Capelli, and M.G. Daidone. 2006. Breast cancer stem cells: an overview. European Journal of Cancer. 42: 1219–24.

Pu, X.X., G.L. Huang, H.Q. Guo, C.C. Guo, H. Li, S. Ye, S. Ling, L. Jiang, Y. Tian, and T.Y. Lin. 2010. Circulating miR-221 directly amplified from plasma is a potential diagnostic and prognostic marker of colorectal cancer and is correlated with p53 expression. Journal of Gastroenterology and Hepatology. 25: 1674–80.

Qian, P., Z. Zuo, Z. Wu, X. Meng, G. Li, W. Zhang, S. Tan, V. Pandey, Y. Yao, P. Wang, L. Zhao, J. Wang, Q. Wu, E. Song, P.E. Lobie, Z. Yin, and T. Zhu. 2011. Pivotal role of reduced let-7g expression in breast cancer invasion and metastasis. Cancer Research. 71: 6463–74.

Rask, L., E. Balslev, S. Jorgensen, J. Eriksen, H. Flyger, S. Moller, E. Hogdall, T. Litman, and B.S. Nielsen. 2011. High expression of miR-21 in tumor stroma correlates with increased cancer cell proliferation in human breast cancer. APMIS: Acta Pathologica, Microbiologica, et Immunologica Scandinavica. 119: 663–73.

Reinhart, B.J., F.J. Slack, M. Basson, A.E. Pasquinelli, J.C. Bettinger, A.E. Rougvie, H.R. Horvitz, and G. Ruvkun. 2000. The 21-nucleotide let-7 RNA regulates developmental timing in Caenorhabditis elegans. Nature. 403: 901–6.

Sanguinetti, A., G. Bistoni, and N. Avenia. 2011. Stem cells and breast cancer, where we are? A concise review of literature. Il Giornale di chirurgia. 32: 438–46.

Scott, G.K., M.D. Mattie, C.E. Berger, S.C. Benz, and C.C. Benz. 2006. Rapid alteration of microRNA levels by histone deacetylase inhibition. Cancer Research. 66: 1277–81.

Sempere, L.F., M. Christensen, A. Silahtaroglu, M. Bak, C.V. Heath, G. Schwartz, W. Wells, S. Kauppinen, and C.N. Cole. 2007. Altered MicroRNA expression confined to specific epithelial cell subpopulations in breast cancer. Cancer Research. 67: 11612–20.

Sethupathy, P., C. Borel, M. Gagnebin, G.R. Grant, S. Deutsch, T. S. Elton, A.G. Hatzigeorgiou, and S.E. Antonarakis. 2007. Human microRNA-155 on chromosome 21 differentially interacts with its polymorphic target in the AGTR1 3′ untranslated region: a mechanism for functional single-nucleotide polymorphisms related to phenotypes. American Journal of Human Genetics. 81: 405–13.

Shah, M.Y. and G.A. Calin. 2011. MicroRNAs miR-221 and miR-222: a new level of regulation in aggressive breast cancer. Genome Medicine. 3: 56.

Si, M.L., S. Zhu, H. Wu, Z. Lu, F. Wu, and Y.Y. Mo. 2007. miR-21-mediated tumor growth. Oncogene. 26: 2799–803.

Silveri, L., G. Tilly, J.L. Vilotte, and F. Le Provost. 2006. MicroRNA involvement in mammary gland development and breast cancer. Reproduction, Nutrition, Development. 46: 549–56.

Stinson, S., M.R. Lackner, A.T. Adai, N. Yu, H.J. Kim, C. O'Brien, J. Spoerke, S. Jhunjhunwala, Z. Boyd, T. Januario, R.J. Newman, P. Yue, R. Bourgon, Z. Modrusan, H.M. Stern, S. Warming, F.J. de Sauvage, L. Amler, R.F. Yeh, and D. Dornan. 2011. miR-221/222 targeting of trichorhinophalangeal 1 (TRPS1) promotes epithelial-to-mesenchymal transition in breast cancer. Science signaling. 4: pt5.

Takamizawa, J., H. Konishi, K. Yanagisawa, S. Tomida, H. Osada, H. Endoh, T. Harano, Y. Yatabe, M. Nagino, Y. Nimura, T. Mitsudomi, and T. Takahashi. 2004. Reduced expression of the let-7 microRNAs in human lung cancers in association with shortened postoperative survival. Cancer Research. 64: 3753–6.

Tavazoie, S.F., C. Alarcon, T. Oskarsson, D. Padua, Q. Wang, P.D. Bos, W.L. Gerald, and J. Massague. 2008. Endogenous human microRNAs that suppress breast cancer metastasis. Nature. 451: 147–52.

Tili, E., C.M. Croce, and J.J. Michaille. 2009. miR-155: on the crosstalk between inflammation and cancer. International Reviews of Immunology. 28: 264–84.

Ueda, T., S. Volinia, H. Okumura, M. Shimizu, C. Taccioli, S. Rossi, H. Alder, C.G. Liu, N. Oue, W. Yasui, K. Yoshida, H. Sasaki, S. Nomura, Y. Seto, M. Kaminishi, G.A. Calin, and C.M. Croce. 2010. Relation between microRNA expression and progression and prognosis of gastric cancer: a microRNA expression analysis. The Lancet Oncology. 11: 136–46.

Volinia, S., G.A. Calin, C.G. Liu, S. Ambs, A. Cimmino, F. Petrocca, R. Visone, M. Iorio, C. Roldo, M. Ferracin, R.L. Prueitt, N. Yanaihara, G. Lanza, A. Scarpa, A. Vecchione, M. Negrini, C.C. Harris, and C.M. Croce. 2006. A microRNA expression signature of human solid tumors defines cancer gene targets. Proceedings of the National Academy of Sciences of the United States of America. 103: 2257–61.

Vrba, L., T.J. Jensen, J.C. Garbe, R.L. Heimark, A.E. Cress, S. Dickinson, M.R. Stampfer, and B.W. Futscher. 2010. Role for DNA methylation in the regulation of miR-200c and miR-141 expression in normal and cancer cells. PloS one. 5: e8697.

Walter, B.A., G. Gomez-Macias, V.A. Valera, M. Sobel, and M.J. Merino. 2011. miR-21 Expression in Pregnancy-Associated Breast Cancer: A Possible Marker of Poor Prognosis. Journal of Cancer. 2: 67–75.

Wang, D., J. Huang, and Z. Hu. 2011. RNA helicase DDX5 regulates MicroRNA expression and contributes to cytoskeletal reorganization in basal breast cancer cells. Molecular & Cellular Proteomics. 11(2): M111.011932.

Wang, F., Z. Zheng, J. Guo, and X. Ding. 2010. Correlation and quantitation of microRNA aberrant expression in tissues and sera from patients with breast tumor. Gynecologic Oncology. 119: 586–93.

Wang, Y., Y. Yu, A. Tsuyada, X. Ren, X. Wu, K. Stubblefield, E.K. Rankin-Gee, and S.E. Wang. 2011. Transforming growth factor-beta regulates the sphere-initiating stem cell-like feature in breast cancer through miRNA-181 and ATM. Oncogene. 30: 1470–80.

Wickramasinghe, N.S., T.T. Manavalan, S.M. Dougherty, K.A. Riggs, Y. Li, and C.M. Klinge. 2009. Estradiol downregulates miR-21 expression and increases miR-21 target gene expression in MCF-7 breast cancer cells. Nucleic Acids Research. 37: 2584–95.

Wu, H., S. Zhu, and Y.Y. Mo. 2009. Suppression of cell growth and invasion by miR-205 in breast cancer. Cell research. 19: 439–48.

Wu, Z.S., Q. Wu, C.Q. Wang, X.N. Wang, J. Huang, J.J. Zhao, S.S. Mao, G.H. Zhang, X.C. Xu, and N. Zhang. 2011. miR-340 inhibition of breast cancer cell migration and invasion through targeting of oncoprotein c-Met. Cancer. 117: 2842–52.

Yang, K., A.M. Handorean, and K.A. Iczkowski. 2009. MicroRNAs 373 and 520c are downregulated in prostate cancer, suppress CD44 translation and enhance invasion of prostate cancer cells *in vitro*. International Journal of Clinical and Experimental Pathology. 2: 361–9.

Yang, M., J. Chen, F. Su, B. Yu, L. Lin, Y. Liu, J. D. Huang, and E. Song. 2011. Microvesicles secreted by macrophages shuttle invasion-potentiating microRNAs into breast cancer cells. Molecular Cancer. 10: 117.

Yang, N., S. Kaur, S. Volinia, J. Greshock, H. Lassus, K. Hasegawa, S. Liang, A. Leminen, S. Deng, L. Smith, C.N. Johnstone, X.M. Chen, C.G. Liu, Q. Huang, D. Katsaros, G.A. Calin, B.L. Weber, R. Butzow, C.M. Croce, G. Coukos, and L. Zhang. 2008. MicroRNA microarray identifies Let-7i as a novel biomarker and therapeutic target in human epithelial ovarian cancer. Cancer Research. 68: 10307–14.

Yu, F., H. Yao, P. Zhu, X. Zhang, Q. Pan, C. Gong, Y. Huang, X. Hu, F. Su, J. Lieberman, and E. Song. 2007. let-7 regulates self renewal and tumorigenicity of breast cancer cells. Cell. 131: 1109–23.

Yu, F., Y. Jiao, Y. Zhu, Y. Wang, J. Zhu, X. Cui, Y. Liu, Y. He, E.Y. Park, H. Zhang, X. Lv, K. Ma, F. Su, J.H. Park, and E. Song. 2012. MicroRNA 34c Gene Down-regulation via DNA Methylation Promotes Self-renewal and Epithelial-Mesenchymal Transition in Breast Tumor-initiating Cells. The Journal of Biological Chemistry. 287: 465–73.

Yu, Z., R. Baserga, L. Chen, C. Wang, M. P. Lisanti, and R.G. Pestell. 2010. microRNA, cell cycle, and human breast cancer. The American Journal of Pathology. 176: 1058–64.

Zhang, L., P.S. Sullivan, J.C. Goodman, P.H. Gunaratne, and D. Marchetti. 2011. MicroRNA-1258 suppresses breast cancer brain metastasis by targeting heparanase. Cancer Research. 71: 645–54.

Zhao, R., J. Wu, W. Jia, C. Gong, F. Yu, Z. Ren, K. Chen, J. He, and F. Su. 2011. Plasma miR-221 as a predictive biomarker for chemoresistance in breast cancer patients who previously received neoadjuvant chemotherapy. Onkologie. 34: 675–80.

Zhu, J., X.Q. Hu, G.L. Guo, Y. Zhang, O.C. Wang, J. You, Q.D. Huang, and X.H. Zhang. 2010. Expression and its clinical significance of miR-155 in human primary breast cancer. Zhonghua wai ke za zhi [Chinese Journal of Surgery]. 48: 205–8.

Zhu, S., M.L. Si, H. Wu, and Y.Y. Mo. 2007. MicroRNA-21 targets the tumor suppressor gene tropomyosin 1 (TPM1). The Journal of Biological Chemistry. 282: 14328–36.

Zhu, Y., F. Yu, Y. Jiao, J. Feng, W. Tang, H. Yao, C. Gong, J. Chen, F. Su, Y. Zhang, and E. Song. 2011. Reduced miR-128 in breast tumor-initiating cells induces chemotherapeutic resistance via Bmi-1 and ABCC5. Clinical cancer research: An Official Journal of the American Association for Cancer Research. 17: 7105–15.

CHAPTER 3

MicroRNAs CONTROLLING INVASION AND METASTASIS IN BREAST CANCER

César López-Camarillo,[1,a,*] Laurence A. Marchat[2]
and José Ali Flores-Perez[1,b]

ABSTRACT

MicroRNAs (miRNAs) are small non-coding RNAs of ~22 nucleotides
that function as negative regulators of gene expression by either
inhibiting translation or inducing degradation of messenger RNAs
(mRNAs). Aberrant expression of miRNAs expression is associated with
initiation and progression of breast cancer where they act as oncogenes
or tumor suppressors contributing to tumorigenesis. Deregulated
miRNAs represent novel prognostic tumor biomarkers and potential
therapeutic targets in cancer. Recently, several miRNAs have been
shown to initiate invasion and metastasis of cancer cells, and thus they
have been denominated as metastamiRs. Here, we present a review of
the current knowledge of metastamiRs in breast cancer.

[1]Genomics Sciences Program, Oncogenomics and Cancer Proteomics Laboratory, Autonomous
University of Mexico City, Mexico.
[a]Email: cesar.lopez@uacm.edu.mx
[b]Email: qfb_ali@hotmail.com
[2]Biotechnology Program, Institutional Program of Molecular Biomedicine, National School of
Medicine and Homeopathy of the National Polytechnic Institute, Mexico.
Email: lmarchat@ipn.mx
*Corresponding author

1. INTRODUCTION

The hallmarks of cancer comprise six essential alterations in cell physiology that dictate malignant growth. Malignant cells acquire self-sufficiency in growth signals, lack of sensitivity to anti-growth signals, evasion of apoptosis, high replicative potential, angiogenesis, and tissue invasion and metastasis (Hanahan and Weinberg 2000, Lazebnik 2010). The capabilities of cancer cells to invade and metastasize to other tissues and organs represent the most deadly hallmark of breast tumors (Schmidt-Kitter et al. 2003). Metastasis results from cancer cells detaching from a primary tumor, consequently adapting to distant tissues and organs, and forming a secondary tumor. It depends on genetic and epigenetic events that are acquired during tumor progression. To successfully metastasize, a tumor cell must complete a complex set of processes, including invasion, survival and arrest in the circulatory system, and colonization of foreign organs. This involves the activation of complex genetic programs that turn on-off specific genes (Steeg 2006).

MicroRNAs (miRNA) are small non-coding single-stranded RNAs of ~22 nucleotides that act at posttranscriptional level as negative regulators of gene expression by inhibiting mRNA translation or inducing its degradation (Bartel 2004). Notably, miRNAs recognize their targets in a sequence specific manner causing down-regulation of individual or multiple gene targets. Now it is widely accepted that miRNAs have important roles in cancer development and progression. An increasing number of studies have shown that miRNAs expression is deregulated in diverse human pathologies including cancer (Croce 2009); thus they have been denominated as "oncomirs", as they might function as oncogenes and tumor-suppressors to inhibit the expression of cancer-related target genes and to suppress or promote tumorigenesis (Zhang et al. 2007). These oncogenic miRNAs usually induce tumor development by inhibiting tumor suppressor genes or genes controlling cell differentiation and death. Deregulation of miRNAs expression in cancer is associated with genetic and epigenetic alterations, including amplification, deletions, mutations, aberrant DNA methylation and histones modifications (Esquela-Kerscher and Slack 2006). The relevance of miRNAs studies relies on the fact that identification of aberrantly expressed miRNAs in cancer and their targets could provide potential diagnostic and prognostic tumor biomarkers and represent new therapeutic targets for cancer therapy (Hernando 2007). In addition, large high-throughput studies in patients revealed that miRNA profiling has the potential to classify tumors and predict patient outcome with high accuracy (Jiang and Chen 2012).

2. METASTAMIRS: MicroRNAs DRIVING METASTASIS IN HUMAN CANCER

Despite advancements in metastasis biology, molecular players and mechanisms are poorly understood. Some studies have reported that miRNAs modulate expression of proteins involved in migration and metastasis, thus they had earlier been denominated as metastamiRs (Hurst et al. 2009a). Metastasis-associated miRNAs have pro- and anti-metastatic effects, promote migration and invasion of cancer cells, and it seems that they do not influence the primary tumor neither development or initiation steps of tumorigenesis. As we have reviewed below, in general metastamiRs did not influence cell growth or proliferation of cancer cells, but they did regulate several steps in the metastatic program and processes, such as epithelial-mesenchymal transition (EMT), apoptosis, angiogenesis, migration, invasion and metastasis *in vitro* and *in vivo* (Nguyen et al. 2009). In this chapter, we describe the current knowledge of the roles of the most prominent miRNAs involved in invasion and metastasis in breast cancer (Table 1).

3. METASTAMIRS IN BREAST CANCER

Breast cancer is the neoplasia with the highest incidence and mortality that affects women worldwide (Ferlay et al. 2010). Breast carcinomas represent a heterogeneous group of diseases that are diverse in behavior, outcome, and response to therapy. Despite advances in screening, diagnosis, and therapies, causes of this disease still remain poorly understood. Most breast cancer deaths are due to the development of metastasis. Because of the existence of suitable metastasis models, the majority of metastamiRs have been identified in breast cancer tumors and cell lines. Consequently, there are an increasing number of reports about the multiple functions of microRNAs in cancer invasion and metastasis.

3.1 The miRNA-10b and HOXD10 Connection

The relevant role of miRNAs in breast cancer metastasis was first established by Ma et al. from Weinberg's group. Authors showed that up-regulation of miR-10b suppressed homeobox D10 (HOXD10) expression, consequently inducing activation of RHOC gene and initiation of breast cancer invasion and metastasis. HOXD10 maintains a differentiated phenotype in epithelial cells and is essential for morphogenesis and differentiation control (Ma et al. 2007). The authors also showed that miR-10b was 50-fold overexpressed in metastatic MDA-MB-231 cells in comparison with non-metastatic MCF7

Table 1. MicroRNAs involved in invasion and metastasis in breast cancer.

miRNA	Target	Deregulation in tumors	Function	References
Pro-metastasis				
miR-10b	HOXD10	Upregulated	Enhanced migration, invasion and metastasis *in vitro and in vivo*	Myers et al. 2002, Wang et al. 2011
miR-373 and -520c	CD44	Upregulated	Promotes migration and *in vivo* metastasis inhibits cell adhesion	Huang et al. 2008 Somesh et al. 2009
miR-21	TPM1, PDCD4, RECK, TIMP3, PTEN,	Upregulated	Invasion and migration are increased and apoptosis is decreased	Zhu et al. 2008, Zhang et al. 2010
miR-155	RHOA	Upregulated	Regulation of EMT and tight junctions	Kong et al. 2008
miR-378	Fus-1, SuFu	Upregulated	Enhances cell survival, reduces caspase-3 activity, and promotes tumor growth and angiogenesis	Lee et al. 2007
Metastasis inhibitors				
miR-196	HOXC8	Downregulated	Abrogated invasion *in vitro* and reduced spontaneous metastasis *in vivo*	Li, Yong et al. 2010
miR-31	FZD3, ITGA5, M-RIP, MMP16, RDX, RHOA	Downregulated	Local invasion, extravasation or initial survival at a distant site, and metastatic colonization	Valastyan et al. 2009
miR-126	CRK1, PIK3R2, SPRED1, VCAM1	Downregulated	Inhibits cell proliferation and tumorigenesis. Remodeling of actin and cell adhesion	Milena et al. 2009
miR-335	SOX4, PTPRN2, MERTK, TNC	Downregulated	Inhibition of cell migration	Tavazoie et al. 2008
miR146 a/b	IRAK1, TRAF6, EGFR	Downregulated	Inhibition of invasion and migration	Hurst et al. 2009
miR-200 family	BMI1, ZEB1, ZEB2	Downregulated	EMT regulation	O'Day et al. 2010
miR-206	NOTCH3, SRC1,SRC3	Downregulated	Inhibition of cancer cell migration and focus formation	Guisheng Song et al. 2009
miR-101	EZH2	Downregulated	Inhibition of cell proliferation and motility epigenetic pathway	Milena et al. 2009
miR-205	ZEB,VEGF, HER3	Downregulated	Inhibition of EMT *in vitro* and metastasis *in vivo*	Gregory et al. 2008, Wang et al. 2011
miR-17/20	cyclin D1	Downregulated	Suppressed breast cancer cell proliferation and tumor colony formation	Yu et al. 2008
miR-448	SATB1	Downregulated	Suppression of EMT	Li et al. 2010

cell line. Importantly, ectopic expression of miR-10b resulted in an increase in migration and invasion in human breast cell lines. In contrast, miR-10b silencing induced a 10-fold reduction of the invasive properties in breast cancer transfected cells. By using *in vivo* models, authors demonstrated that enhanced ectopic expression of miR-10b in non-metastatic tumorigenic cell lines promoted invasion, and lung micro-metastasis. In order to define the mechanisms of miR-10b regulation, they showed that transcription factor TWIST1 activates expression of *mir-10b* gene by specifically binding to its promoter region (Yang et al. 2004). TWIST1-mediated overexpression of miR-10b induced the translation inhibition of HOXD10 mRNA target resulting in an augmented expression of signaling GTPase-protein RHOC (Ras homolog gene family member C). The role of RHOC as a pro-metastatic gene and its dependence on miR-10b were demonstrated by suppression of RHOC expression by small interfering RNAs which caused inhibition on miR-10b induced cell migration and invasion.

3.2 Therapeutic Intervention of miR-10b Induced Metastasis

Efforts to develop drugs that would prevent a primary tumor from spreading to new sites have been hindered by a lack of metastasis-specific targets. In an additional outstanding report, Ma et al. from Weinberg's group reported that antagomirs-based therapeutic silencing of miR-10b in tumor-bearing mice effectively suppressed breast cancer metastasis in a mouse mammary tumor model (Ma et al. 2010). First they showed that silencing of miR-10b in the highly tumorigenic and metastatic 4T1 cells significantly decreases miR-10b levels and inhibits cell migration and invasiveness in trasnswell and matrigel assays. In contrast, cell proliferation was not affected. The antagomiR-10b induced reduction of mature miR-10b correlates with a marked increase of HOXD10 protein whose transcript is targeted by miR-10b (Botas 1993). In addition, knockdown of *Hoxd10* gene using small interfering RNAs reverted the loss of motility and invasiveness observed after antagomir-10b treatments. These *in vitro* results were also recapitulated *in vivo* by systemic administration of antagomir-10b in the mammary fat pad of immunocompetent syngenic BALB/c host mice. Authors did not observe significant difference in primary tumor size between mice treated with PBS or anatago-miR10b as expected given that miR-10b does not affect cell proliferation. Importantly, they observed an 86% decrease in the number of macroscopically visible pulmonary metastasis. Taken together, these results indicate that miR-10b antagomir can be efficiently delivered to rapidly growing tumor cells *in vivo* and can prevent metastasis formation in highly malignant cells.

3.3 The miR-373 and miR-520c

In another hallmark paper, Huang et al., discovered that miR-373 and miR-520c can promote tumor invasion and metastasis by regulating the cell surface receptor for hyaluronan CD44 (Huang et al. 2008). Authors transduced the non-metastatic MCF7 breast cancer cell line with a miRNA-expression library and identified relevant miRNAs through migration assays screening. They found that both miR-373 and miR-520c stimulated cell migration and invasion *in vitro* and *in vivo*. In contrast, both miR-373 and miR-520c did not affect cell proliferation. Authors evidenced that both miR-373 and miR-520c target the 3'UTR of CD44 messenger RNA. Moreover, enhanced expression of a CD44 gene that was unresponsive to miR-373/miR-520c inhibited the migratory activity of MCF7 cells overexpressing miR-373 and miR-520c. In addition, they analyzed matched normal/tumor breast tumors samples and found a significant up-regulation of miR-373, which inversely correlated with CD44 expression in specimens exhibiting lymph node metastasis. These findings indicate that miR-373 and miR-520c are involved in tumor migration and invasion as metastasis-promoting miRNAs.

3.4 The Metastasis-suppressors miR-335, miR-126, and miR-206

MicroRNA profiling is a useful tool to dissect metastasis associated mechanisms. The team led by Joan Massague identified some miRNAs whose expression was specifically lost when human metastatic breast cancer undergoes metastasis in contrast with other studies reporting overexpression of specific miRNAs related to metastasis. Tavazoie and coworkers performed an array-based miRNA profiling in MDA-MB-231 breast cancer cell derivatives that were highly metastatic to bone and lung, and found a signature represented by miR-335, miR-126, miR-206, miR-122a, miR-199a*, and miR-489 genes whose expression was highly repressed in metastatic cells (Tavazoie et al. 2008). Interestingly, restoring the expression of miR-335, miR-126 or miR-206 in highly metastatic to lung LM2 cells decreased the lung colonization. In contrast, restoration of miR-122a, miR-199a* or miR-489 expression inhibited lung colonization at early time points but did not result in a significant decrease in lung metastasis at the end point. These miRNAs seem to operate by distinct mechanisms of metastasis suppression. Authors showed that only miR-126 restoration significantly reduced tumor growth and proliferation, but not apoptosis. Thus, miR-126 suppresses tumorigenesis and metastasis in part by proliferation inhibition. On the contrary, miR-335 or miR-206 did not alter proliferation of LM2 cells *in vitro* and *in vivo*. Overexpression of miR-335 or miR-206 in LM2 cells and CN34-BoM1 cells resulted in a significant reduction of migration and invasive capacities in trans-well and matrigel assays confirming their relevant role in metastasis.

The analysis of miRNAs expression in clinical specimens showed that low expression of miR-335 or miR-126 in primary tumors from patients was associated with poor distal metastasis-free survival. To obtain more insights about the miR-335 downregulation consequences, authors transfected MDA-MB-231 cells with an antagomir-335 and found that inhibition of miR-335 enhanced lung-colonization. They then, profiled LM2 cells overexpressing miR-335 and identified 756 genes whose expression was decreased when compared with control LM2 cells, including genes previously implicated in extracellular matrix and cytoskeleton control and signal transduction, as well as in cell migration, such as the tenascin C (TNC) an extracellular matrix glycoprotein of stem cells niches (Oskarsson et al. 2011) and the SRY-box containing transcription factor SOX4 (van de Wetering et al. 1993). This experimental evidence indicated that miR-335 suppresses metastasis and migration by targeting the progenitor cell transcription factor SOX4 and TNC transcripts. Consequently, loss of miR-335 leads to the activation of SRY-box containing SOX4 and TNC, which are responsible for the acquisition of metastatic properties. Knockdown of SOX4 and TNC using RNA interference diminished *in vitro* invasive ability and *in vivo* metastatic potential, confirming that these proteins are key effectors of metastasis. On the other hand, Tavazoie and coworkers reported that miR-126 suppresses metastatic endothelial cells recruitment, metastatic angiogenesis and metastatic colonization through targeting of insulin-like growth factor binding protein 2 (IGFBP2), PITPNC1 and *c*-Mer tyrosine kinase (MERTK) receptor mRNAs (Png et al. 2011). Taken together, all these data showed that miR-335, miR-126, and miR-206 are metastasis suppressor microRNAs in human breast cancer.

3.5 miR-146a and miR-146b are Suppressors of Metastasis Mediated by NF-κB

Pathways activated by nuclear factor-κB (NF-κB), a proinflamatory transcription factor, have prominent roles in cancer promoting tumorigenesis. In a wide variety of human tumors NF-κB is constitutively activated, suggesting that suppression of NF-κB activation should be effective in the prevention and treatment of cancer (Bharat and Aggarwal 2004). Bhaumik and coworkers showed that miR-146a and miR-146b inhibited invasion and migration of breast cancer cells by down-regulating NF-κB activity through targeting of IRAK1 and TRAF6 two key components of the Toll-like and IL-1 receptor signaling network that activate NF-κB (Bhaumik et al. 2008). They used lentivirus system to introduce miR-146a and miR-146b in the highly metastatic breast cancer cell line MDA-MB-231 which has a constitutively activated NF-κB pathway. Notably, miR-146a/b-expressing cells showed a reduction of IRAK1 and TRAF6 proteins, but mRNA levels were not affected. The authors then searched for the consequences of the inhibition of these essential proteins through measurements of NF-κB activity. Phosphorylation of IκBα is a key measure of NF-κB activation. IκBα inhibits NF-κB by masking the nuclear localization signals of NF-κB and keeping it sequestered in an inactive state in cytoplasm, thus repressing its DNA binding and transcriptional activities. Authors showed that although protein levels of both IκBα and the NF-κB p65/Rel A component were similar between miR-146a/b-expressing and control MDA-MB-231 cells, phosphorylation of IκBα on serine 32, which is essential for its degradation, was reduced in miR-146a and miR-146b cells, respectively. Moreover, by using electrophoretic mobility shift assays and the consensus NF-κB response element, they evidenced that miR-146a/b-expressing cells have a reduced NF-κB DNA binding activity in comparison with control cells. The consequences of this reduced transcriptional activity of NF-κB were assessed in miR-146a/b-expressing and control MDA-MB-231 cells by measuring the expression of its target genes. Results showed that IL-8 expression, a cytokine upregulated in many aggressive cancers, IL-6, and the metalloproteinase MMP-9, all genes activated by NF-κB, were suppressed in miR-146a/b overexpressing cells. Because NF-κB activity is essential for epithelial-mesenchymal transition, migration and metastasis *in vitro* and *in vivo* (F-Huber et al. 2004), authors also examined the functional consequences of NF-κB suppression by miR-146 a/b. Matrigel invasion and migration assays indicated that ability of miR-146a/b overexpressing cells to migrate and invade a basement membrane-like extracellular matrix was significantly reduced in relation to control MDA-MB-231 cells. In contrast, proliferation and apoptosis rates in miR-146a/b expressing cells were not affected.

3.6 BRMS1 Regulated miR-146a and miR-146b Participate in Invasion and Metastasis

Mechanisms regulating microRNAs expression have not been completely understood. In an effort to understand the mechanism regulating miR-146a/b expression, Hurst and coworkers showed that breast cancer metastasis suppressor 1 (BRMS1), a protein that regulates expression of multiple genes leading to suppression of metastasis, significantly up-regulates miR-146a and miR-146b in metastatic breast cancer cells. They overexpressed *brms1* gene in MDA-MB-231 and MDA-MB-435 breast cancer cells and showed that miR-146a was upregulated in both cell lines, while miR-146b was upregulated in MDA-MB-435, but not in MDA-MB-231 cells. To further corroborate that BRMS1 was responsible for miR-146a/b upregulation, they silenced *brms1* gene expression using small interfering RNAs and showed that miR-146a and miR-146b levels were reduced by 20% and 50%, respectively. Interestingly, authors showed that transduction of miR-146a or miR-146b into MDA-MB-231 cells produced a down-regulated expression of epidermal growth factor receptor, inhibited invasion and migration *in vitro*, and suppressed experimental lung metastasis (Hurst et al. 2009b).

3.7 The Metastasis Suppressor miR-31

In order to identify miRNAs that might regulate breast cancer metastasis Valastyan and coworkers from Weinberg's group analyzed 10 selected miRNAs that were previously identified from expression profiling studies of clinical breast tumors, global analysis of miRNA copy-number in breast carcinomas, and localization of cancer-related miRNAs to fragile chromosome sites (Iorio et al. 2005, Calin et al. 2004). They analyzed the expression of these miRNAs in breast cancer lines including normal, tumorigenic but not metastatic, and metastatic tumor cells. Interestingly, miR-31 expression was specifically attenuated in human metastatic MDA-MB-231 and SUM-159 breast cancer cells (Valastyan et al. 2009). Ectopic expression of miR-31 in MDA-MB-231 and SUM-159 did not affect proliferation *in vitro*, but did reduce invasion and motility by 20-fold and 10-fold respectively. In addition, a 60% diminished resistance to anoikis-mediated cell death was observed in miR-31 overexpressing cells. Experiments in mice showed that MDA-MB-231 and SUM-159 cells expressing miR-31 generate larger primary tumors, and were impaired in their capacity to seed lung metastasis. Additional *in vivo* experiments evidenced that miR-31 affects metastatic colonization, and influences local invasion and early postintravasation events. Clinically, miR-31 levels were lower in breast cancer patients with metastasis. The authors then used PicTar and TargetScan programs to predict miR-31 putative targets. They defined more than 200 targets some of them encoding proteins involved in motility related processes, such as cell adhesion, cytoskeletal remodeling, and cell polarity. By using reporter luciferase assays they determined that six genes were repressed in miR-31 expressing cells, including frizzled3 (Fzd3), integrin a5 (ITGA5), myosin phosphatase-Rho interacting protein (M-RIP), matrix metallopeptidase 16 (MMP16), radixin (RDX), and RhoA. Interestingly, in miR-31-expressing cells, Fzd3, ITGA5, RDX, or RhoA reverted, at least partially, miR-31-imposed invasion and motility defects. Contrarily, M-RIP or MMP16 had no effect. Moreover, re-expression of RDX or RhoA completely allowed to rescue miR-31-mediated invasion and motility defects. These data evidenced that Fzd3, ITGA5, RDX, or RhoA are functional effectors of miR-31.

3.8 miR-21: A Master Regulator of Invasion and Metastasis

In another study, it was reported that forced suppression of miR-21 in metastatic MDA-MB-231 breast cancer did not affect cell growth *in vitro*, neither orthotopic growth after 28 days injection of MDA-MB-231 cells in mice.

In contrast, *in vitro* invasion assays showed that invasiveness of cells was significantly inhibited by 60%. To further confirm the role of miR-21 in metastasis, authors injected anti-miR-21-transfected and control MDA-MB-231 cells in tail-vein of nude mice and observed that antagomir-21 significantly reduced the number of lung metastasis. In a search for additional miR-21 targets to explain the anti-metastatic effects, they identified programmed cell death 4 (PDCD4) and maspin. Both genes have been implicated in invasion and metastasis. Moreover, transfection of Myc-tagged PDC4 and Flag-tagged maspin in MDA-MB-231 efficiently suppressed cell invasion in matrigel assays. Taken together, these data suggest that miR-21 may induce tumor invasion and metastasis by simultaneously repressing metastasis-related tumor suppressor genes (Zhu et al. 2008).

It is well known that human epidermal growth factor receptor 2 (HER2/neu) enhances local tumor invasion and lung metastasis. However the miRNAs downstream effectors of HER2/neu signaling have not been identified. Huang and coworkers (Huang et al. 2009) found that miR-21 expression correlates with HER2/neu upregulation, and linked to invasion induced by the membrane receptor. Ectopic expression of HER2/neu in MDA-MB-435 breast cancer cells was sufficient to induce miR-21 upregulation. In addition, authors showed that miR-21 upregulation is mediated via the MAPK (ERK1/2) pathway upon activation of HER2/neu signaling in BT-474 cells. miR-21 promoter contains consensus binding sites for AP-1, STAT3 and ETS-1 transcription factors. Interestingly, knocking down of Ets-1 transcription factor inhibited pri-miR-21 transcription by 40% in HER2/neu-overexpressing BT-474 cells, suggesting that Ets-1 may be regulating miR-21 expression. To test the role of miR-21 in cell invasion, a 550-bp genomic sequence containing the miR-21 pri-cursor was ectopically expressed in MDA-MB-435 cells. Results showed that MDA-MB-435/pri-miR-21 cells exhibited an increased ability to invade in matrigel-coated Boyden chamber assays, indicating that miR-21 upregulation mediated by HER2/neu signaling contributes to the invasiveness *in vitro*. Similarly,

Song and coworkers reported that in MDA-MB-231 breast cancer cell miR-21 regulates invasion in matrigel invasion assays. Using western blot and luciferase assays, they showed that miR-21 targets the tissue inhibitor of metalloproteinase (TIMP3) gene in miR-21 overexpressing MDA-MB-231 and MDA-MB-435 cells (Song et al. 2010).

3.9 miR-145 Tumor Suppressor

Sachdeva et al. previously showed that miR-145 acts as a tumor suppressor and inhibited cell growth in MCF-7 and HCT-116 cells, but not in MDA-MB-231 and LM2-4142 (Sachdeva and Mo 2010). Interestingly, transfection of miR-145 in MDA-MB-231 cells induced morphological changes that include round and flat cells, which seems to be associated with migration failure, suggesting that miR-145 may play a suppressive role in migration and invasion of metastatic cells. These assumptions were corroborated by matrigel chamber assays. Ectopic expression of miR-145 in MDA-MB-231 and LM2-4142 cells caused a reduction of invasion by 50% and 75%, respectively. *In vivo* metastasis studies by tail vein injection in mice revealed that miR-145 reduced the average number of lung tumors nodules by 3-fold in both MDA-MB-231 and LM2-4142 cells. In addition Sachdeva and Mo reported that miR-145 suppresses cell invasion and metastasis by directly targeting *mucin 1* gene which codified the MUC1protein that is frequently upregulated in some types of tumors. Suppression of MUC1 was corroborated by Western blot and immunofluorescence assays. To confirm the role of miR-145 in suppression of invasion through MUC1 inhibition, authors constructed plasmids with *muc1* gene with or without the 3'UTR. Cotransfection of these constructs with miR-145 suppressed the expression of MUC1 protein in clones carrying the but not in construct without 3'UTR, as expected. Moreover, enhanced expression of MUC1 induced important morphological changes in cells that appeared more elongated, compatible with a more invasive phenotype which was reverted by miR-145. Interestingly, suppression of MUC1 by miR-145 causes a reduction of β-catenin and its downstream effector gene cyclin D1. Moreover, silencing of MUC1 by RNA interference mimics the miR-145 function in suppression of invasion, associated with downregulation of β-catenin and cadherin 11. These data evidenced the prominent role of a miRNA function as tumor suppressor and invasion and metastasis suppressor.

4. CONCLUSIONS

Increasing experimental evidence showed that metastamiRs have emerged as new molecular players to regulate invasion and metastasis events in breast cancer (Fig. 1). Some of these miRNAs are regulators of cell motility and invasion (Table 1). Understanding the role of metastamiRs in regulating tumor invasion and metastasis program will provide a promising strategy for the identification of molecular markers for progression and prognosis, for response to chemotherapy, and for the development of novel metastamiRs-based treatments. Further investigations about the role of metastamiRs are required in order to use them as targets for therapy, prognosis and diagnosis in the near future.

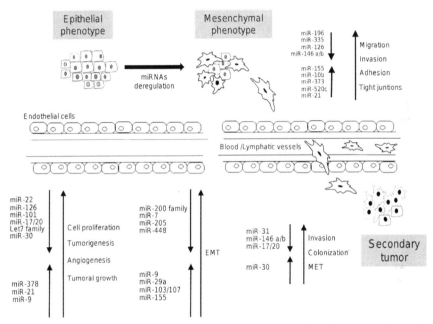

Figure 1. **Regulatory effects of selected miRNAs in cell proliferation, tumorigenesis, angiogenesis, and tumor growth.** MetastasmiRs involved in migration, invasion and metastasis events are outlined. EMT: epithelial-mesenchymal transition.

5. SUMMARY POINTS

- Aberrant expression of miRNAs expression is associated with initiation and progression of breast cancer where they act as oncogenes or tumor suppressors contributing to tumorigenesis.
- miRNAs have been shown to initiate invasion and metastasis, and thus they have been denominated as metastamiRs.
- Because of the existence of suitable metastasis models, the majority of metastamiRs have been identified in breast cancer tumors and cell lines.

- Metastasis-associated miRNAs have pro- and anti-metastatic effects; promote migration and invasion of cancer cells.
- MetastamiRs regulate several steps in the metastatic program and processes, such as epithelial-mesenchymal transition, apoptosis, angiogenesis, migration, invasion and metastasis.
- Successfull therapeutic intervention using antagomirs has been associated to the efficient delivery to growing tumor cells, inhibiting metastasis formation.

ACKNOWLEDGMENTS

Authors gratefully acknowledge the financial support from the National Council of Science and Technology (CONACyT), Mexico (grant 115306). This work was also supported by SIP-IPN (Mexico) and COFAA-IPN (Mexico).

ABBREVIATIONS

BRMS1	:	breast cancer metastasis suppressor 1
EMT	:	epithelial-mesenchymal transition
HER2/neu	:	human epidermal growth factor receptor 2
HOXD10	:	homeobox D10
Fzd3	:	frizzled3
IGFBP2	:	insulin-like growth factor binding protein 2
ITGA5	:	integrin a5
MERTK	:	*c*-Mer tyrosine kinase
metastamiR	:	metastasis associated miRNA
miRNA	:	microRNA
mRNAs	:	messenger RNAs
M-RIP	:	myosin phosphatase-Rho interacting protein
MMP16	:	matrix metallopeptidase 16
NF-κB	:	nuclear factor-κB
OncomiR	:	oncogenic miRNA
PDCD4	:	programmed cell death 4
RDX	:	radixin
RHOC	:	Ras homolog gene family member C
TNC	:	tenascin C
TIMP3	:	tissue inhibitor of metalloproteinase
UTR	:	untranslated region

REFERENCES

Aggarwal, B. Bharat. 2004. Nuclear factor-kappaB: the enemy within. Cancer Cell. 6: 203–61.

Baranwal, Somesh, and Suresh K. Alahari. 2010. miRNA control of tumor cell invasion and metastasis. Int J Cancer. 126: 1283–1290.

Bartel, D.P. 2004. MicroRNAs: Genomics, biogenesis, mechanism, and function. Cell. 116: 281–297.

Bhaumik, D., G.K. Scott, S. Schokrpur, C.K. Patil, J. Campisi, and C.C. Benz. 2008. Expression of microRNA-146 suppresses NF-kappaB activity with reduction of metastatic potential in breast cancer cells. Oncogene. 27: 5643–5647.

Bhaumik, D., G.K. Scott, S. Schokrpur, C.K. Patil, J. Campisi, and C.C. Benz. 2008. Expression of microRNA-146 suppresses NF-κB activity with reduction of metastatic potential in breast cancer cells. Oncogene. 27: 5643–5647.

Botas, J. 1993. Control of morphogenesis and differentiation by HOM/Hox genes. Curr Opin Cell Biol. 5: 1015–1022.

Calin, G.A., C. Sevignani, C.D. Dumitru, T. Hyslop, E. Noch, S. Yendamuri, M. Shimizu, S. Rattan, F. Bullrich, and M. Negrini. 2004. Human microRNA genes are frequently located at fragile sites and genomic regions involved in cancers. Proc Natl Acad Sci USA. 101: 2999–3004.

Croce, C.M. 2009. Causes and consequences of microRNA dysregulation in cancer. Nat Rev Genet. 10: 704–14. Review.

Esquela-Kerscher, A. and F.J. Slack. 2006. Oncomirs-microRNAs with a role in cancer. Nat Rev Cancer. 6: 259–269.

Ferlay, J., H.R. Shin, F. Bray, D. Forman, C. Mathers, and D.M. Parkin. 2010. Estimates of worldwide burden of cancer in 2008: GLOBOCAN 2008. Int J Cancer. 127: 2893–917.

F-Huber, M.A., N. Azoitei, B. Baumann, S. Grünert, A. Sommer, H. Pehamberger, N. Kraut, H. Beug, and T. Wirth. 2004. kappaB is essential for epithelial-mesenchymal transition and metastasis in a model of breast cancer progression. J Clin Invest. 114: 569–81.

Gregory, P.A., A.G. Bert, E.L. Paterson, S.C. Barry, A. Tsykin, G. Farshid, M.A. Vadas, Y. Khew-Goodall, and G.J. Goodall. 2008. The miR-200 family and miR-205 regulate epithelial to mesenchymal transition by targeting ZEB1 and SIP1. Nat Cell Biol. 10: 593–601.

Hanahan, D. and R.A. Weinberg. 2000. The hallmarks of cancer. Cell. 100: 57–70.

Hernando, E. 2007. microRNAs and cancer: Role in tumorigenesis, patient classification and therapy. Clin Transl Oncol. 9: 155–160.

Huang, Q., K. Gumireddy, M. Schrier, C. le Sage, R. Nagel, S. Nair, D.A. Egan, A. Li, G. Huang, and A.J. Klein-Szanto. 2008. The microRNAs miR-373 and miR-520c promote tumour invasion and metastasis. Nat Cell Biol. 10: 202–210.

Huang, T.H., F. Wu, G.B. Loeb, R. Hsu, A. Heidersbach, A. Brincat, D. Horiuchi, R.J. Lebbink, Y.Y. Mo, A. Goga, and M.T. McManus. 2009. Up-regulation of miR-21 by HER2/neu signaling promotes cell invasion. J Biol Chem. 284: 18515–24.

Hurst, D.R., M.D. Edmonds, and D.R. Welch. 2009a. Metastamir: The field of metastasis-regulatory microRNA is spreading. Cancer Res. 69: 7495–749.

Hurst, D.R., M.D. Edmonds, G.K. Scott, C.C. Benz, K.S. Vaidya, and D.R. Welch. 2009b. Breast cancer metastasis suppressor 1 up-regulates miR-146, which suppresses breast cancer metastasis. Cancer Res. 69:1279–1283.

Iorio, M.V., M. Ferracin, C.G. Liu, A. Veronese, R. Spizzo, S. Sabbioni, E. Magri, M. Pedriali, M. Fabbri, and M. Campiglio. 2005. MicroRNA gene expression deregulation in human breast cancer. Cancer Res. 65: 7065–7070.

Jiang, Y.W. and L.A. Chen. 2012. microRNAs as tumor inhibitors, oncogenes, biomarkers for drug efficacy and outcome predictors in lung cancer. Mol Med Report. 5: 890–4.

Kong, W., H. Yang, L. He, J.J. Zhao, D. Coppola, W.S. Dalton, and J.Q. Cheng. 2008. MicroRNA-155 is regulated by the transforming growth factor beta/Smad pathway and contributes to epithelial cell plasticity by targeting RhoA. Mol Cell Biol. 28: 6773–84.

Lazebnik, Y. 2010. What are the hallmarks of cancer? Nat Rev Cancer. 10: 232–233.

Lee, Daniel Y., Zhaoqun Deng, Chia-Hui Wang, and Burton B. Yang. 2007. MicroRNA-378 promotes cell survival, tumor growth, and angiogenesis by targeting SuFu and Fus-1 expression. PNAS. 104: 20350–20355.

Li, Q.Q., Z.Q. Chen, X.X. Cao, J.D. Xu, J.W. Xu, and Y.Y. Chen. 2010. Involvement of NF-kappaB/miR-448 regulatory feedback loop in chemotherapy-induced epithelial-mesenchymal transition of breast cancer cells. Cell Death Differ. 18: 16–25.

Ma, L., J. Teruya-Feldstein, and R.A. Weinberg. 2007. Tumour invasion and metastasis initiated by microRNA-10b in breast cancer. Nature. 449: 682–688.

Ma, L., F. Reinhardt, E. Pan, J. Soutschek, B. Bhat, E.G. Marcusson, J. Teruya-Feldstein, G.W. Bell, and R.A. Weinberg. 2010. Therapeutic silencing of miR-10b inhibits metastasis in a mouse mammary tumor model. Nat Biotechnol. 28: 341–347.

Nguyen, D.X., P.D. Bos, and J. Massagué. 2009. Metastasis: from dissemination to organ-specific colonization. Nat Rev Cancer. 9: 274–84. Review.

Nicoloso, S. Milena, Riccardo Spizzo, Masayoshi Shimizu, Simona Rossi, and George A. Calin. 2009. MicroRNAs—the micro steering wheel of tumour metastases. Nat Rev Cancer. 9: 293–302.

Oskarsson, T., S. Acharyya, X.H. Zhang, S. Vanharanta, S.F. Tavazoie, P.G. Morris, R.J. Downey, K. Manova-Todorova, E. Brogi, and J. Massagué. 2011. Breast cancer cells produce tenascin C as a metastatic niche component to colonize the lungs. Nat Med. 17: 867–874.

O'Day, E. and A. Lal. 2011. MicroRNAs and their target gene networks in breast cancer. Breast Cancer Res. 12: 201–210.

Png, K.J., N. Halberg, M. Yoshida, and S.F. Tavazoie. 2011. A microRNA regulon that mediates endothelial recruitment and metastasis by cancer cells. Nature. 481: 190–4.

Sachdeva, M. and Y.Y. Mo. 2010. MicroRNA-145 suppresses cell invasion and metastasis by directly targeting mucin 1. Cancer Res. 70: 378–387.

Song, B., C. Wang, J. Liu, X. Wang, L. Lv, L. Wei, L. Xie, Y. Zheng, and X. Song. 2010. MicroRNA-21 regulates breast cancer invasion partly by targeting tissue inhibitor of metalloproteinase 3 expression. J Exp Clin Cancer Res. 27: 29:29.

Song Guisheng, Yuxia Zhang, and Li Wang. 2009. MicroRNA-206 Targets notch3, activates apoptosis, and inhibits tumor cell migration and focus formation. The J Biol Chem. 284: 31921–31927.

Schmidt-Kittler, O., T. Ragg, A. Daskalakis, M. Granzow, A. Ahr, T.J. Blankenstein, M. Kaufmann, J. Diebold, H. Arnholdt, P. Muller et al. 2003. From latent disseminated cells to overt metastasis: Genetic analysis of systemic breast cancer progression. Proc Natl Acad Sci USA. 100: 7737–7742.

Steeg, P.S. 2006. Tumor metastasis: Mechanistic insights and clinical challenges. Nat Med. 12: 895–904.

Tavazoie, S.F., C. Alarcon, T. Oskarsson, D. Padua, Q. Wang, P.D. Bos, W.L. Gerald, and J. Massague. 2008. Endogenous human microRNAs that suppress breast cancer metastasis. Nature. 451: 147–152.

Valastyan, S., F. Reinhardt, N. Benaich, D. Calogrias, A.M. Szász, Z.C. Wang, J.E. Brock, A.L. Richardson, and R.A. Weinberg. 2009. A pleiotropically acting microRNA, miR-31, inhibits breast cancer metastasis. Cell. 137: 1032–1046.

Van de Wetering, M., M. Oosterwegel, K. van Norren, and H. Clevers. 1993. Sox-4, an Sry-like HMG box protein, is a transcriptional activator in lymphocytes. EMBO J. 12: 3847–3854.

Wang, L. and J. Wang. 2011. MicroRNA-mediated breast cancer metastasis: from primary site to distant organs. Oncogene. 31: 2499–2511.

Yang, J., S.A. Mani, J.L. Donaher, S. Ramaswamy, R.A. Itzykson, C. Come, P. Savagner, I. Gitelman, A. Richardson, and R.A. Weinberg. 2004. Twist, a master regulator of morphogenesis, plays an essential role in tumor metastasis. Cell. 117: 927–939.

Yong, Li, Maoxiang Zhang, Huijun Chen, Zheng Dong, Vadivel Ganapathy, Muthusamy Thangaraju, and Shuang Huang. 2010. Ratio of miR-196s to HOXC8 mRNA correlates with breast cancer cell migration and metastasis. Cancer Res. 70: 7894–7904.

Yu, Z., C. Wang, M. Wang et al. 2008. A cyclin D1/microRNA 17/20 regulatory feedback loop in control of breast cancer cell proliferation. J Cell Biol. 182: 509–17.

Zhang, H., Y. Li, and M. Lai. 2010. The microRNA network and tumor metastasis. Oncogene. 29: 937–48.

Zhang, B., X. Pan, G.P. Cobb, and T.A. Anderson. 2007. microRNAs as oncogenes and tumor suppressors. Dev Biol. 302: 1–12.

Zhu, S., H. Wu, F. Wu, D. Nie, S. Sheng, and Y.Y. Mo. 2008. MicroRNA-21 targets tumor suppressor genes in invasion and metastasis. Cell Res. 18: 350–359.

CHAPTER 4

ENDOMETRIAL CANCER AND MicroRNAs

Jaime Snowdon[a] and Harriet Feilotter[b,*]

ABSTRACT

Endometrial carcinoma, the most common gynecologic malignancy in Canada, is a heterogeneous disease. Accurate diagnosis and prognosis remains challenging, due both to the generally small amount of biopsy material available for examination, and the limited validation of biomarkers that have been correlated with outcome. As is the case with many human cancers, investigation into the differential expression of miRNAs is now beginning to provide robust and easily measured biomarkers associated with this cancer. This review will focus on the available miRNA literature as it pertains to endometrial carcinoma, including a discussion of miRNAs that appear to be consistently dysregulated in endometrial cancer compared to normal tissues, and on those whose expression correlates with outcome of the disease. Consolidating the available studies on miRNA expression in endometrial carcinoma, we propose to focus on findings that have been replicated in multiple studies, as well as those which appear to echo results from other cancers. We propose to explore potential downstream

Department of Pathology and Molecular Medicine. Queen's University 76 Stuart Street Kingston, Ontario, Canada.
[a]Email: 6jfs@queensu.ca
[b]Email: feilotth@KGH.KARI.NET
*Corresponding author

targets of the key identified miRNAs, and assess their putative role in endometrial carcinogenesis. Finally, we propose to summarize studies in human primary tumors and model systems that provide information about the molecular pathways that may underlie the development of this cancer.

1. INTRODUCTION

Endometrial carcinoma is the most common gynecologic malignancy in North America with an estimated 46,470 new cases and 8,120 resultant deaths in the United States in 2011 (Siegel et al. 2011). Endometrial carcinomas are classified into two types based on microscopic appearance, clinical behavior and epidemiology. Type I endometrial carcinomas, which comprise 70 to 80 percent of newly diagnosed cases, occur at a median age of 60 years and are associated with chronic exposure to unopposed estrogen. They most commonly present with endometrioid histology and are often preceded by atypical endometrial hyperplasia. Type II endometrial carcinomas account for less than 20 percent of all endometrial cancers and tend to occur at an older age, approximately five-10 years later than type I carcinomas. They are clinically more aggressive and have a poorer prognosis than type I tumors. They present with nonendometrioid histology, usually papillary serous or clear cell, and do not appear to be related to estrogen stimulation or endometrial hyperplasia.

Recently, molecular studies have supported this dualistic model through identification of different genetic profiles between the two types of carcinomas (reviewed in Liu 2007, Llobet et al. 2009). Briefly, type I tumors and endometrial hyperplasia are both associated with mutations in PTEN, β-catenin, K-ras, and defects in DNA mismatch repair. Mutations in p53 are rare and occur late in the development of type I cancers. Conversely, type II tumors are associated with early p53 mutations as well as loss of heterozygosity (LOH) on several chromosomes as well as Her-2/neu overexpression, supporting the idea that these tumors arise through a different biological process than do type 1 cancers. This, in turn, has led to the idea that dysregulation of other genes, such as miRNAs, might further delineate these tumors and perhaps shed light on their evolution. This review focuses on these studies, and attempts to summarize what is a rapidly growing literature that points to the critical role that these regulatory molecules play in the development of cancers of the endometrium.

2. MicroRNA DYSREGULATION IN ENDOMETRIAL CANCER

2.1 Global Patterns of microRNA Expression in Endometrial Cancer

There have been several published studies on global miRNA expression levels in type I endometrial carcinoma (Boren et al. 2008, Chung et al. 2009, Wu et al. 2009, Cohn et al. 2010, Ratner et al. 2010, Devor et al. 2011, Lee et al. 2011, Snowdon et al. 2011, Hiroki et al. 2011, Chung et al. 2012). Collectively, 88 up-regulated and 213 down-regulated miRNAs have been identified to date. However, of these, only 34 up-regulated (Table 1) and 80 down-regulated (Table 2) miRNAs were identified in multiple studies, potentially suggesting that these miRNAs might be significant since their dysregulation was evident despite the use of different platforms (microarray, reverse transcription-PCR), different statistical methods (SAM, T-test, ANOVA), and different control samples (atrophic endometrium, proliferative endometrium, combinations). Indeed, several of these miRNAs have been further studied in the context of endometrial carcinoma and other solid tumors, leading to additional information about the role of miRNAs in endometrial tumorigenesis. Alternatively, one must consider that the overall lack of overlap between potentially important miRNAs identified to date might simply reflect the rather arbitrary ways in which global expression data have been filtered.

Fewer global expression studies have been performed on type II endometrial carcinomas, likely reflecting the relative rarity of these tumors. Hiroki et al. (2010) identified 66 up-regulated and 54 down-regulated miRNAs, Hiroki et al. (2011) identified 54 down-regulated miRNAs and Devor et al. (2011) identified 23 up-regulated and nine down-regulated miRNAs in serous carcinomas compared to normal controls. When compared to the miRNAs consistently dysregulated in type I carcinomas (Tables 1 and 2), 20 of 34 up-regulated and 30 of 80 down-regulated miRNAs are shared between the two cancer subtypes (underlined in Tables 1 and 2), suggesting that these miRNAs may play roles in common processes of tumorigenesis. Devor et al. (2011) identified miRNAs significantly underexpressed (seven) or overexpressed (13) in both type I and type II endometrial carcinomas and Hiroki et al. (2011) report 48 miRNAs that are similarly underexpressed in both type I and type II endometrial carcinomas. While study of these miRNAs will likely provide information about general endometrial tumorigenesis, it is the study of differentially dysregulated miRNAs that may ultimately shed light on processes that are specific to the tumor subtypes. In keeping with findings from other tumors, patterns of miRNA expression are much more similar between tumor subtypes than between any subtype and normal controls. However,

Table 1. Summary of up-regulated miRNAs in type I endometrial carcinoma.

miRNA	Reference
hsa-miR-10a	Chung et al. 2009, Wu et al. 2009
hsa-miR-103	Boren et al. 2008, Chung et al. 2009
hsa-miR-106a	Boren et al. 2008, Chung et al. 2009
hsa-miR-107	Boren et al. 2008, Chung et al. 2009, Devor et al. 2011
hsa-miR-135b	Lee et al. 2011, Chung et al. 2012
hsa-miR-141	Chung et al. 2009, Wu et al. 2009, Snowdon et al. 2011, Lee et al. 2011
hsa-miR-146a	Snowdon et al. 2011, Lee et al. 2011, Devor et al. 2011, Chung et al. 2012
hsa-miR-148a	Lee et al. 2011, Devor et al. 2011
hsa-miR-155	Chung et al. 2009, Wu et al. 2009, Lee et al. 2011
hsa-miR-181c	Cohn et al. 2010, Lee et al. 2011
hsa-miR-182	Chung et al. 2009, Wu et al. 2009, Ratner et al. 2010, Lee et al. 2011, Chung et al. 2012
hsa-miR-183	Chung et al. 2009, Ratner et al. 2010, Cohn et al. 2010, Lee et al. 2011, Chung et al. 2012
hsa-miR-192	Lee et al. 2011, Chung et al. 2012
hsa-miR-194	Chung et al. 2009, Lee et al. 2011, Chung et al. 2012
hsa-miR-200a	Chung et al. 2009, Wu et al. 2009, Ratner et al. 2010, Snowdon et al. 2011, Lee et al. 2011
hsa-miR-200b	Wu et al. 2009, Snowdon et al. 2011, Lee et al. 2011
hsa-miR-200c	Chung et al 2009, Wu et al. 2009, Cohn et al. 2010, Snowdon et al. 2011, Lee et al. 2011
hsa-miR-203	Chung et al. 2009, Wu et al. 2009, Snowdon et al. 2011, Lee et al. 2011, Devor et al. 2011
hsa-miR-205	Chung et al. 2009, Wu et al. 2009, Ratner et al. 2010, Cohn et al. 2010, Snowdon et al. 2011, Chung et al. 2012
hsa-miR-210	Boren et al. 2008, Chung et al. 2009, Wu et al. 2009, Snowdon et al. 2011, Devor et al. 2011
hsa-miR-215	Chung et al. 2009, Lee et al. 2011
hsa-miR-223	Chung et al. 2009, Cohn et al. 2010, Chung et al 2012
hsa-miR-31	Wu et al. 2009, Lee et al. 2011
hsa-miR-330	Chung et al. 2009, Lee et al. 2011
hsa-miR-34a	Chung et al. 2009, Ratner et al. 2010, Lee et al. 2011
hsa-miR-423	Boren et al. 2008, Cohn et al. 2010
hsa-miR-425	Cohn et al. 2010, Devor et al. 2011
hsa-miR-429	Wu et al. 2009, Snowdon et al. 2011, Lee et al. 2011, Chung et al. 2012
hsa-miR-449	Wu et al. 2009, Chung et al. 2012
hsa-miR-7	Lee et al. 2011, Chung et al. 2012
hsa-miR-9	Cohn et al. 2010, Snowdon et al. 2011, Devor et al. 2011, Chung et al. 2012
hsa-miR-9*	Snowdon et al. 2011, Devor et al. 2011, Chung et al. 2012
hsa-miR-95	Chung et al. 2009, Chung et al. 2012
hsa-miR-96	Wu et al. 2009, Snowdon et al. 2011

The listed miRNAs have been shown to be up-regulated (p<0.05) in at least two independent studies. The underlined miRNAs are similarly dysregulated in type I and type II endometrial carcinomas.

Table 2. Summary of down-regulated miRNAs in type I endometrial carcinoma.

miRNA	Reference
hsa-let-7a	Cohn et al. 2010, Hiroki et al. 2011
hsa-let-7c	Lee et al. 2011, Devor et al. 2011, Hiroki et al. 2011
hsa-let-7e	Lee et al. 2011, Hiroki et al. 2011
hsa-let-7i	Boren et al. 2008, Lee et al. 2011, Hiroki et al. 2011
hsa-miR-10b*	Snowdon et al. 2011, Hiroki et al. 2011
hsa-miR-100	Snowdon et al. 2011, Lee et al. 2011, Devor et al. 2011, Hiroki et al. 2011
hsa-miR-101	Lee et al. 2011, Hiroki et al. 2011
hsa-miR-10b	Lee et al. 2011, Hiroki et al. 2011
hsa-miR-125b	Lee et al. 2011, Devor et al. 2011, Hiroki et al. 2011
hsa-miR-126	Lee et al. 2011, Hiroki et al. 2011
hsa-miR-126*	Lee et al. 2011, Hiroki et al. 2011
hsa-miR-130a	Lee et al. 2011, Devor et al. 2011, Hiroki et al. 2011
hsa-miR-132	Lee et al. 2011, Hiroki et al. 2011
hsa-miR-133a	Lee et al. 2011, Hiroki et al. 2011
hsa-miR-133b	Wu et al. 2009, Lee et al. 2011, Hiroki et al. 2011
hsa-miR-136	Lee et al. 2011, Hiroki et al. 2011
hsa-miR-143	Lee et al. 2011, Hiroki et al. 2011
hsa-miR-145	Lee et al. 2011, Hiroki et al. 2011
hsa-miR-149	Lee et al. 2011, Hiroki et al. 2011
hsa-miR-152	Boren et al. 2008, Snowdon et al. 2011, Lee et al. 2011, Hiroki et al. 2011
hsa-miR-154	Lee et al. 2011, Hiroki et al. 2011
hsa-miR-193b	Wu et al. 2009, Lee et al. 2011, Hiroki et al. 2011
hsa-miR-195	Lee et al. 2011, Hiroki et al. 2011
hsa-miR-196b	Lee et al. 2011, Hiroki et al. 2011
hsa-miR-199a-5p	Devor et al. 2011, Hiroki et al. 2011
hsa-miR-199b	Lee et al. 2011, Chung et al. 2012
hsa-miR-199b-3p	Snowdon et al. 2011, Hiroki et al. 2011
hsa-miR-199b-5p	Snowdon et al. 2011, Hiroki et al. 2011
hsa-miR-204	Wu et al. 2009, Lee et al. 2011, Hiroki et al. 2011, Chung et al. 2012
hsa-miR-212	Lee et al. 2011, Devor et al. 2011
hsa-miR-214	Lee et al. 2011, Devor et al. 2011, Hiroki et al. 2011
hsa-miR-214*	Devor et al. 2011, Hiroki et al. 2011
hsa-miR-218	Lee et al. 2011, Devor et al. 2011
hsa-miR-22	Lee et al. 2011, Hiroki et al. 2011
hsa-miR-221	Boren et al. 2008, Lee et al. 2011, Hiroki et al. 2011
hsa-miR-222	Lee et al. 2011, Hiroki et al. 2011
hsa-miR-224	Lee et al. 2011, Hiroki et al. 2011
hsa-miR-23b	Lee et al. 2011, Hiroki et al. 2011

Table 2. contd....

Table 2. contd....

miRNA	Reference
hsa-miR-24	Lee et al. 2011, Hiroki et al. 2011
hsa-miR-26a	Lee et al. 2011, Hiroki et al. 2011
hsa-miR-26b	Lee et al. 2011, Hiroki et al. 2011
hsa-miR-27b	Lee et al. 2011, Hiroki et al. 2011
hsa-miR-299-3p	Lee et al. 2011, Hiroki et al. 2011
hsa-miR-299-5p	Lee et al. 2011, Hiroki et al. 2011
hsa-miR-29a	Lee et al. 2011, Hiroki et al. 2011
hsa-miR-30b	Lee et al. 2011, Hiroki et al. 2011
hsa-miR-30c	Boren et al. 2008, Hiroki et al. 2011
hsa-miR-320	Lee et al. 2011, Hiroki et al. 2011
hsa-miR-324-5p	Lee et al. 2011, Hiroki et al. 2011
hsa-miR-368	Wu et al. 2009, Lee et al. 2011
hsa-miR-376a	Snowdon et al. 2011, Lee et al. 2011, Hiroki et al. 2011
hsa-miR-377	Lee et al. 2011, Hiroki et al. 2011
hsa-miR-378	Lee et al. 2011, Hiroki et al. 2011
hsa-miR-379	Lee et al. 2011, Hiroki et al. 2011
hsa-miR-381	Snowdon et al. 2011, Lee et al. 2011, Hiroki et al. 2011
hsa-miR-382	Lee et al. 2011, Hiroki et al. 2011
hsa-miR-409-3p	Lee et al. 2011, Hiroki et al. 2011
hsa-miR-410	Snowdon et al. 2011, Lee et al. 2011, Devor et al. 2011, Hiroki et al. 2011
hsa-miR-411	Wu et al. 2009, Lee et al. 2011, Hiroki et al. 2011
hsa-miR-424	Cohn et al. 2010, Snowdon et al. 2011, Lee et al. 2011, Devor et al. 2011, Hiroki et al. 2011
hsa-miR-424*	Snowdon et al. 2011, Hiroki et al. 2011
hsa-miR-431	Cohn et al. 2010, Snowdon et al. 2011, Lee et al. 2011
hsa-miR-432	Snowdon et al. 2011, Lee et al. 2011, Hiroki et al. 2011
hsa-miR-452	Lee et al. 2011, Hiroki et al. 2011
hsa-miR-487b	Wu et al. 2009, Lee et al. 2011, Hiroki et al. 2011
hsa-miR-495	Lee et al. 2011, Devor et al. 2011, Hiroki et al. 2011
hsa-miR-497	Lee et al. 2011, Hiroki et al. 2011
hsa-miR-502-3p	Devor et al. 2011, Hiroki et al. 2011
hsa-miR-502-5p	Devor et al. 2011, Hiroki et al. 2011
hsa-miR-503	Cohn et al. 2011, Snowdon et al. 2011, Lee et al. 2011, Hiroki et al. 2011
hsa-miR-505	Lee et al. 2011, Hiroki et al. 2011
hsa-miR-532-5p	Devor et al. 2011, Hiroki et al. 2011
hsa-miR-542-3p	Snowdon et al. 2011, Lee et al. 2011, Hiroki et al. 2011
hsa-miR-542-5p	Snowdon et al. 2011, Lee et al. 2011, Hiroki et al. 2011
hsa-miR-598	Lee et al. 2011, Hiroki et al. 2011

Table 2. contd....

Table 2. contd....

miRNA	Reference
hsa-miR-660	Lee et al. 2011, Hiroki et al. 2011
hsa-miR-768-3p	Lee et al. 2011, Hiroki et al. 2011
hsa-miR-98	Lee et al. 2011, Hiroki et al. 2011
hsa-miR-99a	Lee et al. 2011, Devor et al. 2011, Hiroki et al. 2011
hsa-miR-99b	Wu et al. 2009, Lee et al. 2011, Hiroki et al. 2011

The listed miRNAs have been shown to be down-regulated (p<0.05) in at least two independent studies. The underlined miRNAs are similarly dysregulated in type I and type II endometrial carcinomas.

there have been attempts to identify miRNAs whose expression patterns differentiate between tumor subtypes (Devor et al. 2011), resulting in a short list of potentially interesting molecules (Table 3). Similarly, Hiroki et al. (2011) identified six miRNAs (miR-34b, miR-34b*, miR-34c-5p, miR-33a, miR-142-3p, and miR-142-5p) that are down-regulated in type II but not type I endometrial carcinomas. The identification of the more subtle patterns that can differentiate subtypes may hold the key to unlock our understanding of the different clinicopathologic characteristics of these tumor types. A more detailed understanding of the processes in which these miRNAs are involved is a step in that direction.

Table 3. List of miRNAs significantly different (p<0.05) between serous and endometrioid tumors (Devor et al. 2011).

miRNA	Fold Change
hsa-let-7c	3.2
hsa-miR-130a	5.9
hsa-miR-146a	–8.9
hsa-miR-218	5.9
hsa-miR-338-3p	10.9
hsa-miR-375	–232.7
hsa-miR-423-5p	–4.6
hsa-miR-450a	5.4
hsa-miR-490-3p	11.6
hsa-miR-504	8.1
hsa-miR-518e	19.8
hsa-miR-542-3p	5.8
hsa-miR-570	27.6
hsa-miR-675	80.2
hsa-miR-9	–9.8
hsa-miR-9*	–13.6
hsa-miR-99a/100	4.8

The fold change is presented as serous tumors compared to endometrioid tumors.

2.2 MicroRNAs and the Malignant Phenotype

The study of individual miRNAs in the context of endometrial cancer has provided links to key pathways that are likely to underlie tumor development and behavior. While a detailed investigation of any given pathway is beyond the scope of this review, a summary of key findings implicates well known pathways through their association with critical dysregulated miRNAs (summarized in Table 4).

Table 4. Summary of miRNAs that are implicated in endometrial tumorigenesis.

microRNA	Target Pathway	Reference
miR-200c	Epithelial-to-mesenchymal transition (ZEB1), TUBB3, PTEN, BRD7	Cochrane et al. 2009, Cochrane et al. 2010, Cohn et al. 2010, Park et al. 2012
miR-194	BMI-1	Dong et al. 2011
miR-182	PTEN	Cohn et al. 2010
miR-9, miR-27, miR-96, miR-153, miR-182, miR-183, miR-96	FOXO1	Myatt et al. 2010
miR-204	FOXC1	Chung et al. 2012
miR-101	COX2	Hiroki et al. 2010
miR-125b	TP53INPI	Jiang et al. 2011
miR-196a	ANXA1	Luthra et al. 2008
miR-145	OCT4	Wu et al. 2011

2.2.1 MicroRNAs and the epithelial-to-mesenchymal transition

The epithelial-to-mesenchymal transition (EMT) is a form of genetic reprogramming whereby carcinoma cells lose their epithelial characteristics and gain a mesenchymal phenotype which is better suited for invasion and metastasis (Zeisberg et al. 2009). The EMT is characterized by loss of epithelial surface makers such as E-cadherin and acquisition of mesenchymal markers such as N-cadherin and vimentin. Cells that have undergone EMT are often spindled in appearance and gain cell motility relative to their epithelial counterparts. They also secrete proteases and other enzymes that allow them to disrupt extracellular matrix networks.

The EMT process is positively regulated by several transcription factors including ZEB1 and ZEB2 which negatively regulate E-cadherin expression. Studies of normal endometrium and low-grade endometrioid carcinomas show that ZEB1 expression is limited to the stromal cells. However, high grade endometrioid carcinomas as well as serous carcinomas and carcinosarcomas show elevated ZEB1 levels and correspondingly low E-cadherin levels (Spoelstra et al. 2006, Singh et al. 2008).

The miR-200 family has been extensively studied with respect to its role in regulating genes of the EMT. This family consists of five co-ordinately regulated members clustered on chromosome 1 (miR-200c/b/429) and chromosome 12 (miR-200a/141) (Vrba et al. 2010). They share a common seed sequence (Fig. 1) consistent with their ability to regulate similar genes. Members of the miR-200 family, as well as miR-205, have been shown to directly target and regulate ZEB1 and ZEB2 and have thus been implicated in the intricate regulation of EMT in several carcinomas, such as breast (Paterson et al. 2008) and bladder (Adam et al. 2009). In addition to targeting ZEB1 and ZEB2, miR-200c has also been shown to regulate other genes involved in EMT including fibronectin 1, moesin, neurotrophic tyrosine receptor kinase type 2, leptin receptor and Rho GTPase activating protein 19 (Howe et al. 2011).

Interestingly, the miR-200 family has consistently shown increased expression in type I endometrioid adenocarcinomas relative to controls (Table 1), in keeping with observations from other tumors including melanoma (Mueller et al. 2009, Elson-Schwab et al. 2010), ovarian carcinoma (Iorio et al. 2007), and colorectal carcinoma (Xi et al. 2006). Higher levels of miR-200 family members are also observed in the epithelial component of uterine carcinosarcomas (Castilla et al. 2011) consistent with the hypothesis that the miR-200 family may help to maintain an epithelial phenotype, potentially by negatively regulating ZEB1 and ZEB2. This is supported by evidence that restoration of miR-200c expression to endometrial cancer cell lines restores E-cadherin expression and reduces cell migration and invasion potential (Cochrane et al. 2009).

The miR-200 family is known to be regulated in turn through exposure to estrogen (Klinge 2009, Bhat-Nakshatri et al. 2009) and estrogen receptor α binding sites have been identified upstream of the miR-200 promoter (Bhat-Nakshatri et al. 2009). It has been well-established that ER positive endometrial carcinomas are less aggressive than ER negative tumors; 89 percent of low grade endometrioid adenocarcinomas express ER and this

hsa-miR-200b	5′	UAAUACUGCCUGGUAAUGAUGAC	3′
hsa-miR-429	5′	UAAUACUGUCUGGUAAAACCGU	3′
hsa-miR-200c	5′	UAAUACUGCCGGGUAAUGAUGG	3′
hsa-miR-200a	5′	UAACACUGUCUGGUAACGAUGU	3′
hsa-miR-141	5′	UAACACUGUCUGGUAAAGAUGG	3′

Figure 1. Sequence alignment of members of the miR-200 family with the seed sequence (nucleotides 2-7) underlined.

expression decreases with higher grade tumors (Kounelis et al. 2000). It is therefore postulated that the miR-200 family, under the influence of estrogen, helps to maintain an epithelial phenotype in lower grade endometrioid adenocarcinomas. This hypothesis is supported by the morphology of invasive low grade endometrioid adenocarcinomas which still maintain an epithelial phenotype as they penetrate the myometrium (Fig. 2). The association of estrogen, estrogen receptor, the miR-200 family and the EMT requires further investigation.

B lymphoma mouse Moloney leukemia virus insertion region 1 (BMI-1) is overexpressed in a number of cancers including type I endometrial carcinomas (Honig et al. 2010). Studies have demonstrated that BMI-1 is associated with EMT in squamous cell carcinoma (Yang et al. 2010) and endometrial cancer cell lines (Dong et al. 2011) as well as malignant transformation in hepatocellular carcinoma (Sasaki et al. 2008) and breast cancer metastasis (Hoenerhoff et al. 2009). MicroRNA-194 suppresses BMI-1 expression via direct binding to the 3'-untranslated region (3'UTR) of this gene and ectopic miR-194 induces a mesenchymal-to-epithelial phenotype (Dong et al. 2011), suggesting that repression of BMI-1 by miR-194 may have therapeutic applications. It is intriguing that, like miR-200 and miR-205, miR-194 is up-regulated in type I carcinomas (Table 1), potentially assisting cells in maintaining an epithelial phenotype.

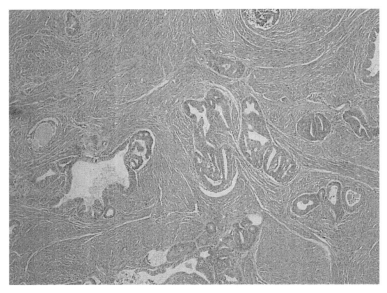

Figure 2. Representative section of a low grade type I endometrial endometrioid adenocarcinoma with extensive myometrial invasion yet with preservation of cell-to-cell adhesion and epithelial phenotype (Hematoxylin-Phloxine-Saffron stain 100x).

2.2.2 MicroRNAs and chemoresistance

In addition to mediating cancer invasion and metastasis, EMT has been linked to the development of chemoresistance in tumors of the lung (Witta et al. 2006) breast (Zhang et al. 2012) and colon (Yang et al. 2006, Buck et al. 2007). In endometrial cancer cell lines, miR-200c has been shown to target a microtubule component, class III β-tubulin (*TUBB3*), normally expressed in neuronal cells (Cochrane et al. 2010). The expression of TUBB3 is a mechanism of resistance to the microtubule-binding chemotherapies that are often employed against endometrial carcinoma. Cochrane et al. (2010) demonstrate that miR-200c can restore sensitivity to such chemotherapeutic agents by directly targeting and repressing TUBB3.

2.2.3 MicroRNAs and tumorigenesis

Phosphatase and Tensin Homologue (PTEN). PTEN is a tumor suppressor gene located on chromosome 10q23 and a key regulator of the PTEN-PI3K-AKT pathway (Steck et al. 1997, Li et al. 1997). PTEN protein dephosphorylates phosphotidylinositol (3,4,5) triphosphate (PI (3,4,5) triphosphate) thereby abrogating the ability of PI (3,4,5) triphosphate to activate Akt through phosphorylation. Loss of PTEN activity leads to the constitutive activation of Akt and the subsequent inhibition of apoptosis and promotion of cell proliferation. Loss of PTEN expression has been described in 40–80 percent of endometrioid adenocarcinomas and in up to 55 percent of the precursor lesion, atypical hyperplasia (Kong et al. 1997, Risinger et al. 1997, Mutter et al. 2000). These results implicate this protein in early carcinogenesis in this tumor type. Loss of the protein can occur through mutations, loss of heterozygosity, and potentially epigenetic mechanisms.

Putative miRNA regulators of the PTEN gene include miR-182 and members of the miR-200 family (Cohn et al. 2010). Cohn et al. (2010) demonstrate that transfection of miR-182 and miR-200c into an endometrial cancer cell line leads to the decreased expression of PTEN as assessed by western blot. Both miR-182 and the miR-200 family have consistently been shown to be up-regulated in endometrioid adenocarcinomas and may therefore be contributing to the early steps of carcinogenesis in this tumor type.

Forkhead box (FOX) proteins. The FOX family proteins are transcription factors that regulate the expression of genes involved in cell growth, proliferation and differentiation (Lehmann et al. 2003, Tuteja and Kaestner 2007a,b). Family members including FOXA1, FOXC1, FOXO1 and FOXP1 have been shown to be aberrantly expressed in endometrial carcinoma (Giatromanolaki et al. 2006, Wong et al. 2007, Goto et al. 2008).

The FOXO family of transcription factors function downstream of the PI3K/AKT pathway and are therefore central to a myriad of cellular functions including proliferation, apoptosis, differentiation, and resistance to oxidative stress (Arden 2008, Burgering 2008, Calnan and Brunet 2008, Gomes et al. 2008). FOXO proteins have been shown to act as tumor suppressor genes in animal studies of thymic lymphomas and haemangiomas (Paik et al. 2007). In humans, FOXO1 is expressed in normal cycling endometrial epithelial cells and decidualized stromal cells (Lam et al. 2012). FOXO1 has been shown to be down-regulated in endometrial cancer compared to normal controls (Goto et al. 2008) and Myatt et al. (2010) demonstrate that this reduced expression is associated with increased expression of several miRNAs predicted to target the 3'-UTR of FOXO1 including miR-9, miR-27, miR-96, miR-153, miR-182, miR-183 and miR-186. MicroRNA-9, miR-96, miR-182 and miR-183 have consistently been shown to be up-regulated in type I endometrial carcinomas (Table 1). Furthermore, miR-182, miR-183, and miR-96 are a family whose expression has been shown to be prognostic in glioma (Jiang et al. 2010) and lung cancer (Zhu et al. 2011) and they have also been implicated in tumorigenesis (Guttilla and White 2009) and sensitivity to chemotherapeutics (Moskwa et al. 2011) in breast carcinoma.

The FOXC genes are important regulators of angiogenesis (Myatt and Lam 2007, Hayashi et al. 2008), are associated with poor prognosis in breast carcinoma (Taube et al. 2010) and are overexpressed in early invasive and basal-like breast cancers (Ray et al. 2010). Chung et al. (2012) demonstrate the miR-204, which is consistently down-regulated in endometrial carcinomas (Table 2), targets FOXC1 through two binding sites in the 3'-UTR of this gene. FOXC1 expression is inversely related to miR-204 expression in endometrial carcinoma and overexpression of miR-204 results in decreased migration and invasion in an endometrial cancer cell line (HEC1A cells). These results highlight the important miRNA-FOX gene interactions in endometrial carcinoma and further studies may reveal important prognostic and therapeutic strategies.

Bromodomain containing 7 (BRD7). Bromodomain containing 7 (BRD7) is a gene with tumor suppressor characteristics in nasopharyngeal carcinoma (Peng et al. 2007, Liu et al. 2008). Park et al. (2012) demonstrate that miR-200c inhibits BRD7 in endometrial carcinoma cell lines but not through direct binding of the 3'-UTR of the BRD7 gene. Rather, miR-200c appears to regulate the translocation of β-catenin from the cytoplasm to the nucleus through BRD7 which results in increased expression of the β-catenin target genes, *cyclin D1* and *c-myc*. This has a tumorigenic effect on cells and this oncogenic capacity of miR-200c is supported by evidence that miR-200c can regulate cell survival, apoptosis and proliferation. The authors propose that an intermediate molecule may exist which is the direct target of miR-200c and mediates the inhibition of BRD7. These results support an oncogenic role for miR-200c which helps to reconcile the consistent up-regulation of this miRNA in endometrial carcinoma (Table 1).

Cyclooxygenase-2 (COX-2). The COX-2 enzyme catalyzes the key step in prostaglandin formation from arachidonic acids and COX-2 levels are normally under tight transcriptional and post-transcriptional control. Loss of this control and the subsequent increase in COX-2 levels has been shown to be tumorigenic in cancers of the colon (Eberhart et al. 1994, Sano et al. 1995, Kutchera et al. 1996), lung (Wolff et al. 1998), pancreas (Tucker et al. 1999), prostate (Yoshimura et al. 2000) and breast (Subbaramaiah et al. 2002). In a mouse model of endometrial carcinoma, loss of PTEN is associated with elevated COX-2 levels and decreased levels of two miRNAs predicted to target COX-2: miR-199a* and miR-101a (Daikoku et al. 2008). *In vitro* studies using HeLa cells show that miR-199a and miR-101a directly target COX-2 mRNA (Chakrabarty et al. 2007) and in human type II endometrial carcinomas, COX-2 expression is inversely correlated with miR-101 levels (Hiroki et al. 2010). The role of miRNAs in the regulation of this key enzyme will undoubtedly be further investigated in several human tumors including tumors of the endometrium.

2.2.4 Tumor protein 53-induced nuclear protein 1 (TP53INP1)

MicroRNA-125b has been shown to be dysregulated in several tumors including prostate (Shi et al. 2007) and liver (Liang et al. 2010) and overexpression of miR-125b has been demonstrated in advanced endometrial carcinoma (Cohn et al. 2010). Jiang et al. (2011) demonstrate that exogenous miR-125b increases proliferation and migration in a type I endometrial cancer cell line while abrogation of miR-125b with synthetic inhibitors suppresses proliferation and migration in a type II endometrial cancer cell line. This study demonstrates the target of miR-125b is the tumor suppressor gene Tumor Protein 53-Induced Nuclear Protein 1 (TP53INP1) which is a proapoptotic stress-induced p53 target gene and which is reduced or lost in cancers of the breast (Ito et al. 2006), stomach (Jiang et al. 2006), and pancreas (Gironella et al. 2007). In type II endometrial cancer, where p53 is often mutated, TP53INP1 may have an important role in antiproliferative and proapoptotic cellular pathways and the down-regulation of TP53INP1 by miR-125b may play an important role in type II endometrial carcinogenesis

Annexin A1 (ANXA1). Annexin A1 is an important mediator of apoptosis (Solito et al. 2001) that has been shown to be dysregulated in several carcinomas including prostate (Kang et al. 2002), esophageal (Paweletz et al. 2000, Xia et al. 2002, Hu et al. 2004), pancreatic (Bai et al. 2004), hepatic (Masaki et al. 1996) and stomach (Sinha et al. 1998). Interestingly, the regulatory role of ANXA1 appears to be tissue- or tumor-specific as some tumors, like those of the prostate and esophagus, show suppression of this protein while those of the pancreas, liver, and stomach show increased expression. Luthra et al. (2008) demonstrate a reciprocal relationship between miR-196a and ANXA1 mRNA levels in endometrial cancer cell lines and also show that transfection of miR-196a mimics results in lower ANXA1 mRNA levels as detected by real time quantitative PCR and western blot. These results suggest that the dysregulation of ANXA1 in tumors may be mediated in part by miRNAs; the significance of this pathway in endometrial carcinoma deserves further characterization.

Octamer-binding Transcription Factor 4 (OCT4). Octamer-binding transcription factor 4 (OCT4), a member of the POU family of transcription factors, is critically involved in the self-renewal of embryonic stem cells (ESC) (Babaie et al. 2007). OCT4 has been implicated as an oncogene (Hochedlinger et al. 2005) and miR-145 has been shown to regulate OCT4 levels in human ESC (Xu et al. 2009). Wu et al. (2011) demonstrate that up-regulating miR-145 in a endometrial cancer cell lines results in the repression of OCT4 and promotes cell differentation. In addition, miR-145 levels are decreased in poorly differentiated endometrial carcinomas concurrent with an overexpression of OCT4. These results support a tumor suppressor role for miR-145, possibly through the regulation of OCT4. In support of this, miR-145 has been shown to be down-regulated in type I carcinomas compared to normal endometrium (Lee et al. 2011).

Mismatch Repair Genes (MMR). Lynch syndrome, an autosomal dominant syndrome, is caused by germline mutations in DNA mismatch repair genes (most commonly *hMLH1, hMSH2, hMSH6* and *hPMS2*) which frequently results in the development of microsatellite instability (MSI) (Peltomaki et al. 1993, Loeb 1994). Patients with Lynch syndrome are at an increased risk for the development of various solid tumors with those of the colon and endometrium occurring most commonly (Vasen et al. 1990, Aarnio et al. 1999). Defects in MMR genes can be hereditary, as in Lynch syndrome, or can be sporadic resulting from promoter methylation. The overall prevalence of MSI in endometrial carcinomas is 20–25 percent and in 75 percent of cases, this results from epigenetic methylation of the *hMLH1* promoter (Esteller et al. 1998, Gurin et al. 1999). In sporadic colon cancer, overexpression of miR-155 is associated with a down-regulation of MMR genes (Valeri et al. 2010). Although the relationship between miR-155 and MMR genes has not been formally studied in endometrial cancer, it should be noted that mir-155 has consistently been shown be overexpressed in both type I and type II endometrial tumors (Table 1). As such, miR-155 may be involved in early tumorigenesis in this tumor type through the regulation of MMR genes and further studies in this area are required.

Clearly, studies on the impact of individual miRNAs on pathways in subtypes of endometrial carcinoma, as well as in other cancers, provides glimpses into potential mechanisms of many aspects of tumor behaviour. Continuation of such studies may provide not only novel insights, but also begin to delineate novel therapeutic targets that may be relevant to increasingly specific subsets of disease, thereby potentially improving detection, diagnosis and outcomes.

3. MECHANISMS OF miRNA DYSREGULATION

What regulates the regulators is a subject of much interest, and provides an additional layer of information about cancer genesis. Studies in endometrial carcinoma as well as other tumor types have begun to shed some light on the mechanisms of miRNA regulation, leading to the identification of additional therapeutic targets.

3.1 DNA Hypermethylation

DNA hypermethylation within CpG islands leads to the inactivation of many tumor suppressor genes (Herman and Baylin 2003) and has also been shown to contribute to the down-regulation of miRNAs (Chuang and Jones 2007). Tsuruta et al. (2011) have shown that miR-152, often down-regulated in type I carcinoma, is epigenetically silenced in this tumor. Restoration of miR-152 expression in cell line models was sufficient to inhibit tumor cell growth *in vitro* and *in vivo,* suggesting that this silencing contributes to the growth potential of the tumors.

Similarly, miR-129-2 is down-regulated in type I carcinomas and this is associated with hypermethylation of the miR-129-2 CpG island in endometrial cancer cell lines and solid tumors (Huang et al. 2009). This silencing of miR-129-2 is associated with an up-regulation of the oncogene SOX4 and restoration of miR-129-2 leads to a decrease in SOX4 expression and reduced proliferation of cell lines. Additionally, the hypermethylation of the miR-129-2 promoter is associated with worse survival in endometrial cancer. Aberrant SOX4 expression may in part be caused by epigenetic alterations in the tumor suppressor miR-129-2.

Members of the miR-34 family, which include miR-34a, miR-34b, and miR-34c, have been shown to be direct targets of p53 and have shown tumor suppressor capabilities (Corney et al. 2007, He et al. 2007, Bommer et al. 2007, Tarasov et al. 2007, Hermeking 2010). The promoters of the miR-34a and miR-34b/c genes have been shown to be subject to inactivation by DNA methylation in tumors of the colon, breast, pancreas, ovary, soft tissue, bladder and kidney (reviewed in Vogt et al. 2011) and the tumorigenic effect of this down-regulation may be mediated by the MET proto-oncogene (He et al. 2007). Hiroki et al. (2011) demonstrate that members of the miR-34 family are down-regulated in type II endometrial tumors and ectopic expression of miR-34b in an endometrial cancer cell line inhibits cell growth, migration and invasion. The promoter for miR-34b was shown to be hypermethylated and treating cells lines with 5-aza-2′ deoxycytidine restored miR-34b expression supporting the epigenetic regulation of this miRNA. The effects of miR-34b suppression may be mediated by MET as the expression of this oncogene was reduced following restoration of miR-34b *in vitro*. These results coupled with the growing literature on the interplay between the miR-34 family, p53 and MET support the role of miRNAs and epigenetics in endometrial carcinoma.

3.2 Dysregulation of MicroRNA Biogenesis

Drosha and Dicer are important regulators of miRNA biogenesis (Bartel 2004, Ambros 2004, and Saito et al. 2005). MicroRNAs are first transcribed to primary miRNAs by polymerase II and then are sequentially cleaved by Drosha and then Dicer before they are incorporated into the functional RISC complex. Alterations in Dicer and Drosha gene expression have been associated with various malignancies including breast (Hinkal et al. 2011), ovarian (Faggad et al. 2010), and endometrial (Torres et al. 2011). Using quantitative RT-PCR, Dicer and Drosha gene levels were decreased by a factor of 1.54 (p=0.009) and 1.4 (p=0.008) respectively in endometrial carcinoma cases compared to normal controls. The down-regulation of these components of miRNA biogenesis is thought to contribute to the miRNA expression alterations observed in solid malignancies. Clearly, additional mechanism whereby key miRNA expression is controlled will be identified as our understanding of the regulation of these molecules increases.

3.3 MicroRNAs and Prognosis

In cases of type II endometrial carcinoma, decreased expression of miR-10b*, miR-29b or miR-455-5p correlates with vascular invasion (Hiroki et al. 2010) while in type I carcinomas, levels of miR-10a, miR-34a and miR-95 correlate with lymph node involvement (Chung et al. 2009). Decreased expression of miR-199a has been shown to be associated with a decreased progression-free interval (13.3 vs. 26.7 months, p=0.048) and overall survival (20 vs 40 months, p=0.0068) in surgically-staged endometrial carcinomas (Cohn et al.

2010). Similarly, decreased expression of miR-152 is a statistically significant independent risk factor for overall survival (p=0.021) and disease-free survival (p=0.010) while decreased expression of miR-101 is a risk factor for disease-free survival (p=0.016) in type II carcinoma (Hiroki et al. 2010). *e* studies (Hiroki et al. 2010) show that increased expression of miR-152 and miR-101 decrease cellular proliferation in serous endometrial carcinomas supporting the importance of these miRNAs in this tumor type.

4. CONCLUSIONS

The study of microRNAs in the context of endometrial carcinoma development and classification is clearly yielding important information about the mechanisms by which these tumors progress, and may potentially provide critical insights into appropriate and effective treatments for carcinomas of all stages or types. Parallels between miRNA expression in endometrial carcinoma and other tumors may also assist in exploring common mechanisms of cellular dysregulation. Clearly, biological targets of the most consistently dysregulated miRNAs will now need to be identified to begin to truly unravel the details of this and other cancers.

5. SUMMARY POINTS

- Global studies have provided some consistency in identification of dysregulated miRNAs, but such studies must be examined in context of controls used, platforms used, filters applied and hormonal influences that were not well controlled.
- The set of identified miRNAs that are either differentially regulated in endometrial vs. normal or between endometrial subtypes can be further examined in the context of their likely targets, implicating key pathways in the development of endometrial carcinomas.
- The use of miRNA expression patterns for diagnosis or prognosis has been fruitful in many other cancers and is achieving similar results in this more rare tumor.

ABBREVIATIONS

BMI	:	B lymphoma mouse moloney leukemia virus insertion region 1
BRD7	:	bromodomain containing 7
COX	:	cyclooxygenase-2
EMT	:	epithelial-to-mesenchymal transition
FOX	:	forkhead box

LOH	:	loss of heterozygosity
mir	:	miRNA: microRNA
mRNA	:	messenger RNA
MMR	:	mismatch repair
OCT4	:	octamer-binding transcription factor 4
PTEN	:	phosphatase and tensin homologue
RISC	:	RNA-induced silencing complex
SAM	:	significance analysis of microarrays
TUBB3	:	class III β-tubulin
UTR	:	untranslated region

REFERENCES

Aarnio, M., R. Sankila, E. Pukkala, R. Salovaara, L.A. Aaltonen, A. de la Chapelle, P. Peltomaki, J.P. Mecklin, and H.J. Jarvinen. 1999. Cancer risk in mutation carriers of DNA-mismatch-repair genes. Int J Cancer. 81: 214–218.

Adam, L., M. Zhong, W. Choi, W. Qi, M. Nicoloso, A. Arora, G. Calin, H. Wang, A. Siefker-Radtke, D. McConkey, M. Bar-Eli, and C. Dinney. 2009. miR-200 expression regulates epithelial-to-mesenchymal transition in bladder cancer cells and reverses resistance to epidermal growth factor receptor therapy. Clin Cancer Res. 15: 5060–5072.

Ambros, V. 2004. The functions of animal microRNAs. Nature. 431: 350–355.

Arden, K.C. 2008. FOXO animal models reveal a variety of diverse roles for FOXO transcription factors. Oncogene. 27: 2345–2350.

Babaie, Y., R. Herwig, B. Greber, T.C. Brink, W. Wruck, D. Groth, H. Lehrach, T. Burdon, and J. Adjaye. 2007. Analysis of Oct4-dependent transcriptional networks regulating self-renewal and pluripotency in human embryonic stem cells. Stem Cells. 25: 500–510.

Bai, X.F., X.G. Ni, P. Zhao, S.M. Liu, H.X. Wang, B. Guo, L.P. Zhou, F. Liu, J.S. Zhang, K. Wang, Y.Q. Xie, Y.F. Shao, and X.H. Zhao. 2004. Overexpression of annexin 1 in pancreatic cancer and its clinical significance. World J Gastroenterol. 10: 1466–1470.

Bartel, D.P. 2004. MicroRNAs: genomics, biogenesis, mechanism, and function. Cell. 116: 281–297.

Bhat-Nakshatri, P., G. Wang, N.R. Collins, M.J. Thomson, T.R. Geistlinger, J.S. Carroll, M. Brown, S. Hammond, E.F. Srour, Y. Liu, and H. Nakshatri. 2009. Estradiol-regulated microRNAs control estradiol response in breast cancer cells. Nucleic Acids Res. 37: 4850–4861.

Bommer, G.T., I. Gerin, Y. Feng, A.J. Kaczorowski, R. Kuick, R.E. Love, Y. Zhai, T.J. Giordano, Z.S. Qin, B.B. Moore, O.A. MacDougald, K.R. Cho, and E.R. Fearon. 2007. p53-mediated activation of miRNA34 candidate tumor-suppressor genes. Curr Biol. 17: 1298–1307.

Boren, T., Y. Xiong, A. Hakam, R. Wenham, S. Apte, Z. Wei, S. Kamath, D.T. Chen, H. Dressman, and J.M. Lancaster. 2008. MicroRNAs and their target messenger RNAs associated with endometrial carcinogenesis. Gynecol Oncol. 110: 206–215.

Buck, E., A. Eyzaguirre, S. Barr, S. Thompson, R. Sennello, D. Young, K.K. Iwata, N.W. Gibson, P. Cagnoni, and J.D. Haley. 2007. Loss of homotypic cell adhesion by epithelial-mesenchymal transition or mutation limits sensitivity to epidermal growth factor receptor inhibition. Mol Cancer Ther. 6: 532–541.

Burgering, B.M. 2008. A brief introduction to FOXOlogy. Oncogene. 27: 2258–2262.

Calnan, D.R. and A. Brunet. 2008. The FoxO code. Oncogene. 27: 2276–2288.

Castilla, M.A., G. Moreno-Bueno, L. Romero-Perez, K. Van De Vijver, M. Biscuola, M.A. Lopez-Garcia, J. Prat, X. Matias-Guiu, A. Cano, E. Oliva, and J. Palacios. 2011. Micro-RNA signature of the epithelial-mesenchymal transition in endometrial carcinosarcoma. J Pathol. 223: 72–80.

Chakrabarty, A., S. Tranguch, T. Daikoku, K. Jensen, H. Furneaux, and S.K. Dey. 2007. MicroRNA regulation of cyclooxygenase-2 during embryo implantation. Proc Natl Acad Sci USA. 104: 15144–15149.

Chuang, J.C. and P.A. Jones. 2007. Epigenetics and microRNAs. Pediatr Res. 61: 24R–29R.

Chung, T.K., T.H. Cheung, N.Y. Huen, K.W. Wong, K.W. Lo, S.F. Yim, N.S. Siu, Y.M. Wong, P.T. Tsang, M.W. Pang, M.Y. Yu, K.F. To, S.C. Mok, V.W. Wang, C. Li, A.Y. Cheung, G. Doran, M.J. Birrer, D.I. Smith, and Y.F. Wong. 2009. Dysregulated microRNAs and their predicted targets associated with endometrioid endometrial adenocarcinoma in Hong Kong women. Int J Cancer. 124: 1358–1365.

Chung, T.K., T.S. Lau, T.H. Cheung, S.F. Yim, K.W. Lo, N.S. Siu, L.K. Chan, M.Y. Yu, J. Kwong, G. Doran, L.M. Barroilhet, A.S. Ng, R.R. Wong, V.W. Wang, S.C. Mok, D.I. Smith, R.S. Berkowitz, and Y.F. Wong. 2012. Dysregulation of microRNA-204 mediates migration and invasion of endometrial cancer by regulating FOXC1. Int J Cancer. 130: 1036–1045.

Cochrane, D.R., N.S. Spoelstra, E.N. Howe, S.K. Nordeen, and J.K. Richer. 2009. MicroRNA-200c mitigates invasiveness and restores sensitivity to microtubule-targeting chemotherapeutic agents. Mol Cancer Ther. 8: 1055–1066.

Cochrane, D.R., E.N. Howe, N.S. Spoelstra, and J.K. Richer. 2010. Loss of miR-200c: A Marker of Aggressiveness and Chemoresistance in Female Reproductive Cancers. J Oncol. 2010: 821717.

Cohn, D.E., M. Fabbri, N. Valeri, H. Alder, I. Ivanov, C.G. Liu, C.M. Croce, and K.E. Resnick. 2010. Comprehensive miRNA profiling of surgically staged endometrial cancer. Am J Obstet Gynecol. 202: 656.e1–656.e8.

Corney, D.C., A. Flesken-Nikitin, A.K. Godwin, W. Wang, and A.Y. Nikitin. 2007. MicroRNA-34b and MicroRNA-34c are targets of p53 and cooperate in control of cell proliferation and adhesion-independent growth. Cancer Res. 67: 8433–8438.

Daikoku, T., Y. Hirota, S. Tranguch, A.R. Joshi, F.J. DeMayo, J.P. Lydon, L.H. Ellenson, and S.K. Dey. 2008. Conditional loss of uterine Pten unfailingly and rapidly induces endometrial cancer in mice. Cancer Res. 68: 5619–5627.

Devor, E.J., A.M. Hovey, M.J. Goodheart, S. Ramachandran, and K.K. Leslie. 2011. microRNA expression profiling of endometrial endometrioid adenocarcinomas and serous adenocarcinomas reveals profiles containing shared, unique and differentiating groups of microRNAs. Oncol Rep. 26: 995–1002.

Dong, P., M. Kaneuchi, H. Watari, J. Hamada, S. Sudo, J. Ju, and N. Sakuragi. 2011. MicroRNA-194 inhibits epithelial to mesenchymal transition of endometrial cancer cells by targeting oncogene BMI-1. Mol Cancer. 10: 99.

Eberhart, C.E., R.J. Coffey, A. Radhika, F.M. Giardiello, S. Ferrenbach, and R.N. DuBois. 1994. Up-regulation of cyclooxygenase 2 gene expression in human colorectal adenomas and adenocarcinomas. Gastroenterology. 107: 1183–1188.

Elson-Schwab, I., A. Lorentzen, and C.J. Marshall. 2010. MicroRNA-200 family members differentially regulate morphological plasticity and mode of melanoma cell invasion. PLoS One. 5: e13176.

Esteller, M., R. Levine, S.B. Baylin, L.H. Ellenson, and J.G. Herman. 1998. MLH1 promoter hypermethylation is associated with the microsatellite instability phenotype in sporadic endometrial carcinomas. Oncogene. 17: 2413–2417.

Faggad, A., J. Budczies, O. Tchernitsa, S. Darb-Esfahani, J. Sehouli, B.M. Muller, R. Wirtz, R. Chekerov, W. Weichert, B. Sinn, C. Mucha, N.E. Elwali, R. Schafer, M. Dietel, and C. Denkert. 2010. Prognostic significance of Dicer expression in ovarian cancer-link to global microRNA changes and oestrogen receptor expression. J Pathol. 220: 382–391.

Giatromanolaki, A., M.I. Koukourakis, E. Sivridis, K.C. Gatter, A.L. Harris, and A.H. Banham. 2006. Loss of expression and nuclear/cytoplasmic localization of the FOXP1 forkhead transcription factor are common events in early endometrial cancer: relationship with estrogen receptors and HIF-1alpha expression. Mod Pathol. 19: 9–16.

Gironella, M., M. Seux, M.J. Xie, C. Cano, R. Tomasini, J. Gommeaux, S. Garcia, J. Nowak, M.L. Yeung, K.T. Jeang, A. Chaix, L. Fazli, Y. Motoo, Q. Wang, P. Rocchi, A. Russo,

M. Gleave, J.C. Dagorn, J.L. Iovanna, A. Carrier, M.J. Pebusque, and N.J. Dusetti. 2007. Tumor protein 53-induced nuclear protein 1 expression is repressed by miR-155, and its restoration inhibits pancreatic tumor development. Proc Natl Acad Sci USA. 104: 16170–16175.

Gomes, A.R., J.J. Brosens, and E.W. Lam. 2008. Resist or die: FOXO transcription factors determine the cellular response to chemotherapy. Cell Cycle. 7: 3133–3136.

Goto, T., M. Takano, A. Albergaria, J. Briese, K.M. Pomeranz, B. Cloke, L. Fusi, F. Feroze-Zaidi, N. Maywald, M. Sajin, R.E. Dina, O. Ishihara, S. Takeda, E.W. Lam, A.M. Bamberger, S. Ghaem-Maghami, and J.J. Brosens. 2008. Mechanism and functional consequences of loss of FOXO1 expression in endometrioid endometrial cancer cells. Oncogene. 27: 9–19.

Gurin, C.C., M.G. Federici, L. Kang, and J. Boyd. 1999. Causes and consequences of microsatellite instability in endometrial carcinoma. Cancer Res. 59: 462–466.

Guttilla, I.K. and B.A. White. 2009. Coordinate regulation of FOXO1 by miR-27a, miR-96, and miR-182 in breast cancer cells. J Biol Chem. 284: 23204–23216.

Hayashi, H., H. Sano, S. Seo, and T. Kume. 2008. The Foxc2 transcription factor regulates angiogenesis via induction of integrin beta3 expression. J Biol Chem. 283: 23791–23800.

He, L., X. He, L.P. Lim, E. de Stanchina, Z. Xuan, Y. Liang, W. Xue, L. Zender, J. Magnus, D. Ridzon, A.L. Jackson, P.S. Linsley, C. Chen, S.W. Lowe, M.A. Cleary, and G.J. Hannon. 2007. A microRNA component of the p53 tumour suppressor network. Nature. 447: 1130–1134.

Herman, J.G. and S.B. Baylin. 2003. Gene silencing in cancer in association with promoter hypermethylation. N Engl J Med. 349: 2042–2054.

Hermeking, H. 2010. The miR-34 family in cancer and apoptosis. Cell Death Differ. 17: 193–199.

Hinkal, G.W., G. Grelier, A. Puisieux, and C. Moyret-Lalle. 2011. Complexity in the regulation of Dicer expression: Dicer variant proteins are differentially expressed in epithelial and mesenchymal breast cancer cells and decreased during EMT. Br J Cancer. 104: 387–388.

Hiroki, E., J. Akahira, F. Suzuki, S. Nagase, K. Ito, T. Suzuki, H. Sasano, and N. Yaegashi. 2010. Changes in microRNA expression levels correlate with clinicopathological features and prognoses in endometrial serous adenocarcinomas. Cancer Sci. 101: 241–249.

Hiroki, E., F. Suzuki, J. Akahira, S. Nagase, K. Ito, J. Sugawara, Y. Miki, T. Suzuki, H. Sasano, and N. Yaegashi. 2011. MicroRNA-34b functions as a potential tumor suppressor in endometrial serous adenocarcinoma. Int J Cancer. 131: 395–404.

Hochedlinger, K., Y. Yamada, C. Beard, and R. Jaenisch. 2005. Ectopic expression of Oct-4 blocks progenitor-cell differentiation and causes dysplasia in epithelial tissues. Cell. 121: 465–477.

Hoenerhoff, M.J., I. Chu, D. Barkan, Z.Y. Liu, S. Datta, G.P. Dimri, and J.E. Green. 2009. BMI1 cooperates with H-RAS to induce an aggressive breast cancer phenotype with brain metastases. Oncogene. 28: 3022–3032.

Honig, A., C. Weidler, S. Hausler, M. Krockenberger, S. Buchholz, F. Koster, S.E. Segerer, J. Dietl, and J.B. Engel. 2010. Overexpression of polycomb protein BMI-1 in human specimens of breast, ovarian, endometrial and cervical cancer. Anticancer Res. 30: 1559–1564.

Howe, E.N., D.R. Cochrane, and J.K. Richer. 2011. Targets of miR-200c mediate suppression of cell motility and anoikis resistance. Breast Cancer Res. 13: R45.

Hu, N., M.J. Flaig, H. Su, J.Z. Shou, M.J. Roth, W.J. Li, C. Wang, A.M. Goldstein, G. Li, M.R. Emmert-Buck, and P.R. Taylor. 2004. Comprehensive characterization of annexin I alterations in esophageal squamous cell carcinoma. Clin Cancer Res. 10: 6013–6022.

Huang, Y.W., J.C. Liu, D.E. Deatherage, J. Luo, D.G. Mutch, P.J. Goodfellow, D.S. Miller, and T.H. Huang. 2009. Epigenetic repression of microRNA-129-2 leads to overexpression of SOX4 oncogene in endometrial cancer. Cancer Res. 69: 9038–9046.

Iorio, M.V., R. Visone, G. Di Leva, V. Donati, F. Petrocca, P. Casalini, C. Taccioli, S. Volinia, C.G. Liu, H. Alder, G.A. Calin, S. Menard, and C.M. Croce. 2007. MicroRNA signatures in human ovarian cancer. Cancer Res. 67: 8699–8707.

Ito, Y., Y. Motoo, H. Yoshida, J.L. Iovanna, Y. Takamura, A. Miya, K. Kuma, and A. Miyauchi. 2006. Decreased expression of tumor protein p53-induced nuclear protein 1 (TP53INP1) in breast carcinoma. Anticancer Res. 26: 4391–4395.

Jiang, F., T. Liu, Y. He, Q. Yan, X. Chen, H. Wang, and X. Wan. 2011. MiR-125b promotes proliferation and migration of type II endometrial carcinoma cells through targeting TP53INP1 tumor suppressor *in vitro* and *in vivo*. BMC Cancer. 11: 425.

Jiang, L., P. Mao, L. Song, J. Wu, J. Huang, C. Lin, J. Yuan, L. Qu, S.Y. Cheng, and J. Li. 2010. miR-182 as a prognostic marker for glioma progression and patient survival. Am J Pathol. 177: 29–38.

Jiang, P.H., Y. Motoo, S. Garcia, J.L. Iovanna, M.J. Pebusque, and N. Sawabu. 2006. Down-expression of tumor protein p53-induced nuclear protein 1 in human gastric cancer. World J Gastroenterol. 12: 691–696.

Kang, J.S., B.F. Calvo, S.J. Maygarden, L.S. Caskey, J.L. Mohler, and D.K. Ornstein. 2002. Dysregulation of annexin I protein expression in high-grade prostatic intraepithelial neoplasia and prostate cancer. Clin Cancer Res. 8: 117–123.

Klinge, C.M. 2009. Estrogen Regulation of MicroRNA Expression. Curr Genomics. 10: 169–183.

Kong, D., A. Suzuki, T.T. Zou, A. Sakurada, L.W. Kemp, S. Wakatsuki, T. Yokoyama, H. Yamakawa, T. Furukawa, M. Sato, N. Ohuchi, S. Sato, J. Yin, S. Wang, J.M. Abraham, R.F. Souza, K.N. Smolinski, S.J. Meltzer, and A. Horii. 1997. PTEN1 is frequently mutated in primary endometrial carcinomas. Nat Genet. 17: 143–144.

Kounelis, S., N. Kapranos, E. Kouri, D. Coppola, H. Papadaki, and M.W. Jones. 2000. Immunohistochemical profile of endometrial adenocarcinoma: a study of 61 cases and review of the literature. Mod Pathol. 13: 379–388.

Kutchera, W., D.A. Jones, N. Matsunami, J. Groden, T.M. McIntyre, G.A. Zimmerman, R.L. White, and S.M. Prescott. 1996. Prostaglandin H synthase 2 is expressed abnormally in human colon cancer: evidence for a transcriptional effect. Proc Natl Acad Sci USA. 93: 4816–4820.

Lam, E.W., K. Shah, and J.J. Brosens. 2012. The diversity of sex steroid action: the role of micro-RNAs and FOXO transcription factors in cycling endometrium and cancer. J Endocrinol. 212: 13–25.

Lee, J.W., Y.A. Park, J.J. Choi, Y.Y. Lee, C.J. Kim, C. Choi, T.J. Kim, N.W. Lee, B.G. Kim, and D.S. Bae. 2011. The expression of the miRNA-200 family in endometrial endometrioid carcinoma. Gynecol Oncol. 120: 56–62.

Lehmann, O.J., J.C. Sowden, P. Carlsson, T. Jordan, and S.S. Bhattacharya. 2003. Fox's in development and disease. Trends Genet. 19: 339–344.

Li, J., C. Yen, D. Liaw, K. Podsypanina, S. Bose, S.I. Wang, J. Puc, C. Miliaresis, L. Rodgers, R. McCombie, S.H. Bigner, B.C. Giovanella, M. Ittmann, B. Tycko, H. Hibshoosh, M.H. Wigler, and R. Parsons. 1997. PTEN, a putative protein tyrosine phosphatase gene mutated in human brain, breast, and prostate cancer. Science. 275: 1943–1947.

Liang, L., C.M. Wong, Q. Ying, D.N. Fan, S. Huang, J. Ding, J. Yao, M. Yan, J. Li, M. Yao, I.O. Ng, and X. He. 2010. MicroRNA-125b suppressesed human liver cancer cell proliferation and metastasis by directly targeting oncogene LIN28B2. Hepatology. 52: 1731–1740.

Liu, F.S. 2007. Molecular carcinogenesis of endometrial cancer. Taiwan. J Obstet Gynecol. 46: 26–32.

Liu, H., L. Zhang, Z. Niu, M. Zhou, C. Peng, X. Li, T. Deng, L. Shi, Y. Tan, and G. Li. 2008. Promoter methylation inhibits BRD7 expression in human nasopharyngeal carcinoma cells. BMC Cancer. 8: 253.

Llobet, D., J. Pallares, A. Yeramian, M. Santacana, N. Eritja, A. Velasco, X. Dolcet, and X. Matias-Guiu. 2009. Molecular pathology of endometrial carcinoma: practical aspects from the diagnostic and therapeutic viewpoints. J Clin Pathol. 62: 777–785.

Loeb, L.A. 1994. Microsatellite instability: marker of a mutator phenotype in cancer. Cancer Res. 54: 5059–5063.

Luthra, R., R.R. Singh, M.G. Luthra, Y.X. Li, C. Hannah, A.M. Romans, B.A. Barkoh, S.S. Chen, J. Ensor, D.M. Maru, R.R. Broaddus, A. Rashid, and C.T. Albarracin. 2008. MicroRNA-196a targets annexin A1: a microRNA-mediated mechanism of annexin A1 downregulation in cancers. Oncogene. 27: 6667–6678.

Masaki, T., M. Tokuda, M. Ohnishi, S. Watanabe, T. Fujimura, K. Miyamoto, T. Itano, H. Matsui, K. Arima, M. Shirai, T. Maeba, K. Sogawa, R. Konishi, K. Taniguchi, Y. Hatanaka, O. Hatase, and M. Nishioka. 1996. Enhanced expression of the protein kinase substrate annexin in human hepatocellular carcinoma. Hepatology. 24: 72–81.

Moskwa, P., F.M. Buffa, Y. Pan, R. Panchakshari, P. Gottipati, R.J. Muschel, J. Beech, R. Kulshrestha, K. Abdelmohsen, D.M. Weinstock, M. Gorospe, A.L. Harris, T. Helleday, and D. Chowdhury. 2011. miR-182-mediated downregulation of BRCA1 impacts DNA repair and sensitivity to PARP inhibitors. Mol Cell. 41: 210–220.

Mueller, D.W., M. Rehli, and A.K. Bosserhoff. 2009. miRNA expression profiling in melanocytes and melanoma cell lines reveals miRNAs associated with formation and progression of malignant melanoma. J Invest Dermatol. 129: 1740–1751.

Mutter, G.L., M.C. Lin, J.T. Fitzgerald, J.B. Kum, J.P. Baak, J.A. Lees, L.P. Weng, and C. Eng. 2000. Altered PTEN expression as a diagnostic marker for the earliest endometrial precancers. J Natl Cancer Inst. 92: 924–930.

Myatt, S.S. and E.W. Lam. 2007. The emerging roles of forkhead box (Fox) proteins in cancer. Nat Rev Cancer. 7: 847–859.

Myatt, S.S., J. Wang, L.J. Monteiro, M. Christian, K.K. Ho, L. Fusi, R.E. Dina, J.J. Brosens, S. Ghaem-Maghami, and E.W. Lam. 2010. Definition of microRNAs that repress expression of the tumor suppressor gene FOXO1 in endometrial cancer. Cancer Res. 70: 367–377.

Paik, J.H., R. Kollipara, G. Chu, H. Ji, Y. Xiao, Z. Ding, L. Miao, Z. Tothova, J.W. Horner, D.R. Carrasco, S. Jiang, D.G. Gilliland, L. Chin, W.H. Wong, D.H. Castrillon, and R.A. DePinho. 2007. FoxOs are lineage-restricted redundant tumor suppressors and regulate endothelial cell homeostasis. Cell. 128: 309–323.

Park, Y.A., J.W. Lee, J.J. Choi, H.K. Jeon, Y. Cho, C. Choi, T.J. Kim, N.W. Lee, B.G. Kim, and D.S. Bae. 2012. The interactions between MicroRNA-200c and BRD7 in endometrial carcinoma. Gynecol Oncol. 124: 125–133.

Paterson, E.L., N. Kolesnikoff, P.A. Gregory, A.G. Bert, Y. Khew-Goodall, and G.J. Goodall. 2008. The microRNA-200 family regulates epithelial to mesenchymal transition. Scientific World Journal. 8: 901–904.

Paweletz, C.P., D.K. Ornstein, M.J. Roth, V.E. Bichsel, J.W. Gillespie, V.S. Calvert, C.D. Vocke, S.M. Hewitt, P.H. Duray, J. Herring, Q.H. Wang, N. Hu, W.M. Linehan, P.R. Taylor, L.A. Liotta, M.R. Emmert-Buck, and E.F. Petricoin 3rd. 2000. Loss of annexin 1 correlates with early onset of tumorigenesis in esophageal and prostate carcinoma. Cancer Res. 60: 6293–6297.

Peltomaki, P., R.A. Lothe, L.A. Aaltonen, L. Pylkkanen, M. Nystrom-Lahti, R. Seruca, L. David, R. Holm, D. Ryberg, and A. Haugen. 1993. Microsatellite instability is associated with tumors that characterize the hereditary non-polyposis colorectal carcinoma syndrome. Cancer Res. 53: 5853–5855.

Peng, C., H.Y. Liu, M. Zhou, L.M. Zhang, X.L. Li, S.R. Shen, and G.Y. Li. 2007. BRD7 suppresses the growth of Nasopharyngeal Carcinoma cells (HNE1) through negatively regulating beta-catenin and ERK pathways. Mol Cell Biochem. 303: 141–149.

Ratner, E.S., D. Tuck, C. Richter, S. Nallur, R.M. Patel, V. Schultz, P. Hui, P.E. Schwartz, T.J. Rutherford, and J.B. Weidhaas. 2010. MicroRNA signatures differentiate uterine cancer tumor subtypes. Gynecol Oncol. 118: 251–257.

Ray, P.S., J. Wang, Y. Qu, M.S. Sim, J. Shamonki, S.P. Bagaria, X. Ye, B. Liu, D. Elashoff, D.S. Hoon, M.A. Walter, J.W. Martens, A.L. Richardson, A.E. Giuliano, and X. Cui. 2010. FOXC1 is a potential prognostic biomarker with functional significance in basal-like breast cancer. Cancer Res. 70: 3870–3876.

Risinger, J.I., A.K. Hayes, A. Berchuck, and J.C. Barrett. 1997. PTEN/MMAC1 mutations in endometrial cancers. Cancer Res. 57: 4736–4738.

Saito, K., A. Ishizuka, H. Siomi, and M.C. Siomi. 2005. Processing of pre-microRNAs by the Dicer-1-Loquacious complex in Drosophila cells. PLoS Biol. 3: e235.

Sano, H., Y. Kawahito, R.L. Wilder, A. Hashiramoto, S. Mukai, K. Asai, S. Kimura, H. Kato, M. Kondo, and T. Hla. 1995. Expression of cyclooxygenase-1 and -2 in human colorectal cancer. Cancer Res. 55: 3785–3789.

Sasaki, M., H. Ikeda, K. Itatsu, J. Yamaguchi, S. Sawada, H. Minato, T. Ohta, and Y. Nakanuma. 2008. The overexpression of polycomb group proteins Bmi1 and EZH2 is associated with the progression and aggressive biological behavior of hepatocellular carcinoma. Lab Invest. 88: 873–882.

Shi, X.B., L. Xue, J. Yang, A.H. Ma, J. Zhao, M. Xu, C.G. Tepper, C.P. Evans, H.J. Kung, and R.W. deVere White. 2007. An androgen-regulated miRNA suppresses Bak1 expression and induces androgen-independent growth of prostate cancer cells. Proc Natl Acad Sci USA. 104: 19983–19988.

Siegel, R., E. Ward, O. Brawley, and A. Jemal. 2011. Cancer statistics, 2011: the impact of eliminating socioeconomic and racial disparities on premature cancer deaths. CA Cancer. J Clin. 61: 212–236.

Singh, M., N.S. Spoelstra, A. Jean, E. Howe, K.C. Torkko, H.R. Clark, D.S. Darling, K.R. Shroyer, K.B. Horwitz, R.R. Broaddus, and J.K. Richer. 2008. ZEB1 expression in type I vs. type II endometrial cancers: a marker of aggressive disease. Mod Pathol. 21: 912–923.

Sinha, P., G. Hutter, E. Kottgen, M. Dietel, D. Schadendorf, and H. Lage. 1998. Increased expression of annexin I and thioredoxin detected by two-dimensional gel electrophoresis of drug resistant human stomach cancer cells. J Biochem Biophys Methods. 37: 105–116.

Snowdon, J., X. Zhang, T. Childs, V.A. Tron, and H. Feilotter. 2011. The microRNA-200 family is upregulated in endometrial carcinoma. PLoS One. 6: e22828.

Solito, E., C. de Coupade, S. Canaider, N.J. Goulding, and M. Perretti. 2001. Transfection of annexin 1 in monocytic cells produces a high degree of spontaneous and stimulated apoptosis associated with caspase-3 activation. Br J Pharmacol. 133: 217–228.

Spoelstra, N.S., N.G. Manning, Y. Higashi, D. Darling, M. Singh, K.R. Shroyer, R.R. Broaddus, K.B. Horwitz, and J.K. Richer. 2006. The transcription factor ZEB1 is aberrantly expressed in aggressive uterine cancers. Cancer Res. 66: 3893–3902.

Steck, P.A., M.A. Pershouse, S.A. Jasser, W.K. Yung, H. Lin, A.H. Ligon, L.A. Langford, M.L. Baumgard, T. Hattier, T. Davis, C. Frye, R. Hu, B. Swedlund, D.H. Teng, and S.V. Tavtigian. 1997. Identification of a candidate tumour suppressor gene, MMAC1, at chromosome 10q23.3 that is mutated in multiple advanced cancers. Nat Genet. 15: 356–362.

Subbaramaiah, K., L. Norton, W. Gerald, and A.J. Dannenberg. 2002. Cyclooxygenase-2 is overexpressed in HER-2/neu-positive breast cancer: evidence for involvement of AP-1 and PEA3. J Biol Chem. 277: 18649–18657.

Tarasov, V., P. Jung, B. Verdoodt, D. Lodygin, A. Epanchintsev, A. Menssen, G. Meister, and H. Hermeking. 2007. Differential regulation of microRNAs by p53 revealed by massively parallel sequencing: miR-34a is a p53 target that induces apoptosis and G1-arrest. Cell Cycle. 6: 1586–1593.

Taube, J.H., J.I. Herschkowitz, K. Komurov, A.Y. Zhou, S. Gupta, J. Yang, K. Hartwell, T.T. Onder, P.B. Gupta, K.W. Evans, B.G. Hollier, P.T. Ram, E.S. Lander, J.M. Rosen, R.A. Weinberg, and S.A. Mani. 2010. Core epithelial-to-mesenchymal transition interactome gene-expression signature is associated with claudin-low and metaplastic breast cancer subtypes. Proc Natl Acad Sci USA. 107: 15449–15454.

Torres, A., K. Torres, T. Paszkowski, B. Jodlowska-Jedrych, T. Radomanski, A. Ksiazek, and R. Maciejewski. 2011. Major regulators of microRNAs biogenesis Dicer and Drosha are down-regulated in endometrial cancer. Tumour Biol. 32: 769–776.

Tsuruta, T., K. Kozaki, A. Uesugi, M. Furuta, A. Hirasawa, I. Imoto, N. Susumu, D. Aoki, and J. Inazawa. 2011. miR-152 is a tumor suppressor microRNA that is silenced by DNA hypermethylation in endometrial cancer. Cancer Res. 71: 6450–6462.

Tucker, O.N., A.J. Dannenberg, E.K. Yang, F. Zhang, L. Teng, J.M. Daly, R.A. Soslow, J.L. Masferrer, B.M. Woerner, A.T. Koki, and T.J. Fahey 3rd. 1999. Cyclooxygenase-2 expression is up-regulated in human pancreatic cancer. Cancer Res. 59: 987–990.

Tuteja, G. and K.H. Kaestner. 2007a. SnapShot: forkhead transcription factors I. Cell. 130: 1160.

Tuteja, G. and K.H. Kaestner. 2007b. Forkhead transcription factors II. Cell. 131: 192.

Valeri, N., P. Gasparini, M. Fabbri, C. Braconi, A. Veronese, F. Lovat, B. Adair, I. Vannini, F. Fanini, A. Bottoni, S. Costinean, S.K. Sandhu, G.J. Nuovo, H. Alder, R. Gafa, F. Calore, M. Ferracin, G. Lanza, S. Volinia, M. Negrini, M.A. McIlhatton, D. Amadori, R. Fishel, and C.M. Croce. 2010. Modulation of mismatch repair and genomic stability by miR-155. Proc Natl Acad Sci USA. 107: 6982–6987.

Vasen, H.F., G.J. Offerhaus, F.C. den Hartog Jager, F.H. Menko, F.M. Nagengast, G. Griffioen, R.B. van Hogezand, and A.P. Heintz. 1990. The tumour spectrum in hereditary non-polyposis colorectal cancer: a study of 24 kindreds in the Netherlands. Int J Cancer 46: 31–34.

Vogt, M., J. Munding, M. Gruner, S.T. Liffers, B. Verdoodt, J. Hauk, L. Steinstraesser, A. Tannapfel, and H. Hermeking. 2011. Frequent concomitant inactivation of miR-34a and miR-34b/c by CpG methylation in colorectal, pancreatic, mammary, ovarian, urothelial, and renal cell carcinomas and soft tissue sarcomas. Virchows Arch. 458: 313–322.

Vrba, L., T.J. Jensen, J.C. Garbe, R.L. Heimark, A.E. Cress, S. Dickinson, M.R. Stampfer, and B.W. Futscher. 2010. Role for DNA methylation in the regulation of miR-200c and miR-141 expression in normal and cancer cells. PLoS One. 5: e8697.

Witta, S.E., R.M. Gemmill, F.R. Hirsch, C.D. Coldren, K. Hedman, L. Ravdel, B. Helfrich, R. Dziadziuszko, D.C. Chan, M. Sugita, Z. Chan, A. Baron, W. Franklin, H.A. Drabkin, L. Girard, A.F. Gazdar, J.D. Minna, and P.A. Bunn Jr. 2006. Restoring E-cadherin expression increases sensitivity to epidermal growth factor receptor inhibitors in lung cancer cell lines. Cancer Res. 66: 944–950.

Wolff, H., K. Saukkonen, S. Anttila, A. Karjalainen, H. Vainio, and A. Ristimaki. 1998. Expression of cyclooxygenase-2 in human lung carcinoma. Cancer Res. 58: 4997–5001.

Wong, Y.F., T.H. Cheung, K.W. Lo, S.F. Yim, N.S. Siu, S.C. Chan, T.W. Ho, K.W. Wong, M.Y. Yu, V.W. Wang, C. Li, G.J. Gardner, T. Bonome, W.B. Johnson, D.I. Smith, T.K. Chung, and M.J. Birrer. 2007. Identification of molecular markers and signaling pathway in endometrial cancer in Hong Kong Chinese women by genome-wide gene expression profiling. Oncogene. 26: 1971–1982.

Wu, W., Z. Lin, Z. Zhuang, and X. Liang. 2009. Expression profile of mammalian microRNAs in endometrioid adenocarcinoma. Eur J Cancer Prev. 18: 50–55.

Wu, Y., S. Liu, H. Xin, J. Jiang, E. Younglai, S. Sun, and H. Wang. 2011. Up-regulation of microRNA-145 promotes differentiation by repressing OCT4 in human endometrial adenocarcinoma cells. Cancer. 117: 3989–3998.

Xi, Y., A. Formentini, M. Chien, D.B. Weir, J.J. Russo, J. Ju, M. Kornmann, and J. Ju. 2006. Prognostic Values of microRNAs in Colorectal Cancer. Biomark Insights. 2: 113–121.

Xia, S.H., L.P. Hu, H. Hu, W.T. Ying, X. Xu, Y. Cai, Y.L. Han, B.S. Chen, F. Wei, X.H. Qian, Y.Y. Cai, Y. Shen, M. Wu, and M.R. Wang. 2002. Three isoforms of annexin I are preferentially expressed in normal esophageal epithelia but down-regulated in esophageal squamous cell carcinomas. Oncogene. 21: 6641–6648.

Xu, N., T. Papagiannakopoulos, G. Pan, J.A. Thomson, and K.S. Kosik. 2009. MicroRNA-145 regulates OCT4, SOX2, and KLF4 and represses pluripotency in human embryonic stem cells. Cell. 137: 647–658.

Yang, A.D., F. Fan, E.R. Camp, G. van Buren, W. Liu, R. Somcio, M.J. Gray, H. Cheng, P.M. Hoff, and L.M. Ellis. 2006. Chronic oxaliplatin resistance induces epithelial-to-mesenchymal transition in colorectal cancer cell lines. Clin Cancer Res. 12: 4147–4153.

Yang, M.H., D.S. Hsu, H.W. Wang, H.J. Wang, H.Y. Lan, W.H. Yang, C.H. Huang, S.Y. Kao, C.H. Tzeng, S.K. Tai, S.Y. Chang, O.K. Lee, and K.J. Wu. 2010. Bmi1 is essential in Twist1-induced epithelial-mesenchymal transition. Nat Cell Biol. 12: 982–992.

Yoshimura, R., H. Sano, C. Masuda, M. Kawamura, Y. Tsubouchi, J. Chargui, N. Yoshimura, T. Hla, and S. Wada. 2000. Expression of cyclooxygenase-2 in prostate carcinoma. Cancer. 89: 589–596.

Zeisberg, M. and E.G. Neilson. 2009. Biomarkers for epithelial-mesenchymal transitions. J Clin Invest. 119: 1429–1437.

Zhang, W., M. Feng, G. Zheng, Y. Chen, X. Wang, B. Pen, J. Yin, Y. Yu, and Z. He. 2012. Chemoresistance to 5-fluorouracil induces epithelial-mesenchymal transition via up-regulation of Snail in MCF7 human breast cancer cells. Biochem. Biophys Res Commun. 417: 679–685.

Zhu, W., X. Liu, J. He, D. Chen, Y. Hunag, and Y.K. Zhang. 2011. Overexpression of members of the microRNA-183 family is a risk factor for lung cancer: a case control study. BMC Cancer. 11: 393.

CHAPTER 5

MicroRNAs IN CERVICAL CANCER: EVIDENCES OF A miRNA DE-REGULATION CAUSED BY HPV

Carlos Pérez-Plasencia,[1,2,3,a,*] Nadia Jacobo-Herrera,[4,b]
Carlos Barba-Ostria,[2,c] Rosa Alvarez-Gómez,[2,d]
David Cantú-De León,[2] Jorge Fernández-Retana,[2]
Elena Arechaga-Ocampo,[4,e] Luis A. Herrera,[5]
César López-Camarillo[6,f] and Oscar Peralta-Zaragoza[6,g]

ABSTRACT

Cervical carcinoma (CC) is one of the most common cancers and a leading cause of mortality in women worldwide. Epidemiologic and experimental studies have clearly confirmed a causal role of high-risk Human Papillomavirus (HR-HPV) types in cervical carcinogenesis, which affect cellular processes by targeting and inactivating p53 and pRB host proteins. CC arises through a multi-step carcinogenesis process mainly induced by the HR-HPV E6 and E7 oncoproteins, that have the ability to deregulate several cellular processes such as apoptosis, cell cycle control, migration, immune evasion and induction of genetic instability, which, among others, promotes accumulation of mutations

Authors' affiliations given at the end of the chapter.

and aneuploidy. The complex interactions between early HR-HPV genes affects important epigenetic mechanisms in host cell by means of direct targeting on histone methyltransferases (HMTs), modulating their enzymatic activities by increasing transcriptional activity on specific histone demethylases (KDM6 and KDM6B) thus leading to aberrant expression of cellular oncogenic and tumor suppressive miRNAs. It has been stated that HPV infection and E6/E7 expression are essential but not sufficient for CC development; hence other genetic and epigenetic factors have to be involved in this complex disease. Recent evidence suggests an important level of interaction between E6/E7 viral proteins and cellular miRNAs and other noncoding RNAs. The aim of the current chapter is to review and analyze recent data pointing to describe the influence of the HPV life cycle on specific noncoding RNAs, leading to a mechanistic view of HPV-induced oncogenesis. Finally, we review the use of RNA interference (RNAi) as a therapeutical alternative, seeking to block E6 and E7 expression to restore p53 and pRb levels.

1. INTRODUCTION

Cervical cancer (CC) is the third most common cancer in women around the world, the seventh overall, and contributes to 9.8% of all female cancers with an estimated 530,000 new cases in 2008. More than 85% of the global burden occurs in developing countries, where it accounts for 13% of all female cancers. High-risk regions are eastern and western Africa (greater than 30 cases per 100,000 inhabitants), southern Africa (26.8 per 100,000), south-central Asia (24.6 per 100,000), South America and middle Africa (23.9 and 23.0 per 100,000 respectively). Rates are lowest in western Asia, North America and Australia/New Zealand (less than 6 per 100,000). Cervical cancer remains the most common cancer in women only in eastern Africa, south-central Asia and Melanesia. Overall, the mortality: incidence ratio is 52%, and cervical cancer is responsible for 275,000 deaths in 2008, about 88% of which occurred in developing countries: 53,000 in Africa, 31,700 in Latin America and the Caribbean, and 159,800 in Asia (Ferlay et al. 2010). Clinical, epidemiological and molecular data points to persistent human papillomavirus (HPV) infection as the main etiological agent in cervical cancer development (zur Hausen 2002). Anogenital cancers including those of the cervix are associated with HPV16 and HPV18, and the presence of high risk HPV (HR-HPV) genomes is associated with 99.7% of all studied cervical carcinomas (Walboomers et al. 1999). During the cervical carcinogenesis processes, E6 and E7 from high-risk HPV (HR-HPV) play a major role in inactivating cellular mechanisms involved in the control of cell cycle; it has, however, been postulated that HPV is a necessary—but not sufficient—cause for developing cervical carcinoma; thus, other types of factors such as cellular, immunological, genetic, epigenetic or environmental ones, can affect the final outcome of the disease. In this work, we discuss the possible effect of microRNA expression on cervical carcinogenesis; moreover, we explore the hypothesis of viral oncoproteins affecting the expression of specific microRNAs that could possibly participate on the establishment of malignant phenotype.

2. HPV CYCLE AND CERVICAL CANCER PROGRESSION

Epidemiological, molecular and clinical evidences have demonstrated a causal relationship between high-risk HPV infection and CC development (Walboomers et al. 1999). However, it has been postulated that HPV is a necessary agent, but not a sufficient cause of cervical intraepithelial neoplasia and CC. Epidemiological studies have shown that HPV is one of the most common sexually transmitted agents, with a prevalence between 10–40 percent in women who have no cytological abnormalities (Jacobs et al. 2000, Molano et al. 2002). There is enough epidemiological evidence suggesting that most women are likely to have an HPV sub-clinical infection, with a higher risk at juvenile ages (younger than 25 years old). This is due to many factors, for instance, older women could have acquired immunity to HPV from previous exposures, likewise, alcohol consumption and an elevated number of sexual partners can increase specific risk (Burk et al. 1996, Kjaer et al. 1997, Ho et al. 1998, Lazcano-Ponce et al. 2001).

Persistent infection with a carcinogenic human papillomavirus (HPV) is a prerequisite for cervical cancer (Steben and Duarte-Franco 2007), and around 20% of women with one-year persistence of HPV cervical infection will develop cervical intraepithelial neoplasia or cervical cancer in the next five years (Castle et al. 2009). However, different cohort and multi-center studies have shown that asymptomatic infection with HPV can go up to 85% in women who have no cytological abnormalities detected by pap-smear (Roteli-Martins et al. 2011); which means that HPV sequences are very common in the genital epithelia of asymptomatic population. Although HPV infection is a necessary cause to develop cervical cancer, the presence of certain co-factors in combination with infection will significantly increase the risk of developing cervical (Ho et al. 1998, Molano et al. 2002, Franco et al. 1999).

With regard to HPV molecular events leading to CC, we can group them in three key events during HPV course of infection: 1) viral DNA integration to host genome, 2) expression of viral proteins (namely E1, E2, E4, E5, E6 and E7), and 3) complex interactions between E2, E6/E7 and cellular proteins (Fig. 1). Cervical cancer is a complex disease caused by the interaction of viral, host, and environmental factors, which influence disease progression from early cervical abnormalities to invasive cancer and thus the appropriate identification of factors involved will lead us to a better knowledge of the natural history of HPV infection.

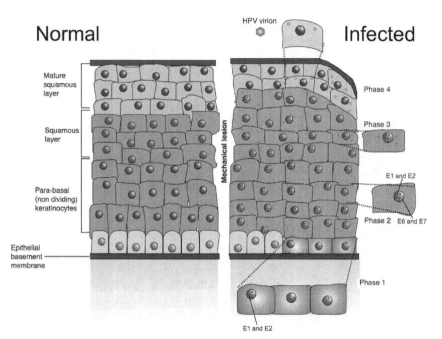

Figure 1. HPV viral cycle and cervical cancer development.
Color image of this figure appears in the color plate section at the end of the book.

After basal cells are infected by HPV, the viral genome starts replication as an episome and genes like E1–E7 are expressed. These genes are important proteins, vital for viral genome replication and viral cycle completion (Matsukura et al. 1989). For instance, the E1 gene is an ATP-dependent DNA helicase, whose principal function is the unwinding of the double-stranded viral DNA and the interaction with the primase subunit of the DNA polymerase, to recruit the replication complex to the viral replication origin (Masterson et al. 1998, Conger et al. 1999). Moreover, E1 acts together with several cyclins and it is known to be phosphorylated by cyclin/CDK complexes (Cueille et al. 1998, Dalton et al. 1995). For the interactions between E1 and cyclins, the consensus RxL cyclin-binding motif, present in the amino-terminal domain of the E1 protein, is required. Mutation of this motif hampers the replication of the viral genome (Ma et al. 1999), indicating that E1 probably regulates HPV genome replication through the interaction with cyclin/CDK complexes (Deng et al. 2004).

Human papillomavirus infects epithelial basal cells through mechanical lesions or by infecting the transformation zone which is a transition from a basal columnar to a squamous epithelium (Phase 1). Following infection, cells express early viral genes E1, E2, E4 and E5 (Phase 2). E2 exerts transcriptional repression on E6 and E7, hence these viral onco-genes are expressed in limited amounts and infected cells acquire a faster cell cycle. Differentiation pushes up infected cells from basal layers to the lumen expressing the late capside genes L1 and L2 (Phase 3). Viral genome is replicated as an episome in sub-clinical infections or low grade intra-epithelial-lesions (LGSIL), productive infections produce viral particles then can infect new zones of epithelium or be sexually transmitted (Phase 4). Only a limited number of infections progress to high-grade intra-epithelial-lesions (HGSIL) and cervical carcinoma (CC). The progression of LGSIL to CC is associated with the integration of the HPV genome into the host genome and the loss of transcriptional repression endeavored by E2.

On the other hand, the full-length E2 protein is a transcription factor that regulates the expression of all HPV genes. This regulation is accomplished through four consensus E2-binding sites (E2-BSs), ACCGN4CGGT, whose locations within the upstream regulatory region (URR) are highly conserved among genital HPVs (Fig. 1, Hegde 2002, Hou et al. 2002). Furthermore, E2 is involved in viral DNA replication via interaction with E1, and it also participates in viral DNA replication via interaction with the E1 protein (Chiang et al. 1992). In this way, the aforementioned functions of the E2 protein could be explained by E2-BSs occupancy in a context-dependent fashion. E2 binding to E2-BS4 can specifically up-regulate viral early gene expression, including the expression of E6 and E7 oncogenes. However, the binding of E2 protein to the promoter-proximal sites E2-BS1 and E2-BS2 leads to transcriptional repression of the early genes, including E6 and E7, whereas the E2-BS3 site is important for viral DNA replication, contributing to the cell cycle control by regulating the expression of E6 and E7 (Steger and Corbach 1997, Stubenrauch and Pfister 1994, Stubenrauch et al. 1998). Besides, E2-mediated repression of HPV early genes appears to involve the displacement of cellular transcription factors from the viral promoter (Tan et al. 1994). Thus, Fig. 2A depicts the relative occupancy of different E2 binding sites, which are finely modulated by the concentration of the E2 protein.

Moreover, E2 is able to induce apoptosis in absence of other HPV open reading frames by its association to the DED motif of caspase-8. Whereas HPV-18 E2 protein induces caspase oligomerization through its amino-terminus motif containing a 27 amino-acid helix, HPV-16 E2 protein induces apoptosis by means of p53 binding (Demeret et al. 2003, Thierry and Demeret 2008). Such interaction has important implications in the viral cycle, where E2 is known to have two main mechanisms of regulating cell cycle progression: 1) by means of apoptosis, either p53-dependent or caspase-induced, and 2) by balancing the expression/repression of oncoviral proteins E6 an E7. The balance executed by E2 is broken upon HPV integration onto the host genome (Webster et al. 2000, Brown et al. 2008). In benign cervical precursor lesions, HPV genome replicates as episome or extrachromosomic molecule; in general, cancer tissues contain both

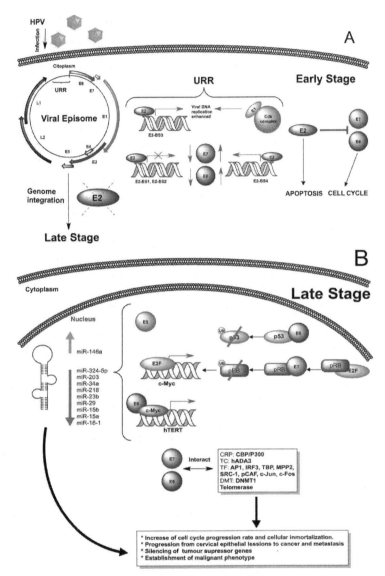

Figure 2. Molecular mechanisms induced by HPV early-expressed proteins. (A) While HPV replicates as an episome, E1 and E2 are actively expressed, as they are essential for viral genome replication and viral cycle completion. E2 is a sequence-specific transcription factor which regulates E6 and E7 expression rate; depending on occupancy of E2-BSs sites. (B) Human papilloma virus E6 and E7 interact with nuclear proteins such as transcription factors (TF), chromatin remodelers (CRP), transcriptional co-activators (TC) and DNA methyl-transferases (DMT) to influence gene expression and miRNA profile, which ultimately affects cellular processes leading to malignant phenotype and tumoral progression.

Color image of this figure appears in the color plate section at the end of the book.

episomal and integrated HPV DNA that has been covalently incorporated into the host cell chromosomal DNA (Cullen et al. 1991, Hudelist et al. 2008). Due to the ring form of the HPV genome, it requires a breakage in the E1–E2 open reading frames region and deletion of E2 and adjacent regions E2–E4, E5, and L2, before integration to the cell genome. Hence, the fine tuning of the E6 and E7 expression levels exerted by E2 is lost and viral oncogenes E6 and E7 are actively expressed in CC tissue (Ueda et al. 2003). The suggested sites for viral integration are cellular genes that might be essential to the enhanced progression risk of HPV-induced premalignant lesions to neoplastic lesions. Thus, the most frequent integration sites are near MYC, NR4A2, hTERT, APM-1, FANCC, and TNFAIP2 (reviewed by Wentzensen et al. 2004).

2.1 Effect of E6 and E7 Protein-protein Interactions with Nuclear Proteins in Regulation of Transcription

The epithelial expression of E6 and E7 proteins, and their interactions with cellular proteins have been at the center of the HPV the biomedical research scenario for probably the last 20 years. The central core of the classic E6/E7 model is the binding and inactivation of the tumor suppressor proteins p53 and pRb respectively, described between late 1980's and early 1990's (Dyson et al. 1989, Scheffner et al. 1990). Currently, it is known that the proliferation capacity of malignant cells and the uncoupling of differentiation through targeting of prominent of cell cycle progression regulators, are activated by the expression of E6 and E7. Also, these proteins interact with a plethora of cellular proteins, both in the nucleus and in the cytoplasm, that participate in molecular pathways involved in the activation and establishment of the malignant phenotype (Moody and Laimins 2010, Lavia et al. 2003). E6 and E7 do not posses DNA-binding domains (Mallon et al. 1987, Grossman et al. 1989), however, in the nucleus these viral products interact with chromatin remodeling proteins, such as the histone acetyl transferase CREB-binding protein CBP/P300; with transcriptional coactivators, such as hADA3; with transcription factors, such as AP1, IRF3, E2F1, TBP, MPP2, SRC-1 and pCAF; DNA-methyl transferases, and with the telomerase (Hwang et al. 2002, Ronco et al. 1998, Antinore et al. 1996, Phillips and Vousden 1997, Maldonado et al. 2002, Lüscher-Firzlaff et al. 1999, Baldwin et al. 2006, Huang and McCance 2002, Burgers et al. 2007, Liu et al. 2009).

The above mentioned protein-protein interactions, participate in important changes in transcriptional regulation by the direct action of these viral oncoproteins upon specific genes. There are many reported examples elsewhere (Howe et al. 1999, Haertel-Wiesmann et al. 2000, Kumar et al. 2002, Zeng et al. 2002, Bernat et al. 2003, Subbaramaiah and Dannenberg

2007, Burgers et al. 2007, Sekaric et al. 2008, Hu et al. 2009, Liu et al. 2011, Balan et al. 2011).

As previously explained, E2 has a functional DNA-binding domain and regulates viral gene expression; it can also regulate the expression of several cellular genes. E2 is able to bind and transactivate to the promoter of the splicing factor SF2/ASF. This factor is involved in the regulation of alternative splicing, and its overexpression mediated by E2 could be related to the production of the viral alternative transcripts of L1 and L2 in the replicative cycle (Mole et al. 2009). E2 is associated to mechanisms that participate in differentiation, growth inhibition and senescence induction (Dowhanick et al. 1995). For instance, E2 interacts with the transcription factor Sp1 in the promoter of hTERT to repress its expression and can interact with C/EBP to promote keratinocyte differentiation (Hadaschick et al. 2003, Lee et al. 2002). The interaction among E2, E6 and E7 with nuclear proteins is relevant due the possibility to constitute a hallmark of transcriptional regulation. These findings are important to understand the molecular pathology of HPV, where cellular and viral proteins work together in the promotion of a malignant phenotype.

3. miRNAs ASSOCIATED TO DEVELOPMENT AND PROGRESSION IN CERVICAL CANCER

CC is a complex disease involving genetic and epigenetic abnormalities that affect the global expression profile of thousands of genes. Molecular evidence during the last three decades has pointed out the alteration of protein-coding oncogenes or tumor suppressor genes as the main drivers to malignant phenotype. However, the general idea of complexity in tumor cells has been significantly increased with the discovery of thousands of genes producing non-coding RNAs (ncRNAs). In this scenario microRNAs (miRNAs) are considered important players in regulating cellular processes such as apoptosis, cell cycle progression, metastasis and drug-resistance. miRNAs have been studied in the context of etiology, progression and prognosis of cancer (Kumar et al. 2007).

The first evidence showing changes in the expression level of miRNAs on tumor cells was done in B-cell chronic lymphocytic leukemia (Calin et al. 2002); since then, expression profiles of miRNAs in tumor tissues and derived cell lines have been detected in different types of cancer. The seminal work of Lu et al. showed that miRNA tumor profiles are informative, reflecting the developmental lineage and differentiation state of the tumor. Moreover, the authors could successfully classify each tumor type by means of miRNA profiling, whereas messenger profiles proved inaccurate to do so (Lu et al. 2005).

Since then, several works have been published. For instance, in the study done by Lee and co-workers, the differential miRNA expression was analyzed by means of TaqMan real-time quantitative PCR arrays, using 157 human mature miRNAs. The work comprised ten normal tumor biopsies, staged as primary invasive squamous cell carcinomas (ISCC), and ten normal tissues. They found that 70 miRNAs were differentially expressed significantly: 68 were up-regulated and 2 were down-regulated. Among them, 10 miRNAs were most over-expressed with—fold changes of more than 100. The top ten over-expressed miRNAs were miR-199-s, miR-9, miR199a*, miR-199a, miR-199b, miR-145, miR-133a, miR-133b, miR-214 and miR-127. On the other hand, the only two down-expressed miRNAs were miR-149 and miR-203. Interestingly, expression of miR-127 was significantly associated to lymph node metastasis. Finally, they chose miR-199a to block its expression by means of anti-miR-199a transfection on CC-derived cell lines, showing an important reduction on cell growth (Lee et al. 2008).

In a larger study with a sample of 102 CC tumor biopsies, the expression profile of 96 cancer-related miRNAs was analyzed by quantitative RT-PCR (Hu et al. 2010). The researchers used a support vector machine selecting 10 miRNAs to predict the overall survival in the patient cohort. Among those selected miRNAs, they found that miR-200a and miR-9 were significantly associated to overall survival. Finally, they individually transfected miR-200a and miR-9 into HeLa cells and analyzed the expression profile of the transfected cells. By means of gene set enrichment and gene ontology-based analyses, they showed that miR-200a could regulate the metastatic potential of cancer cells to migrate to distant sites; hence, miR-200a could control cancer phenotype thorugh regulation of metastasis processes. They also showed that genes associated to miR-9 were involved in metabolic processes, explaining the maintenance of the high metabolic rate that is important for the rapid proliferation of cervical cancer cells.

In the same research line, Pereira et al. studied the differential expression associated to CC progression, employing normal cells, moderate/severe dysplasias and invasive squamous cell carcinomas infected with HPV16 sequences. They identified the altered expression of 10 miRNAs between the cancer and dysplasia tissues compared to the normal cervical tissues (miR-16, miR-21, miR-106b, miR-135b, miR-141, miR-223, miR-301b, and miR-449a significantly over-expressed). Their results showed that miR-21, miR-135b, miR-223, miR-301b and miR-135b were significantly over-expressed in cervical cancer tissue compared to both normal and dysplasia tissues. In the work, it was proposed that the above listed miRNAs might be useful in distinguishing cervical cancer and dysplasia from normal cervical tissue (Pereira et al. 2010).

The analyses of the expression of miR-100 in 125 cervical tissues including normal cervical epithelium, cervical intraepithelial neoplasia

(CIN), and cervical cancer, as well as five cervical cell lines by quantitative real-time reverse transcriptase PCR (qRT-PCR) were recently published. miR-100 expression showed a significantly and gradually reduced tendency from low-grade CIN to high-grade CIN and cervical cancer tissues, as well as a significant decrease in HPV positive cervical cancer cell lines. The modulation of miR-100 expression remarkably influenced cell proliferation, cell cycle and apoptosis, as well as the level of PLK1 protein (Polo-like kinase1), but not mRNA, in *in vitro* experiments. It was also found that PLK1 expression was negatively correlated with miR-100 expression in CIN3 and cervical cancer tissues. Finally, they conclude that miR-100 participates in the development of cervical cancer at least partly through loss of inhibition of the target gene PLK1, which probably occurs in a relative late phase of carcinogenesis (Li et al. 2011).

In another work, the miRNA expression related to CC progression in 18 cervical tissues was analyzed (six for each normal, CIN 2-3 HPV 16 positive and six squamous cell carcinoma HPV 16 positive) by means of an miRNA microarray covering 875 human miRNAs (Li et al. 2011). Interestingly, the findings showed 31 unique miRNAs with a significant increased expression from normal to CC (17 up-regulated and 14 down-modulated). Among these, miR-218 was the most significantly down-regulated from normal to CC, while miR-29 showed over-expression. Furthermore, an important negative correlation of YY1 and CDK6 expression with miR-29 was shown, and it was suggested that HR-HPV E6/E7 could regulate the expression level of miR-29. The expression profile of 1145 miRNAs by means of microarray-based technology was analyzed comparing 2 primary CC samples (staged IIA) versus their adjacent normal counterparts. The authors found that four miRNAs, (miR-886-5p, miR-610, miR-10a* and miR-30a*) were over-expressed in CC tissues; whereas three (miR-302d, miR-346 and miR-518b), were under-expressed, and that miR-886-5p negatively regulates Bax. It is well known that Bax is a downstream transcriptional target of p53 and that it is up-regulated by p53; therefore it is feasible to hypothesize that over-expression of miR-886-5p be a virus-mediated strategy developed by HPV to block host-mediated apoptosis (Li et al. 2011). In addition, in a publication about the regulation of E7 and miR-886-5p, the authors demonstrated that the E7 protein downregulates miR-203 expression upon differentiation, which may occur through the mitogen-activated protein (MAP) kinase/protein kinase C (PKC) pathway. Moreover, they showed that high levels of miR-203 are inhibitory to HPV amplification and that HPV proteins act to suppress expression of this microRNA to allow productive replication in differentiating cells (Melar-New and Laimins 2010). Those results put in context the ability of HPV oncoproteins to regulate the expression of selected miRNA in order to maintain epithelial cells in proliferation, permitting viral replication. Table 1 lists the main miRNAs identified in CC progression.

Table 1. miRNAs involved in the development and progression of cervical cancer.

miRNA	Up/Down	Cellular process	Target gene	Clinical background
miR-34a	Down	p53-dependent pathway (cell cycle progression, cellular senescence and apoptosis)	p18Ink4c, CDK4, CDK6, Cyclin E2, E2F1, E2F3, E2F5, BCL2, BIRC3, and DcR3	↓CIN I, ↓↓CIN II, ↓↓↓CIN III
miR-218	Down		LAMB3	↓CIN III, ↓↓↓CaCu
miR-200a, miR-205	Basal expression	Metastasis (inhibit the epithelial to mesenchymal transition)	ZEB1, ZEB2 and SIP1	CaCu → CaCu metastasis
miR-9	Up	Tumor cell metabolism (ATPase activity, Group transfer coenzyme metabolic process, Glutamine family amino acid metabolic process	Not identified	Cervical cancers
miR-127	Up	Metastasis	Not identified	↓NSE-↑↑ISCCs
miR-199a	Up	Cell growth	Not identified	↓NSE-↑↑ISCCs
miR-372	Down	Cell growth (induced arrest in the S/G2 phases of cell cycle)	CDK2, Cyclin A1	normal → cancer
miR-203	Up	Keratinocyte differentiation/ maintain HPV episomes	p63-family	Normal epithelia → HPV-infected epithelia
miR-143	Down	Cellular Growth and Proliferation	PPAR Signaling	↑Normal, ↓↓CIN, ↓↓CIN III, ↓↓Carcinoma
miR-145	Down	Cellular Movement	IGF-1	↑Normal, ↓CIN, ↓CIN III, ↓Carcinoma
miR-99a, miR-203, miR-513, miR-29a	Down	Cell Death, Tissue Development	IGF-1, BCL2L2, VEGFA and CDK6	↑Normal, ↓CIN, ↓CIN III, ↓Carcinoma
miR-522*	Up	Cell Cycle: G2/M DNA Damage Checkpoint Regulation	Not identified	↑Normal, ↑↑CIN, ↑↑CIN, ↓Carcinoma
miR-148a	Up	Tumor supresor genes	PTEN, TP53INP1 and TP53INP2	↑Normal, ↑↑CIN, ↑↑CIN, ↑↑↑Carcinoma
miR-10a, miR-196a, miR-132	Up	Cell transformation and progression	(HOX) genes	↑Normal, ↑↑↑CIN, ↑↑↑CIN, ↑↑↑Carcinoma
miR-886-5p	Up	Apoptosis	BAX	↑ANTT, ↑↑↑CSCC
miR-100	Down	Growth, cell cycle, and apoptosis	PLK1	↑ Normal, ↓CIN, ↓↓Carcinoma
miR-26a	Down	Cellular Growth and Proliferation	Not identified	↑Normal, ↓CIN, ↓CIN III, ↓Carcinoma

3.1 miRNA Profile Regulated by HR-HPV Oncoproteins

E6 and E7 control, by means of binding and inactivation of cellular proteins p53 and pRb, several important molecular routes that are hallmarks of cancer, including auto-proliferative signaling, growth suppressor evasion, cell death resistance, and replicative immortality, among others. It is well known that complex interactions between HR-HPV E6 and E7 involve the activation of transcription factors, such as E2F and c-Myc, which can promote the transactivation of miRNA expression (El Baroudi et al. 2011, Woods et al. 2007). Hence, it is conceivable that expression of viral oncoproteins modulate the expression levels of miRNAs, which could participate as oncomiRs. In this concern, Au Yeung et al. demonstrated that miR-23b is downregulated by E6 and that miR-23b targets urokinase-type plasminogen activator (uPA), which is over-expressed in CC. Therefore, decreased levels of miR-23b increase the expression of uPA, and thus induce the migration of CC-derived cells. Moreover, a consensus p53-binding site was detected in the promoter region of miR-23b; hence, miR-23b/uPA is involved in HPV-16 E6-associated cervical cancer development (Au Yeung et al. 2001).

miR-34a has been identified as a direct transcriptional target of cellular transcription factor p53 (Chang 2007, He et al. 2007). Transactivation of miR-34a expression is elicited by the binding of p53 to a consensus binding site present in the miR-34a promoter region. Consequently, as HPV E6 oncoprotein destabilizes p53 during virus infection, it is feasible to assume a down-regulation of miR-34a expression in most cervical cancer tissues with oncogenic HPV infection. Thus, miR-34a is down-regulated in productive, pre-malignant HPV infections, cervical cancer tissues, and cervical cancer cells. miR-34a targets multiple cell cycle components, including CDK4, cyclin E2, E2F-1, hepatocyte growth factor receptor MET, and Bcl-2 (Boomer et al. 2007, He et al. 2007, Tazawa et al. 2007, Welch et al. 2007). Therefore, is feasible to postulate miR-34a as a tumor suppressor miRNA whose expression is modulated by the HR-HPV E6 oncoprotein.

A key process during CC progression is the proliferative capacity of differentiating epithelial cells. miR-203 is a critical molecule for regulating the transition of keratinocytes from a proliferative state in undifferentiated basal cells to a nonproliferative status in differentiated supra-basal cells. It has even been suggested to promote epithelial maturation by suppressing the "stemness" potential (Sonkoly et al. 2007). The p63 family of transcription factors, which regulate the balance between epithelial proliferation and differentiation, are the main target of miR-203 (Lena et al. 2008). Table 2 lists the major miRNAs affected by HPV-associated genes.

Table 2. HPV oncoproteins regulate the expression of miRNAs.

Protein	miRNAs	Up/Down	Target gen	Cellular process
E5	mir-146a	Up	ZNF813	Cell adhesion and cell cycle
E5	mir-324-5p	Down	CDH2, CTNNB1	Transendothelial migration
E5	mir-203	Down	p63	Cell juntion, cell migration, and cell motility
E6	mir-34a	Down	p18Ink4c, CDK4, CDK6, Cyclin E2	Cell cycle progression
E6	mir-218	Down	LAMB3	Not identified
E6	mir-23b	Down	uPA	Cell migration
E6/E7	mir-29	Down	YY1 and CDK6	Cell cycle progression restriction and apoptosis induction
E7	mir-15b	Down	CCNA2, CCNB1, CCNB2, MSH6 and MCM7	Recognition of mismatched nucleotides, prior to their repair, and initiation of eukaryotic genome replication
E7	miR-15a/ miR-16-1 and miR-203	Down	c-Myc, c-Myb, PPAR	Cell proliferation, survival, and invasion

4. siRNAs FOR HPV E6 AND E7 ONCOGENES, AS POTENTIAL GENE THERAPY FOR CERVICAL CANCER

siRNAs may silence the expression of genes that encode for tumor antigens or viral oncogenes, in order to repress the specific proliferation of cancerous cells. As a consequence, the silencing of genes with siRNAs is a potential mechanism to inactivate foreign DNA sequences and a successful strategy to silence the expression of HPV oncogenes in cervical cancer. The first studies carried out with synthetic siRNA in order to induce the silencing of HPV16 E6 and E7 oncogene expression were developed by Jiang et al. in 2002. In this study, the authors showed the biological effect of siRNAs in human cells from cervical carcinoma. The administration of siRNAs led to mRNA cleavage and specific silencing of HPV16 E6 and E7 oncogene expression. Besides this, E6 silencing induced the expression of the p53 gene, trans-activation of the p21-inhibiting gene CIP1/WAF1 cyclin-kinase, and decrease of cellular proliferation; whereas silencing of E7 induced cellular death by apoptosis. Thus, the findings reported by this group demonstrated, for the first time, that the expression of HPV E6 and E7 oncogenes might be specifically silenced by siRNAs in human tumoral cervical cells that have been transformed by oncogenic HPVs.

4.1 Silencing of HPV E6 and E7 Bicistron with siRNAs

Attention has been focused on the possibility of using siRNAs for HPV E6 and E7 oncogenes and their ability to silence the HPV E6-E7 bicistron. One example is the work by Yoshinouch et al. (2003) using SiHa cells (HPV16+). Their study showed the effect of a synthetic siRNA for HPV16 E6 oncogene, as well as the silencing of both E6 and E7 oncogenes. In addition, they pointed out some observations regarding the inhibition of cellular proliferation, p53 protein expression, the induction of p21-Cip1/Waf1 gene, and the identification of the cell cycle arrest mediator hypo-phosphorylated form of pRb, which associates with and inactivates E2F transcription factors, in the same cell line. Furthermore, when SiHa cells were inoculated in immune deficient (SCID) mice and treated with siRNAs for the E6 oncogene, a decrease in the ability of cancerous cells to induce tumor formation in animals was detected. Therefore, the findings reported by Yoshinouchi et al. showed that siRNAs for HPV16 E6 oncogene could delete the expression of E6-E7 bicistron *in vitro* as well as *in vivo*.

Another group has reported the use of synthetic siRNAs for HPV18 E6 oncogene (Butz et al. 2003). This study showed the induction of apoptosis of CaSki (HPV16+) cells, the increment of p53 and p21-Cip1/Waf1 expression, and the expression of the pRb. A remarkable result in this work was that siRNAs for HPV18 E6 did not affect HPV18 E7 expression. Initially, this observation seems to contradict what was reported by Jiang et al. who demonstrated that siRNAs for E6 oncogene have an effect on both the E6 and E7 oncogenes (Jiang et al. 2002). Nevertheless, these data are not totally opposite since they can be explained by the design of siRNA nucleotide sequences (Lea et al. 2007). In that work, the design of E7 siRNA sequences affects the expression of the E6-E7 bicistron due to their relative position in the mRNA. On the other hand, siRNAs for E6 were complementary to the mRNA in a sequence that does not influence the silencing of the E6-E7 bicistronic transcript. Again, this evidence supports the fact that the silencing of HPV E6-E7 bicistron expression is dependent on the design of the siRNA sequences and suggests that the alternative splicing of HPV E6 and E7 oncogenes precedes the silencing by siRNAs. This evidence supports the fact that viral E6 and E7 oncoproteins can have anti-apoptotic properties by their influence on the p53 protein function and that, when these oncogenes are silenced with siRNAs, p53 recovers the cellular cycle control functions and apoptosis.

4.2 siRNAs Delivering Molecules for HPV E6 and E7

Evaluation of biological effects of siRNAs *in vitro* as well as *in vivo*, have been developed using liposomes as delivering molecules. CaSki cells were cultivated with liposomes containing E6/E7 siRNAs; silencing of both oncogenes was induced and apoptosis was observed in tumor cells. With this same system, and through the development of a murine tumor model with CaSki cells, the effects of siRNAs for E6 have been evaluated and the silencing of the viral oncogene has been determined, as well as induction of apoptosis of tumor cells and a significant inhibition of the growth of the tumoral mass *in vivo*. Although these findings are significant, when synthetic siRNAs are used and administered by lipofection to mammalian cells, a potential problem that appears is the cleavage of siRNAs by the action of endogenous cellular endonucleases. An alternative design to protect the siRNAs from this cleavage is the synthesis of siRNAs with chemical modifications; however, this may induce undesirable collateral effects (Niu et al. 2006, Jiang et al. 2004). The combination of adhesive biogels and liposomes containing siRNAs for the HPV16 E7 oncogene have resulted in specific silencing of E7 and induction of apoptosis of cancerous cells *in vitro*. In addition, the use of atelocollagen as a vehicle to administer the siRNAs for HPV18 E6 and E7, *in vitro* as well as *in vivo*, has also been reported (Takuma et al. 2006). Although the silencing effects of synthetic siRNAs are evident, the half-life of these molecules after their administration is relatively short, even when they are attached to transporting molecules; this limits their application in pre-clinical or clinical trials. Besides this, the real application of siRNAs for high-risk HPV oncogenes in clinical phase studies requires a better understanding of the development of highly specific siRNAs and greater efficiency in the *in vivo* liberation systems. In this sense, protocols have been developed for the generation of lentivirus as molecular liberation vectors for specific siRNAs for high risk HPV E6 and E7 oncogenes, as well as for their stable transfection and transduction in human cancer cervical cells. The biological effects of silencing with siRNAs, in *in vitro* as well as in *in vivo* models, are being evaluated (Gu et al. 2008).

The effect that E6 and E7 silencing by means of siRNAs has on cellular transcriptome has been analyzed. In 2007 Kuner et al. (2007), analyzed the transcriptome of HeLa cells after viral oncogenes silencing. In this study, 360 cellular genes were identified which had a negative regulation and 288 genes with positive regulation due to the effect of E6 and E7 siRNAs. Most of these genes are involved in relevant biological processes during the development of the tumor cell, such as: apoptosis control, regulation of the cell cycle, formation of the mitotic spindle, processing of mRNA by splicing, metabolism, DNA replication and repair, nuclear transport, cell proliferation and gene regulation by c-Myc (Kuner et al. 2007). Depletion of E6/E7 expression can lead to analyze the major cellular pathways involved in viral transformation.

5. CONCLUSIONS

Cervical Cancer is a complex disease in which progression from pre-neoplasic lesions, metastatic disease, recurrence of cancer and patient clinical outcome, are affected by several factors, not only associated to HR-HPV infection and E6/E7 oncoprotein interaction with cellular components. In this scenario, Wnt/beta-catenin has been postulated as a "second hit" affected during cancer progression (Pérez-Plasencia et al. 2007, Pérez-Plasencia et al. 2008). However, other factors such as miRNAs aberrant expression can contribute to the carcinogenesis process.

miRNAs have emerged as important regulators of carcinogenesis, hence, "OncomiRs" have been described as drivers in carcinogenesis process (Fig. 3). *In vitro* and *in vivo* studies provide exciting information about the biology of OncomiRs and emphasize central aspects that must be considered for the design of an efficient miRNA-based therapy. As siRNA- and miRNA-based therapies begin to be evaluated in clinical approaches, the next few years we will witness the use of drug combinations and siRNA.

Science of miRNAs started as a catalogue of tissue-specific expressed sequences, and then it was clear that their role in carcinogenesis could be important. Nowadays, we know that these tiny molecules are able to orchestrate essential cellular processes that are hallmarks of cancer (Fig. 3).

miRNAs Timeline

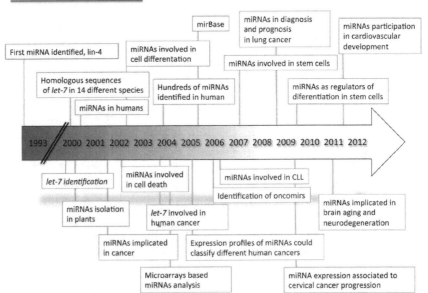

Figure 3. miRNAs Timeline summarizing the main discoveries on cancer miRNAs profile.

Hopefully, in the near future we can bring the discoveries on miRNA science to develop molecular markers associated to cancer progression, clinical outcome, or probably use them not only to block E6 and E7 expression in cervical cancer cells, but also to reprogram the cancer cell.

6. SUMMARY POINTS

- Epidemiological and experimental studies have clearly confirmed a causal role of high-risk Human Papillomavirus (HR-HPV) types in cervical carcinogenesis, which affect the cellular processes by targeting and inactivating p53 and pRB host proteins.
- HR-HPV genes affects important epigenetic mechanisms in host cell by means of direct targeting of histone methyltransferases (HMTs), modulating their enzymatic activities and by increasing transcriptional activity on specific histone demethylases (KDM6 and KDM6B), effectively leading to aberrant expression of cellular oncogenic and tumor-suppressive miRNAs.
- HPV oncoproteins regulate the expression of selected miRNAs in order to maintain epithelial cells in proliferation, permitting viral replication.

ACKNOWLEDGMENTS

This work was supported in part by UNAM (Grant IACOD-TB200111) and by CONACyT (Grant SALUD-2009-01-113948, SALUD-2009-01-111892 and SALUD-2010-01-141907).

REFERENCES

Antinore, M., M.J. Birrer, D. Patel, L. Nader, and D.J. McCance. 1996. The human papillomavirus type 16 E7 gene product interacts with and trans-activates the AP1 family of transcription factors. EMBO J. 15(8): 1950–1960.

Au Yeung, C.L., T.Y. Tsang, P.L. Yau, and T.T. Kwok. 2011. Human papillomavirus type 16 E6 induces cervical cancer cell migration through the p53/microRNA-23b/urokinase-type plasminogen activator pathway. Oncogene. 30(21): 2401–10.

Balan, R., C. Amălinei, S.E. Giuşcă, D. Ditescu, V. Gheorghiţă, E. Crauciuc, and I.D. Căruntu. 2011. Immunohistochemical evaluation of COX-2 expression in HPV-positive cervical squamous intraepithelial lesions. Rom J Morphol Embryol. 52(1): 39–43.

Baldwin, A., H. Kyung-Won, and K. Münger. 2006. Human papillomavirus E7 oncoprotein dysregulates steroid receptor coactivator 1 localization and function. J Virol. 80(13): 6669–77.

Bernat, A., N. Avvakumov, J.S. Mymryk, and L. Banks. 2003. Interaction between the HPV E7 oncoprotein and the transcriptional coactivator p300. Oncogene. 22(39): 7871–81.

Bommer, G.T., I. Gerin, Y. Feng, A.J. Kaczorowski, R. Kuick, R.E. Love, Y. Zhai, T.J. Giordano, Z.S. Qin, B.B. Moore, O.A. MacDougald, K.R. Cho, and E.R. Fearon. 2007. p53-mediated activation of miRNA34 candidate tumorsuppressor genes. Curr Biol. 17: 1298–1307.

Brown, C., A.M. Kowalczyk, A.R. Taylor, I.M. Morgan, and K. Gaston. 2008. P53 represses human papillomavirus type 16 DNA replication via the viral E2 protein. Virol J. 5: 5.

Burgers, W.A., L. Blanchon, S. Pradhan, Y. de Launoit, T. Kouzarides, and F. Fuks. 2007. Viral oncoproteins target the DNA methyltransferases. Oncogene. 26(11): 1650–5.

Burk, R.D., P. Kelly, J. Feldman, J. Bromberg, S.H. Vermund, J.A. DeHovitz, and S.H. Landesman. 1996. Declining prevalence of cervicovaginal human papillomavirus infection with age is independent of other risk factors. Sex Transm Dis. 23: 333–341.

Butz, K., T. Ristriani, A. Hengstermann, C. Denk, M. Scheffner, and F. Hoppe-Seyler. 2003. siRNA targeting of the viral E6 oncogene efficiently kills human papillomavirus-positive cancer cells. Oncogene. 22: 5938–5945.

Calin, G.A., C.D. Dumitru, M. Shimizu, R. Bichi, S. Zupo, E. Noch, H. Aldler, S. Rattan, M. Keating, K. Rai, L. Rassenti, T. Kipps, M. Negrini, F. Bullrich, and C.M. Croce. 2002. Frequent deletions and down-regulation of micro- RNA genes miR15 and miR16 at 13q14 in chronic lymphocytic leukemia. Proc Natl Acad Sci USA. 26; 99(24): 15524–9.

Castle, P.E., A.C. Rodriguez, R.D. Burk, R. Herrero, S. Wacholder, M. Alfaro et al. 2009. Short term persistence of human papillomavirus and risk of cervical precancer and cancer: population based cohort study. BMJ. 339: b2569.

Chang, T.C., E.A. Wentzel, O.A. Kent, K. Ramachandran, M. Mullendore, K.H. Lee, G. Feldmann, M. Yamakuchi, M. Ferlito, C.J. Lowenstein, D.E. Arking, M.A. Beer, A. Maitra, and J.T. Mendell. 2007. Transactivation of miR-34a by p53 broadly influences gene expression and promotes apoptosis. Mol Cell. 26: 745–752.

Chiang, C.M., M. Ustav, A. Stenlund, T.F. Ho, T.R. Broker, and L.T. Chow. 1992. Viral E1 and E2 proteins support replication of homologous and heterologous papillomaviral origins. Proc Natl Acad Sci USA. 89: 5799–803.

Conger, K.L., J.S. Liu, S.R. Kuo, L.T. Chow, and T.S. Wang. 1999. Human papillomavirus DNA replication. Interactions between the viral E1 protein and two subunits of human DNA polymerase alpha/primase. J Biol Chem. 274: 2696–705.

Cueille, N., R. Nougarede, F. Mechali, M. Philippe, and C. Bonne-Andrea. 1998. Functional interaction between the bovine papillomavirus virus type 1 replicative helicase E1 and cyclin E-Cdk2. J Virol. 72: 7255–7262.

Cullen, A.P., R. Reid, M. Campion, and A.T. Lorincz. 1991. Analysis of the physical state of different human papillomavirus DNAs in intraepithelial and invasive cervical neoplasm. J Virol. 65: 606–12.

Dalton, S. and L. Whitbread. 1995. Cell cycle-regulated nuclear import and export of Cdc47, a protein essential for initiation of DNA replication in budding yeast. Proc Natl Acad Sci USA. 92: 2514–2518.

Demeret, C., A. Garcia-Carranca, and F. Thierry. 2003. Transcription-independent triggering of the extrinsic pathway of apoptosis by human papillomavirus 18 E2 protein. Oncogene. 22(2): 168–75.

Deng, W., B.Y. Lin, G. Jin, C.G. Wheeler, T. Ma, J.W. Harper, T.R. Broker, and L.T. Chow. 2004. Cyclin/CDK regulates the nucleocytoplasmic localization of the human papillomavirus E1 DNA helicase. J Virol. 78(24): 13954–65.

Dowhanick, J.J., A.A. McBride, and P.M. Howley. 1995. Suppression of cellular proliferation by the papillomavirus E2 protein. J Virol. 69(12): 7791–9.

Dyson, N., P.M. Howley, K. Munger, and E. Harlow. 1989. The human papillomavirus-16 El oncoprotein is able to bind to the retinoblastoma gene product. Science. 243: 934–940.

El Baroudi, M., D. Corà, C. Bosia, M. Osella, and M. Caselle. 2011. A curated database of miRNA mediated feed-forward loops involving MYC as master regulator. PLoS One. 6(3): e14742.

Ferlay, J., H.R. Shin, F. Bray, D. Forman, C. Mathers, and D.M. Parkin. 2010. Estimates of worldwide burden of cancer in 2008: GLOBOCAN 2008. Int J Cancer. 127(12): 2893–917.

Franco, E.L., L.L. Villa, J.P. Sobrinho, J.M. Prado, M.C. Rousseau, M. Désy, and T.E. Rohan. 1999. Epidemiology of acquisition and clearance of cervical human papillomavirus infection in women from a high-risk area for cervical cancer. J Infect Dis. 180: 1415–23.

Grossman, S.R., R. Mora, and L.A. Laimins. 1989. Intracellular localization and DNA-binding properties of human papillomavirus type 18 E6 protein expressed with a baculovirus vector. J Virol. 63(1): 366–74.

Gu, W., L. Putral, and N. McMillan. 2008. siRNA and shRNA as anticancer agents in a cervical cancer model. Methods Mol Biol. 442: 159–72.

Hadaschik, D., K. Hinterkeuser, M. Oldak, H.J. Pfister, and S. Smola-Hess. 2003. The Papillomavirus E2 protein binds to and synergizes with C/EBP factors involved in keratinocyte differentiation. J Virol. 77(9): 5253–65.

Haertel-Wiesmann, M., Y. Liang, W.J. Fantl, and L.T. Williams. 2000. Regulation of cyclooxygenase-2 and periostin by Wnt-3 in mouse mammary epithelial cells. J Biol Chem. 275(41): 32046–51.

He, L., X. He, L.P. Lim, E. de Stanchina, Z. Xuan, Y. Liang, W. Xue, L. Zender, J. Magnus, D. Ridzon, A.L. Jackson, P.S. Linsley, C. Chen, S.W. Lowe, M.A. Cleary, and G.J. Hannon. 2007. A microRNA component of the p53 tumor suppressor network. Nature. 447: 1130–1134.

Hegde, R.S. 2002. The papillomavirus E2 proteins: structure, function, and biology. Annu Rev Biophys Biomol Struct. 31: 343–60.

Ho, G.Y., R. Bierman, L. Beardsley, C.J. Chang, and R.D. Burk. 1998. Natural history of cervicovaginal papillomavirus infection in young women. N Engl J Med. 338: 423–428.

Hou, S.Y., S.Y. Wu, and C.M. Chiang. 2002. Transcriptional activity among high and low-risk human papillomavirus E2 proteins correlates with E2 DNA binding. J Biol Chem. 277(47): 45619–29.

Howe, L.R., K. Subbaramaiah, W.J. Chung, A.J. Dannenberg, and A.M. Brown. 1999. Transcriptional activation of cyclooxygenase-2 in Wnt-1-transformed mouse mammary epithelial cells. Cancer Res. 59(7): 1572–7.

Hu, X., J.K. Schwarz, J.S. Jr. Lewis, P.C. Huettner, J.S. Rader, J.O. Deasy, P.W. Grigsby, and X. Wang. 2010. MicroRNA expression signature for cervical cancer prognosis. Cancer Res. 70: 1441–1448.

Hu, Y., F. Ye, W. Lu, D. Hong, X. Wan, and X. Xie. 2009. HPV16 E6-induced and E6AP-dependent inhibition of the transcriptional coactivator hADA3 in human cervical carcinoma cells. Cancer Invest. 27(3): 298–306.

Huang, S.M. and D.J. McCance. 2002. Down regulation of the interleukin-8 promoter by human papillomavirus type 16 E6 and E7 through effects on CREB binding protein/p300 and P/CAF. J Virol. 76(17): 8710–21.

Hudelist, G., M. Manavi, K.I. Pischinger, T. Watkins-Riedel, C.F. Singer, E. Kubista, and K.F. Czerwenka. 2008. Physical state and expression of HPV DNA in benign and dysplastic cervical tissue: different levels of viral integration are correlated with lesion grade. Gynecol Oncol. 92: 873–880.

Hwang, S.G., D. Lee, J. Kim, T. Seo, and J. Choe. 2002. Human papillomavirus type 16 E7 binds to E2F1 and activates E2F1-driven transcription in a retinoblastoma protein-independent manner. J Biol Chem. 277: 2923–30.

Jacobs, M.V., J.M. Walboomers, P.J. Snijders, F.J. Voorhorst, R.H. Verheijen, N. Fransen-Daalmeijer, and C.J. Meijer. 2000. Distribution of 37 mucosotropic HPV types in women with cytologically normal cervical smears: the age-related patterns for high-risk and low-risk types. Int J Cancer. 87: 221–7.

Jiang, M. and J. Milner. 2002. Selective silencing of viral gene expression in HPV-positive human cervical carcinoma cells treated with siRNA, a primer of RNA interference. Oncogene. 21: 6041–6048.

Jiang, M., C.P. Rubbi, and J. Milner. 2004. Gel-based application of siRNA to human epithelial cancer cells induces RNAi-dependent apoptosis. Oligonucleotides. 14: 239–248.

Kjaer, S.K., A.J. van den Brule, J.E. Bock, P.A. Poll, G. Engholm, M.E. Sherman, J.M. Walboomers, C.J. Meijer. 1997. Determinants for genital human papillomavirus (HPV) infection in 1000 randomly chosen young Danish women with normal Pap smear: are there different risk profiles for oncogenic and nononcogenic HPV types? Cancer Epidemiol Biomarkers Prev. 6: 799–805.

Kumar, A., Y. Zhao, G. Meng, M. Zeng, S. Srinivasan, L.M. Delmolino, Q. Gao, G. Dimri, F. Weber, D.E. Wazer, H. Band, and V. Band. 2002. Human papillomavirus oncoprotein E6 inactivates the transcriptional coactivator human ADA3. Mol Cell Biol. 22(16): 5801–12

Kumar, M.S., J. Lu, K.L. Mercer, T.R. Golub, and T. Jacks. 2007. Impaired microRNA processing enhances cellular transformation and tumorigenesis. Nature Genet. 39: 673–677.

Kuner, R., M. Vogt, H. Sultmann, A. Buness, S. Dymalla, J. Bulkescher, M. Fellmann, K. Butz, A. Poustka, and F. Hoppe-Seyler. 2007. Identification of cellular targets for the human papillomavirus E6 and E7 oncogenes by RNA interference and transcriptome analyses. J Mol Med. 85: 1253–1262.

Lavia, P., A.M. Mileo, A. Giordano, and M.G. Paggi. 2003. Emerging roles of DNA tumor viruses in cell proliferation: new insights into genomic instability. Oncogene. 22(42): 6508–16.

Lazcano-Ponce, E., R. Herrero, N. Muñoz, A. Cruz, K.V. Shah, P. Alonso, P. Hernández, J. Salmerón, and M. Hernández. 2001. Epidemiology of HPV infection among Mexican women with normal cervical cytology. Int J Cancer. 91: 412–20.

Lea, J.S., N. Sunaga, M. Sato, G. Kalahasti, D.S. Miller, J.D. Minna, and C.Y. Muller. 2007. Silencing of HPV 18 oncoproteins with RNA interference causes growth inhibition of cervical cancer cells. Reprod Sci. 14: 20–28.

Lee, D., H.Z. Kim, K.W. Jeong, Y.S. Shim, I. Horikawa, J.C. Barrett, and J. Choe. 2002. Human papillomavirus E2 down-regulates the human telomerase reverse transcriptase promoter. J Biol Chem. 277(31): 27748–56.

Lee, J.W., C.H. Choi, J.J. Choi, Y.A. Park, S.J. Kim, S.Y. Hwang, W.Y. Kim, T.J. Kim, J.H. Lee, B.G. Kim, and D.S. Bae. 2008. Altered microRNA Expression in Cervical Carcinomas. Clin. Cancer Res. 14: 2535–2542.

Lena, A.M., R. Shalom-Feuerstein, P. Rivetti di Val Cervo, D. Aberdam, R. A. Knight, G. Melino, and E. Candi. 2008. miR-203 represses 'stemness' by repressing DeltaNp63. Cell Death Differ. 15: 1187–1195.

Li, B.H., J.S. Zhou, F. Ye, X.D. Cheng, C.Y. Zhou, W.G. Lu, and X. Xie. 2011. Reduced miR-100 expression in cervical cancer and precursors and its carcinogenic effect through targeting PLK1 protein. Eur J Cancer. 47: 2166–2174.

Li, J.H., X. Xiao, Y.N. Zhang, Y.M. Wang, L.M. Feng, Y.M. Wu, and Y.X. Zhang. 2011. MicroRNA miR-886-5p inhibits apoptosis by down-regulating Bax expression in human cervical carcinoma cells. Gynecol Oncol. 120(1): 145–51.

Liu, H., J. Xiao, Y. Yang, Y. Liu, R. Ma, Y. Li, F. Deng, and Y. Zhang. 2011. COX-2 expression is correlated with VEGF-C, lymphangiogenesis and lymph node metastasis in human cervical cancer. Microvasc Res. 82(2): 131–40.

Liu, X., A. Dakic, Y. Zhang, Y. Dai, R. Chen, and R. Schlegel. 2009. HPV E6 protein interacts physically and functionally with the cellular telomerase complex. Proc Natl Acad Sci USA. 106(44): 18780–5.

Lu, J., G. Getz, E.A. Miska, E. Alvarez-Saavedra, J. Lamb, D. Peck, A. Sweet-Cordero, B.L. Ebert, R.H. Mak, A.A. Ferrando, J.R. Downing, T. Jacks, H.R. Horvitz, and T.R. Golub. 2005. MicroRNA expression profiles classify human cancers. Nature. 435(7043): 834–8.

Lüscher-Firzlaff, J.M., J.M. Westendorf, J. Zwicker, H. Burkhardt, M. Henriksson, R. Müller, F. Pirollet, and B. Lüscher. 1999. Interaction of the fork head domain transcription factor MPP2 with the human papilloma virus 16 E7 protein: enhancement of transformation and transactivation. Oncogene. 18: 5620–30.

Ma, T., N. Zou, B.Y. Lin, L.T. Chow, and J.W. Harper. 1999. Interaction between cyclin-dependent kinases and human papillomavirus replication-initiation protein E1 is required for efficient viral replication. Proc Natl Acad Sci USA. 96: 382–387.

Maldonado, E., M.E. Cabrejos, L. Banks, and J.E. Allende. 2002. Human papillomavirus-16 E7 protein inhibits the DNA interaction of the TATA binding transcription factor. J Cell Biochem. 85: 663–9.

Mallon, R.G., D. Wojciechowicz, and V. Defendi. 1987. DNA-binding activity of papillomavirus proteins. J Virol. 61(5): 1655–60.

Masterson, P.J., M.A. Stanley, A.P. Lewis, and M.A. Romanos. 1998. A C-terminal helicase domain of the human papillomavirus E1 protein binds E2 and the DNA polymerase alpha-primase p68 subunit. J Virol. 72: 7407–19.

Matsukura, T., S. Koi, and M. Sugase. 1989. Both episomal and integrated forms of human papillomavirus type 16 are involved in invasive cervical cancers. Virology. 172: 63–72.

Melar-New, M. and L.A. Laimins. 2010. Human papillomaviruses modulate expression of microRNA 203 upon epithelial differentiation to control levels of p63 proteins. J Virol. 84(10): 5212–21.

Molano, M., H. Posso, E. Weiderpass, A.J. van den Brule, M. Ronderos, S. Franceschi, C.J. Meijer, A. Arslan, and N. Munoz. 2002. Prevalence and determinants of HPV infection among Colombian women with normal cytology. Br J Cancer. 87: 324–33.

Mole, S., S.G. Milligan, and S.V. Graham. 2009. Human papillomavirus type 16 E2 protein transcriptionally activates the promoter of a key cellular splicing factor, SF2/ASF. J Virol. 83(1): 357–67.

Moody, C.A. and L.A. Laimins. 2010. Human papillomavirus oncoproteins: pathways to transformation. Nat Rev Cancer. 10(8): 550–60.

Niu, X.Y., Z.L. Peng, W.Q. Duan, H. Wang, and P. Wang. 2006. Inhibition of HPV 16 E6 oncogene expression by RNA interference in vitro and in vivo. Int J Gynecol Cancer. 16: 743–751.

Pereira, P.M., J.P. Marques, A.R. Soares, L. Carreto, and M.A. Santos. 2010. MicroRNA expression variability in human cervical tissues. PLoS ONE. 5(7): e11780.

Pérez-Plasencia, C., A. Duenas-Gonzalez, and B. Alatorre-Tavera. 2008. Second hit in cervical carcinogenesis process: involvement of wnt/beta catenin pathway. Int Arch Med. 1(1): 10.

Pérez-Plasencia, C., G. Vázquez-Ortiz, R. López-Romero, P. Piña-Sanchez, J. Moreno, and M. Salcedo. 2007. Genome wide expression analysis in HPV16 cervical cancer: identification of altered metabolic pathways. Infect Agent Cancer. 2: 16.

Phillips, A.C. and K.H. Vousden. 1997. Analysis of the interaction between human papillomavirus type 16 E7 and the TATA-binding protein, TBP. J Gen Virol. 78: 905–9.

Ronco, L.V., A.Y. Karpova, M. Vidal, and P.M. Howley. 1998. Human papillomavirus 16 E6 oncoprotein binds to interferon regulatory factor-3 and inhibits its transcriptional activity. Genes Dev. 12: 2061–72.

Roteli-Martins, C.M., N.S. de Carvalho, P. Naud, J. Teixeira, P. Borba, S. Derchain, S. Tyring, S. Gall, A. Diaz, M. Blatter, R.M. Shier, B. Romanowski, W.G. Quint, J. Issam, C. Galindo, A. Schuind, and G. Dubin. 2011. Prevalence of human papillomavirus infection and associated risk factors in young women in Brazil, Canada, and the United States: a multicenter cross-sectional study. Int J Gynecol Pathol. 30(2): 173–84.

Scheffner, M., B.A. Werness, J.M. Huibregtse, A.J. Levine, and P.M. Howley. 1990. The E6 oncoprotein encoded by human papillomavirus types 16 and 18 promotes the degradation of p53. Cell. 63: 1129–1136.

Sekaric, P., J.J. Cherry, and E.J. Androphy. 2008. Binding of human papillomavirus type 16 E6 to E6AP is not required for activation of hTERT. J Virol. 82(1): 71–6.

Sonkoly, E., T. Wei, P.C. Janson, A. Saaf, L. Lundeberg, M. Tengvall-Linder, G. Norstedt, H. Alenius, B. Homey, A. Scheynius, M. Stahle, and A. Pivarcsi. 2007. MicroRNAs: novel regulators involved in the pathogenesis of psoriasis? PLoS One. 2: e610.

Steben, M. and E. Duarte-Franco. 2007. Human papillomavirus infection: epidemiology and pathophysiology. Gynecol Oncol. 107: S2–5.

Steger, G. and S. Corbach. 1997. Dose-dependent regulation of the early promoter of human papillomavirus type 18 by the viral E2 protein. J Virol. 71: 50–58.

Stubenrauch, F. and H. Pfister. 1994. Low-affinity E2-binding site mediates downmodulation of E2 transactivation of the human papillomavirus type 8 late promoter. J Virol. 68: 6959–6966.

Stubenrauch, F., A.M. Colbert, and L.A. Laimins. 1998. Transactivation by the E2 protein of oncogenic human papillomavirus type 31 is not essential for early and late viral functions. J Virol. 72: 8115–23.

Subbaramaiah, K. and A.J. Dannenberg. 2007. Cyclooxygenase-2 transcription is regulated by human papillomavirus 16 E6 and E7 oncoproteins: evidence of a corepressor/coactivator exchange. Cancer Res. 67(8): 3976–85.

Tan, S.H., L.E. Leong, P.A. Walker, and H.U. Bernard. 1994. The human papillomavirus type 16 E2 transcription factor binds with low cooperativity to two flanking sites and represses the E6 promoter through displacement of Sp1 and TFIID. J Virol. 68(10): 6411–20.

Takuma, F., S. Miyuki, I. Eri, O. Takahiro, T. Yoshifumi, H. Shigenori, O. Akiko, H. Nobumaru, N. Masaru, K. Kaneyuki, T. Katsumi, and A. Daisuke. 2006. Intratumor injection of small interfering RNA-targeting human papillomavirus 18 E6 and E7 successfully inhibits the growth of cervical cancer. Int J Oncology. 29: 541–548.

Tazawa, H., N. Tsuchiya, M. Izumiya, and H. Nakagama. 2007. Tumor-suppressive miR-34a induces senescence-like growth arrest through modulation of the E2F pathway in human colon cancer cells. Proc Natl Acad Sci USA. 104: 15472–15477.

Thierry, F. and C. Demeret. 2008. Direct activation of caspase 8 by the proapoptotic E2 protein of HPV18 independent of adaptor proteins. Cell Death Differ. 15(9): 1356–63.

Ueda, Y., T. Enomoto, T. Miyatake, K. Ozaki, T. Yoshizaki, H. Kanao, Y. Ueno, R. Nakashima, K.R. Shroyer, and Y. Murata. 2003. Monoclonal expansion with integration of high-risk type human papillomavirus is an essential step for cervical carcinogenesis: association of clonal status and human papillomavirus infection with clinical outcome in cervical intraepithelial neoplasia. Lab Invest. 83: 1517–1527.

Walboomers, J.M., M.V. Jacobs, M.M. Manos, F.X. Bosch, J.A. Kummer, K.V. Shah, P.J. Snijders, J. Peto, C.J. Meijer, and N. Muñoz. 1999. Human papillomavirus is a necessary cause of invasive cervical cancer worldwide. J Pathol. 189(1): 12–9.

Webster, K., J. Parish, M. Pandya, P.L. Stern, A.R. Clarke, and K. Gaston. 2000. The human papillomavirus HPV-16 E2 protein induces apoptosis in the absence of other HPV proteins and via a p53-dependent pathway. J Biol Chem. 275: 87–94.

Welch, C., Y. Chen, and R.L. Stallings. 2007. MicroRNA-34a functions as a potential tumor suppressor by inducing apoptosis in neuroblastoma cells. Oncogene. 26: 5017–5022.

Wentzensen, N., S. Vinokurova, and M. von Knebel Doeberitz. 2004. Systematic review of genomic integration sites of human papillomavirus genomes in epithelial dysplasia and invasive cancer of the female lower genital tract. Cancer Res. 64: 3878–3884.

Woods, K., J.M. Thomson, and S.M. Hammond. 2007. Direct regulation of an oncogenic micro-RNA cluster by E2F transcription factors. J Biol Chem. 282(4): 2130–4.

Yoshinouchi, M., T. Yamada, M. Kizaki, J. Fen, T. Koseki, Y. Ikeda, T. Nishihara, and K. Yamato. 2003. *In vitro* and *in vivo* growth suppression of human papillomavirus 16-Positive cervical cancer cells by E6 siRNA. Mol Ther. 8: 762–768.

Zeng, M., A. Kumar, G. Meng, Q. Gao, G. Dimri, D. Wazer, H. Band, and V. Band. 2002. Human papilloma virus 16 E6 oncoprotein inhibits retinoic X receptor-mediated transactivation by targeting human ADA3 coactivator. J Biol Chem. 277(47): 45611–8.

zur Hausen, H. 2002. Papillomaviruses and cancer: from basic studies to clinical application. Nat Rev Cancer. 5: 342–50.

[1]Universidad Nacional Autónoma de México UNAM, FES-Iztacala, UBIMED. Tlalnepantla, Estado de México

[2]Instituto Nacional de Cancerología, Laboratorio de Genómica.

[3]Instituto Nacional de Cancerología, Unidad de Genómica y Secuenciación Masiva (UGESEM-INCan).

[4]INCMNSZ, Unidad de Bioquímica.

[5]Unidad de Investigación Biomédica en Cáncer, Instituto Nacional de Cancerología (INCan).

[6]Universidad Autonoma de la Ciudad de Mexico Posgrado en Ciencias Genómicas.

[7]Centro de Investigaciones Sobre Enfermedades Infecciosas. Instituto Nacional de Salud Pública. Avenida Universidad No. 655, Cuernavaca, Morelos, México.

*Corresponding author

[a]Email: carlos.pplas@gmail.com

[b]Email: nadia.jacobo@gmail.com

[c]Email: carlosbarba.ostria@gmail.com

[d]Email: rosamag2@hotmail.com

Email VFO: ontiverosfvero@yahoo.com.mx

Email APT: abraneet@hotmail.com

[e]Email: e_arechaga@hotmail.com

[f]Email: cesar.lopez@uacm.edu.mx

[g]Email: operalta@insp.mx

CHAPTER 6

MicroRNAs IN PROSTATE CANCER

Aya Misawa,[1,a] Ken-ichi Takayama[1, 2,b] and
Satoshi Inoue[1, 2, 3,c,]*

ABSTRACT

Androgen ablation is the mainstay for management of advanced prostate cancer (PCa). However, most tumors relapse to an incurable hormone-independent state. The androgen receptor (AR) plays an essential role in the hormone-dependent and hormone-independent states. Nevertheless, its function in the androgen-independent state and the regulatory mechanisms that cause this transition remain largely unknown.

Although the expression of microRNA (miRNA) in PCa has been extensively studied, a conclusive miRNA profile cannot be deduced yet. In this chapter, we first present an overview of the miRNAs expressed in PCa by drawing on literature data derived from clinical samples. Next, we focus on androgen-responsive miRNAs and on miRNAs that are differentially expressed in castration-resistant prostate cancer (CRPC).

[1]Department of Anti-Aging Medicine.
[2]Department of Geriatric Medicine, Graduate School of Medicine, The University of Tokyo.
[3]Division of Gene Regulation and Signal Transduction, Research Center for Genomic Medicine, Saitama Medical University. 7-3-1 Hongo Bunkyo-ku, Tokyo 113-8655, Japan.
[a]Email: amisawa-tky@umin.ac.jp
[b]Email: ktakayama-tky@umin.ac.jp
[c]Email: INOUE-GER@h.u-tokyo.ac.jp
*Corresponding author

Finally, we describe novel technologies for miRNA expression profiling, including analysis of PCa samples by next generation sequencing. Next generation sequencing has enabled elucidation of the complete set of miRNAs present in an RNA sample, thereby providing information on miRNA expression profiles, identifying sequence isoforms, and predicting novel miRNAs and potential mRNA targets. The technique may facilitate the use of miRNAs in new diagnostic and therapeutic approaches.

1. INTRODUCTION

Prostate cancer (PCa) is the most frequent tumor in men and the second most common cause of cancer-related deaths in the occidental male population (Gronberg 2003, Jemal et al. 2008). Although surgery and radiation are generally effective against clinically localized PCa, metastatic disease invariably remains incurable.

Clinically, PCa is diagnosed as local or advanced. Treatments range from surveillance to radiotherapy, radical prostatectomy, or androgen-deprivation therapy (ADT). Prostate-specific antigen (PSA) has been used as a biological marker for the disease. Diagnosis is currently based on examination of histopathological specimens, obtained by transrectal core biopsy. PCa is graded using the Gleason score, which consists of the sum of 2 scores, each varying from 1 (well differentiated) to 5 (undifferentiated). The 2 scores represent the most prevalent and the second most prevalent differentiation grade within the tumor. PCa local staging (T stage, classified from T1 to T4 depending on malignancy) is determined on the basis of digital rectal examination.

Androgen ablation, the mainstay for management of advanced PCa, reduces symptoms in 70–80% of patients. However, most tumors relapse to an incurable hormone-independent state, which is ultimately responsible for PCa mortality (Damber and Aus 2008). The androgen receptor (AR) is essential in both hormone-dependent and hormone-independent states. Nevertheless, its function in the androgen-independent state and the regulatory mechanisms that cause this transition remain largely unknown.

For early-stage clinically localized disease, radical prostatectomy is generally curative. Nevertheless, treatment is often deferred in favor of active monitoring, particularly in older men. This approach is based on the assumption that many prostate tumors are indolent, and progress slowly relative to the patient's life expectancy (Wu et al. 2004). Selection of the best treatment for clinically localized PCa is not trivial, because radical surgery and radiotherapy do not offer a clear survival advantage in some patients.

Moreover, treatment often results in a poorer quality of life. PSA testing has resulted in increased detection of early-stage PCa. Although some patients are cured of life-threatening disease, there are concerns about over-treatment and related morbidity (Loeb and Catalona 2007). Indeed, the serum PSA level, primary tumor stage, and Gleason score do not reliably predict the outcome for individual patients. Therefore, the detection of aggressiveness indicators would be helpful in guiding therapeutic decisions, by identifying individuals with potentially life-threatening disease for whom treatment is actually necessary. Diagnostic, prognostic, and therapeutic molecules are required to predict disease severity, select from among available treatments, and establish more effective therapies for advanced PCa. In this complex scenario, miRNAs may have a role to play (Kasinski et al. 2011).

Most studies indicate that miRNAs are aberrantly expressed in PCa. The miRNA expression profile of PCa has been extensively studied, but with inconsistent and controversial results. Thus, there is no consensus on which miRNAs are specific biomarkers for PCa. Conflicting data in the literature may reflect diverse experimental variables (e.g. tumor stage and previous treatment), selection of control samples (benign prostatic hyperplasia, non-neoplastic tissue surrounding the tumor, or prostate tissue from non-cancer individuals), specimen preservation and preparation (frozen, formalin-fixed paraffin-embedded, bulk, or microdissected tissues) and RNA purification methods (total RNA or small RNA samples).

Gene-expression microarray analysis is a useful technology for transcriptome analysis. However, it relies on sequence-specific probe hybridization, and therefore measures only the relative abundances of transcripts. Furthermore, it suffers from background and cross-hybridization problems. Thus, it is not surprising that published miRNA expression profiles are inconsistent. Results obtained using different measurement platforms may not perfectly correlate. Thus, when translating from high-throughput screening to single candidate miRNAs for development as potential biomarkers, validation of data by an independent technique (e.g. quantitative reverse transcription PCR [qRT-PCR] analysis, RNase protection assay, or northern blotting) is required.

The variety of miRNAs implicated in prostate carcinogenesis reflects not only differences in profiling strategies and study design, but also disease heterogeneity. miRNA expression is dynamic and differs according to the phases of prostate carcinogenesis (initiation, progression, or metastasis), dependence upon androgens for growth, treatment exposure (radiation therapy or chemotherapy), and the tumor-specific molecular pathway. The introduction of deep-sequencing technology has enabled the simultaneous sequencing of up to millions of different DNA and RNA molecules. This tag-based sequencing method measures absolute abundance and is not limited by array content, providing a comprehensive miRNA landscape and new challenges for data analysis.

In this chapter, we first present an overview of the miRNAs expressed in PCa, by drawing on data derived using clinical samples. Next, we focus on androgen-responsive miRNAs and on miRNAs that are differentially expressed in castration-resistant prostate cancer (CRPC). Finally, we describe novel technologies for miRNA expression profiling, including analysis of PCa samples by next generation sequencing. The technique represents an approach for rational validation of candidate miRNAs, which may be clinically relevant in the management of PCa.

2. MicroRNA EXPRESSION IN CLINICAL PROSTATE CANCER SPECIMENS

2.1 miRNA Expression Profiles

Many miRNAs are differentially expressed between normal and malignant tissues. Thus, miRNA expression profiles represent potential tools for cancer diagnosis and prognosis. In comparison to mRNAs, miRNAs may be used more accurately to classify and diagnose even poorly differentiated tumor samples (Lu et al. 2005).

miRNA expression profiles have recently been extensively investigated in a large cohort of human cancers. The first miRNA expression profiling study to include PCa specimens was performed in 2005, using a bead-based flow cytometric expression profiling method (Lu et al. 2005). Systematic expression analysis of 217 mammalian miRNAs in different human cancers, including 6 prostate tumors and 8 normal prostate tissues, revealed that miRNA expression profiles reflect the developmental lineage and differentiation stage of tumors. Moreover, in comparison with normal tissues, tumors showed a general downregulation of miRNAs.

In 2006, large-scale microarray analysis on solid tumors of different histotypes identified a common miRNA signature, which was mainly composed of overexpressed miRNAs (Volinia et al. 2006). In most of the tumors analyzed, miR-17-5p, miR-20a, miR-21, miR-92, miR-106a, and miR-155 were shown to be overexpressed. In the subset of prostate specimens, including 56 carcinomas and 7 normal tissues, 84% of differentially expressed miRNAs were upregulated.

The first systematic profiling report detailing miRNA expression in PCa was published in 2007 (Porkka et al. 2007). Using a self-synthesized microarray to compare 9 prostate carcinomas and 4 benign prostatic hyperplasia samples, the authors identified 37 upregulated and 14 downregulated miRNAs. Hierarchical clustering by means of miRNA expression accurately separated carcinomas from benign prostate cells, and further classified the tumors according to androgen dependence.

In 2008, a similar study was conducted using 16 clinically localized PCa specimens and 10 benign prostatic hyperplasia samples (Ozen et al. 2008). Analysis of the expression of 480 miRNAs revealed a widespread (up to 86%) downregulation of PCa miRNAs, including miR-125b, miR-145, and let-7c. These expression data were further validated by qRT-PCR analysis. In the same year, genome-wide expression profiling of miRNAs and mRNAs in 60 primary PCa samples and 16 surrounding non-tumor tissues was performed (Ambs et al. 2008). Key components of miRNA processing, and several genes carrying miRNA coding sequences in their introns, (e.g. *MCM7* and *C9orf5*), were shown to be significantly upregulated in tumors. In comparison with non-tumor tissues, PCa samples consistently expressed the miR-106b-25 cluster and miR-32 (which map to intron 13 on *MCM7* and intron 14 of *C9orf5* respectively) at significantly higher levels. Furthermore, miR-32 was found to inhibit the expression of BIM-1, a proapoptotic member of the BCL2 family, whose downregulation can contribute to resistance of tumor cells to death stimuli (Ambs et al. 2008). The miR-106b cluster was found to play an antiapoptotic growth-inhibitory activity, by suppressing E2F1 and p21/WAF1 (Ambs et al. 2008). E2F1 is also a target of miR-17-5p and miR-20a, which belong to the miR-17-92 cluster, and were found to be upregulated in many tumors, including PCa (Volinia et al. 2006, Sylvestre et al. 2007). Changes in miRNA abundance were also observed between organ-confined tumors and tumors with extraprostatic extension. In a further study of a series of 57 primary PCa specimens, the same research group observed differential expression of 19 miRNAs in tumors with and without perineural invasion (the dominant pathway for local invasion in PCa), with higher expression of miR-224 in tumors with perineural invasion (Prueitt et al. 2008).

The first paired analysis of miRNA expression in microdissected malignant and uninvolved areas of 40 prostatectomy specimens from state T2a/b PCa patients was carried out in 2009 (Tong et al. 2009). A high-throughput liquid-phase hybridization technique and validation by qRT-PCR revealed that 5 miRNAs, including miR-23b, miR-100, miR-145, miR-221, and miR-222, were significantly downregulated in cancerous tissues. Moreover, in individuals exhibiting elevated PSA within 2 years of surgery, the expression profiles of 16 miRNAs differed from those of non-relapsing patients, suggesting that aberrant miRNA expression features may reflect a tendency for early biochemical failure. miR-23b can act as a tumor suppressor, by inhibiting the expression of the glutaminase-encoding mRNA. Low miR-23b expression leads to an increase in glutaminase expression, and thereby to pro-oncogenic effects (Gao et al. 2009).

In 2010, analysis of miRNA expression in matched tumor and adjacent normal tissues obtained from 76 patients revealed 15 differentially expressed miRNAs: miR-16, miR-31, miR-125b, miR-145, miR-149, miR-181b, miR-184, miR-205, miR-221, and miR-222 were downregulated, whereas miR-96, miR-182, miR-182*, miR-183, and miR-375 were upregulated. Expression of 5 miRNAs correlated with the Gleason score (miR-31, miR-96, and miR-205) and/or pathological tumor stage (miR-125b, miR-205, and miR-222). In addition, miR-205 was found to be the best discriminating miRNA, providing a correct overall classification of 72% (Schaefer et al. 2010). miR-205 was reported to be downregulated in PCa cell lines compared with the untransformed RWPE-1 cell line and in 23 out of 31 prostate carcinomas compared with normal prostate tissues (Gandellini et al. 2009). miR-205 restoration in DU145 PCa cells resulted in marked morphological changes, resembling a reverse epithelial–mesenchymal transition (EMT). miR-205 reintroduction significantly reduced migratory and invasive capabilities of DU145 and PC3 cells. These effects were partially explained by miR-205 targeting of PKCε (Wu and Terrian 2002) and ZEB2 (Peinado et al. 2007) mRNAs.

A set of miRNAs showing consistently different expression profiles between malignant prostate and adjacent normal tissues was recently identified in 20 randomly selected patients from the Swedish Watchful Waiting cohort (Carlsson et al. 2011). Of 667 unique miRNAs analyzed by miRNA array, 30 were found to be differentially expressed ($p < 0.0001$). Five downregulated (miR-26A, miR-29A*, miR-34a*, miR-126*, miR-195) and 4 upregulated (miR-30D, miR-342-3P, miR-425* and miR-622) miRNAs were differentially expressed at a more stringent level ($p < 0.00001$). Six of these (miR-26A, miR-126*, miR-195, miR-30D, miR-29A*, and miR-342-3P) were previously reported to be involved in the development of PCa. Deficiency of miR-126* in PCa cells has been shown to be caused by poor expression of its host gene, the EGF-like domain 7. The absence of miR-126* results in high levels of its natural target, prostein (a prostate antigen), the reduction of which impairs LNCaP cell migration and invasion (Musiyenko et al. 2008).

3. EXPRESSION OF CIRCULATING miRNAs IN SERUM OR BLOOD PLASMA

Several groups have evaluated miRNA expression in the clinical context. The structure and size of miRNAs protects them from RNase attack and delays their degradation. Moreover, they are secreted relatively abundantly into bodily fluids, within protective exosomal packages (Weber et al. 2010). Thus, circulating miRNAs are emerging as promising non-invasive biomarkers of disease.

Mitchell and coworkers were the first to demonstrate a correlation between plasma miRNAs and the presence of PCa. They showed that miRNAs originating from human PCa xenografts can enter the circulation, be measured in the plasma, and thus robustly be used to distinguish xenografted mice from controls. Evaluation of the expression of miR-141 in a case-control cohort of serum samples collected from 25 individuals with metastatic PCa, and 25 healthy age-matched male control individuals, revealed markedly higher expression of miR-141 (as detected by qRT-PCR) in the PCa serum samples. Moreover, measurement of serum miR-141 levels allowed individuals with cancer to be detected with high accuracy (Mitchell et al. 2008). Lodes and coworkers screened 21 sera from 5 different cancer entities (colon, ovarian, breast, lung, and prostate). In total, 6 PCa samples were studied, of which 5 were from patients in advanced PCa stages, and 1 was from a patient in a non-advanced stage. miR-141 was revealed to be the only miRNA associated with advanced PCa stages (Lodes et al. 2009). A strong correlation between miR-141 values and clinical course was also observed in a study designed to evaluate the efficacy of circulating miR-141 as a potential biomarker of therapeutic response in PCa patients. This correlation was similar to those observed for other clinically validated biomarkers, such as PSA, circulating tumor cells (CTC), and lactate dehydrogenase (LDH) (Gonzales et al. 2011).

Serum screening for 667 miRNAs in a small sample of localized ($n = 14$) and metastatic ($n = 7$) cancer patients revealed that circulating miR-141 and miR-375 were the most pronounced markers for high-risk tumors. These miRNAs may therefore represent robust diagnostic and prognostic biomarkers (Brase et al. 2011). miR-141 and miR-375 were also identified in a global miRNA profiling study, using a mouse model of PCa as a tool to detect serum miRNAs that could be assessed in a clinical setting (Selth et al. 2011). In this study, 46 miRNAs with significantly altered levels ($p \leq 0.05$) were identified in the serum of Transgenic Adenocarcinoma of Mouse Prostate (TRAMP) mice with advanced PCa. A subset of these miRNAs with known human homologues was validated in an independent cohort of mice, and subsequently measured in serum from men with metastatic CRPC (mCRPC; $n = 25$), and serum from healthy individuals ($n = 25$) (Selth et al. 2011).

miR-21 has also been shown to be closely correlated with serum PSA (Zhang et al. 2011). Specific qRT-PCR quantification of miR-21 from the serum samples of 56 patients, including 20 patients with localized PCa, 20 patients with androgen-dependent PCa (ADPC), 10 patients with hormone-refractory PCa (HRPC), and 6 patients with benign prostatic hyperplasia (BPH), revealed that the serum miR-21 level correlated with the serum PSA level in patients with ADPC and HRPC. In these patients, higher levels of miR-21 were detected when the PSA level exceeded 4 ng/mL.

A diagnostic panel based on a novel multiplex qRT-PCR method, was utilized to define different miRNAs to those previously identified (Moltzahn et al. 2011). By this method, 384 human miRNAs were evaluated from 4 sets of patients (12 patients in each set) with differing risk-stratifying "cancer of the prostate risk assessment" (CAPRA) scores. miRNAs known to correlate with groupings have frequently been shown to possess oncogenic or tumor-suppressive functions in different cancer contexts. Thus, altered miRNA levels in the serum may reflect functional roles within the tumors. Moltzahn and coworkers identified 10 miRNAs that differed substantially between healthy and malignant samples. Four of these (miR-223, miR-26b, miR-30c, and miR-24) were downregulated in the cancer group, while the remaining 6 (miR-20b, miR-874, miR-1274a, miR-1207-5p, miR-93, and miR-106a) were upregulated. Two miRNAs (miR-19a and miR-451) differed significantly between the healthy and high-risk groups. With the exception of miR-19a and miR-20b, the screening results correlated with individual PCR data for all other miRNAs (miR-24, miR-26b, miR-30c, miR-93, miR-106a, miR-223, miR-451, miR-874, miR-1207-5p, and miR-1274a). These signatures may offer a means of identifying patients within different risk groups and may also provide additional prognostic information (Moltzahn et al. 2011).

In a study of circulating oncogenic miRNAs conducted using serum samples from patients with localized PCa, patients with benign prostate hyperplasia (BPH), and healthy individuals (HI), miR-26a, miR-32, miR-195, and miR-let7i allowed sensitive (89%) discrimination of PCa and BPH patients with moderate specificity (Mahn et al. 2011). Tissue miRNA levels were correlated with preprostatectomy serum miRNA levels, while serum miRNA levels decreased following prostatectomy, thereby indicating tumor-associated release of miRNA.

In another study, the plasma levels of miR-21, miR-141, and miR-221 were shown to vary between localized/local advanced PCa or metastatic PCa patients and healthy individuals (Yaman Agaoglu et al. 2011). miRNA levels (in particular, the level of miR-21) were higher in PCa patients. In comparison to patients with localized/local advanced PCa, patients with metastatic PCa exhibited significantly higher levels of all the 3 miRNAs (especially miR-141).

miR-20a is a member of the miR-17-92 cluster, which is reported to decrease apoptosis in PC-3 PCa cells, through repression of E2F2 and E2F3 mRNA translation. By contrast, inhibition by antisense oligonucleotide results in increased cell death following doxorubicin treatment (Sylvestre et al. 2007). Large-scale microarray analysis, using solid tumors of different histotypes to identify a common miRNA signature, revealed that miR-20a (and also miR-17-5p, miR-21, miR-92, miR-106a, and miR-155) were commonly overexpressed in most of the tumors analyzed (Volinia et al. 2006). In a study to investigate the relationship between miRNA expression and the clinicopathological features of PCa, miR-20a expression was found to be significantly higher in patients with a high Gleason score (7–10) than in patients with a low Gleason score (0–6) (Pesta et al. 2010). Taken together, these results indicate that dedifferentiated PCa cells exhibit higher expression of miR-20a, which supports the oncogenic role of miR-20a in PCa carcinogenesis.

Hagman and coworkers performed qRT-PCR on 49 transurethral resections of the prostate (TURP) specimens from PCa patients, and on 25 TURP specimens from patients with benign prostatic hyperplasia. miR-34c was shown to be downregulated in PCa patients (Hagman et al. 2010). Moreover, miR-34c expression was inversely correlated with tumor aggressiveness, WHO grade, PSA level, and occurrence of metastases. Kaplan-Meier analysis of patient survival divided patients into high- and low-risk groups, based on low (<50th percentile) and high (>50th percentile) miR-34c expression. Ectopic expression of miR-34c resulted in reduced cell growth, because of a decrease in cellular proliferation and an increase in apoptosis, leading to suppressed migration and invasion. In concordance with this, miR-34c was shown to negatively regulate the oncogenes E2F3 and BCL-2. Conversely, blocking miR-34c *in vitro* resulted in increased cell growth.

In a study to identify miRNAs involved in PCa progression, Leite and coworkers analyzed the expression profiles of 14 miRNAs in localized high-grade carcinoma and bone metastasis, by using RT-PCR. miR-let7c, miR-100, and miR-218 were shown to be significantly overexpressed in localized high Gleason Score, pT3 PCa compared with metastatic carcinoma, suggesting that these miRNAs may be involved in the process of PCa metastasization (Leite et al. 2011b). miR-100 expression was previously reported to be related to the biochemical recurrence of localized PCa in patients treated with radical prostatectomy (Leite et al. 2011a).

miR-373 and miR-520c were identified as metastasis-promoting miRNAs, which stimulate cancer cell migration and invasion through the suppression of CD44 (Huang et al. 2008). Expression of miR-373 and miR-520c was decreased in PCa cell lines and in matched benign and malignant patient prostate tissues. Migration and invasion were stimulated by exogenous expression, and cells also exhibited CD44 3′UTR binding (Yang et al. 2009).

Saini and coworkers performed miRNA expression profiling of human prostate cell lines, to identify dysregulated miRNA components of advanced PCa. They identified miR-203 as an "antimetastatic" miRNA, which acts at multiple steps of the PCa metastatic cascade, through the repression of a cohort of prometastatic targets (Saini et al. 2011). miR-203 expression was specifically attenuated in bone metastatic PCa. Moreover, reintroduction of miR-203 suppressed metastasis by inhibiting several critical steps of the metastatic cascade, including EMT, invasion, and motility. Ectopic miR-203 significantly attenuated the development of metastasis in a PCa bone metastatic model, by regulating a cohort of pro-metastatic genes (e.g. ZEB2, Bmi, survivin) and bone-specific effectors (e.g. Runx2, a master regulator of bone metastasis).

An analysis of polycomb repressive complexes 1 and 2 (PRC1 and PRC2) repressing endogenous miRNAs in a cohort of benign prostate, localized, and metastatic PCa cells revealed that miR-181a, miR-181b, miR-200a, miR-200b, miR-200c, miR-203, miR-101, and miR-217 were lowest in metastatic PCa tissues and highest in benign prostate tissues (Cao et al. 2011, Varambally et al. 2008). Immunoblot analyses revealed that BMI1 and RING2 (PRC1 components), and EZH2 (a PRC2 component), which are targets of these miRNAs, were increased in metastatic PCa compared with benign tissues and localized cancer samples. Ubiquitination of histone H2A at lysine 119 (uH2A), mediated by PRC1, was also increased in metastatic samples. PRC2 was shown to repress miR-181a, miR-181b, miR-200b, miR-200c, and miR-203 expression epigenetically by facilitating H3K27me3 trimethylation at these loci. Furthermore, endogenous overexpression of these miRNAs attenuated growth, invasiveness, and self-renewal of PCa cells *in vitro* and overexpression of miR-181b, miR-200b, and miR-203 suppressed prostate tumor formation and growth in mouse xenografts.

4. ANDROGEN-RESPONSIVE MicroRNAs

Androgen signaling plays a pivotal role in the biology of the normal prostate and also in PCa. The action of androgen is mediated through AR, which is a ligand-dependent transcription factor and a member of the hormone nuclear receptor superfamily. In the normal prostate, AR is expressed in stromal and epithelial cells. In epithelial cells, AR inhibits cell proliferation, drives differentiation, and regulates prostate metabolic and secretory functions. In the stromal compartment, AR promotes proliferation of epithelial cells. However, in cancer, AR signaling is modified to facilitate growth and survival in an autocrine fashion (Vander et al. 2010).

AR is expressed during all stages of PCa progression, in androgen-dependent and also in castration-resistant tumors. In CRPC, AR is known to be overexpressed. This overexpression of AR can partly be explained by amplification of the AR gene (Visakorpi et al. 1995, Linja et al. 2001), and appears to be sufficient to transform cancer cells from androgen-dependent to androgen-independent cells *in vitro* and also *in vivo* (Chen et al. 2004, Waltering et al. 2009). In addition to gene amplification, mutations of the AR gene leading to promiscuous ligand utilization have been found in 10–30% of anti-androgen-treated PCa cases (Taplin et al. 2003). Well-known AR-regulated androgen target genes in normal prostate epithelia and PCa include PSA (or kallikrein-related peptidase 3, KLK3), transmembrane protease, serine 2 (TMPRSS2), and prostate acidic phosphatase (PAP). AR-signaling pathways have also been shown to upregulate the expression of many enzymes involved in steroidogenesis and the cell cycle (Waltering et al. 2009). It appears that the TMPRSS2:ERG fusion gene, arising from a common genetic rearrangement, is one critical downstream gene (Tomlins et al. 2005, Saramaki et al. 2008). However, this fusion is found at the same frequency in untreated PCa and CRPC, and therefore probably does not explain the emergence of androgen-independent tumors (Saramaki et al. 2008). Eliminating AR function continues to be a focus of therapeutic investigation (Vander et al. 2010). Nevertheless, AR regulatory mechanisms remain poorly understood and the many AR target genes involved in the development and progression of PCa remain to be identified (Takayama et al. 2007).

The relationship between miRNA and androgen signaling forms a complex feedback loop, with androgen-responsive miRNAs and miRNAs modulating the androgen pathway. Importantly, androgen-mediated miRNA expression is affected by the integrity of the androgen pathway. Several studies support a role for AR in the transcriptional regulation of miRNA expression. However, the nature of this role is virtually unknown. One report revealed that miRNA expression was associated with androgen status in PCa cell lines and xenografts (Porkka et al. 2007), while another demonstrated that ectopic expression of AR in AR-null PC3 cells induced the differential expression of 11 miRNAs (Lin et al. 2008). There is also evidence that AR signaling directly regulates the expression of certain miRNAs (Shi et al. 2007).

miR-125b was reported to be upregulated by androgen, and overexpressed in androgen-independent LNCaP sublines relative to parental cells (Shi et al. 2007). These controversial findings suggest that androgen-independent regulation of miR-125b may also exist (see "miRNA Profiling in Castration-Resistant Prostate Cancer"). In the absence of androgens, reduced levels of miR-125b were detected in LNCaP PCa cell lines, while androgen treatment stimulated a marked increase in miR-125b. In addition, transfection of AR into AR-negative prostate epithelial cells, followed by treatment with androgen, increased the abundance of miR-125b. A typical TATA box and potential androgen-responsive elements (AREs) were identified in the 5′ end DNA region of this miRNA. Chromatin immunoprecipitation (ChIP) analysis revealed that AR was indeed recruited to the 5′ end DNA region of the miR-125b locus in an androgen-dependent manner. Taken together, these findings suggest that AR signaling mediates the regulation of miR-125b in PCa cells, thereby facilitating androgen-independent growth in LNCaP cells, by targeting BAK1, BBC3, and p53 (Shi et al. 2011).

High-throughput microarray analysis of 2 androgen-responsive PCa cell lines revealed the existence of a set of androgen-regulated miRNAs. Among the 16 candidate AR-dependent miRNAs, miR-21 was confirmed by northern blot as androgen responsive in several AR-positive PCa cell lines. Furthermore, androgen-dependent binding of AR to the reported promoter of miR-21 was confirmed by ChIP assay (Ribas et al. 2009).

miR-148a, and also miR-141 and miR-200a, were identified as androgen-upregulated miRNAs, by short RNA sequencing and qRT-PCR in androgen-sensitive LNCaP PCa cell lines (Murata et al. 2010). Transfection of miRNA precursors into LNCaP PCa cell lines revealed that these miRNAs had growth-promoting effects, and this was confirmed by knockdown experiments with each miRNA inhibitor. Furthermore, cullin-associated protein, CAND1 (also known as TIP120A), was identified as a target of miR-148a. Reduction of CAND1 expression by siRNA accelerated LNCaP cell growth, suggesting that androgen-dependent upregulation of miR-148a could promote PCa growth by reducing its target CAND1.

A recent profiling study employing clinical specimens of untreated and CRPC, an AR-overexpressing LNCaP-derived model, the VCaP cell line, and 13 intact-castrated PCa xenograft pairs analyzed the expression of androgen-responsive miRNAs in cells and xenografts by microarray and qRT-PCR (Waltering et al. 2011). Of 55 candidates, miR-141 showed similar androgen regulation following dihydrotestosterone (DHT) treatment in the cell lines and after castration in AR-positive xenografts. miR-141 was found to be overexpressed in PCa and CRPC, compared to benign prostate hyperplasia. Overexpression of miR-141 enhanced the growth of parental LNCaP cells, while inhibition of miR-141 by anti-miR-141 suppressed the growth of the AR-overexpressing LNCaP subline (Waltering et al. 2011).

5. miRNA PROFILING IN CASTRATION-RESISTANT PROSTATE CANCER

The most common treatment for PCa patients with advanced and metastatic disease consists of androgen ablation by castration, or administration of AR antagonists (e.g. flutamide, bicalutamide, nilutamide). However, following an initially favorable response to this therapy, most tumors relapse to an incurable androgen-independent state. Such CRPC has been shown to be unresponsive to further ADT, and yet to have an active AR-signaling pathway. This pathway operates through a variety of mechanisms that alter AR sensitivity or specificity, including AR mutations, alternative splicing, amplification, alterations of coregulators, sensitization by growth factors and cytokines, and response to intracrine androgens.

Elucidation of the factors that regulate AR expression and activation will enhance our understanding of the mechanisms leading to castration resistance, and will therefore play a key role in developing therapeutic strategies to combat this CRPC. Given the important roles of miRNAs in post-transcriptional regulation, identification of differentially expressed and novel miRNAs will facilitate identification of the molecular mechanisms underlying the progression of androgen-independent PCa.

miR-21, which has been documented as an androgen-induced miRNA, has been shown to enhance PCa tumor growth *in vivo*, and to be sufficient for androgen-dependent tumors to overcome castration-mediated growth arrest. Thus, elevated miR-21 expression alone can impart castration resistance (Ribas et al. 2009). miR-21 was found to be elevated in PC3 and DU145 androgen-independent PCa cell lines (Li et al. 2009). Blocking of miR-21 by antisense oligonucleotides did not affect cell proliferation, but sensitized cells to staurosporine-induced apoptosis, and impaired cell motility and invasion. Myristoylated alanine-rich protein kinase C substrate (MARCKS) was identified as a new miR-21 target in PCa cells. Given the involvement of MARCKS in cellular processes such as cell adhesion, spreading, membrane trafficking, and cell motility (Arbuzova et al. 2002), it is conceivable that the observed effects were caused, at least in part, by modified MARCKS levels. However, investigation of other cancers suggests that a wider spectrum of genes may be altered in miR-21-overexpressing cells. Validated targets of miR-21 enclose several genes implicated mainly in suppressing cell migration and invasion, including PDCD4 (Asangani et al. 2008, Frankel et al. 2008), PTEN (Meng et al. 2007), TPM1 (Zhu et al. 2007), SPRY2 (Sayed et al. 2008), and the metalloprotease inhibitors TIMP3 and RECK (Gabriely et al. 2008). Zhang and coworkers observed that miR-21 was closely correlated with serum PSA and docetaxel resistance. Serum miR-21 levels were significantly higher in patients who were resistant to docetaxel-based chemotherapy, than in those who were sensitive to chemotherapy. Thus, miR-21 may be applicable as a marker to indicate the transformation to hormone refractory disease and as a potential predictor for the efficacy of docetaxel-based chemotherapy (Zhang et al. 2011).

miR-125b, another androgen-regulated miRNA, was observed to stimulate androgen-independent growth of PCa cells, and also repress its target Bak1 (Shi et al. 2007). miR-125b was overexpressed in androgen-independent LNCaP sublines relative to parental cells, suggesting that androgen-independent regulation of miR-125b may exist. Transfection of synthetic miR-125b stimulated the growth of LNCaP cells in the absence of androgens, whereas anti-miR-125b inhibited the growth of androgen-independent cells. The pro-apoptotic, Bak1, was identified as a target. Bak1 expression is reduced in hormone-refractory tumors (Yoshino et al. 2006), and has also been reported to be associated with the progression of

PCa. However, 3 miRNA expression profiles have documented miR-125b as downregulated in PCa (Porkka et al. 2007, Ozen et al. 2008, Tong et al. 2009), and therefore its oncogenic role in PCa progression requires further investigation.

Genome-wide expression profiling of miRNAs revealed that expression levels of several miRNAs, in particular miR-221 and miR-222, were significantly increased in CRPC cells (the LNCaP-derived cell line LNCaP-Abl), compared with those in the androgen-dependent PCa cell line (LNCaP) (Sun et al. 2009). Overexpression of miR-221 or miR-222 in LNCaP or LAPC-4, another androgen-dependent cell line, significantly reduced the level of DHT-induced upregulation of PSA expression, and increased androgen-independent growth of LNCaP cells. Knocking down the expression level of miR-221 and miR-222 with antagonist miRNAs in the LNCaP-Abl cell line restored the response to the DHT-induced upregulation of PSA transcription, and also increased the growth response of the LNCaP-Abl cells to androgen treatment. Changing the expression level of p27/kip1, a known target of miR-221 and miR-222, in LNCaP cells affected DHT-independent cell growth, but did not significantly influence the response of PSA transcription to DHT treatment. These data suggest the involvement of miR-221 and miR-222 in the development or maintenance of the CRPC phenotype (Sun et al. 2009, Zheng et al. 2011). Further studies are required to elucidate the role of miR-221/miR-222 in progression to androgen independence. Upregulation of miR-221 has been reported in several cancers, including pancreatic, breast, papillary thyroid, glioblastoma, and chronic lymphocytic leukemia. By contrast, in PCa, downregulation of miR-221 has been repeatedly confirmed by miRNA expression studies (Lu et al. 2005, Porkka et al. 2007, Ozen et al. 2008, Ambs et al. 2008, Tong et al. 2009). In the scenario of an early, hormone-dependent disease, androgens may play a protective role by keeping miR-221/miR-222 levels low. In a large and diverse cohort of prostate carcinoma samples, miR-221 was found to be downregulated in patients with tumors bearing TMPRSS2:ERG fusion transcripts. The aberrant expression of the oncogene, the TMPRSS2:ERG fusion gene, is unique to PCa and is found in more than 50% of PCa patients. Taken together, these findings indicate a link between miRNA and gene fusion expression (Gordanpour et al. 2011).

Quantitative polymerase chain reaction (qPCR) analysis of 20 prostate tumor-derived primary cultures and *in situ* hybridization analysis of 15 prostate tumor biopsies revealed extensive downregulation of miR-15a/miR-16 levels (Bonci et al. 2008). A stable lentivirus-mediated sequestering of miR-15a/miR-16 in RWPE-1 cells induced an increase in proliferation and migration *in vitro*, and allowed these cells to form small tumor masses in non-obese, diabetes/severe combined immunodeficiency (NOD/SCID) mice. By contrast, lentivirus-mediated miR-15a/miR-16 reconstitution in

LNCaP cells resulted in a dramatic apoptotic effect *in vitro* and considerable regression of tumor xenografts *in vivo*. miR-15a and miR-16 are able post-transcriptionally to repress the expression of BCL2, CCND1, and WNT3A, thus interfering with multiple oncogenic activities. BCL2 upregulation has been described as a common event, especially in hormone-refractory PCa, where it can support survival of cells in the absence of androgens (McDonnell et al. 1992, Colombel et al. 1993). WNT3A has been implicated to play a role in PCa pathogenesis by enhancing β-catenin stabilization and its transcriptional activity and by activating pivotal oncogenic pathways (e.g. AKT and MAPK) (Almeida et al. 2005, Yun et al. 2005). β-Catenin stabilization leads to the transcription of several oncogenic genes (e.g. CCND1, MYC, MMPs, and endothelin). Moreover, by acting as an AR co-activator, it promotes AR activity in the presence of low androgen levels (Schweizer et al. 2008). AKT and MAPK activation can also provide PCa cells with survival and proliferation signals in a hormone-deprived environment (Yeh et al. 1999, Graff et al. 2000).

Recently, miR-let-7c was identified as a key regulator of AR expression. let-7c suppresses AR expression and activity in human PCa cells, by targeting AR transcription via c-Myc. Suppression of AR by let-7c resulted in decreased proliferation of PCa cells. Downregulation of let-7c in PCa specimens was inversely correlated with AR expression, whereas the expression of Lin28 (a repressor of let-7) was positively correlated with AR expression. These findings indicate that let-7c plays an important role in the regulation of androgen signaling in PCa, by downregulating AR expression. (Nadiminty et al. 2011). Let-7c has been previously reported to be induced by androgen in a high-throughput genome analysis combining 5′-cap analysis of gene expression (CAGE) and chromatin immunoprecipitation (ChIP) on array (ChIP-chip) analysis (Takayama et al. 2011), which suggests that the relationship between let-7c expression and androgen signaling may form a complex feedback loop.

Some miRNAs, such as miR-146, appear to play opposing roles in tumorigenesis, depending on the cellular context. Expression levels of the miR-146 family members, miR-146a and miR-146b-5p (which are encoded by separate genes) are elevated in breast, prostate, endocrine pancreatic, cervical, thyroid, and ovarian carcinomas (He et al. 2005, Volinia et al. 2006, Dahiya et al. 2008, Wang et al. 2008). Whether this elevation drives tumorigenesis, or whether it acts as a cellular safeguard to prevent tumorigenesis, remains to be elucidated. In PCa, downregulation of miR-146a was observed in androgen-independent human prostate cancer cells compared to those in androgen-dependent cells (Lin et al. 2008). Indeed, miR-146a targets ROCK1, a key kinase for the activation of hyaluronan (HA)-mediated CRPC transformation *in vivo* and in PC3 cells, and thus, partly responsible for the development of CRPC.

Screening for miRNAs that are downregulated during the EMT has identified 5 members of the miR-200 family (miR-200a, miR-200b, miR-200c, miR-141, and miR-429) (Gregory et al. 2008, Park et al. 2008). miR-200 loss is common in aggressive lung, prostate, and pancreatic cancers (Gibbons et al. 2009, Wellner et al. 2009, Kong et al. 2010). These findings suggest an association between EMT and the loss of miR-200. Profiling of miRNAs in 4 PCa cell lines, by means of a simple but effective array platform, revealed that a set of miRNAs was differentially expressed between androgen-dependent and androgen-independent metastatic PCa cells (Tang et al. 2011). The differential expression of miR-205 and miR-200c was further validated by northern blot analysis.

Genetic variations in miRNAs and miRNA target sites may be associated with the efficacy of ADT in men with PCa. Bao and coworkers systematically evaluated 61 common single nucleotide polymorphisms (SNPs) inside miRNAs and miRNA target sites, in a cohort of 601 men diagnosed with advanced PCa and receiving ADT (Bao et al. 2011). The prognostic significance of these SNPs on disease progression, PCa-specific mortality (PCSM), and all-cause mortality (ACM) following ADT was assessed by the Kaplan-Meier method and Cox regression model. Univariate analysis revealed that 4, 7, and 4 SNPs were significantly associated with disease progression, PCSM, and ACM, respectively. A strong combined genotype effect on disease progression and PCSM was also observed. Patients with a higher number of unfavorable genotypes exhibited a shorter time to progression, and poorer PCa-specific survival. Thus, SNPs inside miRNAs and miRNA target sites may play a role in improving the outcome prediction in PCa patients receiving ADT. Moreover, novel approaches such as direct sequencing offer the potential to identify modestly expressed miRNAs, including SNPs between samples, with improved accuracy.

6. MicroRNA EXPRESSION ANALYSIS BY MEANS OF NEXT GENERATION SEQUENCING

The emergence of next generation sequencing technologies has dramatically and cost-effectively altered the speed of sequencing to permit an unbiased, quantitative, and in-depth investigation of small RNA transcriptomes.

For miRNA sequencing, the "wet-lab" portion comprises several steps. Short RNA is purified from the sample, and then ligated to the sequencing adapters and reverse transcribed to cDNA. The resulting cDNA library is run on a polyacrylamide agarose gel, and the band containing the molecules corresponding to the miRNA fragments with ligated adapters (11–40 nucleotides) is cut out for subsequent sequencing. The output of a next generation miRNA sequencing experiment will typically contain

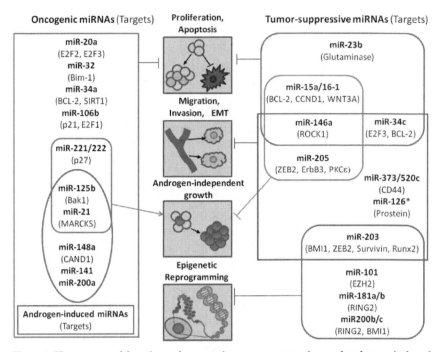

Figure 1. The targets and functions of oncogenic, tumor-suppressive, and androgen-induced miRNAs in prostate cancer. Oncogenic miRNAs enhance the cancer cell phenotype (cell proliferation and apoptosis resistance). Among the oncogenic miRNAs, miR-125b, miR-148a, miR-141, and miR-21 are induced by androgens, whereas miR-221/222, miR-125b, and miR-21 are involved in androgen-independent growth. miR-23b, miR-15/16-1, miR-34c, and miR-146a are known as tumor-suppressive miRNAs that decrease cellular proliferation and increase apoptosis. miR-34c, miR-146a, miR-373/520c, miR-126*, miR-205, and miR-203 suppress migration and invasion. miR-15a/16-1, miR-146a, and miR-205 are involved in androgen-independent growth. miR-203, miR-101, miR-181a/b, and miR-200b/c attenuate cancer cell phenotypes through epigenetic mechanisms.

Color image of this figure appears in the color plate section at the end of the book.

millions of short reads. Before specific research questions can be addressed, several basic data preprocessing steps must be performed to extract the relevant information from these raw data. The reads from a next generation sequencing experiment are commonly provided as text files in FASTQ format.

Next generation sequencing of short RNAs identified a set of androgen-responsive miRNAs in the androgen-sensitive LNCaP PCa cell line treated with R1881 (10 nM) (Murata et al. 2010). Following the construction of short RNA libraries (Taft et al. 2009), 784,315, 512,503, and 470,693 tag sequences were obtained for cells treated with R1881 for 0 hr, six hr, and 24 hr, respectively. Comparison of the sequence frequencies between 24 hr and 0 hr revealed 1809 sequences significantly upregulated by androgen.

A BLAST search identified 431 sequences as androgen-upregulated short RNAs, among 631 RNA sequences uniquely mapped to the human genome. From these short RNAs, miRNAs with known sequences were selected based on miRBase 9.1. miRNAs in a list of the 10 highest frequencies were selected, and the androgen responsiveness (>1.5-fold over basal) of 10 miRNAs (miR-22, miR-29b, miR-30b, miR-141, miR-148a, miR-17-5p, miR-27a, miR-30a-5p, miR106b, and miR-200a) was validated by qPCR, using miR-125b as a positive control.

The miRNA profiles of primary PCa and non-cancer prostate tissues were compared by next generation sequencing (Szczyrba et al. 2010). 33 miRNAs were found to be upregulated or downregulated >1.5-fold. The deregulation of selected miRNAs in established PCa cell lines and clinical tissue samples was confirmed by northern blotting and by qRT-PCR. A computational search identified the 3'-untranslated region (UTR) of the mRNA for myosin VI (MYO6) as a potential target for miR-143 and miR-145, the expression of which was reduced in tumor tissues.

Xu and coworkers used high-throughput Illumina next generation sequencing to comprehensively represent the entire complement of individual small RNAs and to characterize miRNA expression profiles between androgen-dependent and androgen-independent PCa cell lines (Xu et al. 2010). The study employed the androgen-dependent LNCaP PCa cell line and its androgen-independent derivatives obtained by long-term androgen deprivation by culturing the LNCaP cell line in an androgen-depleted medium supplemented with antiandrogen flutamide. At least 83 miRNAs were found to be significantly differentially expressed. Of these, 41 were upregulated and 42 were downregulated, indicating that these miRNAs may be involved in the transition of PCa to an androgen-independent phenotype. In addition, 43 novel miRNAs were identified from the androgen-dependent and androgen-independent PCa libraries, of which 3 were specific to the androgen-independent PCa. Function annotation of target genes indicated that most of the differentially expressed miRNAs tended to target genes involved in signal transduction and cell communication, especially the MAPK signaling pathway. The small RNA transcriptomes obtained are of considerable value in facilitating a better understanding of the expression and function of small RNAs in the development of androgen-independent PCa.

miRNA profiling using next generation sequencing and locked nucleic acid miRNA microarrays revealed that miR-125b and the miR-99 family members (miR-99a, miR-99b, and miR-100) were downregulated in advanced PCa cell lines relative to parental cell lines (Sun et al. 2011). All 4 miRNAs were also downregulated in human prostate tumor tissue relative to normal prostate tissue. Transfection of miR-99a, miR-99b, or miR-100 inhibited the growth of PCa cells, and decreased the expression of PSA, indicating potential roles as tumor suppressors. A combined approach using computational prediction of potential targets, and experimental validation by microarray and polyribosomal loading analysis, identified 3 direct targets of the miR-99 family—the chromatin-remodeling factors SMARCA5 and SMARCD1, and the growth regulatory kinase mTOR. Taken together, these findings demonstrate a post-transcriptional regulation of PSA by the miR-99 family members, at least in part, through repression of SMARCA5.

In order to systematically characterize AR regulatory mechanisms involving miRNAs, paired-end RNA sequencing was conducted on a cDNA library prepared with fragmented RNA from the VCaP PCa cell line, using the Illumina next generation sequencing (Östling et al. 2011). The 3'-UTR of AR (and other genes) was demonstrated to be much longer than those currently used in miRNA target prediction programs. The data predicted that miRNA regulation of AR would primarily target an extended 6-kb 3'-UTR. 3'-UTR-binding assays validated 13 miRNAs (miR-135b, miR-185, miR-297, miR-299-3p, miR-34a, miR-34c, miR-371-3p, miR-421, miR-449a, miR-449b, miR-634, miR-654-5p, and miR-9) able to regulate this long 3'-UTR. miR-34a is a known oncogenic miRNA, which is reported to inhibit BCL-2 and SIRT1, thereby promoting survival under genotoxic and oxidative stress (Yamakuchi et al. 2008, Fujita et al. 2008, Kojima et al. 2010).

An integrative approach combining primary transcription, genome-wide DNA methylation, and H3K9Ac marks with miRNA expression, identified miRNA genes that were epigenetically modified in cancer (Hulf et al. 2011). miR-205, miR-21, and miR-196b were demonstrated to be epigenetically repressed, whereas miR-615 was epigenetically activated, in LNCaP and PrEC PCa cell lines.

Martens-Uzunova and coworkers employed Illumina sequencing to analyze the composition of the entire small transcriptome, in order to evaluate the diversity and abundance of small non-coding RNAs (ncRNAs) in PCa (Martens-Uzunova et al. 2011). The miRNA expression signatures of 102 fresh-frozen PCa patient samples were further analyzed by miRNA microarrays. The 2 platforms were cross-validated by qRT-PCR. Besides altered expression of several miRNAs, deep-sequencing analyses revealed strong differential expression of small nucleolar RNAs (snoRNAs) and transfer RNAs (tRNAs). Using microarray analysis, a miRNA diagnostic classifier able accurately to distinguish normal from cancer samples was identified. Furthermore, a PCa prognostic predictor able independently to forecast the postoperative outcome was constructed. Importantly, the majority of miRNAs included in the predictor exhibited high sequence counts and concordant differential expression in Illumina PCa samples (as validated by qRT-PCR). Thus, miRNA expression signatures may serve as an accurate tool for the diagnosis and prognosis of PCa.

Androgen-regulated coding genes and ncRNAs (including miRNAs) were determined in a genome-wide screening of androgen target genes (Takayama et al. 2011). Androgen-regulated transcription start sites (TSSs) were determined by 5′-cap analysis of gene expression (CAGE), while AR-binding sites (ARBSs) and histone H3 acetylated (AcH3) sites were identified by chromatin immunoprecipitation (ChIP) on array (ChIP-chip) analysis. CAGE determined 13,110 distinct, androgen-regulated TSSs, while ChIP-chip analysis identified 2872 androgen-dependent ARBSs and 25,945 AcH3 sites. Besides prototypic androgen-regulated TSSs in annotated gene promoter regions, many androgen-dependent TSSs widely distributed throughout the genome were identified, including those in the antisense direction of RefSeq genes or TSSs of miRNA primary transcripts. A cluster of androgen-inducible miRNAs (miR-125b, miR-99a and let-7c) in the intronic region of *C21orf34* gene was identified by the integration of CAGE and ChIP-chip analyses. These data, combining integrated high-throughput genome analyses of CAGE and ChIP-chip analysis, will facilitate elucidation of the AR-mediated transcriptional network, which contributes to the development and progression of PCa.

Next generation sequencing technologies represent a highly robust, accurate, and scalable system, which sets a new standard for the rapid, productive, and cost-effective investigation of the miRNA transcriptome. Such innovative tools will contribute towards improved miRNA profiling, and will offer an enhanced global view of the expression and function of miRNAs in PCa.

7. CONCLUSIONS

An important challenge in PCa research is to develop efficient predictors of tumor recurrence following surgery, in order to determine whether immediate adjuvant therapy is warranted (Fendler et al. 2011, Long et al. 2011). Currently, no biomarker can accurately predict the risk of relapse at the time of surgery. However, miRNAs are emerging as useful disease markers. Few studies on PCa tissue have compared miRNA expression levels in more than 10 benign and malignant samples (Ambs et al. 2008, Ozen et al. 2008, Porkka et al. 2007, Volinia et al. 2006). Nevertheless, no overlapping subset between downregulated or upregulated miRNAs has been detected.

High-throughput sequencing offers the potential to determine the length variation of mature miRNA and possible enzymatic modifications. Moreover, it allows the successful detection of novel miRNAs. Direct sequencing further allows the identification of modestly and differentially expressed miRNAs, such as SNPs inside miRNAs, with improved accuracy.

Technological advances have enabled elucidation of the role of miRNAs in human disease. As technology becomes more available and cost-effective, further knowledge will be gained. miRNAs represent promising biomarkers and therapeutic targets, and may be clinically relevant in the management of PCa.

8. SUMMARY POINTS

- miRNA expression profiles have been investigated in a large cohort of human cancers, including PCa. The role exerted by specific miRNAs in the initiation and progression of PCa remains to be elucidated. Several studies evaluating PCa-related miRNA expression in serum or blood samples of patients have shown that certain miRNAs, such as miR-21, miR-141, and miR-375, may be potential diagnostic and prognostic biomarkers.
- Some miRNAs have been identified as responsive to AR signaling, which facilitates the growth and survival of PCa cells. miR-21, miR-125b, miR-141, miR-148a, and miR-200a are reportedly induced by an androgen, and their overexpression supports the growth of PCa cell lines. miR-221/222, miR-125b, and miR-21 were shown to be involved in androgen-independent growth. miR-23b, miR-15/16-1, miR-34c, and miR-146a, known as tumor-suppressive miRNAs, decrease cellular proliferation and increase apoptosis. miR-34c, miR-146a, miR-373/520c, miR-126*, miR-205, and miR-203 were shown to suppress migration and invasion. miR-15a/16-1, miR-146a, and miR-205 are reported to be involved in androgen-independent growth. Moreover, miR-203, miR-101, miR-181a/b, and miR-200b/c attenuate cancer cell phenotypes through epigenetic mechanisms.
- Next generation sequencing has enabled the sequencing of the complete set of miRNAs present in an RNA sample, thus, providing information on miRNA expression profiling, identifying sequence isoforms, and predicting novel miRNAs and potential mRNA targets. This novel technique will facilitate potential applications of miRNAs towards new diagnostic and therapeutic approaches.

ACKNOWLEDGEMENTS

Grant support: Grants of the Cell Innovation Program (S. I.) and the P-Direct (S. I.) from the MEXT, Japan, the Grants (A. M., K. T. and S. I.) from the JSPS, Japan, and the Program for Promotion of Fundamental Studies in Health Sciences (S. I.), NIBIO, Japan. We thank Drs. K. Horie and T. Miyazaki for critical reading of the manuscript.

ABBREVIATIONS

ADPC	:	Androgen-dependent prostate cancer
ADT	:	Androgen deprivation therapy
AR	:	Androgen receptor
ARBS	:	Androgen receptor binding site
CAGE	:	5'-cap analysis of gene expression
ChIP	:	Chromatin immunoprecipitation
CRPC	:	Castration resistant prostate cancer
HRPC	:	Hormone-refractory prostate cancer
miRNA	:	microRNA
ncRNA	:	non-coding RNA
PCa	:	Prostate cancer
PSA	:	Prostate specific antigen
qPCR	:	Quantitative Polymerase Chain Reaction
qRT-PCR	:	Quantitative Reverse Transcription Polymerase Chain Reaction
snoRNA	:	small nucleolar RNA
SNPs	:	Single nucleotide polymorphisms

REFERENCES

Almeida, R.C., D.G. Souza, R.C. Soletti, M.G. Lopez, A.L. Rodrigues, and N.H. Gabilan. 2006. Involvement of PKA, MAPK/ERK and CaMKII, but not PKC in the acute antidepressant-like effect of memantine in mice. Neurosci Lett. 395(2): 93–97.

Ambs, S., R.L. Prueitt, M. Yi, R.S. Hudson, T.M. Howe, F. Petrocca, T.A. Wallace, C.G. Liu, S. Volinia, G.A. Calin, H.G. Yfantis, R.M. Stephens, and C.M. Croce. 2008. Genomic profiling of microRNA and messenger RNA reveals deregulated microRNA expression in prostate cancer. Cancer Res. 68(15): 6162–6170.

Arbuzova, A., A.A. Schmitz, and G. Vergeres. 2002. Cross-talk unfolded: MARCKS proteins. Biochem J. 362(Pt 1): 1–12.

Asangani, I. A., S.A. Rasheed, D.A. Nikolova, J.H. Leupold, N.H. Colburn, S. Post, and H. Allgayer. 2008. MicroRNA-21 (miR-21) post-transcriptionally downregulates tumor suppressor Pdcd4 and stimulates invasion, intravasation and metastasis in colorectal cancer. Oncogene. 27(15): 2128–2136.

Bao, B.Y., J.B. Pao, C.N. Huang, Y.S. Pu, T.Y. Chang, Y.H. Lan, T.L. Lu, H.Z. Lee, S.H. Juang, L.M. Chen, C.J. Hsieh, and S.P. Huang. 2011. Polymorphisms inside microRNAs and microRNA target sites predict clinical outcomes in prostate cancer patients receiving androgen-deprivation therapy. Clin Cancer Res. 17(4): 928–936.

Bonci, D., V. Coppola, M. Musumeci, A. Addario, R. Giuffrida, L. Memeo, L. D'Urso, A. Pagliuca, M. Biffoni, C. Labbaye, M. Bartucci, G. Muto, C. Peschle, and R. De Maria. 2008. The miR-15a-miR-16-1 cluster controls prostate cancer by targeting multiple oncogenic activities. Nat Med. 14(11): 1271–1277.

Brase, J.C., M. Johannes, T. Schlomm, M. Falth, A. Haese, T. Steuber, T. Beissbarth, R. Kuner, and H. Sultmann. 2011. Circulating miRNAs are correlated with tumor progression in prostate cancer. Int J Cancer. 128(3): 608–616.

Carlsson, J., S. Davidsson, G. Helenius, M. Karlsson, Z. Lubovac, O. Andren, B. Olsson, and K. Klinga-Levan. 2011. A miRNA expression signature that separates between normal and malignant prostate tissues. Cancer Cell Int. 11(1): 14.

Chen, C.D., D.S. Welsbie, C. Tran, S.H. Baek, R. Chen, R. Vessella, M.G. Rosenfeld, and C.L. Sawyers. 2004. Molecular determinants of resistance to antiandrogen therapy. Nat Med. 10(1): 33–39.

Colombel, M., F. Symmans, S. Gil, K.M. O'Toole, D. Chopin, M. Benson, C.A. Olsson, S. Korsmeyer, and R. Buttyan. 1993. Detection of the apoptosis-suppressing oncoprotein bcl-2 in hormone-refractory human prostate cancers. Am J Pathol. 143(2): 390–400.

Dahiya, N., C.A. Sherman-Baust, T.L. Wang, B. Davidson, Shih IeM, Y. Zhang, W. Wood 3rd, K.G. Becker, and P.J. Morin. 2008. MicroRNA expression and identification of putative miRNA targets in ovarian cancer. PLoS One. 3(6): e2436.

Damber, J.E. and G. Aus. 2008. Prostate cancer. Lancet. 371(9625): 1710–1721.

Fendler, A., M. Jung, C. Stephan, R.J. Honey, R.J. Stewart, K.T. Pace, A. Erbersdobler, S. Samaan, K. Jung, and G.M. Yousef. 2011. miRNAs can predict prostate cancer biochemical relapse and are involved in tumor progression. Int J Oncol. 39(5): 1183–1192.

Fujita, Y., K. Kojima, N. Hamada, R. Ohhashi, Y. Akao, Y. Nozawa, T. Deguchi, and M. Ito. 2008. Effects of miR-34a on cell growth and chemoresistance in prostate cancer PC3 cells. Biochem Biophys Res Commun. Dec 5; 377(1):114–119.

Frankel, L.B., N.R. Christoffersen, A. Jacobsen, M. Lindow, A. Krogh, and A.H. Lund. 2008. Programmed cell death 4 (PDCD4) is an important functional target of the microRNA miR-21 in breast cancer cells. J Biol Chem. 283(2): 1026–1033.

Gabriely, G., T. Wurdinger, S. Kesari, C.C. Esau, J. Burchard, P.S. Linsley, and A.M. Krichevsky. 2008. MicroRNA 21 promotes glioma invasion by targeting matrix metalloproteinase regulators. Mol Cell Biol. 28(17): 5369–5380.

Gao, P., I. Tchernyshyov, T.C. Chang, Y.S. Lee, K. Kita, T. Ochi, K.I. Zeller, A.M. De Marzo, J.E. Van Eyk, J.T. Mendell, et al. 2009. c-Myc suppression of miR-23a/b enhances mitochondrial glutaminase expression and glutamine metabolism. Nature. 458: 762–765.

Gibbons, D.L., W. Lin, C.J. Creighton, Z.H. Rizvi, P.A. Gregory, G.J. Goodall, N. Thilaganathan, L. Du, Y. Zhang, A. Pertsemlidis, and J.M. Kurie. 2009. Contextual extracellular cues promote tumor cell EMT and metastasis by regulating miR-200 family expression. Genes Dev. 23(18): 2140–2151.

Gordanpour, A., A. Stanimirovic, R.K. Nam, C.S. Moreno, C. Sherman, L. Sugar, and A. Seth. 2011. miR-221 Is down-regulated in TMPRSS2:ERG fusion-positive prostate cancer. Anticancer Res. 31(2): 403–410.

Graff, J.R., B.W. Konicek, A.M. McNulty, Z. Wang, K. Houck, S. Allen, J.D. Paul, A. Hbaiu, R.G. Goode, G.E. Sandusky, R.L. Vessella, and B.L. Neubauer. 2000. Increased AKT activity contributes to prostate cancer progression by dramatically accelerating prostate tumor growth and diminishing p27Kip1 expression. J Biol Chem. 275(32): 24500–24505.

Gregory, P.A., A.G. Bert, E.L. Paterson, S.C. Barry, A. Tsykin, G. Farshid, M.A. Vadas, Y. Khew-Goodall, and G.J. Goodall. 2008. The miR-200 family and miR-205 regulate epithelial to mesenchymal transition by targeting ZEB1 and SIP1. Nat Cell Biol. 10(5): 593–601.

Gronberg, H. 2003. Prostate cancer epidemiology. Lancet. 361(9360): 859–864.

Hagman, Z., O. Larne, A. Edsjo, A. Bjartell, R.A. Ehrnstrom, D. Ulmert, H. Lilja, and Y. Ceder. 2010. miR-34c is down regulated in prostate cancer and exerts tumor suppressive functions. Int J Cancer. 127(12): 2768–2776.

He, H., K. Jazdzewski, W. Li, S. Liyanarachchi, R. Nagy, S. Volinia, G.A. Calin, C.G. Liu, K. Franssila, S. Suster, R.T. Kloos, C.M. Croce, and A. de la Chapelle. 2005. The role of microRNA genes in papillary thyroid carcinoma. Proc Natl Acad Sci USA. 102(52): 19075–19080.

Huang, Q., K. Gumireddy, M. Schrier, C. le Sage, R. Nagel, S. Nair, D.A. Egan, A. Li, G. Huang, A.J. Klein-Szanto, P.A. Gimotty, D. Katsaros, G. Coukos, L. Zhang, E. Puré, and R. Agami.

2008. The microRNAs miR-373 and miR-520c promote tumour invasion and metastasis. Nat Cell Biol. 10(2): 202–210.

Hulf, T., T. Sibbritt, E.D. Wiklund, S. Bert, D. Strbenac, A.L. Statham, M.D. Robinson, and S.J. Clark. 2011. Discovery pipeline for epigenetically deregulated miRNAs in cancer: integration of primary miRNA transcription. BMC Genomics. 12: 54.

Jemal, A., R. Siegel, E. Ward, Y. Hao, J. Xu, T. Murray, and M.J. Thun. 2008. Cancer statistics, 2008. CA Cancer J Clin. 58(2): 71–96.

Kasinski, A.L. and F.J. Slack. 2011. MicroRNAs en route to the clinic: progress in validating and targeting microRNAs for cancer therapy. Nat Rev Cancer. 11(12): 849–864.

Kojima, K., Y. Fujita, Y. Nozawa, T. Deguchi, and M. Ito. 2010. MiR-34a attenuates paclitaxel-resistance of hormone-refractory prostate cancer PC3 cells through direct and indirect mechanisms. Prostate. 1;70(14): 1501–1512.

Kong, D., S. Banerjee, A. Ahmad, Y. Li, Z. Wang, S. Sethi, and F.H. Sarkar. 2010. Epithelial to mesenchymal transition is mechanistically linked with stem cell signatures in prostate cancer cells. PLoS One. 5(8): e12445.

Leite, K.R., J.M. Sousa-Canavez, S.T. Reis, A.H. Tomiyama, L.H. Camara-Lopes, A. Sanudo, A.A. Antunes, and M. Srougi. 2011a. Change in expression of miR-let7c, miR-100, and miR-218 from high grade localized prostate cancer to metastasis. Urol Oncol. 29(3): 265–269.

Leite, K.R., A. Tomiyama, S.T. Reis, J.M. Sousa-Canavez, A. Sanudo, M.F. Dall'Oglio, L.H. Camara-Lopes, and M. Srougi. 2011b. MicroRNA-100 expression is independently related to biochemical recurrence of prostate cancer. J Urol. 185(3): 1118–1122.

Li, T., D. Li, J. Sha, P. Sun, and Y. Huang. 2009. MicroRNA-21 directly targets MARCKS and promotes apoptosis resistance and invasion in prostate cancer cells. Biochem Biophys Res Commun. 383(3): 280–285.

Lin, S.L., A. Chiang, D. Chang, and S.Y. Ying. 2008. Loss of miR-146a function in hormone-refractory prostate cancer. RNA. 14(3): 417–424.

Linja, M.J., K.J. Savinainen, O.R. Saramaki, T.L. Tammela, R.L. Vessella, and T. Visakorpi. 2001. Amplification and overexpression of androgen receptor gene in hormone-refractory prostate cancer. Cancer Res. 61(9): 3550–3555.

Lodes, M.J., M. Caraballo, D. Suciu, S. Munro, A. Kumar, and B. Anderson. 2009. Detection of cancer with serum miRNAs on an oligonucleotide microarray. PLoS One. 4(7): e6229.

Loeb, S. and W.J. Catalona. 2007. Prostate-specific antigen in clinical practice. Cancer Lett. 249(1): 30–39.

Long, Q., B.A. Johnson, A.O. Osunkoya, Y.H. Lai, W. Zhou, M. Abramovitz, M. Xia, M.B. Bouzyk, R.K. Nam, L. Sugar, A. Stanimirovic, D.J. Williams, B.R. Leyland-Jones, A.K. Seth, J.A. Petros, and C.S. Moreno. 2011. Protein-coding and microRNA biomarkers of recurrence of prostate cancer following radical prostatectomy. Am J Pathol. 179(1): 46–54.

Lu, J., G. Getz, E.A. Miska, E. Alvarez-Saavedra, J. Lamb, D. Peck, A. Sweet-Cordero, B.L. Ebert, R.H. Mak, A.A. Ferrando, J.R. Downing, T. Jacks, H.R. Horvitz, and T.R. Golub. 2005. MicroRNA expression profiles classify human cancers. Nature. 435(7043): 834–838.

Lu, Z., M. Liu, V. Stribinskis, C.M. Klinge, K.S. Ramos, N.H. Colburn, and Y. Li. 2008. MicroRNA-21 promotes cell transformation by targeting the programmed cell death 4 gene. Oncogene. 27(31): 4373–4379.

Mahn, R., L.C. Heukamp, S. Rogenhofer, A. von Ruecker, S.C. Muller, and J. Ellinger. 2011. Circulating microRNAs (miRNA) in serum of patients with prostate cancer. Urology. 77(5): 1265.e9–1265.e16.

Martens-Uzunova, E.S., S.E. Jalava, N.F. Dits, G.J. van Leenders, S. Moller, J. Trapman, C.H. Bangma, T. Litman, T. Visakorpi, and G. Jenster. 2011. Diagnostic and prognostic signatures from the small non-coding RNA transcriptome in prostate cancer. Oncogene. 31(8): 978–991.

McDonnell, T.J., P. Troncoso, S.M. Brisbay, C. Logothetis, L.W. Chung, J.T. Hsieh, S.M. Tu, and M.L. Campbell. 1992. Expression of the protooncogene bcl-2 in the prostate and its association with emergence of androgen-independent prostate cancer. Cancer Res. 52(24): 6940–6944.

Meng, F., R. Henson, H. Wehbe-Janek, K. Ghoshal, S.T. Jacob, and T. Patel. 2007. MicroRNA-21 regulates expression of the PTEN tumor suppressor gene in human hepatocellular cancer. Gastroenterology. 133(2): 647–658.

Mitchell, P.S., R.K. Parkin, E.M. Kroh, B.R. Fritz, S.K. Wyman, E.L. Pogosova-Agadjanyan, A. Peterson, J. Noteboom, K.C. O'Briant, A. Allen, D.W. Lin, N. Urban, C.W. Drescher, B.S. Knudsen, D.L. Stirewalt, R. Gentleman, R.L. Vessella, P.S. Nelson, D.B. Martin, and M. Tewari. 2008. Circulating microRNAs as stable blood-based markers for cancer detection. Proc Natl Acad Sci USA. 105(30): 10513–10518.

Moltzahn, F., A.B. Olshen, L. Baehner, A. Peek, L. Fong, H. Stoppler, J. Simko, J.F. Hilton, P. Carroll, and R. Blelloch. 2011. Microfluidic-based multiplex qRT-PCR identifies diagnostic and prognostic microRNA signatures in the sera of prostate cancer patients. Cancer Res. 71(2): 550–560.

Murata, T., K. Takayama, S. Katayama, T. Urano, K. Horie-Inoue, K. Ikeda, S. Takahashi, C. Kawazu, A. Hasegawa, Y. Ouchi, Y. Homma, Y. Hayashizaki, and S. Inoue. 2010. miR-148a is an androgen-responsive microRNA that promotes LNCaP prostate cell growth by repressing its target CAND1 expression. Prostate Cancer Prostatic Dis. 13(4): 356–361.

Musiyenko, A., V. Bitko, and S. Barik. 2008. Ectopic expression of miR-126*, an intronic product of the vascular endothelial EGF-like 7 gene, regulates prostein translation and invasiveness of prostate cancer LNCaP cells. Journal of Molecular Medicine. 86: 313–322.

Nadiminty, N., R. Tummala, W. Lou, Y. Zhu, J. Zhang, X. Chen, R.W. Devere White, H.J. Kung, C.P. Evans, and A.C. Gao. 2011. MicroRNA let-7c suppresses androgen receptor expression and activity via regulation of Myc expression in prostate cancer cells. J Biol Chem. 287(2): 1527–1537.

Ostling, P., S.K. Leivonen, A. Aakula, P. Kohonen, R. Makela, Z. Hagman, A. Edsjo, S. Kangaspeska, H. Edgren, D. Nicorici, A. Bjartell, Y. Ceder, M. Perala, and O. Kallioniemi. 2011. Systematic analysis of microRNAs targeting the androgen receptor in prostate cancer cells. Cancer Res. 71(5): 1956–1967.

Ozen, M., C.J. Creighton, M. Ozdemir, and M. Ittmann. 2008. Widespread deregulation of microRNA expression in human prostate cancer. Oncogene. 27(12): 1788–1793.

Park, S.M., A.B. Gaur, E. Lengyel, and M.E. Peter. 2008. The miR-200 family determines the epithelial phenotype of cancer cells by targeting the E-cadherin repressors ZEB1 and ZEB2. Genes Dev. 22: 894–907.

Peinado, H., D. Olmeda, and A. Cano. 2007. Snail, Zeb and bHLH factors in tumour progression: an alliance against the epithelial phenotype? Nature Reviews. Cancer. 7: 415–428.

Pesta, M., J. Klecka, V. Kulda, O. Topolcan, M. Hora, V. Eret, M. Ludvikova, M. Babjuk, K. Novak, J. Stolz, and L. Holubec. 2010. Importance of miR-20a expression in prostate cancer tissue. Anticancer Res. 30(9): 3579–3583.

Porkka, K.P., M.J. Pfeiffer, K.K. Waltering, R.L. Vessella, T.L. Tammela, and T. Visakorpi. 2007. MicroRNA expression profiling in prostate cancer. Cancer Res. 67(13): 6130–6135.

Prueitt, R.L., M. Yi, R.S. Hudson, T.A. Wallace, T.M. Howe, H.G. Yfantis, D.H. Lee, R.M. Stephens, C.G. Liu, G.A. Calin, C.M. Croce, and S. Ambs. 2008. Expression of microRNAs and protein-coding genes associated with perineural invasion in prostate cancer. Prostate. 68(11): 1152–1164.

Ribas, J., X. Ni, M. Haffner, E.A. Wentzel, A.H. Salmasi, W.H. Chowdhury, T.A. Kudrolli, S. Yegnasubramanian, J. Luo, R. Rodriguez, J.T. Mendell, and S.E. Lupold. 2009. miR-21: an androgen receptor-regulated microRNA that promotes hormone-dependent and hormone-independent prostate cancer growth. Cancer Res. 69(18): 7165–7169.

Saini, S., S. Majid, S. Yamamura, L. Tabatabai, S.O. Suh, V. Shahryari, Y. Chen, G. Deng, Y. Tanaka, and R. Dahiya. 2011. Regulatory Role of mir-203 in Prostate Cancer Progression and Metastasis. Clin Cancer Res. 17(16): 5287–5298.

Saramaki, O.R., A.E. Harjula, P.M. Martikainen, R.L. Vessella, T.L. Tammela, and T. Visakorpi. 2008. TMPRSS2:ERG fusion identifies a subgroup of prostate cancers with a favorable prognosis. Clin Cancer Res. 14(11): 3395–3400.

Sayed, D., S. Rane, J. Lypowy, M. He, I.Y. Chen, H. Vashistha, L. Yan, A. Malhotra, D. Vatner, and M. Abdellatif. 2008. MicroRNA-21 targets Sprouty2 and promotes cellular outgrowths. Mol Biol Cell. 19(8): 3272–3282.

Schaefer, A., M. Jung, H.J. Mollenkopf, I. Wagner, C. Stephan, F. Jentzmik, K. Miller, M. Lein, G. Kristiansen, and K. Jung. 2010. Diagnostic and prognostic implications of microRNA profiling in prostate carcinoma. Int J Cancer. 126(5): 1166–1176.

Schweizer, L., C.A. Rizzo, T.E. Spires, J.S. Platero, Q. Wu, T.A. Lin, M.M. Gottardis, and R.M. Attar. 2008. The androgen receptor can signal through Wnt/beta-Catenin in prostate cancer cells as an adaptation mechanism to castration levels of androgens. BMC Cell Biol. 9: 4.

Selth, L.A., S. Townley, J.L. Gillis, A.M. Ochnik, K. Murti, R.J. Macfarlane, K.N. Chi, V.R. Marshall, W.D. Tilley, and L.M. Butler. 2011. Discovery of circulating microRNAs associated with human prostate cancer using a mouse model of disease. Int J Cancer. 131(3): 652–661.

Shi, X.B., L. Xue, J. Yang, A.H. Ma, J. Zhao, M. Xu, C.G. Tepper, C.P. Evans, H.J. Kung, and R.W. deVere White. 2007. An androgen-regulated miRNA suppresses Bak1 expression and induces androgen-independent growth of prostate cancer cells. Proc Natl Acad Sci USA. 104(50): 19983–19988.

Sun, D., Y.S. Lee, A. Malhotra, H.K. Kim, M. Matecic, C. Evans, R.V. Jensen, C.A. Moskaluk, and A. Dutta. 2011. miR-99 family of MicroRNAs suppresses the expression of prostate-specific antigen and prostate cancer cell proliferation. Cancer Res. 71(4): 1313–1324.

Sun, T., Q. Wang, S. Balk, M. Brown, G.S. Lee, and P. Kantoff. 2009. The role of microRNA-221 and microRNA-222 in androgen-independent prostate cancer cell lines. Cancer Res. 69(8): 3356–3363.

Sylvestre, Y., V. De Guire, E. Querido, U.K. Mukhopadhyay, V. Bourdeau, F. Major, G. Ferbeyre, and P. Chartrand. 2007. An E2F/miR-20a autoregulatory feedback loop. J Biol Chem. 282(4): 2135–2143.

Szczyrba, J., E. Loprich, S. Wach, V. Jung, G. Unteregger, S. Barth, R. Grobholz, W. Wieland, R. Stohr, A. Hartmann, B. Wullich, and F. Grasser. 2010. The microRNA profile of prostate carcinoma obtained by deep sequencing. Mol Cancer Res. 8(4): 529–538.

Taft, R.J., E.A. Glazov, N. Cloonan, C. Simons, S. Stephen, G.J. Faulkner, T. Lassmann, A.R. Forrest, S.M. Grimmond, K. Schroder, K. Irvine, T. Arakawa, M. Nakamura, A. Kubosaki, K. Hayashida, C. Kawazu, M. Murata, H. Nishiyori, S. Fukuda, J. Kawai, C.O. Daub, D.A. Hume, H. Suzuki, V. Orlando, P. Carninci, Y. Hayashizaki, and J.S. Mattick. 2009. Tiny RNAs associated with transcription start sites in animals. Nat Genet. 41(5): 572–578.

Takayama, K., K. Kaneshiro, S. Tsutsumi, K. Horie-Inoue, K. Ikeda, T. Urano, N. Ijichi, Y. Ouchi, K. Shirahige, H. Aburatani, and S. Inoue. 2007. Identification of novel androgen response genes in prostate cancer cells by coupling chromatin immunoprecipitation and genomic microarray analysis. Oncogene. 26(30): 4453–4463.

Takayama, K., S. Tsutsumi, S. Katayama, T. Okayama, K. Horie-Inoue, K. Ikeda, T. Urano, C. Kawazu, A. Hasegawa, K. Ikeo, T. Gojyobori, Y. Ouchi, Y. Hayashizaki, H. Aburatani, and S. Inoue. 2011. Integration of cap analysis of gene expression and chromatin immunoprecipitation analysis on array reveals genome-wide androgen receptor signaling in prostate cancer cells. Oncogene. 30(5): 619–630.

Tang, X., J. Gal, N. Kyprianou, H. Zhu, and G. Tang. 2011. Detection of microRNAs in prostate cancer cells by microRNA array. Methods Mol Biol. 732: 69–88.

Taplin, M.E., B. Rajeshkumar, S. Halabi, C.P. Werner, B.A. Woda, J. Picus, W. Stadler, D.F. Hayes, P.W. Kantoff, N.J. Vogelzang, and E.J. Small. 2003. Androgen receptor mutations in androgen-independent prostate cancer: Cancer and Leukemia Group B Study 9663. J Clin Oncol. 21(14): 2673–2678.

Tomlins, S.A., D.R. Rhodes, S. Perner, S.M. Dhanasekaran, R. Mehra, X.W. Sun, S. Varambally, X. Cao, J. Tchinda, R. Kuefer, C. Lee, J.E. Montie, R.B. Shah, K.J. Pienta, M.A. Rubin, and A.M. Chinnaiyan. 2005. Recurrent fusion of TMPRSS2 and ETS transcription factor genes in prostate cancer. Science. 310(5748): 644–648.

Tong, A.W., P. Fulgham, C. Jay, P. Chen, I. Khalil, S. Liu, N. Senzer, A.C. Eklund, J. Han, and J. Nemunaitis. 2009. MicroRNA profile analysis of human prostate cancers. Cancer Gene Ther. 16(3): 206–216.

Vander Griend, D.J., J. D'Antonio, B. Gurel, L. Antony, A.M. Demarzo, and J.T. Isaacs. 2010. Cell-autonomous intracellular androgen receptor signaling drives the growth of human prostate cancer initiating cells. Prostate. 70(1): 90–99.

Varambally, S., Q. Cao, R.S. Mani, S. Shankar, X. Wang, B. Ateeq, B. Laxman, X. Cao, X. Jing, K. Ramnarayanan, J.C. Brenner, J. Yu, J.H. Kim, B. Han, P. Tan, C. Kumar-Sinha, R.J. Lonigro, N. Palanisamy, C.A. Maher, and A.M. Chinnaiyan. 2008. Genomic loss of microRNA-101 leads to overexpression of histone methyltransferase EZH2 in cancer. Science. 322(5908): 1695–1699.

Visakorpi, T., E. Hyytinen, P. Koivisto, M. Tanner, R. Keinanen, C. Palmberg, A. Palotie, T. Tammela, J. Isola, and O.P. Kallioniemi. 1995. *In vivo* amplification of the androgen receptor gene and progression of human prostate cancer. Nat Genet. 9(4): 401–406.

Volinia, S., G.A. Calin, C.G. Liu, S. Ambs, A. Cimmino, F. Petrocca, R. Visone, M. Iorio, C. Roldo, M. Ferracin, R.L. Prueitt, N. Yanaihara, G. Lanza, A. Scarpa, A. Vecchione, M. Negrini, C.C. Harris, and C.M. Croce. 2006. A microRNA expression signature of human solid tumors defines cancer gene targets. Proc Natl Acad Sci USA. 103(7): 2257–2261.

Waltering, K.K., M.A. Helenius, B. Sahu, V. Manni, M.J. Linja, O.A. Janne, and T. Visakorpi. 2009. Increased expression of androgen receptor sensitizes prostate cancer cells to low levels of androgens. Cancer Res. 69(20): 8141–8149.

Waltering, K.K., K.P. Porkka, S.E. Jalava, A. Urbanucci, P.J. Kohonen, L.M. Latonen, O.P. Kallioniemi, G. Jenster, and T. Visakorpi. 2011. Androgen regulation of micro-RNAs in prostate cancer. Prostate. 71(6): 604–614.

Wang, X., S. Tang, S.Y. Le, R. Lu, J.S. Rader, C. Meyers, and Z.M. Zheng. 2008. Aberrant expression of oncogenic and tumor-suppressive microRNAs in cervical cancer is required for cancer cell growth. PLoS One. 3(7): e2557.

Weber, J.A., D.H. Baxter, S. Zhang, D.Y. Huang, K.H. Huang, M.J. Lee, D.J. Galas, and K. Wang. 2010. The microRNA spectrum in 12 body fluids. Clin Chem. 56(11): 1733–1741.

Wellner, U., J. Schubert, U.C. Burk, O. Schmalhofer, F. Zhu, A. Sonntag, B. Waldvogel, C. Vannier, D. Darling, A. zur Hausen, V.G. Brunton, J. Morton, O. Sansom, J. Schüler, M.P. Stemmler, C. Herzberger, U. Hopt, T. Keck, S. Brabletz, and T. Brabletz. 2009. The EMT-activator ZEB1 promotes tumorigenicity by repressing stemness-inhibiting microRNAs. Nat Cell Biol. 11(12): 1487–1495.

Wu, D. and D.M. Terrian. 2002. Regulation of caveolin-1 expression and secretion by a protein kinase C epsilon signaling pathway in human prostate cancer cells. J Biol Chem. 277(43): 40449–40455.

Wu, H., L. Sun, J.W. Moul, H.Y. Wu, D.G. McLeod, C. Amling, R. Lance, L. Kusuda, T. Donahue, J. Foley, A. Chung, W. Sexton, and D. Soderdahl. 2004. Watchful waiting and factors predictive of secondary treatment of localized prostate cancer. J Urol. 171(3): 1111–1116.

Xu, G., J. Wu, L. Zhou, B. Chen, Z. Sun, F. Zhao, and Z. Tao. 2010. Characterization of the small RNA transcriptomes of androgen dependent and independent prostate cancer cell line by deep sequencing. PLoS One. 5(11): e15519.

Yaman Agaoglu, F., M. Kovancilar, Y. Dizdar, E. Darendeliler, S. Holdenrieder, N. Dalay, and U. Gezer. 2011. Investigation of miR-21, miR-141, and miR-221 in blood circulation of patients with prostate cancer. Tumour Biol. 32(3): 583–588.

Yamakuchi, M., M. Ferlito, and C.J. Lowenstein. 2008. miR-34a repression of SIRT1 regulates apoptosis. PNAS. 105: 13421–13426.

Yang, K., A.M. Handorean, and K.A. Iczkowski. 2009. MicroRNAs 373 and 520c are downregulated in prostate cancer, suppress CD44 translation and enhance invasion of prostate cancer cells *in vitro*. Int J Clin Exp Pathol. 2(4): 361–9.

Yeh, S., H.K. Lin, H.Y. Kang, T.H. Thin, M.F. Lin, and C. Chang. 1999. From HER2/Neu signal cascade to androgen receptor and its coactivators: a novel pathway by induction of androgen target genes through MAP kinase in prostate cancer cells. Proc Natl Acad Sci USA. 96(10): 5458–5463.

Yoshino, T., H. Shiina, S. Urakami, N. Kikuno, T. Yoneda, K. Shigeno, and M. Igawa. 2006. Bcl-2 expression as a predictive marker of hormone-refractory prostate cancer treated with taxane-based chemotherapy. Clin Cancer Res. 12(20 Pt 1): 6116–6124.

Yun, M.S., S.E. Kim, S.H. Jeon, J.S. Lee, and K.Y. Choi. 2005. Both ERK and Wnt/beta-catenin pathways are involved in Wnt3a-induced proliferation. J Cell Sci. 118(Pt 2): 313–322.

Zhang, H.L., L.F. Yang, Y. Zhu, X.D. Yao, S.L. Zhang, B. Dai, Y.P. Zhu, Y.J. Shen, G.H. Shi, and D.W. Ye. 2011. Serum miRNA-21: elevated levels in patients with metastatic hormone-refractory prostate cancer and potential predictive factor for the efficacy of docetaxel-based chemotherapy. Prostate. 71(3): 326–331.

Zheng, C., S. Yinghao, and J. Li. 2011. MiR-221 expression affects invasion potential of human prostate carcinoma cell lines by targeting DVL2. Med Oncol. 29(2): 815–822.

Zhu, S., M.L. Si, H. Wu, and Y.Y. Mo. 2007. MicroRNA-21 targets the tumor suppressor gene tropomyosin 1 (TPM1). J Biol Chem. 282(19): 14328–14336.

CHAPTER 7

MicroRNA DEREGULATIONS IN GASTRIC CANCER

Jiayi Yao,[1] Tusheng Song[2,b] and Chen Huang[2,3,*]

ABSTRACT

Gastric cancer is the second most common cause of cancer death. Elucidating the molecular mechanisms of gastric cancer may illuminate a prognostic strategy to block gastric cancer development. MicroRNAs (miRNAs) are highly conserved small non-coding RNAs that trigger translational repression and/or mRNA degradation mostly through complementary binding to the 3'-untranslated (3'-UTR) regions of target mRNAs. Numerous miRNAs are deregulated in human gastric cancers, and experimental evidence indicates that they can play roles as oncogenes or tumor suppressor genes to affect many cellular processes that are routinely altered in cancer, such as differentiation, cell-cycle progression, proliferation, apoptosis, metastasis, invasiveness and telomere maintenance. miRNA expression can be regulated by different epigenetic mechanisms, including changes in DNA methylation and histone modification in promoter regions. This article reviews the

[1]Cardiovascular Research Center, Xi'an Jiaotong University, College of Medicine, Xi'an 710061 People's Republic of China.
[2]Department of Genetics and Molecular Biology, Medical School of Xi'an Jiaotong University, Xi'an 710061, People's Republic of China.
[a]Email: yaojiayi@mail.xjtu.edu.cn
[b]Email: tusheng@mail.xjtu.edu.cn
[3]Key Laboratory of Environment and Genes Related to Diseases, Xi'an Jiaotong University, Ministry of Education, Xi'an 710061, People's Republic of China.
Email: hchen@mail.xjtu.edu.cn
*Corresponding author

studies that have shown the importance of these small molecules in human gastric cancer, and mechanisms of these microRNAs that can affect or be affected by genetic and epigenetic alterations.

1. INTRODUCTION

Gastric cancer is the second most common cause of cancer-related death in the world, and it remains difficult to cure worldwide, because most patients present advanced disease (Danaei et al. 2005). Although recent advances have been made, surgery still remains to be the most effective treatment for gastric cancer patients. However, patients with locally advanced disease show high rates of distant recurrence even after potentially curative resections (Scartozzi et al. 2010). Even patients who present the most favorable condition and who undergo curative surgical resection often die of recurrent disease. The 5-year survival rate for curative surgical resection ranges from 30% to 50% for patients with stage II disease or stage III disease, respectively (Lee et al. 2009). Stomach cancer can metastasize to the esophagus or the small intestine, and can extend through the stomach wall to nearby lymph nodes and organs such as liver, pancreas, and colon (Zheng et al. 2003). In the past decades, studies have focused on investigating the molecular mechanism underlying the development and progression of gastric cancer. The progression is thought to involve the deregulation of genes that are critical to cellular processes such as cell cycle control, cell growth, apoptosis, metastasis and invasion (Wang et al. 2008). Recently, an increasing number of reports have described a new class of small regulatory RNA molecules termed microRNAs (miRNAs) that are implicated in gastric cancer development (Ueda et al. 2010). Moreover, compelling evidence demonstrates that miRNAs have important roles in gastric cancer progression and directly contribute to the cell proliferation, avoidance of apoptosis, and metastasis of gastric cancer (Wu et al. 2010b). miRNA carries out its biological functions by repressing or activating the expression of its target genes through base-pairing with endogenous mRNAs. In this consideration, miRNA genes have been characterized as novel proto-oncogenes or tumor suppressors in gastric carcinogenesis. Identifying the miRNAs and their targets that are essential for gastric cancer progression may provide promising therapeutic opportunities. In this review, we discuss recent advances in our knowledge of miRNAs and validated targets in gastric cancer progression and the potential future perspectives.

2. FUNCTION AND BIOGENESIS OF miRNAs

MicroRNAs (miRNAs) are noncoding RNAs of ~22nt that function as post-transcriptional regulators to control the efficiency of translation by base-pairing with the complementary sites in the 3' untranslated region (UTR) of mRNA (Bartel 2004). Approximately 40% of total human miRNA loci are located in <3kb from the adjacent miRNA locus in clusters (Altuvia et al. 2005). Furthermore, experimental evidence demonstrated that clustered miRNA loci form an operon-like gene structure and that they are transcribed from a common promoter (Lee et al. 2002, Lee et al. 2004). Most miRNA genes are located in intergenic regions or in antisense orientation of certain genes, which contain their own promoters and regulatory elements. miRNA gene is transcribed by RNA polymerase II in the nucleus, producing primary miRNA that contains a 5' cap, at least one 70-nucleotide hairpin loop structure and a 3' poly (A) tail. After transcription, primary miRNA is processed by a complex consisting of Drosha and its co-factor DGCR8 into precursor miRNA that is deprived of the 5' cap, the 3' poly (A) tail and sequences flanking the hairpin loop structure (Lee et al. 2006). Pre-miRNAs are exported from the nucleus by exportin-5 (Yi et al. 2003). In the cytoplasm, the RNase III enzyme Dicer cleaves pre-miRNAs further to generate their mature forms (Lee et al. 2006, Yi et al. 2003). Finally, one strand of the miRNA duplex is incorporated into the RNA-induced silencing complex (RISC) that mediates the gene suppression effect (Gregory et al. 2004). In this matter, Dicer, along with the transactivating response RNA-binding protein (TRBP), PACT, and an Argonaute protein, contribute to the formation of the RISC loading complex (Kim et al. 2009a). The mature miRNA then guides the RNA-induced silencing complex to its target mRNAs to induce silencing (Tijsterman and Plasterk 2004).

3. miRNAs IN HUMAN CANCER

As a post-translation system, miRNAs are capable of regulating many targets to permit control of crucial developmental and physiological decisions (Bartel 2004). In the same pattern, miRNAs have been proved to exert a profound impact on aberrant processes during malignant transformation, including cell proliferation, apoptosis, metastasis, invasion, stress responses, and metabolism (Bhattacharyya et al. 2006). The relevance of miRNAs in cancer is suggested by changes in their expression patterns (Iorio et al. 2005, Volinia et al. 2006), recurrent amplification and deletion of miRNA genes in tumors (Akao et al. 2006, Calin et al. 2004). This has been demonstrated in tumor initiation and progression using mouse models and cultured tumor cells. For example, *miR-145* in carcinoma originates in the colon, breast, lung or prostate (Arndt et al. 2009, Shi et al. 2007, Villadsen et al. 2011), *miR-15a* and *miR-16-1* in chronic lymphocytic leukemia, *mir-221* in papillary thyroid carcinoma (He et al. 2005), the *miR-17-92* polycistronic cluster in lung carcinoma and *miR-21* in glioblastoma (Hayashita et al. 2005, Ren et al. 2010).

In addition to roles for several individual miRNAs in tumour initiation and progression, global alterations in miRNA expression patterns are a common feature of tumor cells (Lu et al. 2005, Volinia et al. 2006). Recently, studies have emerged directly implicating miRNAs in cancer and thus giving rise to a new molecular taxonomy of human cancers based on miRNA profiling (Kent and Mendell 2006). A comprehensive analysis of the miRNA expression in diverse neoplasms showed an even higher accuracy for tumor diagnosis using the miRNA genetic fingerprint than using a profile of more than 16,000 messenger RNAs (Lu et al. 2005). miRNA profiles are excellently suited not only for the classification and diagnosis of human malignancies, but also for the prognosis of patients with certain carcinomas (Takamizawa et al. 2004, Yanaihara et al. 2006). Even more, miRNA signatures have been shown to correlate with the degree of histological tumor differentiation in tumors or with specific pathologic features such as estrogen and progesterone receptor expression in breast carcinomas (Iorio et al. 2005, Murakami et al. 2006).

4. miRNAs IN HUMAN GASTRIC CANCER

Gastric cancer development is thought to develop in a multi-step process requiring the accumulation of several structural and genomic alterations and affecting many different pathways. Previous studies revealed a broad variety of oncogenes and tumor suppressor genes regulated by different miRNAs in gastric cancer cells. Some of these miRNAs modulate expression of known oncogenes or tumor suppressor genes, whereas others function as so-called onco-miRs or tumor-suppressor miRs (Wu et al. 2010b). Furthermore, evidence for regulation of carcinogenesis by miRNAs has been obtained, including promotion of proliferation, migration, invasion and anti-apoptosis (Li et al. 2011a).

4.1 miRNAs and the Cell Cycle

It is well documented that a single defect in cell cycle control may cause a series of consequences in the development and progression of human cancer. In gastric cancers, a number of oncogenes and tumor suppressors expressions have been identified to be aberrant in cell cycle regulation, thereby promoting gastric cancer cell initiation and progression (Shao et al. 2003). In the cell cycle regulation process, cyclin-dependent kinase complexes, which are composed by the association of cyclin with cyclin-dependent kinase, are able to phosphorylate other compounds under the consumption of ATP. Depending on the types of cyclins attached to CDKs, they serve as unique markers in each phase, which are integral for the correct sequence of events, as well as facilitating the transition to the next phase. In the G1 phase, G1 cyclin-CDK complexes receive a pro-mitotic extracellular signal to become active to prepare the cell for S phase, promoting the expression of transcription factors that in turn promote the expression of S cyclins and of enzymes required for DNA replication. Active S cyclin-CDK complexes phosphorylate proteins that make up the pre-replication complexes assembled during G1 phase on DNA replication origins. Mitotic cyclin-CDK complexes, which are synthesized but inactivated during S and G2 phases, promote the initiation of mitosis by stimulating downstream proteins involved in chromosome condensation and mitotic spindle assembly (Satyanarayana and Kaldis 2009). In mammalian cells, two families of CDK inhibitors are responsible for regulating different CDKs. Members of the Ink4 family (p15Ink4b, p16Ink4a, p18Ink4c and p19Ink4d) bind to CDK4 and 6/cyclin D complexes, thereby inhibiting progression through the G1 restriction point. Members of the Cip/Kip family (p21Cip1, p27Kip1 and p57Kip2) share homology in their N-terminal regions and affect the complexes of CDK2, 4 and 6 with cyclin A, D and E. Because these genes are crucial in prevention of tumor formation, they are known as tumor suppressors. The Cip/Kip proteins are also involved in many aspects of cellular physiology, including apoptosis, transcriptional regulation, cell fate determination, cell migration and cytoskeletal dynamics (Besson et al. 2008).

Recent reports showed that some miRNAs could modulate the major proliferation pathways through interaction with critical cell cycle regulators such as CDK complexes and Cip/Kip family. In G1-S phase, cell-cycle progression in mammalian cells is mainly governed by cyclins D and E with cyclin-dependent kinase (CDK) 2, 4, and 6. The kinase activities of these CDKs are negatively regulated by the cyclin-dependent kinase inhibitors (CDKIs) p16 and p27 (Besson et al. 2008). *miR-148* has been verified to be downregulated frequently and dramatically in human advanced gastric cancer tissues by targeting at p27, thus promoting gastric cancer cell proliferation (Guo et al. 2011). Cyclin dependent kinase 6 (CDK6) is found to be upregulated in gastric cancer and has been implicated in tumor initiation and progression. Feng et al. have identified *miR-107* as a potential regulator of CDK6 expression. Ectopic expression of *miR-107* reduced both mRNA and protein expression levels of CDK6, inhibited proliferation, induced G1 cell cycle arrest. Thus, *miR-107* may have a tumor suppressor function by directly targeting CDK6 to inhibit the proliferation in gastric cancer cells (Feng et al. 2011). *miR-106b, miR-93, miR-222-221* and *miR-106b-25* clusters, which are upregulated in gastric cancer and are downstream targets of the oncogenic transcription factor E2F-1, interfer with the expression of CDKN1A (p21Waf1/Cip1) and BCL2L11 (Bim) (Petrocca et al. 2008, Satyanarayana and Kaldis 2009). Ectopic expression of *miR-222-221* and *miR-106b-25* clusters results in activation of CDK2 activity and facilitates the G1/S phase transition (Cho et al. 2009). Overexpression of *miR-331-3p* blocked G1/S transition in gastric cancer cell lines by targeting cell cycle-related molecule E2F1 (Guo et al. 2010). In addition, *miR-106b* and *miR-93* control p21, whereas *miR-222* and *miR-221* downregulate both p27 and p57. AE1 is an erythroid-specific membrane protein, which is unexpectedly expressed in gastric cancer cells. AE1 sequesters p16 in the cytoplasm and thus promotes cell proliferation. The expression of AE1 is modulated by *miR-24*, which directly binds to its 3′-UTR. Transfection of *miR-24* leads the return of AE1-sequestered p16 to the nucleus and inhibition of cell proliferation (Chan et al. 2010). Song et al. found that *miR-199a* regulated gastric cancer cell proliferation through directly targeting the mitogen-activated protein kinase (Song et al. 2011). Another study identified *miR-27a* as an oncogenic miRNA in a gastric cancer cell line, in which it targeted the tumor suppressor prohibitin, an evolutionary conserved and ubiquitous protein interacting with pRb and its family members (Liu et al. 2009). These findings suggest that aberrant miRNA expression may enhance cell-cycle progression through direct or indirect regulation of cell-cycle regulators.

4.2 miRNAs and Apoptosis

Apoptosis is a tightly regulated and highly efficient cell death program, which requires the interplay of a multitude of factors. The components of the apoptotic signaling network are genetically encoded and are considered to be usually in place in a nucleated cell ready to be activated by a death-inducing stimulus (Muschel et al. 1995). Apoptosis occurs via two main signaling pathways: 1) the intrinsic pathway, and 2) the extrinsic pathway, both of which are potential anticancer therapeutic targets. The intrinsic pathway is triggered by developmental cues or severe cell stress, such as DNA damage. The extrinsic pathway is activated when a pro-apoptotic ligand, such as endogenous Fas ligand or Apo2 ligand/tumor necrosis factor (TNF)-related apoptosis-inducing ligand (Apo2L/TRAIL), binds to pro-apoptotic death receptors, such as Fas, or death receptor 4 (DR4) and death receptor 5 (DR5) (Fiers et al. 1995, Gulbins et al. 1995). Intracellular protease enzymes called caspases that, upon activation by the intrinsic and/or extrinsic pathways, destroy cellular proteins that are vital for cell survival ultimately carry out destruction of the cell. The activation or restoration of apoptosis is emerging as a key strategy to target cancer and other diseases.

miRNAs have been reported to be involved in the regulation of apoptotic process. Bcl2 suppresses apoptosis by controlling the mitochondrial membrane permeability and inhibits caspase activity either by preventing the release of cytochrome c from the mitochondria and/or by binding to the apoptosis-activating factor (APAF-1). Xia et al. reported that microRNA-15b and -16 are involved in the development of multidrug resistance in gastric cancer cells, at least in part by modulating apoptosis (Xia et al. 2008). *MiR-129-2* also regulates apoptosis probably through regulating the relative abundance of proapoptotic and antiapoptotic members of Bcl-2 family (Shen et al. 2010). The p53 tumor suppressor protein and the Akt/PKB kinase play important roles in the transduction of pro-apoptotic and anti-apoptotic signals, respectively. Akt and p53 are integrated via negative feedback between the two pathways. On the one hand, the combination of ionizing radiation and survival factor deprivation, which leads to rapid apoptosis of IL-3 dependent DA-1 cells, entails a caspase- and p53-dependent destruction of Akt. On the other hand upon serum stimulation, when Akt becomes active and enhances cell survival, phosphorylation occurs at an Akt consensus site (serine 166) within the Mdm2 protein, a key regulator of p53 function (Gottlieb et al. 2002). *miR-34* had been demonstrated to restore p53 function in p53-deficient human gastric cancer cells to inhibit cell growth and induce apoptosis. Even more, overexpression of *miR-34* increased caspase-3 activation and impaired tumor formation and growth in p53-mutant gastric cancer cells. The mechanism of *miR-34*-mediated suppression of self-renewal appears to be related to the direct modulation of downstream targets Bcl-2, Notch, and HMGA2, indicating that *miR-34* may be involved in gastric cancer stem cell self-renewal/differentiation decision-making (Ji et al. 2008). Ectopic expression of *miRNA-375* (*miR-375*) in gastric carcinoma cells markedly reduced cell viability via the caspase-mediated apoptosis pathway. Moreover, expression of *miR-375* inhibited expression of PDK1, which is a direct target of *miR-375*, followed by suppression of Akt phosphorylation. Further analysis by gene expression microarray revealed that 14-3-3ζ, a potent antiapoptotic gene, was significantly downregulated at both the mRNA and protein levels in cells transfected with *miR-375* (Tsukamoto et al. 2010).

Transforming growth factor β (TGF-β) induced apoptosis is a well-documented phenomenon in many cell types *in vitro* and is known to involve TIEG1, DAXX, DAPK (death-associated protein kinase), ARTS, SHIP, truncated Bad, AP1, GADD45b, Fas, and Bim, as well as components of the canonical TGF-β signaling pathway, such as Smad3, Smad4, Smad7, and mitogen-activated protein kinases (MAP kinases), such as Jun N-terminal protein kinase and p38 MAP kinase. *miRNA-200c* can indirectly regulate apoptosis through E-cadherin in gastric cancer cells, which may be a possible mechanism of *miRNA-200c* in promoting apoptosis (Chen et al. 2010). Smad3 regulates, at the transcriptional level, *miR-200* family members, which themselves regulate ZEB1 and ZEB2, known transcriptional repressors of E-cadherin, at the posttranscriptional level in a TGF-β-independent manner (Ahn et al. 2011). *In vitro* assays indicated that exogenous *miR-370* expression enhanced the oncogenic potential of gastric cancer cells by downregulating TGFβ-RII (Lo et al. 2011). Recently, it is reported that RUNX3 forms complexes with Smads that regulate target gene expression; thus, RUNX3 is a downstream target of the TGF-β signaling pathway and it is responsible for transcriptional upregulation of *Bim* in TGF-β-induced apoptosis (Yano et al. 2006). Moreover, RUNX3 is closely involved in DNA damage-dependent phosphorylation of tumor suppressor p53 at Ser-15 and acts as a co-activator for p53 (Yamada et al. 2010). Bioinformatics, gene and protein expression analysis in eight gastric cell lines identified *miR-130b* as the top candidate miRNA for RUNX3 binding. Overexpression of *miR-130b* increased cell viability, reduced cell death and decreased expression of *Bim* in TGF-beta mediated apoptosis, subsequent to the downregulation of RUNX3 protein expression. In 15 gastric tumours, *miR-130b* expression was significantly higher compared to match normal tissue, and was inversely associated with RUNX3 hypermethylation (Lai et al. 2010). *miR-150* was overexpressed in gastric cancer cell lines and tissues. Ectopic expression of *miR-150* promoted tumorigenesis and proliferation of gastric cancer cells by targeting of EGR2 (Wu et al. 2010a). Furthermore, *miR-106b*, *miR-93*, and *miR-25* overexpression not only caused a decreased response of gastric cancer cells to TGFβ interfering with the synthesis of p21 and Bim, but also controlled p21 while *miR-222* and *miR-221* regulated both p27 and p57 (Kim et al. 2009b). The antiapoptotic activity of the *miR-17-92* cluster and related miRNAs in other settings, including TGFβ-stimulated gastric cancer cells, may similarly involve downregulation of Bim (Petrocca et al. 2008).

The other important mechanism of cell survival is the activation of transcription of different anti-apoptotic proteins by tumor necrosis factors (TNFs) via the nuclear factors of kappa light polypeptide in B-cells (NFκB) signaling cascade. The members of the tumor necrosis factor ligand family (TNFs) may induce both apoptotic and anti-apoptotic pathways. TNFs transduce cellular responses through activation of different TNF-receptors (TNFRs) (Orange et al. 2005). *miR-218* expression was reduced significantly in gastric cancer tissues, in *H. pylori*-infected gastric mucosa, and in *H. pylori*-infected human gastric cancer cell line AGS cells. Overexpression of *miR-218* inhibited cell proliferation and increased apoptosis *in vitro*. Epidermal growth factor receptor-coamplified and overexpressed protein (ECOP), which regulates nuclear factor kappa B transcriptional activity and is associated with apoptotic response, was a direct target of *miR-218*. Overexpression of *miR-218* also inhibited NFκB transcriptional activation and transcription of cyclooxygenase-2, a proliferative gene regulated by NFκB (Gao et al. 2010). Furthermore, *miR-486* was identified as a significantly downregulated miRNA in primary gastric cancer (GC) tissues and gastric cancer cell lines. Restoring *miR-486* expression in GC cells decreased endogenous OLFM4 mRNA and protein levels, and also inhibited expression of luciferase reporters containing an OLFM4 3′UTR with predicted *miR-486* binding sites (Oh et al. 2011). Besides, NFκB1 was reported to be regulated negatively by *miR-9* at the post-transcriptional level. It may suggest that *miR-9* plays important roles in diverse biological processes by regulating NFκB1 (Wan et al. 2010). More interestingly, NFκB can also regulate *miR-16* and *miR-21* expression in gastric cancer cells, and yet nicotine-promoted cell proliferation is mediated via EP2/4 receptors. *MiR-16* and *miR-21* were upregulated upon nicotine-treated gastric cancer cells. NFκB-binding sites were located in both *miR-16* and *miR-21* gene transcriptional elements and nicotine enhanced the binding of NFκB to the promoters of *miR-16* and *miR-21* (Shin et al. 2011).

4.3 miRNAs and Invasion and Metastasis

Invasion and metastasis are responsible for >90% of cancer-related mortality. As tumors progress with increased malignancy, cells within them develop the ability to invade surrounding normal tissues and through tissue boundaries to form new tumors at sites distinct from the primary tumor. The molecular mechanisms involved in this process are associated with cell–cell and cell-matrix adhesion, with the degradation of extracellular matrix, and with the initiation and maintenance of early growth at the new site (Nomura and Katunuma 2005). Gastric cancer cells produce various growth factors and express their receptors to bring about biological events, including cancer-stromal interactions. These factors induce not only cell growth but also extracellular matrix degradation and angiogenesis that facilitate tumor invasion and metastasis. Among the best-known angiogenic factors of gastric cancer cells are vascular endothelial growth factor (VEGF), interleukin (IL-8), basic fibroblast growth factor (bFGF) and platelet-derived endothelial cell growth factor (PDECGF). Most gastric cancers express IL-8/receptor systems, and IL-8 levels correlate closely with neovascularity. Well differentiated gastric cancer expresses VEGF at high levels, while poorly differentiated scirrhous cancer expresses strongly bFGF (Farinati et al. 2008). Simultaneous expression of EGF/TGF-alpha and EGFR is associated with cancer progression and correlates with poor patient prognosis (Yasui et al. 2006).

During the progression of an invasive cancer, epithelial cells lose expression of proteins that promote cell–cell contact, and acquire mesenchymal markers, which promote cell migration and invasion. These events bear extensive similarities to the process of epithelial to mesenchymal transition (EMT), which has been recognized for several decades as critical feature of embryogenesis. EMT describes the molecular reprogramming and phenotypic changes that characterize the conversion of polarized immotile epithelial cells to motile mesenchymal cells (Zhang et al. 2010a). During EMT, cells lose epithelial features, including the cell–cell adhesion and E-cadherin expression, and acquire mesenchymal characteristics, including increased vimentin expression, concurrent with the manifestation of a migratory phenotype. This process is required by primary tumor cells to escape from the primary organ to a distant site. Ectopic expression of *miR-146a* inhibited migration and invasion and downregulated EGFR and IRAK1 expression in gastric cancer cells (Kogo et al. 2011). Twist, a transcription factor of the helix–loop–helix class, regulates cancer metastasis and induces EMT. Furthermore, loss of Twist expression prevents the intravasation of metastatic tumor cells into the blood circulation (Yang et al. 2006). Twist-enhanced cancer metastasis occurs in gastric cancer. The *miR-200* family has emerged as a major player in the control of EMT. The *miR-200* family inhibits EMT by maintaining the epithelial phenotype through direct targeting of the transcriptional repressors of E-cadherin, and zinc finger E-box-binding homeo box 1 and 2 (ZEB1 [also known as δEF1] and ZEB2 [also known as SIP1]), which are implicated in EMT and metastasis. *miR-223* is overexpressed only in metastatic gastric cancer cells and stimulated nonmetastatic gastric cancer cells migration and invasion. Mechanistically, *miR-223*, induced by the transcription factor Twist, posttranscriptionally downregulates EPB41L3 expression by directly targeting its 3′-untranslated regions. Significantly, overexpression of *miR-223* in primary gastric carcinomas is associated with poor metastasis-free survival (Li et al. 2011c). In addition, ectopic expression of *miR-200a* and *miR-200c* upregulated E-cadherin in cancer cell lines and reduced their motility, while the inhibition of *miR-200* reduced E-cadherin expression, increased expression of vimemtin, and induced EMT (Shinozaki et al. 2010). Crk plays important roles in VEGF-induced cell migration which has been proved to be a target of *miR-126* (Stoletov et al. 2004). *miR-126* was significantly down-regulated in gastric cancer tissues compared with matched normal tissues and was associated with local invasion and tumor-node-metastasis (TNM) stage (Feng et al. 2010).

The NFκB family of transcription factors that represses an epithelial phenotype also plays pivotal roles in EMT by taking control of mesenchymal genes critical for promoting and maintaining an invasive phenotype (Min et al. 2008). ING4 overexpression strongly suppressed the growth of human umbilical vein endothelial cells (HUVEC) and their ability to form tubular structure *in vitro*. ING4 is induced by BRMS1 and that it inhibits cancer angiogenesis by suppressing NFκB activity and IL-6 expression (Li and Li

2010). *miR-650* is involved in lymphatic and distant metastasis in human gastric cancer by targeting ING4. Collectively, further study demonstrates that over-expression of *miR-650* in gastric cancer may promote proliferation and growth of cancer cells, at least partially through directly targeting ING4 (Zhang et al. 2010b). Another study showed that mitogen-activated protein kinase 11 was significantly down regulated by *miR-199a* at the post-transcriptional level, and the level of *miR-199a* expression in gastric cancer significantly correlated with clinical progression. These findings suggested *miR-199a* promoted proliferation and metastasis of gastric cancer cells through a regulatory pathway in gastric cancer (Song et al. 2010). *microRNA-107 (miR-107)* is overexpressed in gastric cancer tissues compared with the matched normal tissues. *miR-409-3p* was found to be downregulated frequently in human GCs, and its expression was significantly associated with tumor-node-metastasis (TNM) stage and lymph node metastasis. *miR-409* suppresses GC cell invasion and metastasis by directly targeting RDX and that patients with down regulated *miR-409-3p* are prone to TNM (Zheng et al. 2011a).

Exosomes also play a major role in cell-to-cell communication, targeting cells to transfer exosomal molecules including proteins, mRNAs, and miRNAs by an endocytosis-like pathway. The enrichment of *let-7* miRNA family in the extracellular fractions, particularly, in the exosomes from AZ-P7a cells may reflect their oncogenic characteristics including tumorigenesis and metastasis. Since *let-7* miRNAs generally play a tumor-suppressive role as targeting oncogenes such as RAS and HMGA2, the results suggest that AZ-P7a cells release *let-7* miRNAs via exosomes into the extracellular environment to maintain their oncogenesis (Ohshima et al. 2010). SLIT-family proteins activate Robo1, which is a member of the immunoglobulin gene superfamily and encodes an integral membrane protein, resulting in a repulsive effect on glioma cell guidance in the developing brain. Decreased *miR-218* levels eliminate Robo1 repression, which activates the Slit-Robo1 pathway through the interaction between Robo1 and Slit2, thus triggering tumor metastasis. The restoration of *miR-218* suppresses Robo1 expression and inhibits tumor cell invasion and metastasis *in vitro* and *in vivo*. *miR-218* expression is decreased along with the expression of one of its host genes, Slit3 in metastatic GC. However, Robo1, one of several Slit receptors, is negatively regulated by *miR-218*, thus establishing a negative feedback loop. Decreased *miR-218* levels eliminate Robo1 repression, which activates the Slit-Robo1 pathway through the interaction between Robo1 and Slit2, thus triggering tumor metastasis (Tie et al. 2010). *Let-7f* was downregulated in the highly metastatic potential gastric cancer cell lines GC9811-P and SGC7901-M, when compared with their parental cell lines, GC9811 and SGC7901-NM. Luciferase reporter assays demonstrated that let-7f directly binds to the 3'UTR of MYH9, which codes for myosin IIA

(Liang et al. 2011). Low expression of *miR-335* was significantly associated with lymph-node metastasis, poor pT stage, poor pN stage and invasion of lymphatic vessels. Overexpression of *miR-335* suppressed gastric cancer cell invasion and metastasis *in vitro* and *in vivo*, but has no significant effects on cell proliferation. Furthermore, *miR-335* might suppress gastric cancer invasion and metastasis by targeting Bcl-w and specificity protein 1 (SP1) (Xu et al. 2011). *miR-148a* functions as a tumor metastasis suppressor in gastric cancer, and downregulation of *miR-148a* contributes to gastric cancer lymph node-metastasis and progression. Luciferase assays confirmed that *miR-148a* could directly bind to the 2 sites of 3' untranslated region of ROCK1. Moreover, in gastric cancer tissues, an inverse correlation between *miR-148a* and ROCK1 expression was further observed. Knockdown of ROCK1 significantly inhibited gastric cancer cell migration and invasion resembling that of *miR-148a* overexpression (Zheng et al. 2011b). Even more, DICER1 is found to be a direct target of *miR-107* in which upregulation of DICER1 resulted in a dramatic reduction of *in vitro* migration, invasion, liver metastasis of nude mice, which is similar to that occurs with the silencing of *miR-107*. Furthermore, the restoration of DICER1 can inhibit *miR-107*-induced gastric cancer cell invasion and metastasis, indicating that DICER1 functions as a metastasis suppressor in gastric cancer (Li et al. 2011b).

5. EPIGENETIC MODULATION OF miRNAs IN HUMAN GASTRIC CANCER

Epigenetic modifications control gene expression mainly by DNA methylation or histone modulation. DNA methylation occurs in specific genomic areas called CpG-islands, which are commonly present in the promoter area of the gene. Methylation of CpG-island is triggered by DNA methyltransferases (DNMTs) and histone modifications are catalyzed by histone deacetylases (HDACs) and histone methyltransferases (HMTs). Tumor genes are globally hypomethylated compared with those of normal tissues, and methylation of CpG islands in the gene promoter area results in inactivation of tumor suppressor genes (Herman and Baylin 2003). Thus, epigenetic modifications could be involved in carcinogenesis, in addition to other well-defined genetic mechanisms, such as gene mutations and loss of deficiency of heterozygosity. It was demonstrated recently that certain genes, in particular those with hypermethylated promoters, require Dicer to maintain the epigenetic status (Herman and Baylin 2003). As mentioned above, Dicer is a key enzyme in microRNA biogenesis (Davalos and Esteller 2010). MicroRNAs act as negative regulators of gene expression, and the altered expression of microRNAs by epigenetic mechanisms is strongly implicated in carcinogenesis. To identify epigenetically silenced miRNAs in gastric cancer (GC), miRNAs were screened by the treatment with 5-aza-2'-deoxycytidine (5-Aza-CdR) and 4-phenylbutyrate. Hirumo et al. found that *miR-34b* and *miR-34c* were epigenetically silenced in GC and that their downregulation is associated with hypermethylation of the neighboring CpG island (Suzuki et al. 2010). The methylation-silenced expression of *miR-34b, miR-127-3p, miR-129-3p* and *miR-409* could be reactivated in gastric cancer cells by treatment with 5-Aza-CdR and trichostatin A in a time-dependent manner. Analysis of the methylation status of these miRNAs showed that the upstream CpG-rich regions of *mir-34b* and *mir-129-2* are frequently methylated in gastric cancer tissues compared to adjacent normal tissues, and their methylation status correlated inversely with their expression patterns (Tsai et al. 2011). By the same method, five miRNAs were found to be upregulated (>3 fold) after 5-aza-CdR treatment compared with untreated cells. Among them, *miR-181c* and *miR-432AS* exhibited CpG islands in their upstream sequences on computational analysis, and their upregulation was verified by reverse transcription–polymerase chain reaction analyses (Hashimoto et al. 2010). By using a methylation array and bisulfate pyrosequencing analysis, Kim et al. found that *miR-10b* promoter CpGs were heavily methylated in gastric cancers. After 5-Aza-CdR treatment of gastric cancer cells, *miR-10b* methylation was significantly decreased, and expression of *miR-10b* and HOXD4, which is 1 kb downstream of *miR-10b*, was greatly restored. Moreover, decreased MAPRE1 expression coincided with increased *miR-10b* expression, suggesting that *miR-10b* targets MAPRE1 transcription (Kim et al. 2011).

Saito et al. found that DNA demethylation and HDAC inhibition at Alu repeats would activate silenced *miR-512-5p* by RNA polymerase II. In addition, activation of *miR-512-5p* by epigenetic treatment induces suppression of Mcl-1, resulting in apoptosis of gastric cancer cells, suggesting that chromatin remodeling at Alu repeats plays critical roles in the regulation of miRNA expression and that epigenetic activation of silenced Alu-associated miRNAs could be a novel therapeutic approach for gastric cancer (Saito et al. 2009). Enhancer of zeste homolog 2 (EZH2) is a mammalian histone methyltransferase that contributes to the epigenetic silencing of target genes and regulates the survival and metastasis of cancer cells. Analysis of human prostate tumors revealed that *miR-101* expression decreases during cancer progression, paralleling an increase in EZH2 expression. One or both of the 2 genomic loci encoding *miR-101* were somatically lost in 37.5% of clinically localized prostate cancer cells (6 of 16) and 66.7% of metastatic disease cells (22 of 33). Genomic loss of *miR-101* in cancer leads to overexpression of EZH2 and concomitant deregulation of epigenetic pathways, resulting in cancer progression (Varambally et al. 2008). Human epidermal growth factor receptor 2 (HER2/neu, ERBB2) is overexpressed on many types of cancer cells, including gastric cancer cells; HER2 overexpression has been associated with metastasis and poor prognosis. Bao et al. found that CD44 binds directly to HER2, which up-regulates the expression of metastasis-associated protein-1, induces deacetylation of histone H3 lysine 9, and suppresses transcription of *microRNA139* (miR-139) to inhibit expression of its target gene, C-X-C chemokine receptor type 4 (CXCR4). Cultures of different types of metastatic cancer cells with histone deacetylase inhibitors and/or DNA methyltransferase resulted in upregulation of *miR-139*. HER2 interaction with CD44 up-regulates CXCR4 by inhibiting expression of *miR-139*, at the epigenetic level, in gastric cancer cells. These findings indicate HER2 signaling might promote gastric tumor progression and metastasis (Bao et al. 2011). SOX4 was up-regulated in gastric cancer compared to benign gastric tissues. To further elucidate the molecular mechanisms underlying upregulation of SOX4 in gastric cancers, Ruizhe et al. analyzed the expression of *miR-129-2* gene, the epigenetic repression of which leads to overexpression of SOX4 in endometrial cancer. Upregulation of SOX4 was inversely associated with the epigenetic silencing of *miR-129-2* in gastric cancer, and restoration of *miR-129-2* down-regulated SOX4 expression (Shen et al. 2010).

6. CONCLUSIONS

MicroRNAs (MiRNAs) are short (20 to 23-nucleotide), non-coding RNAs that control protein expression through various mechanisms in human gastric cancer development and progression. The mechanisms by which miRNAs takes part in tumor promotion and progression are various and complex including cell proliferation, cell cycle regulation, cell apoptosis, metastasis and invasion. Gene profiling studies have demonstrated a number of significantly deregulated miRNAs and identified signatures of both diagnostic and prognostic value in gastric cancer. Mapping of all human miRNAs and the availability of complete miRNAs microarrays will be required to address the question that whether miRNAs profiling can be used for cancer prognosis. miRNAs expression can be regulated by different epigenetic mechanisms, including changes in DNA methylation and histone modification in promoter regions. It is possible that understanding of miRNA deregulation and the associated abnormalities in cellular signaling in human gastric cancer may provide a strategy for the better understanding of cancer treatment and improve the survival of patients.

7. SUMMARY POINTS

- Gastric cancer is the second most common cause of cancer death. Elucidating the molecular mechanisms of gastric cancer may illuminate a prognostic strategy to block gastric cancer development.
- MicroRNAs are highly conserved small non-coding RNAs that trigger translational repression and/or mRNA degradation mostly through complementary binding to the 3'-untranslated (3'-UTR) regions of target mRNAs.
- Recently, microRNAs have been characterized as novel proto-oncogenes or tumor suppressors in carcinogenesis. Identifying the miRNAs and their targets that are essential for cancer progression may provide promising therapeutic opportunities.
- MicroRNAs affect many cellular processes that are routinely altered in cancer, such as differentiation, cell-cycle progression, proliferation, apoptosis, metastasis, invasiveness and telomere maintenance.
- MicroRNA expression can be regulated by different epigenetic mechanisms, including changes in DNA methylation and histone modification in promoter regions.

ACKNOWLEDGEMENTS

This work was funded by Key Science and Technology Major Program of Shaanxi Province, China (2010ZDKG-50), National Natural Science Foundation of China (31100921), and The Fundamental Research Funds for the Central Universities (08142006).

ABBREVIATIONS

GC	:	gastric cancer
miRNAs	:	microRNAs
UTR	:	untranslated region
RISC	:	RNA-induced silencing complex
SP1	:	specificity protein 1
CDK	:	cyclin-dependent kinase
TNM	:	tumor-node-metastasis
5-Aza-CdR	:	5-aza-2'-deoxycytidine
EMT	:	epithelial to mesenchymal transition
DNMTs	:	DNA methyltransferases
HDACs	:	histone deacetylases
HMTs	:	histone methyltransferases
APAF	:	apoptosis-activating factor
TGF-β	:	transforming growth factor β

REFERENCES

Ahn, S.M., J.Y. Cha, J. Kim, D. Kim, H.T. Trang, Y.M. Kim, Y.H. Cho, D. Park, and S. Hong. 2012. Smad3 regulates E-cadherin via miRNA-200 pathway. Oncogene. 31(25): 3051–3059.

Akao, Y., Y. Nakagawa, and T. Naoe. 2006. MicroRNAs 143 and 145 are possible common onco-microRNAs in human cancers. Oncology reports. 16: 845–850.

Altuvia, Y., T. Landgraf, G. Lithwick, N. Elefant, S. Pfeffer, A. Aravin, M.J. Brownstein, T. Tuschl, and H. Margalit. 2005. Clustering and conservation patterns of human microRNAs. Nucleic Acids Research. 33: 2697–2706.

Arndt, G.M., L. Dossey, L.M. Cullen, A. Lai, R. Druker, M. Eisbacher, C. Zhang, N. Tran, H. Fan, K. Retzlaff et al. 2009. Characterization of global microRNA expression reveals oncogenic potential of miR-145 in metastatic colorectal cancer. BMC Cancer. 9: 374.

Bao, W., H.J. Fu, Q.S. Xie, L. Wang, R. Zhang, Z.Y. Guo, J. Zhao, Y.L. Meng, X.L.Ren, T. Wang et al. 2011. HER2 interacts with CD44 to up-regulate CXCR4 via epigenetic silencing of microRNA-139 in gastric cancer cells. Gastroenterology. 141: 2076–2087 e2076.

Bartel, D.P. 2004. MicroRNAs: genomics, biogenesis, mechanism, and function. Cell. 116: 281–297.

Besson, A., S.F. Dowdy, and J.M. Roberts. 2008. CDK inhibitors: cell cycle regulators and beyond. Developmental Cell. 14: 159–169.

Bhattacharyya, S.N., R.Habermacher, U. Martine, E.I. Closs, and W. Filipowicz. 2006. Stress-induced reversal of microRNA repression and mRNA P-body localization in human cells. Cold Spring Harbor Symposia on Quantitative Biology. 71: 513–521.

Calin, G.A., C. Sevignani, C.D. Dumitru, T. Hyslop, E. Noch, S. Yendamuri, M. Shimizu, S. Rattan, F. Bullrich, M. Negrini et al. 2004. Human microRNA genes are frequently located at fragile sites and genomic regions involved in cancers. Proceedings of the National Academy of Sciences of the United States of America. 101: 2999–3004.

Chan, M.C., A.C. Hilyard, C. Wu, B.N. Davis, N.S. Hill, A. Lal, J. Lieberman, G. Lagna, and A. Hata. 2010. Molecular basis for antagonism between PDGF and the TGFbeta family of signalling pathways by control of miR-24 expression. The EMBO Journal. 29: 559–573.

Chen, Y., J. Zuo, Y. Liu, H. Gao, and W. Liu. 2010. Inhibitory effects of miRNA-200c on chemotherapy-resistance and cell proliferation of gastric cancer SGC7901/DDP cells. Chinese Journal of Cancer. 29: 1006–1011.

Cho, W.J., J.M. Shin, J.S. Kim, M.R. Lee, K.S. Hong, J.H. Lee, K.H. Koo, J.W. Park, and K.S. Kim. 2009. miR-372 regulates cell cycle and apoptosis of ags human gastric cancer cell line through direct regulation of LATS2. Molecules and Cells 28: 521–527.

Danaei, G., S. Vander Hoorn, A.D. Lopez, C.J. Murray, and M. Ezzati. 2005. Causes of cancer in the world: comparative risk assessment of nine behavioural and environmental risk factors. Lancet. 366: 1784–1793.

Davalos, V. and M. Esteller. 2010. MicroRNAs and cancer epigenetics: a macrorevolution. Current Opinion in Oncology. 22: 35–45.

Farinati, F., R. Cardin, M. Cassaro, M. Bortolami, D. Nitti, C. Tieppo, G. Zaninotto, and M. Rugge. 2008. Helicobacter pylori, inflammation, oxidative damage and gastric cancer: a morphological, biological and molecular pathway. Eur J Cancer Prev. 17: 195–200.

Feng, L., Y. Xie, H. Zhang, and Y. Wu. 2012. miR-107 targets cyclin-dependent kinase 6 expression, induces cell cycle G1 arrest and inhibits invasion in gastric cancer cells. Med Oncol. 29(2): 856–863.

Feng, R., X. Chen, Y. Yu, L. Su, B. Yu, J. Li, Q. Cai, M. Yan, B. Liu, and Z. Zhu. 2010. miR-126 functions as a tumour suppressor in human gastric cancer. Cancer Letters. 298: 50–63.

Fiers, W., R. Beyaert, E. Boone, S. Cornelis, W. Declercq, E. Decoster, G. Denecker, B. Depuydt, D. De Valck, G. De Wilde et al. 1995. TNF-induced intracellular signaling leading to gene induction or to cytotoxicity by necrosis or by apoptosis. Journal of Inflammation. 47: 67–75.

Gao, C., Z. Zhang, W. Liu, S. Xiao, W. Gu, and H. Lu. 2010. Reduced microRNA-218 expression is associated with high nuclear factor kappa B activation in gastric cancer. Cancer. 116: 41–49.

Gottlieb, T.M., J.F. Leal, R. Seger, Y. Taya, and M. Oren. 2002. Cross-talk between Akt, p53 and Mdm2: possible implications for the regulation of apoptosis. Oncogene. 21: 1299–1303.

Gregory, R.I., K.P. Yan, G. Amuthan, T. Chendrimada, B. Doratotaj, N. Cooch, and R. Shiekhattar. 2004. The Microprocessor complex mediates the genesis of microRNAs. Nature. 432: 235–240.

Gulbins, E., R. Bissonnette, A. Mahboubi, S. Martin, W. Nishioka, T. Brunner, G. Baier, G. Baier-Bitterlich, C. Byrd, F. Lang et al. 1995. FAS-induced apoptosis is mediated via a ceramide-initiated RAS signaling pathway. Immunity. 2: 341–351.

Guo, S.L., Z. Peng, X. Yang, K.J. Fan, H. Ye, Z.H. Li, Y. Wang, X.L. Xu, J. Li, Y.L. Wang et al. 2011. miR-148a promoted cell proliferation by targeting p27 in gastric cancer cells. International Journal of Biological Sciences. 7: 567–574.

Guo, X., L. Guo, J. Ji, J. Zhang, J. Zhang, X. Chen, Q. Cai, J. Li, Q. Gu, B. Liu et al. 2010. miRNA-331-3p directly targets E2F1 and induces growth arrest in human gastric cancer. Biochemical and Biophysical Research Communications. 398: 1–6.

Hashimoto, Y., Y. Akiyama, T. Otsubo, S. Shimada, and Y. Yuasa. 2010. Involvement of epigenetically silenced microRNA-181c in gastric carcinogenesis. Carcinogenesis. 31: 777–784.

Hayashita, Y., H. Osada, Y. Tatematsu, H. Yamada, K. Yanagisawa, S. Tomida, Y. Yatabe, K. Kawahara, Y. Sekido, and T. Takahashi. 2005. A polycistronic microRNA cluster, miR-17-92, is overexpressed in human lung cancers and enhances cell proliferation. Cancer Research. 65: 9628–9632.

He, H., K. Jazdzewski, W. Li, S. Liyanarachchi, R. Nagy, S. Volinia, G.A. Calin, C.G. Liu, K. Franssila, S. Suster et al. 2005. The role of microRNA genes in papillary thyroid carcinoma. Proceedings of the National Academy of Sciences of the United States of America. 102: 19075–19080.

Herman, J.G. and S.B. Baylin. 2003. Gene silencing in cancer in association with promoter hypermethylation. The New England Journal of Medicine. 349: 2042–2054.

Iorio, M.V., M. Ferracin, C.G. Liu, A. Veronese, R. Spizzo, S. Sabbioni, E. Magri, M. Pedriali, M. Fabbri, M. Campiglio et al. 2005. MicroRNA gene expression deregulation in human breast cancer. Cancer Research. 65: 7065–7070.

Ji, Q., X. Hao, Y. Meng, M. Zhang, J. Desano, D. Fan, and L. Xu. 2008. Restoration of tumor suppressor miR-34 inhibits human p53-mutant gastric cancer tumorspheres. BMC Cancer. 8: 266.

Kent, O.A. and J.T. Mendell. 2006. A small piece in the cancer puzzle: microRNAs as tumor suppressors and oncogenes. Oncogene. 25: 6188–6196.

Kim, K., H.C. Lee, J.L. Park, M. Kim, S.Y. Kim, S.M. Noh, K.S. Song, J.C. Kim, and Y.S. Kim. 2011. Epigenetic regulation of microRNA-10b and targeting of oncogenic MAPRE1 in gastric cancer. Epigenetics: Official Journal of the DNA Methylation Society. 6: 740–751.

Kim, V.N., J. Han, and M.C. Siomi. 2009a. Biogenesis of small RNAs in animals. Nature Reviews Molecular Cell Biology. 10: 126–139.

Kim, Y.K., J. Yu, T.S. Han, S.Y. Park, B. Namkoong, D.H. Kim, K. Hur, M.W. Yoo, H.J. Lee, H.K. Yang et al. 2009b. Functional links between clustered microRNAs: suppression of cell-cycle inhibitors by microRNA clusters in gastric cancer. Nucleic Acids Research. 37: 1672–1681.

Kogo, R., K. Mimori, F. Tanaka, S. Komune, and M. Mori. 2011. Clinical significance of miR-146a in gastric cancer cases. Clinical cancer research: An Official Journal of the American Association for Cancer Research. 17: 4277–4284.

Lai, K.W., K.X. Koh, M. Loh, K. Tada, M.M. Subramaniam, X.Y. Lim, A. Vaithilingam, M. Salto-Tellez, B. Iacopetta, Y. Ito et al. 2010. MicroRNA-130b regulates the tumour suppressor RUNX3 in gastric cancer. Eur J Cancer. 46: 1456–1463.

Lee, S.E., K.W. Ryu, B.H. Nam, J.H. Lee, I.J. Choi, M.C. Kook, S.R. Park, and Y.W. Kim. 2009. Prognostic significance of intraoperatively estimated surgical stage in curatively resected gastric cancer patients. J Am Coll Surg. 209: 461–467.

Lee, Y., K. Jeon, J.T. Lee, S. Kim, and V.N. Kim. 2002. MicroRNA maturation: stepwise processing and subcellular localization. The EMBO Journal. 21: 4663–4670.

Lee, Y., M. Kim, J. Han, K.H. Yeom, S. Lee, S.H. Baek, and V.N. Kim. 2004. MicroRNA genes are transcribed by RNA polymerase II. The EMBO Journal. 23: 4051–4060.

Lee, Y., J. Han, K.H. Yeom, H. Jin, and V.N. Kim. 2006. Drosha in primary microRNA processing. Cold Spring Harbor Symposia on Quantitative Biology. 71: 51–57.

Li, J. and G. Li. 2010. Cell cycle regulator ING4 is a suppressor of melanoma angiogenesis that is regulated by the metastasis suppressor BRMS1. Cancer Research. 70: 10445–10453.

Li, X., F. Luo, Q. Li, M. Xu, D. Feng, G. Zhang, and W. Wu. 2011a. Identification of new aberrantly expressed miRNAs in intestinal-type. Oncology Reports. 26: 1431–1439.

Li, X., Y. Zhang, Y. Shi, G. Dong, J. Liang, Y. Han, X. Wang, Q. Zhao, J. Ding, K. Wu et al. 2011b. MicroRNA-107, an oncogene microRNA that regulates tumour invasion and metastasis by targeting DICER1 in gastric cancer. Journal of Cellular and Molecular Medicine. 15: 1887–1895.

Li, X., Y. Zhang, H. Zhang, X. Liu, T. Gong, M. Li, L. Sun, G. Ji, Y. Shi, Z. Han et al. 2011c. miRNA-223 promotes gastric cancer invasion and metastasis by targeting tumor suppressor EPB41L3. Molecular Cancer Research: MCR. 9: 824–833.

Liang, S., L. He, X. Zhao, Y. Miao, Y. Gu, C. Guo, Z. Xue, W. Dou, F. Hu, K. Wu et al. 2011. MicroRNA let-7f inhibits tumor invasion and metastasis by targeting MYH9 in human gastric cancer. PloS one. 6: e18409.

Liu, T., H. Tang, Y. Lang, M. Liu, and X. Li. 2009. MicroRNA-27a functions as an oncogene in gastric adenocarcinoma by targeting prohibitin. Cancer Letters. 273: 233–242.

Lo, S.S., P.S. Hung, J.H. Chen, H.F. Tu, W.L. Fang, C.Y. Chen, W.T. Chen, N.R. Gong, and C.W. Wu. 2011. Overexpression of miR-370 and downregulation of its novel target TGFbeta-RII contribute to the progression of gastric carcinoma. Oncogene. 31: 226–237.

Lu, J., G. Getz, E.A. Miska, E. Alvarez-Saavedra, J. Lamb, D. Peck, A. Sweet-Cordero, B.L. Ebert, R.H. Mak, A.A. Ferrando et al. 2005. MicroRNA expression profiles classify human cancers. Nature. 435: 834–838.

Min, C., S.F. Eddy, D.H. Sherr, and G.E. Sonenshein. 2008. NF-kappaB and epithelial to mesenchymal transition of cancer. Journal of Cellular Biochemistry. 104: 733–744.

Murakami, Y., T. Yasuda, K. Saigo, T. Urashima, H. Toyoda, T. Okanoue, and K. Shimotohno. 2006. Comprehensive analysis of microRNA expression patterns in hepatocellular carcinoma and non-tumorous tissues. Oncogene. 25: 2537–2545.

Muschel, R.J., E.J. Bernhard, L. Garza, W.G. McKenna, and C.J. Koch. 1995. Induction of apoptosis at different oxygen tensions: evidence that oxygen radicals do not mediate apoptotic signaling. Cancer Research. 55: 995–998.

Nomura, T. and N. Katunuma. 2005. Involvement of cathepsins in the invasion, metastasis and proliferation of cancer cells. The Journal of Medical Investigation: JMI. 52: 1–9.

Oh, H.K., A.L. Tan, K. Das, C.H. Ooi, N.T. Deng, I.B. Tan, E. Beillard, J. Lee, K. Ramnarayanan, S.Y. Rha et al. 2011. Genomic loss of miR-486 regulates tumor progression and the OLFM4 antiapoptotic factor in gastric cancer. Clinical cancer research: An Official Journal of the American Association for Cancer Research. 17: 2657–2667.

Ohshima, K., K. Inoue, A. Fujiwara, K. Hatakeyama, K. Kanto, Y. Watanabe, K. Muramatsu, Y. Fukuda, S. Ogura, K. Yamaguchi et al. 2010. Let-7 microRNA family is selectively secreted into the extracellular environment via exosomes in a metastatic gastric cancer cell line. PloS one. 5: e13247.

Orange, J.S., O. Levy, and R.S. Geha. 2005. Human disease resulting from gene mutations that interfere with appropriate nuclear factor-kappaB activation. Immunological Reviews. 203: 21–37.

Petrocca, F., R. Visone, M.R. Onelli, M.H. Shah, M.S. Nicoloso, I. de Martino, D. Iliopoulos, E. Pilozzi, C.G. Liu, M. Negrini et al. 2008. E2F1-regulated microRNAs impair TGFbeta-dependent cell-cycle arrest and apoptosis in gastric cancer. Cancer Cell. 13: 272–286.

Ren, Y., X. Zhou, M. Mei, X.B. Yuan, L. Han, G.X. Wang, Z.F. Jia, P. Xu, P.Y. Pu, and C.S. Kang. 2010. MicroRNA-21 inhibitor sensitizes human glioblastoma cells U251 (PTEN-mutant) and LN229 (PTEN-wild type) to taxol. BMC Cancer. 10: 27.

Saito, Y., H. Suzuki, H. Tsugawa, I. Nakagawa, J. Matsuzaki, Y. Kanai, and T. Hibi 2009. Chromatin remodeling at Alu repeats by epigenetic treatment activates silenced microRNA-512-5p with downregulation of Mcl-1 in human gastric cancer cells. Oncogene. 28: 2738–2744.

Satyanarayana, A. and P. Kaldis. 2009. Mammalian cell-cycle regulation: several Cdks, numerous cyclins and diverse compensatory mechanisms. Oncogene. 28: 2925–2939.

Scartozzi, M., M. Pistelli, A. Bittoni, R. Giampieri, E. Galizia, R. Berardi, L. Faloppi, M. Del Prete, and S. Cascinu. 2010. Novel perspectives for the treatment of gastric cancer: from a global approach to a personalized strategy. Current Oncology Reports. 12: 175–185.

Shao, J.C., J.F. Wu, D.B. Wang, R. Qin, and H. Zhang. 2003. Relationship between the expression of human telomerase reverse transcriptase gene and cell cycle regulators in gastric cancer and its significance. World Journal of Gastroenterology: WJG 9: 427–431.

Shen, R., S. Pan, S. Qi, X. Lin, and S. Cheng. 2010. Epigenetic repression of microRNA-129-2 leads to overexpression of SOX4 in gastric cancer. Biochemical and Biophysical Research Communications. 394: 1047–1052.

Shi, B., L. Sepp-Lorenzino, M. Prisco, P. Linsley, T. deAngelis, and R. Baserga. 2007. MicroRNA 145 targets the insulin receptor substrate-1 and inhibits the growth of colon cancer cells. The Journal of Biological Chemistry. 282: 32582–32590.

Shin, V.Y., H. Jin, E.K. Ng, A.S. Cheng, W.W. Chong, C.Y. Wong, W.K. Leung, J.J. Sung, and K.M. Chu. 2011. NF-kappaB targets miR-16 and miR-21 in gastric cancer: involvement of prostaglandin E receptors. Carcinogenesis. 32: 240–245.

Shinozaki, A., T. Sakatani, T. Ushiku, R. Hino, M. Isogai, S. Ishikawa, H. Uozaki, K. Takada, and M. Fukayama. 2010. Downregulation of microRNA-200 in EBV-associated gastric carcinoma. Cancer Research. 70: 4719–4727.

Song, G., H. Zeng, J. Li, L. Xiao, Y. He, Y. Tang, and Y. Li. 2010. miR-199a regulates the tumor suppressor mitogen-activated protein kinase kinase kinase 11 in gastric cancer. Biological & Pharmaceutical Bulletin. 33: 1822–1827.

Song, Y.X., Z.Y. Yue, Z.N. Wang, Y.Y. Xu, Y. Luo, H.M. Xu, X. Zhang, L. Jiang, C.Z. Xing, and Y. Zhang. 2011. MicroRNA-148b is frequently down-regulated in gastric cancer and acts as a tumor suppressor by inhibiting cell proliferation. Molecular Cancer. 10: 1.

Stoletov, K.V., C. Gong, and B.I. Terman. 2004. Nck and Crk mediate distinct VEGF-induced signaling pathways that serve overlapping functions in focal adhesion turnover and integrin activation. Experimental Cell Research. 295: 258–268.

Suzuki, H., E. Yamamoto, M. Nojima, M. Kai, H.O. Yamano, K. Yoshikawa, T. Kimura, T. Kudo, E. Harada, T. Sugai et al. 2010. Methylation-associated silencing of microRNA-34b/c in gastric cancer and its involvement in an epigenetic field defect. Carcinogenesis. 31: 2066–2073.

Takamizawa, J., H. Konishi, K. Yanagisawa, S. Tomida, H. Osada, T. Endoh, T. Harano, Y. Yatabe, M. Nagino, Y. Nimura et al. 2004. Reduced expression of the let-7 microRNAs in human lung cancers in association with shortened postoperative survival. Cancer Research. 64: 3753–3756.

Tie, J., Y. Pan, L. Zhao, K. Wu, J. Liu, S. Sun, X. Guo, B. Wang, Y. Gang, Y. Zhang et al. 2010. MiR-218 inhibits invasion and metastasis of gastric cancer by targeting the Robo1 receptor. PLoS Genetics 6: e1000879.

Tijsterman, M. and R.H. Plasterk. 2004. Dicers at RISC; the mechanism of RNAi. Cell. 117: 1–3.

Tsai, K.W., C.W. Wu, L.Y. Hu, S.C. Li, Y.L. Liao, C.H. Lai, H.W. Kao, W.L. Fang, K.H. Huang, W.C. Chan et al. 2011. Epigenetic regulation of miR-34b and miR-129 expression in gastric cancer. International Journal of Cancer Journal International du Cancer. 129: 2600–2610.

Tsukamoto, Y., C. Nakada, T. Noguchi, M. Tanigawa, L.T. Nguyen, T. Uchida, N. Hijiya, K. Matsuura, T. Fujioka, M. Seto et al. 2010. MicroRNA-375 is downregulated in gastric carcinomas and regulates cell survival by targeting PDK1 and 14-3-3zeta. Cancer Research. 70: 2339–2349.

Ueda, T., S. Volinia, H. Okumura, M. Shimizu, C. Taccioli, S. Rossi, H. Alder, C.G. Liu, N. Oue, W. Yasui et al. 2010. Relation between microRNA expression and progression and prognosis of gastric cancer: a microRNA expression analysis. The Lancet Oncology. 11: 136–146.

Varambally, S., Q. Cao, R.S. Mani, S. Shankar, X. Wang, B. Ateeq, B. Laxman, X. Cao, X. Jing, K. Ramnarayanan et al. 2008. Genomic loss of microRNA-101 leads to overexpression of histone methyltransferase EZH2 in cancer. Science. 322: 1695–1699.

Villadsen, S.B., J.B. Bramsen, M.S. Ostenfeld, E.D. Wiklund, N. Fristrup, S. Gao, T.B. Hansen, T.I. Jensen, M. Borre, T.F. Orntoft et al. 2011. The miR-143/-145 cluster regulates plasminogen activator inhibitor-1 in bladder cancer. British Journal of Cancer. 106: 366–374.

Volinia, S., G.A. Calin, C.G. Liu, S. Ambs, A. Cimmino, F. Petrocca, R. Visone, M. Iorio, C. Roldo, M. Ferracin et al. 2006. A microRNA expression signature of human solid tumors defines cancer gene targets. Proceedings of the National Academy of Sciences of the United States of America. 103: 2257–2261.

Wan, H.Y., L.M. Guo, T. Liu, M. Liu, X. Li, and H. Tang. 2010. Regulation of the transcription factor NF-kappaB1 by microRNA-9 in human gastric adenocarcinoma. Molecular Cancer. 9: 16.

Wang, Z., Y.L. He, S.R. Cai, W.H. Zhan, Z.R. Li, B.H. Zhu, C.Q. Chen, J.P. Ma, Z.X. Chen, W. Li et al. 2008. Expression and prognostic impact of PRL-3 in lymph node metastasis of gastric cancer: its molecular mechanism was investigated using artificial microRNA interference. International Journal of Cancer Journal International du Cancer. 123: 1439–1447.

Wu, Q., H. Jin, Z. Yang, G. Luo, Y. Lu, K. Li, G. Ren, T. Su, Y. Pan, B. Feng et al. 2010a. MiR-150 promotes gastric cancer proliferation by negatively regulating the pro-apoptotic gene EGR2. Biochemical and Biophysical Research Communications. 392: 340–345.

Wu, W.K., C.W. Lee, C.H. Cho, D. Fan, K. Wu, J. Yu, and J.J. Sung. 2010b. MicroRNA dysregulation in gastric cancer: a new player enters the game. Oncogene. 29: 5761–5771.

Xia, L., D. Zhang, R. Du, Y. Pan, L. Zhao, S. Sun, L. Hong, J. Liu, and D. Fan 2008. miR-15b and miR-16 modulate multidrug resistance by targeting BCL2 in human gastric cancer cells. International Journal of Cancer Journal International du Cancer. 123: 372–379.

Xu, Y., F. Zhao, Z. Wang, Y. Song, Y. Luo, X. Zhang, L. Jiang, Z. Sun, Z. Miao, and H. Xu. 2012. MicroRNA-335 acts as a metastasis suppressor in gastric cancer by targeting Bcl-w and specificity protein 1. Oncogene. 31(11): 1398–1407.

Yamada, C., T. Ozaki, K. Ando, Y. Suenaga, K. Inoue, Y. Ito, R. Okoshi, H. Kageyama, H. Kimura, M. Miyazaki et al. 2010. RUNX3 modulates DNA damage-mediated phosphorylation of tumor suppressor p53 at Ser-15 and acts as a co-activator for p53. The Journal of Biological Chemistry. 285: 16693–16703.

Yanaihara, N., N. Caplen, E. Bowman, M. Seike, K. Kumamoto, M. Yi, R.M. Stephens, A. Okamoto, J. Yokota, T. Tanaka et al. 2006. Unique microRNA molecular profiles in lung cancer diagnosis and prognosis. Cancer Cell. 9: 189–198.

Yang, J., S.A. Mani, and R.A. Weinberg. 2006. Exploring a new twist on tumor metastasis. Cancer Research. 66: 4549–4552.

Yano, T., K. Ito, H. Fukamachi, X.Z. Chi, H.J. Wee, K. Inoue, H. Ida, P. Bouillet, A. Strasser, S.C. Bae et al. 2006. The RUNX3 tumor suppressor upregulates Bim in gastric epithelial cells undergoing transforming growth factor beta-induced apoptosis. Molecular and Cellular Biology. 26: 4474–4488.

Yasui, W., K. Sentani, J. Motoshita, and H. Nakayama. 2006. Molecular pathobiology of gastric cancer. Scandinavian journal of surgery: SJS: Official Organ for the Finnish Surgical Society and the Scandinavian Surgical Society. 95: 225–231.

Yi, R., Y. Qin, I.G. Macara, and B.R. Cullen. 2003. Exportin-5 mediates the nuclear export of pre-microRNAs and short hairpin RNAs. Genes & development. 17: 3011–3016.

Zhang, H., Y. Li, and M. Lai. 2010a. The microRNA network and tumor metastasis. Oncogene. 29: 937–948.

Zhang, X., W. Zhu, J. Zhang, S. Huo, L. Zhou, Z. Gu, and M. Zhang. 2010b. MicroRNA-650 targets ING4 to promote gastric cancer tumorigenicity. Biochemical and Biophysical Research Communications. 395: 275–280.

Zheng, B., L. Liang, S. Huang, R. Zha, L. Liu, D. Jia, Q. Tian, Q. Wang, C. Wang, Z. Long et al. 2011a. MicroRNA-409 suppresses tumour cell invasion and metastasis by directly targeting radixin in gastric cancers. Oncogene. Dec 19. doi: 10.1038/onc.2011.581.

Zheng, B., L. Liang, C. Wang, S. Huang, X. Cao, R. Zha, L. Liu, D. Jia, Q. Tian, J. Wu et al. 2011b. MicroRNA-148a Suppresses Tumor Cell Invasion and Metastasis by Downregulating ROCK1 in Gastric Cancer. Clinical cancer research: An Official Journal of the American Association for Cancer Research. 17: 7574–7583.

Zheng, H.C., Y.L. Li, J.M. Sun, X.F. Yang, Li, W.G. Jiang, Y.C. Zhang, and Y. Xin. 2003. Growth, invasion, metastasis, differentiation, angiogenesis and apoptosis of gastric cancer regulated by expression of PTEN encoding products. World Journal of Gastroenterology: WJG. 9: 1662–1666.

CHAPTER 8

MicroRNAs IN COLORECTAL CANCER

K. Valencia,[1] E. Bandres[2,]* and J. Garcia-Foncillas[3]

ABSTRACT

In 1993, Andrew Fire and Craig Mello discovered that some DNA genes rendered RNA molecules which surprisingly did not produce proteins. Such molecules were termed non-coding RNAs (ncRNAs) and were initially considered "junk-DNA". Later, investigators found that some of these molecules could generate certain structures and give rise to a new class of genes called microRNAs (miRNAs) with important regulatory functions. miRNAs are small ncRNAs that regulate gene expression during crucial cell processes. Changes in the expression profiles of miRNAs have been observed in human tumors, including colorectal cancer (CRC). Functional studies indicate that miRNAs act as tumor suppressors and oncogenes. These findings have been extended to Vogelstein's model of CRC carcinogenesis. In this chapter,

[1]Oncology Division, Center for Applied Medical Research (CIMA), University of Navarra, Pamplona, Spain.
Email: kvalencia@alumni.unav.es
[2]Immunology Unit, Hematology Service, Complejo Hospitalario de Navarra, Pamplona, Spain.
Email: ebandres@hotmail.es
[3]Department of Oncology, University Hospital "Fundacion Jimenez Diaz", Autonomous University of Madrid, Madrid, Spain.
Email: jgfoncillas@fjd.es
*Corresponding author

we discuss the expression profile of the miRNAs associated with CRC clinical phenotypes including recurrence, metastasis and therapeutic outcomes. Moreover, we recapitulate the cutting edge progress in the discovery of miRNAs profiles in plasma and feces associated to CRC as a non-invasive method of early diagnosis discussing their utility as novel prognostic and predictive markers. In conclusion, we summarize what is known about the role of miRNAs in CRC while emphasizing their significance in pathogenesis signalling pathways and their potential as disease biomarkers and novel therapeutic targets.

1. INTRODUCTION

1.1 miRNAs: Genomics, Biogenesis and Function

miRNAs represent an abundant class of small ncRNAs involved in regulating many cellular processes, including proliferation, differentiation, apoptosis and development. They regulate gene expression by targeting messenger RNAs (mRNA) through sequence-specific base pairing with the 3'-untranslated regions (3'-UTR) of mRNAs, resulting in RNA degradation and/or translational repression (Ambros 2004). In the latest version of the Sanger Institute miRBase (version 18) (http://microrna.sanger.ac.uk/sequences/index.shtml), to date, 1527 human mature miRNAs sequences have been deposited.

miRNAs genes represent around 1% of the genome of different species. It has been estimated that approximately 30% of the genes are regulated by at least one miRNA (Bartel 2004). Whereas some miRNAs regulate specific individual targets, others can be central regulators of a process, so key miRNAs regulate the expression levels of hundreds of genes simultaneously, and many types of miRNAs cooperate on the regulation of their targets (Esteller 2011).

miRNAs genes are located in all human chromosomes except the Y chromosome. miRNAs genes are evolutionarily conserved. They may be located either within the introns or exons of protein-coding genes (70%) or in intergenic areas (30%). Many miRNAs are encoded as clusters that range from 2 to 19 miRNAs hairpins in close proximity. The genomic location of a miRNA affects its transcriptional units and regulation. Intronic miRNAs located within a host gene, in the same orientation, are transcribed by the same promoter. In contrast, intergenic miRNAs rely on their own promoters (Macfarlane and Murphy 2010). Remarkably, many miRNAs are mapped at or near fragile sites, common breakpoints, and other sites of genomic rearrangements associated with human cancer (Calin et al. 2004).

The mechanism of miRNAs biosynthesis is evolutionarily conserved. miRNAs are transcribed by RNA polymerase-II or III into primary (pri)-miRNAs, which are processed by Drosha/DGCR8 (DiGeorge syndrome critical region) genes into precursor (pre)-miRNAs. The pre-miRNAs is transported from the nucleus into the cytoplasm by Exportin 5, where it is then processed by Dicer/TRBP (Dicer-TAR RNA binding protein) into a miRNA duplex. The duplex is unwound by a helicase and the mature, more stable strand, is incorporated into the RNA-induced silencing complex (RISC). In this complex, the mature miRNAs is able to bind, though partial complementarity, to the 3'UTR of its mRNA target, regulating gene expression at the post-transcriptional level. Depending on miRNAs complementarity to its target mRNA, the RISC mediates downregulation of gene expression by either translational repression or mRNA degradation. If the miRNAs and the mRNA exhibit perfect complementarity, the target mRNA is cleaved by RISC. This is the predominant mechanism for miRNAs functions in plants. Imperfect base pairing between a miRNA and its target, as occurs with most mammalian miRNAs, leads to translational silencing of the target. However, imperfectly complementarity miRNAs can also reduce the abundance of mRNAs (Fig. 1).

Prediction of miRNAs targets is one of the most relevant fields in miRNAs research, given that miRNAs act by regulating target mRNAs (Dai and Zhou 2010). The specificity of miRNAs–mRNA interaction is principally conferred by the first eight nucleotides of a miRNAs (known as the *seed* sequence). However, the likelihood that a predicted target is a *bona fide* target is influenced not only by seed pairing but also by other factors such as the number of target sites, the context of surrounding sequence in mRNAs, and the occlusion of target sites by RNA-binding proteins. Nowadays, several algorithms have been developed to predict miRNAs target interactions. They pay attention to some major features such as the hairpin-shaped stem loop structure, high evolutionary conservation, and high minimal folding free energy to computationally identify miRNAs targets. These programs indicate that each miRNAs potentially regulates hundreds of target mRNAs, and it seems reasonable that most mRNAs are post-transcriptionally regulated by miRNAs. However, the major problem of these computational algorithms remains target over-prediction. Many targets predicted by *in silico* analyses are not confirmed as real targets in biological assays. Thus, the most reliable assay for miRNAs target identification is the demonstration of repression of a luciferase gene reporter as result of miRNAs-3'UTR interaction of the predicted target and that this repression is abrogated by point mutation in the 3'UTR of the target sequence.

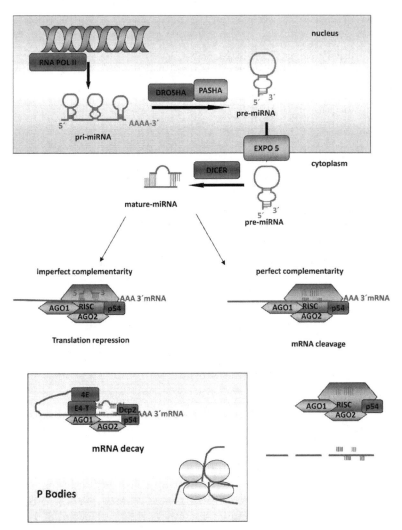

Figure 1. miRNAs biogenesis pathway. miRNAs genes are transcribed by RNA polymerase II in the nucleus to form large pri-miRNAs transcripts. These pri-miRNAs transcripts are processed by the RNase III enzyme Drosha and its co-factor, Pasha, to release the 70-nucleotide pre-miRNAs precursor product. RAN–GTP and exportin 5 transport the pre-miRNAs into the cytoplasm. Subsequently, Dicer processes the pre-miRNAs to generate a transient 22-nucleotide miRNA: miRNA* duplex. This duplex is then loaded into the miRNAs-associated multiprotein RNA-induced silencing complex (miRISC). The mature miRNAs then binds to complementary sites in the mRNA target to negatively regulate gene expression in one of the two ways that depend on the degree of complementarity between the miRNAs and its target. miRNAs that bind to mRNA targets with imperfect complementarity block target gene expression at the level of protein translation. miRNAs that bind to their mRNA targets with perfect complementarity induce target-mRNA cleavage.

Color image of this figure appears in the color plate section at the end of the book.

1.2 Colorectal Cancer (CRC)

CRC is the third most commonly diagnosed cancer and the third leading cause of cancer death in both men and women in the western countries. The 5-year relative survival rate for CRC has increased in the last decade, due to the improvement of the treatment and early detection. However, CRC should be regarded as a heterogeneous, multi-pathway disease, and histologically identical tumors may have different prognoses and not similar response to therapy.

Briefly, the main CRC challenges to date can be summarized as:

1. Survival rate depends on CRC diagnostic stage, so development of new biomarkers for CRC early detection is necessary.
2. New staging system integrating molecular cancer signature are warranted to better stratify patients with different risk of disease recurrence and/or death.
3. Improving the use of current available drugs.
4. Identify cancer targets and develop new therapeutic strategies.

Finally, these critical challenging in CRC showed that the discovery and validation of new cancer biomarkers becomes a decisive challenge in oncology research offering the opportunity of early detection, patient's prognostic stratification and personalized therapy selection.

Dysregulation of miRNAs expression levels emerges as an important mechanism that triggers the loss or gain of functions in cancer cells. These findings highlight the diagnostic and prognostic value of miRNAs and their potential role as therapeutic molecules. We review the findings that, to date, provide evidence of miRNAs therapeutic intervention and, in particular, we emphasize the potential role of miRNAs in CRC.

2. METHODS FOR miRNAs QUANTIFICATION

Some miRNAs features make them technically challenging to obtain miRNAs expression profiles with high sensitivity and specificity. The choice of the most suitable methods for miRNAs detection mainly depends on the specific experimental setting. However, an ideal miRNAs profiling method should be sensitive (to provide quantitative analysis of miRNAs levels, even with small amounts of starting material), specific (to detect reproducible single nucleotide differences between miRNAs), and easy to perform without needing expensive reagents or equipment (Takada and Mano 2007).

Most miRNAs detection methods rely on the Watson–Crick base-pairing between complementary chains of nucleotides, and require hybridization between a strand of nucleic acid and the miRNAs. Thus, there must be a way to translate the hybridization event into a measurable signal. These signals can be detected in several ways, such as electrochemically or by measuring fluorescence and bioluminescence intensities. There are two main hybridization methods: solid-phase (a capture probe is adsorbed or bound to a surface for hybridization to target) and solution-phase (target hybridization occurs in solution methods). While the solid-phase methods are adaptable to high throughput screening and possess high sensitivity, they are often time-consuming. The solution-phase methods are very sensitive, and fast; however, they cannot be applied in high-throughput settings. The current gaps of miRNAs detection include the ability to perform multiplex, sensitive detection of miRNAs with single-nucleotide specificity along with the standardization of these new methods (Cissell and Deo 2009).

Principal technologies used for miRNAs expression profiling include northern blot analysis, microarray methods, quantitative real-time polymerase chain reaction (qRT-PCR) and *in situ* hybridization (Table 1). It should be kept in mind that different methods used to detect miRNAs expression may produce different profiles.

2.1 RNA Extraction

The first step in the analysis of miRNAs is extraction and purification of the RNA from a biological sample. Most RNA isolation kits were developed to recover messenger RNA, while disregarding smaller molecules such as miRNAs. Isolation of miRNAs begins when total RNA, including the small RNA fraction, is isolated from the samples of interest. However, not all isolation methods retain the small RNA fraction. The standard glass fiber filter (GFF) or silicate adsorption method, employed by most RNA isolation kits is inefficient at recovering small RNAs. Therefore, it is important to use RNA isolation methods specifically adapted to retain small RNAs. RNA extraction method may be a source of bias, as demonstrated in a study comparing the expression of miRNAs in samples obtained by three different extraction methods. Although similar expression levels were obtained for most miRNAs analyzed by different extraction methods, the reported data suggest that the method used to isolate RNA may influence the results and this effect appears to be more pronounced in the context of a microarray platform than in northern blots or q-RT-PCR assays (de Planell-Saguer and Rodicio 2011).

Table 1. Comparison of main miRNAs detection methods.

	Northern blot	qRT-PCR	Microarray	ISH
Assay type	Solid-phase	Solution-phase	Solid-phase	
Detection limit	Nanomolar range	Femtomolar range 10^{-22} mol	Femtomolar range 10^{-15}–10^{-18} mol	
Advantages	Quantitative ↑specificity	Quantitative ↑sensitivity ↑accuracy Fast and easy 10 ng–700 ng RNA required	↑throughput Multi-sample test 100 ng–10,000 ng RNA required Fast Computational analysis	↑resolution of a single cell level or subcellular levels
Disadvantages	↓sensitivity ↑ amounts of RNA required Time-consuming ↓throughput	Costly ↓throughput Normalization problems	Non-quantitative ↓ sensitivity ↓specificity	Slow analysis Semi-quantitative Technically difficult ↓throughput
Application	Examination of their target Determination of their sizes	Relative and absolute Quantification Profiling approaches validation MiRNAs expression from cells and tissues Clinical application	High-throughput miRNA Profiling Differential miRNAs expression MiRNA biomarker identification	Direct detection of miRNA in living cells or FFPE tissues Subcellular location information

2.2 Northern Blot

The first used miRNAs detection method was northern blotting. Initially, the sample containing RNA is run on an electrophoresis gel. Next, the RNA is transferred to a nitrocellulose membrane, and then soaked in a solution containing a fluorescent or radiolabeled oligonucleotide probe which is complementary to the target miRNAs for hybridization to occur. This method, however, is time-consuming and has low sensitivity (Cissell and Deo 2009). Mature miRNAs molecules are very short and their prevalence in total RNA is also very low. This fact is translated into a poor sensitivity of routine northern blot analysis. To improve the sensitivity and total assay time for northern blotting, Varallyay et al. introduced LNA (locked-nucleic acid) probes into the complementary sequence. LNAs contain a 2'-O-4'C methylene bridge on the furanose ring, which results in a more rigid and thermally stable double-stranded nucleic acid structure. This could prevent the dissociation of the hybrid between target miRNAs and antisense miRNAs. Substituting LNAs every third nucleotide in the complementary sequence greatly increased the sensitivity of miRNAs detection (Varallyay et al. 2008). With this technology, the assay was improved from 16hr to 2hr hybridization time. Furthermore, due to the thermal stability, LNA probes have reached increased hybridization temperature (50°C) as compared with the 37°C for the DNA oligonucleotide probes. In this study, the authors showed that the optimized high resolution northern blotting can be used to analyze endogenous and exogenous RNAs in the range of 20–70 nt. They demonstrated the usefulness of this technique to investigate miRNAs biogenesis as well as to characterize RNAi technology reagents (Koscianska et al. 2011).

2.3 Microarrays

Microarray is a high throughput technology largely used (Calin and Croce 2006). In a microarray, oligonucleotides, spotted in a solid-phase, complementary with the target miRNAs are used to detect miRNAs levels from a sample. The cDNA is produced by reverse transcription using biotinylated-labeled primers or fluorescent-labeled primers. Once the labeled cDNA samples have been, hybridized, microarrays are washed to remove any unhybridized DNA. Afterwards, if the hybridized cDNA is biotinylated, a streptavidin-labeled fluorophore can be added. If the cDNA has already been labeled with a fluorophore, the fluorescence intensity of each well can be measured. The emission wavelength and intensity of each spot determines the level of expression of each target miRNAs from the initial RNA sample. Although microarrays can analyze thousands of samples in a day, they are quite expensive (Curry et al. 2009). Signal intensities of miRNAs microarray experiments may be several ways of bias caused by differences in sample RNA preparation, labeling, hybridization and washing efficiency, print tip characteristics, spatial or hybridization specific effects or pre-amplification of extracted RNA. Thus, distinct normalization approaches have been applied to decrease the technical variations and to minimize over all variance (Podolska et al. 2011). Data processing includes pre-processing of miRNAs detection experiments and normalization. Pre-processing includes the adjustments and corrections required for each method, and normalization is necessary to minimize technical variations. Appropriate normalization method must be use to avoid false conclusions. Although many normalization strategies have been used in miRNAs detection studies, there is no a standard method, and the strategies must be selected on the basis of the detection method and the characteristics of the data set (Meyer et al. 2010).

2.4 Quantitative-real-time-PCR

The extremely small size of mature miRNAs implies that most conventional biological amplification tools become inefficient due to an unstable binding between the primers and the tiny templates. Moreover, regular real-time PCR methods can only be used to measure miRNAs precursors. In addition, miRNAs are heterogeneous in their GC content, which causes a large interval of melting temperatures of nucleic acid duplexes and furthermore, limits the detection of multiple miRNAs. Challenges to the specificity of miRNAs detection come from the close sequence similarity among miRNAs from the same family and from the target sequence, which is present in pri-miRNAs, pre-miRNAs and mature miRNAs (de Planell-Saguer and Rodicio 2011).

Actually, a novel miRNAs quantification method has been developed using retrotranscription (RT) primers containing a partial stem-loop structure (which increases efficiency and specificity), followed by TaqMan PCR analysis. TaqMan miRNAs assays are specific for mature miRNAs and discriminate between related miRNAs that differ in one nucleotide. Furthermore, they are not affected by genomic DNA contamination. Precise quantification is achieved routinely with 25 pg of total RNA for most miRNAs. The stem-loop primer improves the specificity for only mature miRNA targets, and formation of a RT primer-mature miRNAs chimera, extending the 5′-end of the miRNAs. The q-RT-PCR reaction involves a forward primer, a reverse primer, and a TaqMan probe, which quantifies the number of mature miRNAs molecules present in a sample by fluorescent emission of a reporter dye. The q-RT-PCR miRNAs assays are able to distinguish between the hairpin structure of precursor miRNA and the short mature miRNA molecules. Thus, the q-RT-PCR miRNAs assays detect and quantify only mature miRNAs molecules.

Similar to microarray data, q-RT-PCR data requires to be normalized for accurate gene expression analysis. Several normalization strategies have been developed including those based on small, stable, reference ncRNAs, plate normalizing factor or global mean expression.

Normalization based on predefined invariant endogenous controls, reference miRNAs or other small ncRNAs such as small nuclear, small nucleolar RNA or 5 s rRNA, is a commonly used approach in miRNAs q-RT-PCR profiling data analysis. However, the use of small ncRNAs, different from miRNAs, does not contemplate the physicochemical properties of miRNAs molecules and it has been discussed the preferential use of RNA belonging to the same class as better normalizator gene. Using non-miRNAs reference genes for q-RT-PCR normalization is not appropriate when the overall abundance of miRNAs varies e.g. in experiments affecting the miRNAs processing machinery, or in comparisons involving multiple tissues or combinations of tissues and cell lines. Selection of invariant miRNAs identified by algorithms specifically developed for reference gene evaluation was superior over small ncRNA based normalization. Data driven invariant selection of miRNAs whose expression pattern is similar to the global mean expression has been suggested. Basically, the use of more than one reference gene increases the precision of quantification compared to the use of a single reference gene.

For large scale miRNAs expression profiling studies, the mean expression value normalization outperformed the current normalization strategy that makes use of stable small RNA controls, such as snoRNAs, proposed by manufacturers, in terms of better reduction of technical variation. However, the selection of a limited number of miRNAs or small RNA controls that resemble the mean expression value can be successfully used for normalization in follow-up studies where only a limited number of miRNAs molecules are profiled to allow a more accurate assessment of relevant biological variation from a miRNAs q-RT-PCR profiling experiment (Meyer et al. 2010).

2.5 *In situ* Hybridization

In situ hybridization is specifically used to detect miRNAs accumulation in tissue and subcellular levels. It is a technique that uses labeled complementary nucleic acids sequences (DNA or RNA) to detect a single strand of acid nucleotide (DNA or RNA) in tissue sections or fixed cells, thus conserving the morphology. It shows which type of cells express specific miRNAs. In addition, it also provides semiquantitative analysis of miRNAs expression. However, the small size of miRNAs makes their detection by this method technically difficult. This renders low expression levels and low affinity for targets, limiting the technique. The introduction of LNA-probes by Alenius and colleagues has greatly improved the performance of this technique (Obernosterer et al. 2007). LNA probes show higher affinity for their complimentary RNAs than conventional RNA or DNA based probes, which enables detection of miRNAs in tissue sections in a very tighten hybridization, thus increasing the specificity and sensitivity of miRNAs detection in relation to standard DNA probes. Recently, Schmittgen and colleagues have developed a novel method for *in situ* detection of mature miRNAs molecule after its hybridization to an approximately 100 nucleotide-long ultramer template in formalin-fixed paraffin-embedded (FFPE) tissues (Nuovo et al. 2009).

2.6 Deep Sequencing

Deep sequencing technology enables a simultaneous sequencing of up to millions of DNA or RNA molecules. Using deep sequencing to comprehensively profile mRNA expression remains rather expensive; however, profiling the small RNA fraction that contains miRNAs is more achievable. Small RNA cloning methods were used to identify novel miRNAs, but sequencing technology has evolved rapidly and next-generation sequencing (NGS) appears to be very promising for miRNAs detection, because it provides the main advantages of high-throughput sequencing with very high speed and reduced cost. Several studies have successfully used NGS to discover novel miRNAs, especially for those that are difficult to detect because of their low abundance (Creighton et al. 2009).

Although the number of methods for miRNAs analysis grows every day, they are not standardized. Standardized methods must be reproducible, accurate, and robust. Also, standardized methods must have established normal levels of miRNAs in the cell to determine whether the miRNAs are upregulated or downregulated (Cissell and Deo 2009). To ensure standardization of miRNAs detection results, a set of common technical methods are needed: use of the same miRNAs extraction procedure (spin columns, phenol/chloroform, Trizol, etc.), standard technology platforms (microarray, q-RT-PCR analysis, deep sequencing approaches, etc.), the type of miRNAs to be analyzed (pre-miRNAs, mature miRNAs, etc.), and the sort of specimen to be analyzed (plasma, serum, whole blood, etc.) must all be standardized. It is crucial to use appropriate and, if possible, generally applicable normalization controls. Additional important sources of variability could be the method of specimen storage, blood sample collection, and the use of appropriate statistical methods. Consistent methods of data processing and robust statistical methods to analyse separate groups of patients will further reduce the variability of the reported result.

3. miRNAs EXPRESSION PROFILE OF CRC: IMPLICATION FOR DIAGNOSIS AND PROGNOSIS

Alterations in miRNAs expression profiles have been successfully detected in many types of human tumors. The discovery of miRNAs associated with cancer began with high throughput systematic miRNAs expression profiling. From a large-scale analysis of miRNAs expression profiles on 540 samples of solid cancers, including CRC, Volinia et al. identified a solid cancer miRNAs signature composed by a large portion of dysregulated miRNAs (Volinia et al. 2006).

In CRC, the first report to analyze dysregulated miRNAs was published in 2003. Michael et al. using cloning technology followed by northern blotting observed a consistently reduced expression of the specific mature miR-143 and miR-145 in the adenomatous and carcinoma stages of colorectal neoplasia (Michael et al. 2003). Later, Velculescu's group developed an experimental approach called miRNAs serial analysis of gene expression (miRAGE) and used it to perform one of the largest experimental analyses of human miRNAs (Cummins et al. 2006). Sequence analysis of small RNA tags from human colorectal cells allowed them to identify 200 known mature miRNAs, 133 novel miRNAs candidates, and 112 previously uncharacterized miRNAs forms. Simultaneously, our group examined by q-RT-PCR the expression of 156 mature miRNAs in colorectal tumors and adjacent non-neoplastic tissues from patients and CRC cell lines. This permitted us to identify a group of 13 miRNAs whose expression is significantly altered in this type of tumor. The most significantly dysregulated miRNAs were miR-31, miR-96, miR-135b, miR-183 (up-regulated in tumors and CRC cell-lines) and miR-133b, miR-145 (downregulated) (Bandres et al. 2006). These results, achieved through a standardized real-time PCR method, suggest that miRNAs expression profile could have relevance to the biological and clinical behaviour of colorectal neoplasia.

In another profiling study, Lanza et al. identified 14 differentially expressed miRNAs between CRC samples characterized by microsatellite stability (MSS) or by high levels of microsatellite instability (MSI) (Lanza et al. 2007). Slattery et al. quantified the miRNAs expression levels based on tumor location. The tumors were further categorized into several subtypes based on microsatellite instability, CpG island methylator phenotype, KRAS and p53 mutations. Each subtype had a unique miRNAs profile, indicating a potential of different genetic factors to influence miRNAs expression levels. Comparing these with normal colon tissue, they identified 49 miRNAs significantly dysregulated in tumor tissue. The presence of the KRAS mutation was associated with up-regulation of miR-127-3p, miR-92a, and miR-486-3p and downregulation of miR-378 (Slattery et al. 2011).

A global miRNAs expression was performed in a study including 315 samples that included normal colonic mucosa, tubulovillous adenomas, adenocarcinomas with proficient DNA mismatch repair (pMMR), and adenocarcinomas with defective DNA mismatch repair (dMMR) selected for sporadic and inherited colon cancer. Unsupervised cluster analysis demonstrated that normal colon tissue, adenomas, pMMR carcinomas and dMMR carcinomas were all clearly discernible. The majority of miRNAs that were differentially expressed between normal and colon polyps were also differentially expressed with a similar magnitude in the comparison of normal to both the pMMR and dMMR tumor groups, suggesting a stepwise progression for transformation from normal colon to carcinoma. Among the miRNAs demonstrating the largest fold up- or down-regulated changes were miR-31, miR-1, miR-9 and miR-99a, miR-137 and miR-135b. The comparison between pMMR and dMMR tumors identified four miRNAs (miR-31, miR-552, miR-592 and miR-224) with statistically significant expression differences (Oberg et al. 2011).

Monzo et al. assessed the expression of mature miRNAs in human embryonic colon tissue, as well as in CRC and paired normal colon tissue. Overlapping miRNAs expression was detected between embryonic colonic mucosa and CRC. The miR-17-92 cluster and its target, E2F1, exhibit a similar pattern of expression in human colon development and in colonic carcinogenesis (Monzo et al. 2008).

From a diagnostic point of view, miRNAs expression profiles might also contribute significantly to the further determination of the tissue origin of the cancer of unknown primary sites. Identifying miRNAs differentially expressed between CRC and adenocarcinoma of unknown primary site may improve prognosis. In general, there are several advantages of using miRNAs expression profiling instead of its mRNA counterpart for biomarker identification. As a consequence of the fact that miRNAs target mRNAs with an imperfect sequence complementarity, a single miRNA can regulate the expression of more than 100 mRNAs simultaneously. This might explain why microarrays of 217 miRNAs have a much higher information content to distinguish tissues and tumors than 16000 mRNAs (Lu et al. 2005). A further advantage is that, due to their small size and stem-loop structure, miRNAs are relatively more stable and less subjected to degradation during fixation and sample processing. That is why is possible to obtain relatively easy miRNAs from formalin-fixed paraffin-embedded tissues (FFPE). Due to their small size miRNAs are minimally affected by high temperatures and remarkably well protected from endogenous degradation by ribonucleases. One recent examination compared the miRNAs expression profiles from fresh frozen *versus* FFPE CRC tissues. A good correlation coefficient of 0.86–0.89 was observed (Xi et al. 2007). This can also be of benefit for large retrospective studies based on archived FFPE samples.

Accumulating evidence shows that miRNAs expression patterns have potential to be used as prognostic and predictive factors in clinical routine. Several studies on miRNAs have shown that the expression levels of different miRNAs, such as miR-21, miR-320, miR-498, miR-106a and miR-200c correlate with the probability of recurrence-free survival in CRC. Schetter et al. compared miRNAs expression patterns in stage II colonic adenocarcinoma and adjacent normal tissue using a test set and two validation cohorts. In one of the cohorts, a high tumor to normal expression ratio of miR-20a, miR-21, miR-106a, miR-181b and miR-203 was associated with poor survival. In the second validation set, including stage III, survival analysis showed that higher miR-21 expression predicted poorer survival in treated colonic cancer and poor responsiveness to adjuvant chemotherapy and this association was independent of age, sex, and tumor location (Schetter et al. 2008). Similar associations have been found by other research group. Schepeler and colleagues showed that miR-320 or miR-498 expression was significantly associated with progression-free survival in stage II CRC. The biomarker based on miRNAs expression profiles could predict recurrence of disease with an overall performance accuracy of 81%, thus indicating a potential role for miRNAs in determining tumor aggressiveness. These miRNAs were found to be independent predictors of recurrence-free survival when stratified for age, sex, tumor stage, differentiation and histological grade. Low expression of miR-106a in 110 CRC samples was correlated with advanced disease stage and overall survival (OS). Moreover, miR-106a expression levels were inversely correlated with E2F1 expression levels, supporting the idea that miR-106a act as a tumor suppressor (Schepeler et al. 2008). Three independent groups have reported up-regulation of miR-31 in CRC, but the clinical significance is unclear, because no mRNA targets have been verified for miR-31. Overexpression was associated with advanced tumor stage in two reports (Bandres et al. 2006, Wang et al. 2009a), while in a third study, no correlation was observed between miR-31 expression and tumor stage (Slaby et al. 2007). Another study found that miR-21 was significantly up-regulated, and miR-143 and miR-145 down-regulated in tumors compared with the normal counterparts. High expression of miR-21 was associated with lymph node positivity and distant metastasis and tumors >50 mm in maximal tumor diameter were related to lower expression of miR-143 and miR-145 (Slaby et al. 2007). In addition, miR-18a was found to imply a poorer clinical prognosis (Motoyama et al. 2009). Xi et al. found that tumors expressing high levels of miR-200c are correlated with poorer prognosis, regardless of tumor stage (Xi et al. 2006). The expression of miR-17-5p, miR-106a, and miR-126 was assessed in 110 CRC patients and correlated with survival and the levels of their target mRNAs. Altered expression of miR-17-5p, miR-106a, and EGFL7 was associated with pathological tumor features of poor prognosis. miR-17-5p correlated with disease free survival (DFS) only at early stages and miR-106a downregulation was revealed as a marker of DFS and OS independent of tumor stage (Diaz et al. 2008).

Although these results are promising, larger studies will be needed to prove whether miRNAs really have significant potential to extend prognostic information to practical clinic.

In the Table 2 we summarized some of the most important miRNAs found associated to CRC clinic-pathological features and clinic outcome.

Regulation of miRNAs expression remains largely unknown. miRNAs dysregulation could be due to several mechanisms, including genomic alteration (chromosomal abnormalities), mutation, polymorphisms (SNPs), epigenetic changes, and/or miRNAs processing alterations (altered Drosha or Dicer activity). miRNAs expression in human cancer can be altered

Table 2. Relevant miRNAs dysregulated in CRC.

miRNA	Dysregulation	Clinical-related phenotype	Reference
let-7g	Upregulated in CRC	High level associated with poor S-1 response	(Nakajima et al. 2006)
miR-18 a	Upregulated in CRC	High level associated with poor overall survival	(Motoyama et al. 2009)
miR-21	Upregulated in adenoma, CRC and liver metastasis	High levels associated with lymph node positivity. Poor survival, poor therapeutic outcomes, rapid recurrence. Shorter disease-free survival.	(Slaby et al. 2007, Schetter et al. 2008)
miR-31	Upregulated in CRC	High level associated with higher TNM stages and local invasion	(Slaby et al. 2007, Wang et al. 2009a, Bandres et al. 2006)
miR-106 a	Upregulated in colon cancer	High level associated with shorter disease-free survival and overall survival Downregulation of miR-106a predicted shortened disease-free survival and overall survival	(Schetter et al. 2008) (Diaz et al. 2008)
miR-143	Downregulated in colon cancer and liver metastasis	Low level associated with larger tumor size and shorter disease-free interval	(Slaby et al. 2007, Motoyama et al. 2009, Schepeler et al. 2008)
miR-145	Downregulated in CRC	Low level associated with larger tumor size; related with tumor location	(Schepeler et al. 2008, Wang et al. 2009a)
miR-181b	Upregulated in CRC	High level associated with poor S-1 response	(Nakajima et al. 2006)
miR-320	Downregulated in MSS (microsatellite stable) tumor	Low level associated with shorter progression-free interval	(Schepeler et al. 2008)
miR-498	Downregulated in MSS tumor	Low level associated with shorter progression-free interval	(Schepeler et al. 2008)

by any of the mechanisms described above since an extensive analysis of genomic sequences of miRNAs genes have shown that miRNAs are frequently located in regions of the genome involved in alterations in cancer. Moreover, approximately half of them are associated with CpG islands, suggesting that they could be subjected to this mechanism of regulation (Lujambio and Esteller 2007). In fact, several evidences have proved that an altered methylation status can be responsible for the dysregulated expression of miRNAs in cancer, as the silencing of putative tumor suppressor miRNAs.

In CRC, Lujambio et al. showed the first evidence that one mechanism accounting for the observed downregulation of miR-124a is CpG island hypermethylation (Lujambio et al. 2007). Later, we have showed that suppression of miR-9, miR-129 and miR-137 genes in CRC is mediated at least in part by epigenetic mechanisms including DNA hypermethylation and histone acetylation. Moreover, the frequent hypermethylation of these miRNAs genes in CRC and the correlation between methylation and clinic-pathological support the concept that epigenetic instability is an important event in CRC (Bandres et al. 2009a).

These results were validated by Balaguer et al. who determined the methylation status of miR-137 CpG island in a panel of six CRC cell lines and 409 colorectal tissues, including normal colonic mucosa from healthy individuals (N-N), primary CRC tissues and their corresponding normal mucosa (N-C) and adenomas. Methylation of the miR-137 CpG island was a cancer specific event, and was frequently observed in CRC cell lines (100%), adenomas (82.3%) and CRC (81.4%) but not in normal colon tissues. Expression of miR-137 was restricted to the colonocytes in normal mucosa, and inversely correlated with the level of methylation. Transfection of miR-137 precursor in CRC cells significantly inhibited cell proliferation and LSD-1 gene was validated as a specific miR-137 target. These data indicate that miR-137 acts as a tumor suppressor in the colon and is frequently silenced by promoter hypermethylation (Balaguer et al. 2011).

By microarray analysis, Tang et al. analyze the differential expression of miRNAs after 5-aza-2'-deoxycitidine (5-aza-dC) treatment in CRC cell lines. The expression of mir-345 was significantly increased after 5-aza-dC treatment and DNA methylation analyses of mir-345 showed high methylation levels in colon tumor *versus* normal tissues. Expression of mir-345 was significantly down-regulated in 51.6% of CRC tissues compared with corresponding non-cancerous tissues and low expression of mir-345 was associated with lymph node metastasis and worse histological type. Increased mir-345 function was sufficient to suppress colon cancer cell proliferation and invasiveness *in vitro* (Tang et al. 2011).

Recently, a comprehensive genome wide analysis has been conducted in CRC by high-resolution chromatin immunoprecipitation (ChIP)-sequencing approach. It allowed the identification of 47 miRNAs encoded in 37 primary transcription units as potential targets of epigenetic silencing. Among the miRNAs dysregulated, miR-1-1 was methylated frequently in early and advanced CRC in which it may act as a tumor suppressor (Suzuki et al. 2011).

4. miRNAs EXPRESSION IN CIRCULATING BLOOD AND FECAL SAMPLES: IMPLICATION FOR EARLY DETECTION OF CRC

Although colonoscopy is the most reliable method for early detection of CRC available to date, the invasive nature and the expensive cost have impeded its widespread application. The fecal occult blood test (FOBT), which is the most widely used non-invasive screening tool so far, is limited by its low sensitivity, especially with respect to detection of pre-neoplastic lesions. Stool DNA tests may be a promising alternative in the future, but widespread application is limited by labor-intensive handling and high costs. Other methods have been reported for early detection of CRC, including those based on the detection of mutated DNA, cancer-related methylation analysis, and DNA integrity in fecal samples. However, there is an imperious requirement for new non-invasive biomarkers to improve early detection of CRC.

An ideal screening method should have a high sensitivity and specificity for early-stage cancers and precancerous lesions; it may be safe and affordable. An optimal screening test may reduce the necessity for an invasive procedure, reduce the cost, have better screening compliance and could more precisely select those individuals who require the colonoscopic removal of a neoplastic lesion. The detection of different types and levels of miRNAs in samples is critical in diagnostics. miRNAs can be detected in blood and body fluids as well as in tissues, thus making non-invasive collection of samples possible.

4.1 miRNA Expression in Fecal Samples of CRC Patients

Several studies have assessed to detect CRC by using reverse transcriptase-PCR in fecal samples. However, there is no evidence that the stool RNA test is useful for CRC screening. The main reason for this imprecision is that the nucleic acids in feces are derived from a huge number and variety of bacteria. However, it has described the presence of viable cancer cells in the feces which can be isolated from naturally evacuated feces by using cell isolation methods (Matsushita et al. 2005). Moreover, CRC-related gene mutation, and analysis of CRC-related gene expression have been reported in feces samples. miRNAs are richly present in stool and can be easily and reproducibly detected in stool specimens. Turchinovich and colleagues showed that miRNAs remains stable in the extracellular space for at least one month, suggesting the possibility of using extracellular miRNAs as a biomarker for cancer (Turchinovich et al. 2011). miRNAs may be in stool due to the cell exfoliation and by the accumulation of exosomes from the cells of the gastrointestinal tract, in the same way as proposed for contribution of miRNAs signals by exosomes in blood. Although a blood-based test might be more practical, considering the increased number of exfoliated colonocytes shed in the colon from CRC patients, it is likely that the earliest detectable neoplastic changes in the expression pattern of specific miRNAs may be in feces rather than in blood.

Thus, miRNAs expression profile in CRC tissue and exfoliated colonocytes in feces have been investigated as a new method for early detection of CRC (Koga et al. 2010). The sensitivity and specificity of miRNAs expression assay was 74% and 79%, respectively. This means that miRNAs, in exfoliated colonocytes, may be conserved in greater number compared with mRNA and the miRNAs expression assay should be more useful as a CRC screening method compared with the gene expression assay. Moreover, it has been reported that miRNAs expression analysis in the stool is able to detect patients with resectable CRC as well as those with advanced CRC. Ahmed et al. have demonstrated that the detection of specific miRNAs in fecal specimens clearly discriminated between sporadic human colon cancer and active ulcerative colitis (Ahmed et al. 2009). Furthermore, Link et al. proposed the use of fecal miRNAs as novel biomarkers for CRC screening. They showed that miR-21 and miR-106a are overexpressed in adenomas and CRC compared with individuals who do not have colorectal neoplasia (Link et al. 2010). On the other hand, Koga et al. showed that the expression of the miR-17-92 cluster and miR-135 was significantly higher in feces of CRC than in healthy volunteers (Koga et al. 2010).

Kalimutho et al. found miR-144* overexpressed in feces of CRC patients, suggesting that it could be a potent candidate diagnostic marker for CRC detection, with a sensitivity of 74% and a specificity of 87%. To further investigate the pre-analytical steps involved in fecal collection and transportation, they analyzed miR-16 stability profiles in samples stored at two different temperatures (4°C and 25°C) over a period of six days. They found that feces kept at 4°C are more stable than feces stored at room temperature. They detected a slight degradation pattern of endogenous miR-16 in feces stored at room temperature during the fourth and sixth days relative to 4°C. Thus, these data indicate that endogenous miRNAs are relatively time and temperature stable in feces (Kalimutho et al. 2011a).

Recently, Kalimutho et al. found that miR-34b/c could be an ideal candidate target for CRC screening in fecal microenvironment, whereas the involvement of epigenetic silencing factor in miR-148a correlates to poor prognosis (Kalimutho et al. 2011b).

Therefore, miRNAs evaluation in feces is a promising and non-invasive approach to CRC screening, although further validation of pre-analytical sampling and miRNAs stability is required. Few reports have been published so far on the presence of miRNAs in feces and, further investigations about the properties of miRNAs in feces are important to validate their role as biomarkers of CRC.

4.2 miRNA Expression in Serum and Plama of CRC Patients

It has been shown that the levels of miRNAs in serum and plasma are exceptionally stable, reproducible, and consistent among individuals of the same species. This enables miRNAs to serve as stable biomarkers for cancer and other diseases.

Tumor-derived miRNAs was firstly described in plasma by Mitchell et al. in year 2008 (Mitchell et al. 2008). To directly determine whether miRNAs are present in human plasma, they isolated the 18- to 24-nt RNA fraction from a human plasma sample from a healthy donor and used 5′ and 3′RNA–RNA linker ligations followed by RT-PCR amplification to generate a small RNA cDNA library. Given that plasma has been reported to contain high levels of RNase activity, Tsui et al. sought to determine whether the stability of miRNAs was intrinsic to their small size or chemical structure, or whether it was caused by additional extrinsic factors. They confirm the presence of RNase activity in plasma and that blood miRNAs exist in a form that is resistant to plasma RNase activity (Tsui et al. 2002).

Ng et al. revealed that miR-92 is significantly elevated in plasma of CRC patients and could be a potential non-invasive molecular marker for CRC detection. Elevation of miR-92 differentiates CRC from gastric cancer, inflammatory bowel disease (IBD) and normal subjects, suggesting that miR-92 be a potential non-invasive molecular marker for CRC screening (Ng et al. 2009a). Huang et al. examined the expression levels of 12 miRNAs by using q-RT-PCR in 40 plasma samples (20 CRC and 20 controls), and found that miR-29a and 92a were significantly elevated in CRC plasma when compared with normal controls. Importantly, these two miRNAs also discriminated advanced adenomas and healthy controls suggesting their potential value for early detection of CRC and prognosis marker (Huang et al. 2010). Pu et al. showed that the elevated plasma miR-221 level is a significant prognostic factor for poor overall survival in CRC patients. Patients with high miR-221 levels in plasma had a dramatically lower survival rate than that in the low expression group. The immunohistochemistry analysis demonstrated a significant correlation between plasma miR-221 level and p53 expression (Pu et al. 2010).

Further research is crucial to dilucidate the role of miRNAs in plasma/serum to use them as a reliable clinical tool. A common problem in research on circulating miRNAs is that no consensus housekeeper miRNAs or endogenous controls have been established. An accurate need for a small housekeeping miRNAs still exists because other classes of small RNAs, such as the snoRNA RNU6B, cannot be used for normalization due to the instability of this RNA in serum. Chang and colleagues reported that miR-16 and miR-345 were identified as the most stably expressed reference gene (Chang et al. 2010) but further standardization methods are necessary in order to obtain reproducible results.

Although the clinical significance of these findings has not been clarified yet, these results demonstrated that circulating miRNAs could be non-invasive diagnostic or prognostic markers for cancer.

5. INVOLVEMENT OF miRNAs IN COLORECTAL PATHOGENESIS: IMPLICATION FOR A FUTURE THERAPEUTIC APPROACH IN CRC

In general, dysregulation of miRNAs can influence carcinogenesis if their mRNA targets are encoded by tumor suppressor genes or oncogenes. Both, overexpression and silencing or switching off of specific miRNAs, have been described in the carcinogenesis of CRC. Like a protein-coding gene, a miRNA can act as a tumor suppressor when its function loss can initiate or contribute to the malignant transformation of a normal cell because regulates an oncogenic protein. On contrary, the list of miRNAs that function as oncogenes is short, but the evidence for their role in cancer is very strong. The putative targets for these miRNAs are tumor suppressor genes. In Table 3 we reported the main miRNAs with putative targets associated with CRC identified by bioinformatic analysis.

However, analyses of these miRNAs in mechanistic studies are crucial to better knowledge CRC pathogenesis and to identify novel therapeutic targets.

Many proteins involved in key signaling pathways of CRC, such as members of the Wnt/β-catenin and phosphatidyl-inositol-3-kinase (PI3K) pathways, KRAS, p53, extracellular matrix regulators as well as epithelial-mesenchymal transition (EMT) transcription factors and angiogenesis factors are altered and seemed to be affected by miRNAs regulation in CRC. In Table 4 we present the major validate targets for miRNAs involved in CRC disease.

Table 3. Putative targets of CRC dysregulated miRNAs.

Putative target	miRNAs
AKT1	miR-219, miR-370
AKT2	miR-203
APC	miR-142-3p, miR-145, miR-26b, miR-141
BAX	miR-133a, miR-214, miR-370
B-CATENIN	miR-214, miR-299
CDKN1A	miR-299
CDKN2A	miR-337
CDKN2B	miR-214
EGFR	miR-214
IGFR I	miR-125b, miR-182, miR-215
K-RAS	miR-133a, miR-133b, miR-205, miR-96, let-7g, let-7a, miR-143
MLH1	miR-154*, miR-272, miR-200b, miR-200c
MSH2	miR-205, miR-323, miR-9, miR-135a, miR-135b, miR-141, miR-200a
P53	miR-214
PTEN	miR-205
SMAD2	miR-372, miR-200c
SMAD4	miR-205
TCF4	miR-9*
TGFR II	let-7 g, miR-130a, miR-145, miR-214, miR-372, miR-137
VEGF	miR-125b
VEGFR	miR-125b

Adenomatous polyposis coli (APC) gene inactivation, leading to stimulation of the Wnt pathway via free β-catenin, is a central initiating event in colorectal carcinogenesis, and is observed in more than 60% of colorectal adenomas and carcinomas. According to a study performed by Nagel et al. miRNAs represent a novel mechanism for APC regulation in CRC. miR-135a and miR-135b were found to be upregulated *in vivo* in colorectal adenomas and carcinomas and were correlated with low APC levels. *In vitro* studies showed that miR-135a and miR-135b decrease translation of the APC transcript (Nagel et al. 2008).

In CRC, KRAS and PI3K, downstream molecules in the EGF receptor pathway, promote tumor growth, survival, angiogenesis and metastasis. KRAS oncogene has been reported to be a direct target of the let-7 miRNA family (Johnson et al. 2005). When let-7 low-expressing DLD-1 colon cancer cells were transfected with let-7a-1 precursor, significant growth suppression

Table 4. miRNAs validated targets in CRC.

miRNA	Genome location	Validated targets	Cellular process	Reference
let-7	3p21.1	KRAS	Cell cycle	(Johnson et al. 2005)
miR-1	20q13.33	MET	Invasion	(Migliore et al. 2012)
miR-101	1p31.3	COX2	Inflamation	(Strillacci et al. 2009)
miR-107	10q23.31	HIF-1b	Angiogenesis	(Yamakuchi et al. 2010)
miR-126	9q34.3	P85b (PI3K)	Proliferation	(Guo et al. 2008)
miR-133b	6p12.2	MET	Invasion	(Hu et al. 2010)
miR-135a	3p21.2	APC	Proliferation	(Nagel et al. 2008)
miR-135b	1q32.1	APC	Proliferation	(Nagel et al. 2008)
miR-141	12p13.31	ZEB1	EMT	(Wellner et al. 2009)
		ZEB2	EMT	(Wellner et al. 2009)
miR-143	5q32	K-RAS	Cell cycle	(Chen et al. 2009)
		DNMT3A	Methylation	(Ng et al. 2009b)
miR-18a*	13q31.3	K-RAS	Cell cycle	(Tsang and Kwok 2009)
miR-194	1q41	THBS1	Angiogenesis	(Sundaram et al. 2011)
miR-200a	1p36.33	ZEB1	EMT	(Wellner et al. 2009)
		ZEB2	EMT	(Wellner et al. 2009)
miR-200b	1p36.33	ZEB1	EMT	(Wellner et al. 2009)
		ZEB2	EMT	(Wellner et al. 2009)
miR-200c	12p13.31	ZEB1	EMT	(Wellner et al. 2009)
		ZEB2	EMT	(Wellner et al. 2009)
miR-21	17q23.1	PTEN	Apoptosis	(Iliopoulos et al. 2010)
		CDC25A	Cell cycle	(Wang et al. 2009b)
		MSH2	DNA repair	(Valeri et al. 2010)
		MSH6	DNA repair	(Valeri et al. 2010)
miR-22	17p13.3	HIF-1α	Angiogenesis	(Yamakuchi et al. 2011)
miR-29	7q32.3	MMP2	Invasion	(Ding et al. 2011)
miR-30a-5p	6q13	DTL	Cell cycle	(Baraniskin et al. 2012)
miR-34a	1p36.22	FRA-1	Invasion	(Wu et al. 2012)
		SNAIL	Methylation	(Siemens et al. 2011)
		SIRT1	Cell cycle	(Yamakuchi et al. 2008)
		P53	Cell cycle	(Yamakuchi and Lowenstein 2009)

Table 4. contd....

Table 4. contd....

miRNA	Genome location	Validated targets	Cellular process	Reference
miR-365	1p34.3	MYBL2	Differentiation	(Papetti and Augenlicht 2011)
miR-429	1p36.33	ZEB1	EMT	(Wellner et al. 2009)
		ZEB2	EMT	(Wellner et al. 2009)
miR-451	17q11.2	MIF	Inflamation	(Bandres et al. 2009b)
		ABCB1	Drug resistance	(Bitarte et al. 2011)
miR-499-5p	20q11.22	FOXO4	Invasion	(Liu et al. 2011)
		PDCD4	Invasion	(Liu et al. 2011)
miR-506	Xq27.3	PPARα	Drug resistance	(Tong et al. 2011)
miR-542-3p	Xq26.3	ILK	Invasion	(Oneyama et al. 2011)
miR-92a	13q31.3	BIM	Apoptosis	(Tsuchida et al. 2011)
miR-93	7q22.1	HDAC8	Epigenetic regulation	(Yu et al. 2011)
		TLE4	Epigenetic regulation	(Yu et al. 2011)
miR-192	11q13.1	TYMS	Proliferation	(Boni et al. 2010)
miR-24	9q22.32	DHFR	Drug resistance	(Mishra et al. 2007)
miR-519c	19q13.42	ABCG2	Drug resistance	(To et al. 2008)

with concurrent decrease of the KRAS protein levels was observed. Another miRNA associated with KRAS regulation in CRC is miR-143. KRAS was deduced to be a miR-143 target not only by computational prediction but also by noting the inverse correlation between miR-143 and KRAS protein level in clinical samples. KRAS expression *in vitro* was significantly abolished by treatment with miR-143 precursor, whereas miR-143 inhibitor increased the KRAS protein level (Chen et al. 2009). Recently, miR-18a* was observed to directly regulate KRAS in the colon adenocarcinoma HT-29 cells. MiR-18a* repression by transfection with anti-miR-18a* inhibitor increased the KRAS expression as well as the luciferase activity of a reporter construct containing the 3'-UTR of KRAS mRNA. Furthermore, the miR-18a* repression also increased the cell proliferation and promoted the anchorage-independent growth in soft agar of HT-29 cells (Tsang and Kwok 2009).

PI3K signal is stabilized and propagated by the p85β regulatory subunit, which has been recognized as a miR-126 target. Reconstitution of miR-126 in colon cancer cells resulted in a significant growth reduction as evidenced in clonogenic assays. A search for miR-126 gene targets revealed p85β, a regulatory subunit involved in stabilizing and propagating the PI3K signal, as one of the potential substrates. Restoration of miR-126 in cancer cells induced a significant reduction in p85β protein levels, with no concomitant change in p85α, a gene that is functionally related to p85β but not a supposed target of miR-126. Additionally, using reporter constructs, the authors showed that the p85β-3' untranslated region is directly targeted by miR-126. Furthermore, this miR-126 mediated reduction of p85β expression was accompanied by a substantial reduction in phosphorylated AKT levels in the cancer cells, suggesting impairment in PI3K signaling. Finally, in a panel of matched normal colon and primary colon tumors, each of the tumors demonstrated miR-126 downregulation together with an increase in the p85β protein level (Guo et al. 2008). Similarly, the suppression of PTEN by miR-21 is associated with augmented PI3K signaling and tumor progression (Iliopoulos et al. 2010).

A well-known tumor suppressor gene, p53 is mutated in about 50–75% of all CRC. p53 responds to DNA damage or deregulation of mitogenic oncogenes through the induction of cell cycle checkpoints, apoptosis, or cellular senescence. The conserved miR-34a-c family was found to be direct transcriptional targets of p53. miRNAs expression patterns were analyzed in wild-type p53+/+ and p53–/– mutant HCT-116 colon cancer cell lines after treatment with DNA damaging agents. Several miRNAs were induced in the wild type but not in the p53–/– mutant cells. miR-34a showed the strongest induction. Expression of miR-34a was sufficient to induce apoptosis through p53-dependent and independent mechanisms. miR-34a responsive genes are highly enriched for those that regulate cell cycle progression, cellular proliferation, apoptosis, DNA repair and angiogenesis (Paris et al. 2008). By overexpression of miR-34a, p53 effects like cell-cycle arrest and apoptosis could be achieved. miR-34a regulates silent information regulator 1 (SIRT1) expression, too. miR-34a inhibition of SIRT1 leads to an increase in acetylated p53 and expression of p21 and PUMA, transcriptional targets of p53 that regulate the cell cycle and apoptosis, respectively. Furthermore, miR-34a suppression of SIRT1 ultimately leads to apoptosis in WT human colon cancer cells but not in human colon cancer cells lacking p53 (Yamakuchi et al. 2008). Finally, miR-34a itself regulates p53 levels, suggesting a positive feedback loop between p53 and miR-34a (Yamakuchi and Lowenstein 2009).

Other miRNA involved to regulate epigenetic process is miR-143. miR-143 is frequently downregulated in 87.5% of CRC tissues and DNMT3A was shown to be a direct target of miR-143 by luciferase gene reporter assay. Restoration of the miR-143 expression in colon cell lines decreased tumour cell growth and soft-agar colony formation, and downregulated the DNMT3A expression in both mRNA and protein levels. Furthermore, the miR-143 expression was observed to be inversely correlated with DNMT3A mRNA and protein expression in CRC tissues (Ng et al. 2009b).

Aberrant activation of the Wnt signaling pathway is causally involved in the formation of most CRC. Schepeler et al. identified changes of the miRNAs transcriptome following disruption of beta-catenin/TCF4 activity in DLD1 CRC cells. A set of miRNAs induced by abrogated Wnt signaling *in vitro* was downregulated in two independent series of human primary CRC relative to normal adjacent mucosa. Several of these miRNAs (miR-145, miR-126, miR-30e-3p and miR-139-5p) markedly inhibited CRC cell growth *in vitro* when ectopically expressed. By using an integrative approach of proteomics and expression microarrays, the authors found numerous mRNAs and proteins to be affected by ectopic miR-30e-3p levels. Between them, they identified HELZ and PIK3C2A as direct targets, confirmed by luciferase gene reporter assays (Schepeler et al. 2011).

Overexpressed cyclooxygenase-2 (COX-2) strongly contributes to the growth and invasiveness of tumor cells in patients with CRC. It has been demonstrated that COX-2 overexpression depends upon various cellular pathways involving both transcriptional and post-transcriptional regulations. An inverse correlation was reported between COX-2 and miR-101 expression in CRC cell lines. It was demonstrated *in vitro* that the direct translational inhibition of COX-2 mRNA is mediated by miR-101. Moreover, this correlation was supported by data collected *ex vivo*, in which colon cancer tissues and liver metastases derived from CRC patients were analyzed (Strillacci et al. 2009).

Macrophage migration inhibitory factor (MIF) is an innate cytokine which plays a critical role in the host control of inflammation and immunity, and MIF could inhibit p53 tumor suppressor activity. Interestingly, MIF has been found to play an important role in the colorectal carcinogenesis and hypoxia-induced apoptosis. Recently, our group showed that MIF is a potential target of miR-451. Overexpression of miR-451 in gastric cancer and CRC cells resulted in reduction of cell proliferation, increased their susceptibility to radiotherapy, down-regulated expression of MIF at both mRNA and protein level. Furthermore, an inverse connection between miR-451 and MIF expression in biopsies of gastric tumors was observed, which suggested a role of miR-451 as a tumor suppressor (Bandres et al. 2009b). On the whole, miRNAs play a role in the signal pathway linking inflammation and tumorigenesis.

Epithelial-mesenchymal transition (EMT) is the conversion of an epithelial cell into a mesenchymal cell. Morphologically, EMT is characterized by a decrease of E-cadherin, loss of cell adhesion, and increased cell motility leading to promotion of metastatic behavior of cancer cells (including CRC). The transcriptional repressor zinc-finger E-box binding homeobox 1 (ZEB1) is a crucial inducer of EMT in various human tumours, and it recently was shown to promote invasion and metastasis of tumour cells. Functional assays link miR-200 family (miR-200a, miR-200b, miR-200c, miR-141 and miR-429) to EMT. ZEB1 directly suppresses transcription of miRNAs-200 family members miR-141 and miR-200c, which strongly activate epithelial differentiation in pancreatic, colorectal and breast cancer cells (Wellner et al. 2009).

Metastases process is the major cause of death in patients with CRC. Currently, the genes and molecular mechanisms that are functionally critical in modulating CRC metastasis remain unclear. Several miRNAs have been associated with CRC metastasis and could be considered as potential targets for diagnosis and treatment.

Recently, a study using functional selection in an orthotopic mouse model of CRC identified a set of genes that play an important role in mediating CRC liver metastasis. These genes included APOBEC3G, CD133, LIPC, and S100P. Clinically, these genes are highly expressed in a cohort of human hepatic metastasis and their primary colorectal tumors, suggesting that it might be possible to use these genes to predict the likelihood of hepatic metastasis. The authors have further revealed what we believe to be a novel mechanism in which APOBEC3G promotes colorectal cancer hepatic metastasis through inhibition of miR-29 mediated suppression of MMP2, involving a miRNA in CRC hepatic metastasis (Ding et al. 2011).

In another study, miR-499-5p was detected differentially expressed between SW480 (primary) and SW620 (metastasic) cells lines. The authors confirmed increased miR-499-5p levels in highly invasive CRC cell lines and lymph node-positive CRC specimens. Furthermore, enhancing the expression of miR-499-5p promoted CRC cell migration and invasion *in vitro* and lung and liver metastasis *in vivo*, while silencing its expression resulted in reduced migration and invasion. Additionally, they identified FOXO4 and PDCD4 as direct and functional targets of miR-499-5p. Collectively, these findings suggested that miR-499-5p promoted metastasis of CRC cells and may be useful as a new potential therapeutic target for CRC (Liu et al. 2011).

The tyrosine kinase c-Src is upregulated in various human cancers and mediated tumor progression. Downregulation of miR-542-3p is tightly associated with tumor progression via c-Src-related oncogenic pathways. Integrin-linked kinase (ILK) was identified as a conserved target of miR-542-3p. ILK upregulation promotes cell adhesion and invasion by activating the integrin-focal adhesion kinase (FAK)/c-Src pathway, and can also contribute to tumor growth via the AKT and glycogen synthase kinase 3β pathways. MiR-542-3p expression is downregulated by the activation of c-Src-related signaling molecules, including EGFR, KRAS and RAS/RAF/MAPK pathway. In human colon cancer tissues, downregulation of miR-542-3p is significantly correlated with the upregulation of c-Src and ILK, suggesting that the novel c-Src-miR-542-3p-ILK-FAK circuit plays a crucial role in controlling tumor progression of CRC (Oneyama et al. 2011).

MET, the tyrosine kinase receptor for hepatocyte growth factor is frequently overexpressed in colon cancers with high metastatic tendency. Two studies have demonstrated that miRNAs are involved to regulate MET expression in CRC. In the first study, Hu et al. demonstrated that miR-133b significantly suppressed a luciferase gene reporter containing the MET-3′-UTR. miR-133b expression was shown to be greatly downregulated in human CRC cells and CRC tissues. In the CRC cell lines SW-620 and HT-29, ectopic expression of miR-133b affected tumor cell proliferation and apoptosis *in vitro* and *in vivo* by direct targeting of the receptor tyrosine kinase MET (Hu et al. 2010). Recently, Migliore et al. showed that miR-1 was downregulated in 85% of CRC and its decrease significantly correlated with MET overexpression, particularly in metastatic tumors. They found that concurrent metastasis-associated in colon cancer 1 (MACC1) gene upregulation and miR-1 downregulation are required to elicit the highest increase of MET expression. Consistent with a suppressive role of miR-1, its forced *in vitro* expression in colon cancer cells reduced MET levels and impaired MET-induced invasive growth (Migliore et al. 2012).

Hypoxia is a common feature of solid tumors and represents a critical factor in their progression. Hipoxia process has been demonstrated to be regulated by miRNAs. miR-22 expression in human colon cancer is lower than in normal colon tissue and miR-22 controls hypoxia inducible factor 1α (HIF-1α) expression in the HCT116 colon cancer cell line. Overexpression of miR-22 inhibits HIF-1α expression, repressing vascular endothelial growth factor (VEGF) production during hypoxia. Conversely, knockdown of endogenous miR-22 enhances hypoxia induced expression of HIF-1α and VEGF. The conditioned media from cells over-expressing miR-22 contain less VEGF protein than control cells, and also induce less endothelial cell growth and invasion, suggesting miR-22 in adjacent cells influences endothelial cell function. These data suggest that miR-22 might have an anti-angiogenic effect in colon cancer (Yamakuchi et al. 2011). Similarly, miR-107 and miR-21 has been involved to regulate hypoxic process in CRC. miR-107 is a microRNA expressed by human colon cancer specimens and regulated by p53. miR-107 decreases hypoxia signaling by suppressing expression of HIF-1β. Knockdown of endogenous miR-107 enhances HIF-1β expression and hypoxic signaling in human colon cancer cells. Conversely, overexpression of miR-107 inhibits HIF-1β expression and hypoxic signaling. Furthermore, overexpression of miR-107 in tumor cells suppresses tumor angiogenesis, tumor growth, and tumor VEGF expression in mice. Finally, in human colon cancer specimens, expression of miR-107 is inversely associated with expression of HIF-1β (Yamakuchi et al. 2010). It has been demonstrated that miR-21 is induced by hypoxic process, serum starvation and DNA damage, and negatively regulates G1-S transition, and participates in DNA damage-induced G2-M checkpoint through downregulation of CDC25A. In contrast, miR-21 deficiency did not affect apoptosis induced by a variety of commonly used anticancer agents or cell proliferation under normal cell culture conditions. Furthermore, miR-21 was found to be underexpressed in a subset of CDC25A-overexpressing colon cancers. These data show a role of miR-21 in modulating cell cycle progression following stress, providing a novel mechanism of CDC25A regulation and a potential explanation of miR-21 in CRC tumorogenesis (Wang et al. 2009b).

These functional studies indicate that miRNAs act as tumor suppressors and oncogenes in CRC. These findings significantly extend Vogelstein's model of CRC pathogenesis and have shown great potential for miRNAs as a novel class of therapeutic targets (Fig. 2).

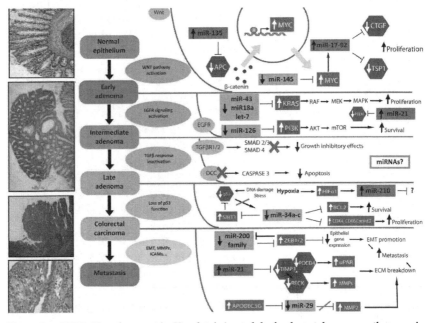

Figure 2. miRNAs' involvement in Vogelstein's model of colorectal cancer pathogenesis. According to the model proposed by Vogelstein different genes are involved in each step of the progression from normal mucosa to metastatic CRC. The knowledge of the mechanism of action of miRNAs in CRC adapts the model with the addition of specific miRNAs that modulate selective targets at each step of this model.

Color image of this figure appears in the color plate section at the end of the book.

6. INVOLVEMENT OF miRNAs IN DRUGS RESISTANCE: IMPLICATION AS PREDICTIVE BIOMARKERS FOR CHEMOTHERAPY RESPONSE IN CRC

Another important question for management of CRC patients is the possibility of predicting therapy response. Several individual miRNAs have been suggested to be of importance for drug sensitivity in CRC.

The role of miRNAs in drug resistance/sensitivity is now recognized; so, miRNAs are a new field of pharmacogenomics. The concept of miRNAs pharmacogenomics was introduced for the first time by Bertino et al. and defined as the study of miRNAs and miRNAs-SNPs affecting drug behavior (Bertino et al. 2007).

miRNAs can be associated with inter-individual differences in drug response by two major mechanisms:

1) miRNAs expression changes can alter mRNA target levels involved in drug response.
2) Sequence variations in miRNAs target regions can alter miRNAs-mRNA target interaction and consequently the target repression regulation.

6.1 Evidences of miRNAs-SNPs in Drug Resistance

Functional miRNAs-SNPs are defined as SNPs in *seed* regions and binding targets of mature miRNAs that can affect the pairing between miRNAs and the target mRNAs, leading to loss of miRNAs function. miRNAs-SNPs affecting expression of genes involved in pathways of drug absorption, metabolism, disposition and targets that participate in the overall clinical effect of a drug may contribute to drug resistance. miRNAs-SNPs can interfere with a miRNAs expression and function, resulting in loss of the miRNAs-mediated regulation of a drug target gene that induce drug resistance. The knowledge of miR-SNPs in a patient's genome may allow oncologists to understand whether the patient will respond to a certain chemotherapeutic drug or to predict toxicity. Many such functional miRNAs-SNPs exist in the human genome that will result in loss or gain of miRNAs function and lead to changes in protein expression, and if a drug target is affected, may induce drug resistance.

In 2007, Mishra et al. confirmed for the first time that a single nucleotide mutation in miR-24 binding targets changed cell response to methotrexate. The presence of a miR-24 miRNA binding site SNP 829 C→T in dihydrofolate reductase 3'UTR lead to loss of miR-24 function which result in dihydrofolate reductase (DHFR) overexpression and methotrexate (MTX) resistance. Deletions in the 3'UTR of the target mRNA also lead to the loss of miRNA binding sites and therefore loss of miRNA function (Mishra et al. 2007). To et al. demonstrated that in S1 drug-resistant cells, there was a ~1500-bp deletion in the 3'UTR of ABCG2 mRNA, where a putative miR-519c binding site was located. ABCG2 mRNA adopts a longer 3'UTR in the parental S1 colon cancer cell line than its mitoxantrone-resistant counterpart S1. miR-519c decreases ABCG2 mRNA and protein levels by acting through a putative binding site located within the longer 3'UTR in parental cells, but the binding site for this miRNA is lost in the shorter 3'UTR in the resistant cells, therefore mRNA degradation and/or repression on protein translation contributes to the overexpression of ABCG2. These findings suggest that escaping from miRNAs-mediated translational repression and mRNA degradation could lead to overexpression of ABCG2 in drug-resistant colon cancer cells (To et al. 2008). Recently, KRAS mutation status has been demonstrated to be a predictive marker of clinical benefit to monoclonal antibodies targeting EGFR. An *in vitro* study demonstrated that a miRNA polymorphism in the let-7 miRNA complementary-binding site (lcs6) of the 3'UTR of KRAS gene was associated with increased KRAS expression (Chin et al. 2008). Zhang et al. studied this miRNA-SNP in 130 mCRC patients who were refractory to fluoropyrimidine, irinotecan, and oxaliplatin, and treated with cetuximab as monotherapy. They found a significant association of the miRNA-SNP with the objective response rate suggesting the SNP as an additional biomarker in KRAS wild-type patients for better predicting cetuximab responsiveness (Zhang et al. 2011b).

6.2 miRNAs Expression Associated with Drug Resistance

Different approaches can be used for identifying miRNAs involved in drug resistance. miRNAs profiles of drug resistance cell lines comparing with parental sensitive cell lines, as well as miRNAs profiling of responders *versus* no responders patients. Finally, functional studies using a candidate gene approach for identifying miRNAs targeting critical drug response genes have been successful used.

Meng et al. provided evidence that miR-21 regulated apoptosis and increased sensitivity to gemcitabine in malignant human cholangiocytes by PTEN-dependent activation of PI 3-kinase signaling. The authors also investigate miRNAs expression to chemotherapeutic responses *in vivo* by profiling miRNAs expression in tumor cell xenografts in nude mice randomized to treatment with either gemcitabine or control diluent. Gemcitabine treatment was found to determine a greater than 2-fold increase in the expression of a group of miRNAs that included miR-202, miR-320, miR-374, miR-1-2, miR-16-1, miR-24-1, let7f-2, let7a-1, and miR-374. Treatment of tumor cell xenografts with systemic gemcitabine altered the expression of a significant number of miRNAs suggesting that miRNAs might be involved in survival cell response (Meng et al. 2006). Ragusa et al. found that miR-146b-3p and miR-486-5p were more abundant in mutant KRAS patients compared with wild-type ones suggesting miRNAs involvement in EGFR pathway. They also investigated miRNAs profiling in 2 human CRC cell lines, one sensitive and the other resistant to cetuximab (Caco-2 and HCT-116, respectively). Caco-2 and HCT-116 miRNAs profile was also studied after treatment with cetuximab. The authors suggested that the downregulation of let-7b and let-7e and up-regulation of miR-17-3p were potential predictive markers of cetuximab resistance. However, at today, no clinical data confirm these finding (Ragusa et al. 2010).

Let-7g and miR-181b were also shown to be associated with chemosensitivity to S-1 (a pro-drug of 5-fluorouracil, 5-FU) based chemotherapy in colon cancer. The roles of let-7g and miR-181b in chemosensitivity are associated with their effect on several genes such as RAS, cyclin D, C-MYC, E2F1 and cytochrome c. These genes have been shown to be important for the transduction of cell signals, the control of cell cycle and chemosensitivity. However, the detailed molecular and cellular mechanisms of let-7g and miR-181b in mediating translational control will require further studies to elucidate the link of these miRNAs in chemosensitivity to fluoropyrimidine-based drugs in colon cancer.

MiR-34a was identified as one of the down-regulated miRNAs in human colorectal cancer 5-FU-resistant DLD-1 cells compared with those in the parental DLD-1 cells. miR-34a was also found down-regulated in drug resistant prostate cancer cells and ectopic miR-34a expression resulted in cell cycle arrest and growth inhibition and attenuated chemoresistance to the anticancer drug camptothecin (Akao et al. 2011).

Valeri et al. demonstrate that miR-21 targets and down-regulates the core MMR recognition protein complex, human mutS homolog 2 (hMSH2) and 6 (hMSH6). Colorectal tumors that express a high level of miR-21 display reduced hMSH2 protein expression. Mismatch Repair system (MMR) impairment appears to cause reduced incorporation of 5-FU metabolites into DNA, leading to reduced G2/M arrest and apoptosis after 5-FU treatment. Cells that overexpress miR-21 exhibit significantly reduced 5-FU-induced G2/M damage arrest and apoptosis that is characteristic of defects in the core MMR component. These results were confirmed in xenograft studies demonstrating that miR-21 overexpression dramatically reduces the therapeutic efficacy of 5-FU. These studies suggest that the downregulation of the MMR mutator gene associated with miR-21 overexpression may be another important clinical indicator of therapeutic efficacy of 5-FU in CRC (Valeri et al. 2010).

Another study determined that 5-FU and oxaliplatin (L-OHP) down-regulated the expression of miR-197, miR-191, miR-92a, miR-93, miR-222 and miR-826 in HCT-8 and HCT-116 colon cancer cells. These results indicate that 5-FU and L-OHP mechanism of action could rely in part on their influence on the down-regulated miRNAs expression providing novel molecular markers (Zhou et al. 2010).

Svoboda et al. reported that median levels of miR-125b and miR-137 were upregulated in rectal cancer patients after a short-course of capecitabine-based chemoradiotherapy, and higher induction of miR-125b and miR-137 were associated with worse response to the treatment. For the first time, miRNAs modification expression was investigated during the therapy process, in patients with rectal cancer undergoing chemoradiotherapy with capecitabine. Importantly, while a number of these miRNAs showed distinct variation 2 weeks after starting therapy, showing profound inter-tumoral variability, miR-125b and miR-137 demonstrated a significant induction and similar expression trends. The increased levels of both miRNAs correlated with minor response to therapy and with higher post-surgery tumor stage suggesting that higher induced levels of miR-125b and miR-137 might be associated with worse response to radiotherapy with capecitabine (Svoboda et al. 2008, Boni et al. 2010). Recently, Della Vittoria Scarpati et al. studied miRNAs expression in tumor biopsies of patients with rectal cancer to identify a specific "signature" correlating with pathological complete response (pCR) in 38 rectal cancer patients receiving capecitabine-oxaliplatin and radiotherapy followed by surgery. miR-622 and miR-630 had a 100% sensitivity and specificity in selecting patients with complete pathological response and may represent specific predictors of response to chemoradiotherapy in rectal cancer patients (Della Vittoria Scarpati et al. 2011).

On the other hand, we showed recently the functional effects of miR-192/215 on cell cycle, likely due to their pleiotropic mechanism of action, reduce the 5-fluorouracil (5-FU) activity. 5-FU is an S-phase specific drug and previous studies underline the determinant role of cell cycle regulation in 5-FU sensitivity. Moreover, we found that miR-192/215 acts increasing p21 expression. In agreement with our results CRC cell lines resistant to 5-FU had 3-fold higher p21 expression compared to sensitive cell lines. Our findings also suggest a potential role of these miRNAs as biomarkers in CRC. According with recent studies, we found an anti-proliferative effect of miR-192/215 overexpression in CRC cell lines (Boni et al. 2010). Braun et al. found that the expression of miRNAs-192/215 was down-regulated in cancer tissue compared to normal mucosa (Braun et al. 2008). Moreover, two studies found a downregulation of miR-192 in CRC tissues when compared to the normal colon. This evidence highlights the important role of miR-192 in CRC development and supports the idea that miR-192 might carry out a tumor suppression function. Moreover, cell cycle arrest in response to DNA damage is an important anti-tumorigenic mechanism.

miRNAs microarrays of sensitive and MTX-resistant HT29 colon cancer cells identified miR-224 as differentially expressed. Downexpression of miR-224 was also observed in CaCo-2 and K562 cells resistant to MTX. Functional assays using an anti-miR against miR-224 desensitized the cells towards MTX, mimicking the resistant phenotype (Mencia et al. 2011). Similarly, miR-140 has been identifies with osteosarcoma and colon MTX-resistance. In this study, high-throughput miRNAs expression analysis revealed that the expression of miR-140 has been related with chemosensitivity in osteosarcoma tumor xenografts. Tumor cells ectopically transfected with miR-140 were more resistant to methotrexate and 5-fluorouracil (5-FU). Blocking endogenous miR-140 by LNA-modified anti-miR partially sensitized resistant colon cancer stem-like cells to 5-FU treatment. miR-140 induced p53 and p21 expression accompanied with G1 and G2 phase arrest only in cell lines containing wild-type p53. Histone deacetylase 4 (HDAC4) was confirmed to be one of the important targets of miR-140 (Song et al. 2009).

Increasing evidence has suggested cancer stem cells (CSCs) are considered to be responsible for cancer formation, recurrence, and metastasis. Recently, many studies have also revealed that miRNAs are strongly implicated in regulating self-renewal and tumorigenicity of CSCs in human cancers. However, with respect to colon cancer, the role of miRNAs in stemness maintenance and tumorigenicity of CSCs still remains to be unknown. Zhang et al. isolated a population of colon CSCs expressing a CD133 surface phenotype from human HT29 colonic adenocarcinoma cell line by flow cytometry cell sorting. The CD133(+) cells possess a greater tumor sphere-forming efficiency *in vitro* and higher tumorigenic potential *in vivo*. The authors identified a colon CSCs miRNAs signature comprising 11 overexpressed and 8 underexpressed miRNAs, such as miR-429, miR-155, and miR-320d, some of which may be involved in regulation of stem cell differentiation (Zhang et al. 2011a). Similarly, we compare miRNAs expression in colon spheres and differentiated colorectal cancer cell lines. The phenotype of colon spheres was confirmed by the expression of several CSCs markers, by their high capacity to initiate xenograft tumors and by their chemoresistance. miR-451 was down-regulated in colon spheres and their restoration reduced the number and size of colon spheres. Moreover, miR-451 overexpression in colon spheres decreased the colon spheres tumorogenity and increased the sensitivity to SN38. We identified COX-2 as a mir-451 related gene involved in colon spheres growth and the ATP-binding cassette (ABCB1) drug transporter as a target of miR-451 directly involved in SN38 sensitization (Bitarte et al. 2011).

LIN28B is a homologue of LIN28 that induces pluripotency when expressed in conjunction with OCT4, SOX2, and KLF4 in somatic fibroblasts. LIN28B represses biogenesis of let-7 microRNAs and is implicated in both development and tumorigenesis. It has been demonstrated that LIN28B is overexpressed in colon tumors and correlates with reduced patient survival and increased probability of tumor recurrence. Constitutive overexpression of LIN28B in colon cancer cells was associated to increased expression of colonic stem cell markers LGR5 and PROM1, mucinous differentiation, and metastasis (King et al. 2011).

Many efforts have been exerted in analyzing the role of miRNAs in the development of drug resistance in a variety of malignancies, including CRC. Several research groups have shown that the expressions of miRNAs in chemoresistant cancer cells and their parental chemosensitive ones are different. The molecular targets and mechanisms of chemosensitivity and chemoresistance are also successfully studies. These results suggest a great potential for incorporate miRNAs to individualized treatment.

7. CONCLUSIONS

miRNAs has emerged as important players in the regulation of gene expression. The microRNome dysregulation is a common feature of several diseases, especially cancer. To date, many efforts have focused on studying miRNAs expression patterns, as well as miRNAs target validation. Strong evidence of their consistent dysregulation in human neoplasms through mechanisms linked to genomic alterations, oncogene activation or epigenetic events, indicate that they may act as oncogenes or tumor suppressor genes.

Luo et al. provide an overview of miRNAs expression studies in CRC published to date. They summarized the results of 23 studies which reported more than 700 miRNAs in 1525 cases of CRC/adenomas and 1102 controls. Among them, 160 miRNAs were found to be significantly dysregulated in at least one study. Out of them, miR-31 (upregulated) and miR-145 (downregulated) were most often described as dysregulated in CRC (both in 8 different studies) (Luo et al. 2011). This review pointed out the requirement to standarization of miRNAs expression analysis in order to translate basic investigation to clinic.

In this sense, serum or stool-based miRNAs has emerged as new potential non-invasive methods to identify diagnostic, prognostic and predictive biomarkers in CRC. Most likely, a signature of miRNAs rather than a single miRNA will be needed to reach enough sensitivity and specificity for an early detection or predictive tests.

Further research exploring the outcome of therapeutic treatment associated with different miRNAs profiles could provide valuable information to advance drug regimen selection. As we gain understanding of specific miRNAs mechanisms and function it could become evident that certain chemotherapeutic drugs might be favorable based on their mode of action in relation to a particular miRNAs profile.

Finally, the fact that miRNAs dysregulation in CRC has a pathogenic effect, provides the rationale for using miRNAs as potential therapeutic targets in CRC cancer. miRNAs-based cancer therapy is a very interesting field of investigation that offers the appeal of targeting multiple gene networks controlled by a single miRNA.

8. SUMMARY POINTS

- miRNAs expression is dysregulated in CRC and play an essential role in carcinogenesis process by regulating multiple gene targets.
- Several miRNAs detection methods have been developed in the last decade but standardization methods are necessary to obtain reproducible results.
- miRNAs are good candidates as non-invasive biomarkers in plasma and feces samples in CRC.
- Essential cellular process involved in key signaling pathways of CRC, including cell proliferation, invasion, angiogenesis and metastasis process are regulated by miRNAs.
- Identification of miRNAs target genes in CRC is essential to identify new therapeutic targets and to overcome drug resistance.

ABBREVIATIONS

5-aza-dC	:	5-aza-2′-deoxycitidine
ABCB1	:	ATP-binding cassette, sub-family B (MDR/TAP), member 1
ABCG2	:	ATP-binding cassette sub-family G member 2
AKT	:	v-akt murine thymoma viral oncogene homolog
APC	:	adenomatous polyposis coli
APOBEC3G	:	apolipoprotein B mRNA-editing, enzyme-catalytic, polypeptide-like 3G
ChIP	:	chromatin immunoprecipitation
DGCR8	:	DiGeorge syndrome critical region 8
DHFR	:	dihydrofolate reductase

DNMT3A	:	DNA (cytosine-5)-methyltransferase 3A
EGFL7	:	EGF-like domain-containing protein 7
FAK	:	integrin-focal adhesion kinase
FOBT	:	fecal occult blood test
FOXO4	:	forkhead box protein O4
GFF	:	glass fiber filter
HDAC4	:	histone deacetylase 4
HELZ	:	helicase with zinc finger
HIF-1α	:	hypoxia inducible factor 1α
hMSH2	:	human mutS homolog 2
hMSH6	:	human mutS homolog 6
ILK	:	integrin-linked kinase
KLF4	:	Kruppel-like factor 4
LGR5	:	Leucine-rich repeat-containing G-protein coupled receptor 5
LIPC	:	hepatic lipase
LNA	:	locked nucleic acid
L-OHP	:	oxaliplatin
LSD-1	:	lysine-specific demethylase 1
MACC1	:	metastasis-associated in colon cancer 1
MET	:	tyrosine kinase receptor for hepatocyte growth factor
MIF	:	macrophage migration inhibitory factor
miRAGE	:	miRNA serial analysis of gene expression
MMP2	:	metalloproteinase-2
MSI	:	microsatellite instability
MTX	:	Methotrexate
MYC	:	myelocytomatosis viral oncogene homolog
NGS	:	next-generation sequencing
OCT4	:	octamer-binding transcription factor 4
PDCD4	:	programmed death cell protein 4
PI3K	:	phosphatidylinositol-3 kinase
PIK3C2A	:	Phosphatidylinositol-4-phosphate 3-kinase C2 domain-containing alpha polypeptide
PROM1(CD133)	:	prominin 1
PTEN	:	phosphatase and tensin homolog
PUMA	:	p53 upregulated modulator of apoptosis
RISC	:	RNA-induced silencing complex
S-1	:	pro-drug of 5-FU
S100P	:	S100 calcium binding protein
SIRT1	:	silent information regulator 1
SOX2	:	sex determining Region Y-box 2
TCF4	:	transcription factor 4

TRBP : TAR RNA binding protein
VEGF : vascular endothelial growth factor
WRL-68 : fetal hepatic cell line
ZEB1 : zinc-finger E-box binding homeobox 1

REFERENCES

Ahmed, F.E., C.D. Jeffries, P.W. Vos, G. Flake, G.J. Nuovo, D.R. Sinar, W. Naziri, and S.P. Marcuard. 2009. Diagnostic microRNA markers for screening sporadic human colon cancer and active ulcerative colitis in stool and tissue. Cancer Genomics Proteomics. 6: 281–95.

Akao, Y., S. Noguchi, A. Iio, K. Kojima, T. Takagi, and T. Naoe. 2011. Dysregulation of microRNA-34a expression causes drug-resistance to 5-FU in human colon cancer DLD-1 cells. Cancer Lett. 300: 197–204.

Ambros, V. 2004. The functions of animal microRNAs. Nature. 431: 350–5.

Balaguer, F., A. Link, J.J. Lozano, M. Cuatrecasas, T. Nagasaka, C.R. Boland, and A. Goel. 2011. Epigenetic silencing of miR-137 is an early event in colorectal carcinogenesis. Cancer Res. 70: 6609–18.

Bandres, E., X. Agirre, N. Bitarte, N. Ramirez, R. Zarate, J. Roman-Gomez, F. Prosper, and J. Garcia-Foncillas. 2009a. Epigenetic regulation of microRNA expression in colorectal cancer. Int J Cancer. 125: 2737–43.

Bandres, E., N. Bitarte, F. Arias, J. Agorreta, P. Fortes, X. Agirre, R. Zarate, J.A. Diaz-Gonzalez, N. Ramirez, J.J. Sola, P. Jimenez, J. Rodriguez, and J. Garcia-Foncillas. 2009b. microRNA-451 regulates macrophage migration inhibitory factor production and proliferation of gastrointestinal cancer cells. Clin Cancer Res. 15: 2281–90.

Bandres, E., E. Cubedo, X. Agirre, R. Malumbres, Zarate, N. Ramirez, A. Abajo, A. Navarro, I. Moreno, M. Monzo, and J. Garcia-Foncillas. 2006. Identification by Real-time PCR of 13 mature microRNAs differentially expressed in colorectal cancer and non-tumoral tissues. Mol Cancer. 5: 29.

Baraniskin, A., K. Birkenkamp-Demtroder, A. Maghnouj, H. Zollner, J. Munding, S. Klein-Scory, A. Reinacher-Schick, I. Schwarte-Waldhoff, W. Schmiegel, and S.A. Hahn. 2012. MiR-30a-5p suppresses tumor growth in colon carcinoma by targeting DTL. Carcinogenesis. 33: 732–9.

Bartel, D.P. 2004. MicroRNAs: genomics, biogenesis, mechanism, and function. Cell. 116: 281–97.

Bertino, J.R., D. Banerjee, and P.J. Mishra. 2007. Pharmacogenomics of microRNA: a miRSNP towards individualized therapy. Pharmacogenomics. 8: 1625–7.

Bitarte, N., E. Bandres, V. Boni, R. Zarate, J. Rodriguez, M. Gonzalez-Huarriz, I. Lopez, J. Javier Sola, M.M. Alonso, P. Fortes, and J. Garcia-Foncillas. 2011 MicroRNA-451 is involved in the self-renewal, tumorigenicity, and chemoresistance of colorectal cancer stem cells. Stem Cells. 29: 1661–71.

Boni, V., N. Bitarte, I. Cristobal, R. Zarate, J. Rodriguez, E. Maiello, J. Garcia-Foncillas, and E. Bandres. 2010. miR-192/miR-215 influence 5-fluorouracil resistance through cell cycle-mediated mechanisms complementary to its post-transcriptional thymidilate synthase regulation. Mol Cancer Ther. 9: 2265–75.

Braun, C. J., X. Zhang, I. Savelyeva, S. Wolff, U.M. Moll, T. Schepeler, T.F. Orntoft, C.L. Andersen, and M. Dobbelstein. 2008. p53-Responsive microRNAs 192 and 215 are capable of inducing cell cycle arrest. Cancer Res. 68: 10094–104.

Calin, G.A., C. Sevignani, C.D. Dumitru, T. Hyslop, E. Noch, S. Yendamuri, M. Shimizu, S. Rattan, F. Bullrich, M. Negrini, and C.M. Croce. 2004. Human microRNA genes are frequently located at fragile sites and genomic regions involved in cancers. Proc Natl Acad Sci USA. 101: 2999–3004.

Calin, G.A. and C.M. Croce. 2006. MicroRNA signatures in human cancers. Nat Rev Cancer. 6: 857–66.

Cissell, K.A. and S.K. Deo. 2009. Trends in microRNA detection. Anal Bioanal Chem. 394: 1109–16.

Creighton, C.J., J.G. Reid, and P.H. Gunaratne. 2009. Expression profiling of microRNAs by deep sequencing. Brief Bioinform. 10: 490–7.

Cummins, J.M., Y. He, R.J. Leary, R. Pagliarini, L.A. Diaz, Jr., T. Sjoblom, O. Barad, Z. Bentwich, A.E. Szafranska, E. Labourier, C.K. Raymond, B.S. Roberts, H. Juhl, K.W. Kinzler, B. Vogelstein, and V.E. Velculescu. 2006. The colorectal microRNAome. Proc Natl Acad Sci USA. 103: 3687–92.

Curry, E., S.E. Ellis, and S.L. Pratt. 2009. Detection of porcine sperm microRNAs using a heterologous microRNA microarray and reverse transcriptase polymerase chain reaction. Mol Reprod Dev. 76: 218–9.

Chang, K.H., P. Mestdagh, J. Vandesompele, M.J. Kerin, and N. Miller. 2010. MicroRNA expression profiling to identify and validate reference genes for relative quantification in colorectal cancer. BMC Cancer. 10: 173.

Chen, X., X. Guo, H. Zhang, Y. Xiang, J. Chen, Y. Yin, X. Cai, K. Wang, G. Wang, Y. Ba, L. Zhu, J. Wang, R. Yang, Y. Zhang, Z. Ren, K. Zen, J. Zhang, and C.Y. Zhang. 2009. Role of miR-143 targeting KRAS in colorectal tumorigenesis. Oncogene. 28: 1385–92.

Chin, L.J., E. Ratner, S. Leng, R. Zhai, S. Nallur, I. Babar, R.U. Muller, E. Straka, L. Su, E.A. Burki, R.E. Crowell, R. Patel, T. Kulkarni, R. Homer, D. Zelterman, K.K. Kidd, Y. Zhu, D.C. Christiani, S.A. Belinsky, F.J. Slack, and J.B. Weidhaas. 2008. A SNP in a let-7 microRNA complementary site in the KRAS 3' untranslated region increases non-small cell lung cancer risk. Cancer Res. 68: 8535–40.

Dai, Y. and X. Zhou. 2010. Computational methods for the identification of microRNA targets. Open Access Bioinformatics. 2: 29–39.

De Planell-Saguer, M. and M.C. Rodicio. 2011. Analytical aspects of microRNA in diagnostics: a review. Anal Chim Acta. 699: 134–52.

Della Vittoria Scarpati, G., F. Falcetta, C. Carlomagno, P. Ubezio, S. Marchini, A. De Stefano, V.K. Singh, M. D'incalci, S. De Placido, and S. Pepe. 2011. A specific miRNA signature correlates with complete pathological response to neoadjuvant chemoradiotherapy in locally advanced rectal cancer. Int J Radiat Oncol Biol Phys. 83: 1113–9.

Diaz, R., J. Silva, J.M. Garcia, Y. Lorenzo, V. Garcia, C. Pena, R. Rodriguez, C. Munoz, F. Garcia, F. Bonilla, and G. Dominguez. 2008. Deregulated expression of miR-106a predicts survival in human colon cancer patients. Genes Chromosomes Cancer. 47: 794–802.

Ding, Q., C.J. Chang, X. Xie, W. Xia, J.Y. Yang, S.C. Wang, Y. Wang, J. Xia, L. Chen, C. Cai, H. Li, C.J. Yen, H.P. Kuo, D.F. Lee, J. Lang, L. Huo, X. Cheng, Y.J. Chen, C.W. Li, L.B. Jeng, J.L. Hsu, L.Y. Li, A. Tan, S.A. Curley, L.M. Ellis, R.N. Dubois, and M.C. Hung. 2011. APOBEC3G promotes liver metastasis in an orthotopic mouse model of colorectal cancer and predicts human hepatic metastasis. J Clin Invest. 121: 4526–36.

Esteller, M. 2011. Non-coding RNAs in human disease. Nat Rev Genet. 12: 861–74.

Guo, C., J.F. Sah, L. Beard, J.K. Willson, S.D. Markowitz, and K. Guda. 2008. The noncoding RNA, miR-126, suppresses the growth of neoplastic cells by targeting phosphatidylinositol 3-kinase signaling and is frequently lost in colon cancers. Genes Chromosomes Cancer. 47: 939–46.

Hu, G., D. Chen, X. Li, K. Yang, H. Wang, and W. Wu. 2010. miR-133b regulates the MET proto-oncogene and inhibits the growth of colorectal cancer cells *in vitro* and *in vivo*. Cancer Biol Ther. 10: 190–7.

Huang, Z., D. Huang, S. Ni, Z. Peng, W. Sheng, and X. Du. 2010. Plasma microRNAs are promising novel biomarkers for early detection of colorectal cancer. Int J Cancer. 127: 118–26.

Iliopoulos, D., S.A. Jaeger, H.A. Hirsch, M.L. Bulyk, and K. Struhl. 2010. STAT3 activation of miR-21 and miR-181b-1 via PTEN and CYLD are part of the epigenetic switch linking inflammation to cancer. Mol Cell. 39: 493–506.

Johnson, S.M., H. Grosshans, J. Shingara, M. Byrom, R. Jarvis, A. Cheng, E. Labourier, K.L. Reinert, D. Brown, and F.J. Slack. 2005. RAS is regulated by the let-7 microRNA family. Cell. 120: 635–47.

Kalimutho, M., G. Del Vecchio Blanco, S. Di Cecilia, P. Sileri, M. Cretella, F. Pallone, G. Federici, and S. Bernardini. 2011a. Differential expression of miR-144* as a novel fecal-based diagnostic marker for colorectal cancer. J Gastroenterol. 46: 1391–402.

Kalimutho, M., S. Di Cecilia, G. Del Vecchio Blanco, F. Roviello, P. Sileri, M. Cretella, A. Formosa, G. Corso, D. Marrelli, F. Pallone, G. Federici, and S. Bernardini. 2011b. Epigenetically silenced miR-34b/c as a novel faecal-based screening marker for colorectal cancer. Br J Cancer. 104: 1770–8.

King, C.E., M. Cuatrecasas, A. Castells, A.R. Sepulveda, J.S. Lee, and A.K. Rustgi. 2011. LIN28B promotes colon cancer progression and metastasis. Cancer Res. 71: 4260–8.

Koga, Y., M. Yasunaga, A. Takahashi, J. Kuroda, Y. Moriya, T. Akasu, S. Fujita, S. Yamamoto, H. Baba, and Y. Matsumura. 2010. MicroRNA expression profiling of exfoliated colonocytes isolated from feces for colorectal cancer screening. Cancer Prev Res Phila. 3: 1435–42.

Koscianska, E., J. Starega-Roslan, L.J. Sznajder, M. Olejniczak, P. Galka-Marciniak, and W.J. Krzyzosiak. 2011. Northern blotting analysis of microRNAs, their precursors and RNA interference triggers. BMC Mol Biol. 12: 14.

Lanza, G., M. Ferracin, R. Gafa, A. Veronese, R. Spizzo, F. Pichiorri, C.G. Liu, G.A. Calin, C.M. Croce, and M. Negrini. 2007. mRNA/microRNA gene expression profile in microsatellite unstable colorectal cancer. Mol Cancer. 6: 54.

Link, A., F. Balaguer, Y. Shen, T. Nagasaka, J.J. Lozano, C.R. Boland, and A. Goel. 2010. Fecal MicroRNAs as novel biomarkers for colon cancer screening. Cancer Epidemiol Biomarkers Prev. 19: 1766–74.

Liu, X., Z. Zhang, L. Sun, N. Chai, S. Tang, J. Jin, H. Hu, Y. Nie, X. Wang, K. Wu, H. Jin, and D. Fan. 2011. MicroRNA-499-5p promotes cellular invasion and tumor metastasis in colorectal cancer by targeting FOXO4 and PDCD4. Carcinogenesis. 32: 1798–805.

Lu, J., G. Getz, E.A. Miska, E. Alvarez-Saavedra, J. Lamb, D. Peck, A. Sweet-Cordero, B.L. Ebert, R.H. Mak, A.A. Ferrando, J.R. Downing, T. Jacks, H.R. Horvitz, and T.R. Golub. 2005. MicroRNA expression profiles classify human cancers. Nature. 435: 834–8.

Lujambio, A. and M. Esteller. 2007. CpG island hypermethylation of tumor suppressor microRNAs in human cancer. Cell Cycle. 6: 1455–9.

Lujambio, A., S. Ropero, E. Ballestar, M.F. Fraga, C. Cerrato, F. Setien, S. Casado, A. Suarez-Gauthier, M. Sanchez-Cespedes, A. Git, I. Spiteri, P.P. Das, C. Caldas, E. Miska, and M. Esteller. 2007. Genetic unmasking of an epigenetically silenced microRNA in human cancer cells. Cancer Res. 67: 1424–9.

Luo, X., B. Burwinkel, S. Tao, and H. Brenner. 2011. MicroRNA signatures: novel biomarker for colorectal cancer? Cancer Epidemiol Biomarkers Prev. 20: 1272–86.

MacFarlane, L.A. and P.R. Murphy. 2010. MicroRNA: Biogenesis, Function and Role in Cancer. Curr Genomics. 11: 537–61.

Matsushita, H., Y. Matsumura, Y. Moriya, T. Akasu, S. Fujita, S. Yamamoto, S. Onouchi, N. Saito, M. Sugito, M. Ito, T. Kozu, T. Minowa, S. Nomura, H. Tsunoda, and T. Kakizoe. 2005. A new method for isolating colonocytes from naturally evacuated feces and its clinical application to colorectal cancer diagnosis. Gastroenterology. 129: 1918–27.

Mencia, N., E. Selga, V. Noe, and C.J. Ciudad. 2011. Underexpression of miR-224 in methotrexate resistant human colon cancer cells. Biochem Pharmacol. 82: 1572–82.

Meng, F., R. Henson, M. Lang, H. Wehbe, A. Maheshwari, J.T. Mendell, J. Jiang, T.D. Schmittgen, and T. Patel. 2006. Involvement of human micro-RNA in growth and response to chemotherapy in human cholangiocarcinoma cell lines. Gastroenterology. 130: 2113–29.

Meyer, S.U., M.W. Pfaffl, and S.E. Ulbrich. 2010. Normalization strategies for microRNA profiling experiments: a 'normal' way to a hidden layer of complexity? Biotechnol Lett. 32: 1777–88.

Michael, M.Z., O.C. Sm, N.G. Van Holst Pellekaan, G.P. Young, and R.J. James. 2003. Reduced accumulation of specific microRNAs in colorectal neoplasia. Mol Cancer Res. 1: 882–91.

Migliore, C., V. Martin, V.P. Leoni, A. Restivo, L. Atzori, A. Petrelli, C. Isella, L. Zorcolo, I. Sarotto, G. Casula, P.M. Comoglio, A. Columbano, and S. Giordano. 2012. MiR-1 Downregulation Cooperates with MACC1 in Promoting MET Overexpression in Human Colon Cancer. Clin Cancer Res. 18: 737–47.

Mishra, P.J., R. Humeniuk, G.S. Longo-Sorbello, D. Banerjee, and J.R. Bertino. 2007. A miR-24 microRNA binding-site polymorphism in dihydrofolate reductase gene leads to methotrexate resistance. Proc Natl Acad Sci USA. 104: 13513–8.

Mitchell, P.S., R.K. Parkin, E.M. Kroh, B.R. Fritz, S.K. Wyman, E.L. Pogosova-Agadjanyan, A. Peterson, J. Noteboom, K.C. O'briant, A. Allen, D.W. Lin, N. Urban, C.W. Drescher, B.S. Knudsen, D.L. Stirewalt, R. Gentleman, R.L. Vessella, P.S. Nelson, D.B. Martin, and M. Tewari. 2008. Circulating microRNAs as stable blood-based markers for cancer detection. Proc Natl Acad Sci USA. 105: 10513–8.

Monzo, M., A. Navarro, E. Bandres, R. Artells, I. Moreno, B. Gel, R. Ibeas, J. Moreno, F. Martinez, T. Diaz, A. Martinez, O. Balague, and J. Garcia-Foncillas. 2008. Overlapping expression of microRNAs in human embryonic colon and colorectal cancer. Cell Res. 18: 823–33.

Motoyama, K., H. Inoue, Y. Takatsuno, F. Tanaka, K. Mimori, H. Uetake, K. Sugihara, and M. Mori. 2009. Over- and under-expressed microRNAs in human colorectal cancer. Int J Oncol. 34: 1069–75.

Nagel, R., C. Le Sage, B. Diosdado, M. Van Der Waal, J.A. Oude Vrielink, A. Bolijn, G.A. Meijer, and R. Agami. 2008. Regulation of the adenomatous polyposis coli gene by the miR-135 family in colorectal cancer. Cancer Res. 68: 5795–802.

Nakajima, G., K. Hayashi, Y. Xi, K. Kudo, K. Uchida, K. Takasaki, M. Yamamoto, and J. Ju. 2006. Non-coding MicroRNAs hsa-let-7g and hsa-miR-181b are Associated with Chemoresponse to S-1 in Colon Cancer. Cancer Genomics Proteomics. 3: 317–324.

Ng, E.K., W.W. Chong, H. Jin, E.K. Lam, V.Y. Shin, J. Yu, T.C. Poon, S.S. Ng, and J.J. Sung. 2009a. Differential expression of microRNAs in plasma of patients with colorectal cancer: a potential marker for colorectal cancer screening. Gut. 58: 1375–81.

Ng, E.K., W.P. Tsang, S.S. Ng, H.C. Jin, J. Yu, J.J. Li, C. Rocken, M.P. Ebert, T.T. Kwok, and J.J. Sung. 2009b. MicroRNA-143 targets DNA methyltransferases 3A in colorectal cancer. Br J Cancer. 101: 699–706.

Nuovo, G., E.J. Lee, S. Lawler, J. Godlewski, and T. Schmittgen. 2009. *In situ* detection of mature microRNAs by labeled extension on ultramer templates. Biotechniques. 46: 115–26.

Oberg, A.L., A.J. French, A.L. Sarver, S. Subramanian, B.W. Morlan, S.M. Riska, P.M. Borralho, J.M. Cunningham, L.A. Boardman, L. Wang, T.C. Smyrk, Y. Asmann, C.J. Steer, and S.N. Thibodeau. 2011. miRNA expression in colon polyps provides evidence for a multihit model of colon cancer. PLoS One. 6: e20465.

Obernosterer, G., J. Martinez, and M. Alenius. 2007. Locked nucleic acid-based *in situ* detection of microRNAs in mouse tissue sections. Nat Protoc. 2: 1508–14.

Oneyama, C., E. Morii, D. Okuzaki, Y. Takahashi, J. Ikeda, N. Wakabayashi, H. Akamatsu, M. Tsujimoto, T. Nishida, K. Aozasa, and M. Okada. 2011. MicroRNA-mediated upregulation of integrin-linked kinase promotes Src-induced tumor progression. Oncogene. 31: 1623–35.

Papetti, M. and L.H. Augenlicht. 2011. Mybl2, downregulated during colon epithelial cell maturation, is suppressed by miR-365. Am J Physiol Gastrointest Liver Physiol. 301: G508–18.

Paris, R., R.E. Henry, S.J. Stephens, M. Mcbryde, and J.M. Espinosa. 2008. Multiple p53-independent gene silencing mechanisms define the cellular response to p53 activation. Cell Cycle. 7: 2427–33.

Podolska, A., B. Kaczkowski, T. Litman, M. Fredholm, and S. Cirera. 2011. How the RNA isolation method can affect microRNA microarray results. Acta Biochim Pol. 58: 535–40.

Pu, X.X., G.L. Huang, H.Q. Guo, C.C. Guo, H. Li, S. Ye, S. Ling, L. Jiang, Y. Tian, and T.Y. Lin. 2010. Circulating miR-221 directly amplified from plasma is a potential diagnostic and prognostic marker of colorectal cancer and is correlated with p53 expression. J Gastroenterol Hepatol. 25: 1674–80.

Ragusa, M., A. Majorana, L. Statello, M. Maugeri, L. Salito, D. Barbagallo, M.R. Guglielmino, L.R. Duro, R. Angelica, R. Caltabiano, A. Biondi, M. Di Vita, G. Privitera, M. Scalia, A. Cappellani, E. Vasquez, S. Lanzafame, F. Basile, C. Di Pietro, and M. Purrello. 2010. Specific alterations of microRNA transcriptome and global network structure in colorectal carcinoma after cetuximab treatment. Mol Cancer Ther. 9: 3396–409.

Schepeler, T., J.T. Reinert, M.S. Ostenfeld, L.L. Christensen, A.N. Silahtaroglu, L. Dyrskjot, C. Wiuf, F.J. Sorensen, M. Kruhoffer, S. Laurberg, S. Kauppinen, T.F. Orntoft, and C.L. Andersen. 2008. Diagnostic and prognostic microRNAs in stage II colon cancer. Cancer Res. 68: 6416–24.

Schepeler, T., A. Holm, P. Halvey, I. Nordentoft, P. Lamy, E.M. Riising, L.L. Christensen, K. Thorsen, D.C. Liebler, K. Helin, T.F. Orntoft, and C.L. Andersen. 2012. Attenuation of the beta-catenin/TCF4 complex in colorectal cancer cells induces several growth-suppressive microRNAs that target cancer promoting genes. Oncogene. 31: 2750–60.

Schetter, A.J., S.Y. Leung, J.J. Sohn, K.A. Zanetti, E.D. Bowman, N. Yanaihara, S.T. Yuen, T.L. Chan, D.L. Kwong, G.K. Au, C.G. Liu, G.A. Calin, C.M. Croce, and C.C. Harris. 2008. MicroRNA expression profiles associated with prognosis and therapeutic outcome in colon adenocarcinoma. JAMA. 299: 425–36.

Siemens, H., R. Jackstadt, S. Hunten, M. Kaller, A. Menssen, U. Gotz, and H. Hermeking. 2011. miR-34 and SNAIL form a double-negative feedback loop to regulate epithelial-mesenchymal transitions. Cell Cycle. 10: 4256–71.

Slaby, O., M. Svoboda, P. Fabian, T. Smerdova, D. Knoflickova, M. Bednarikova, R. Nenutil, and R. Vyzula. 2007. Altered expression of miR-21, miR-31, miR-143 and miR-145 is related to clinicopathologic features of colorectal cancer. Oncology. 72: 397–402.

Slattery, M.L., E. Wolff, M.D. Hoffman, D.F. Pellatt, B. Milash, and R.K. Wolff. 2011. MicroRNAs and colon and rectal cancer: differential expression by tumor location and subtype. Genes Chromosomes Cancer. 50: 196–206.

Song, B., Y. Wang, Y. Xi, K. Kudo, S. Bruheim, G.I. Botchkina, E. Gavin, Y. Wan, A. Formentini, M. Kornmann, O. Fodstad, and J. Ju. 2009. Mechanism of chemoresistance mediated by miR-140 in human osteosarcoma and colon cancer cells. Oncogene. 28: 4065–74.

Strillacci, A., C. Griffoni, P. Sansone, P. Paterini, G. Piazzi, G. Lazzarini, E. Spisni, M.A. Pantaleo, G. Biasco, and V. Tomasi. 2009. MiR-101 downregulation is involved in cyclooxygenase-2 overexpression in human colon cancer cells. Exp Cell Res. 315: 1439–47.

Sundaram, P., S. Hultine, L.M. Smith, M. Dews, J.L. Fox, D. Biyashev, J.M. Schelter, Q. Huang, M.A. Cleary, O.V. Volpert, and A. Thomas-Tikhonenko. 2011. p53-responsive miR-194 inhibits thrombospondin-1 and promotes angiogenesis in colon cancers. Cancer Res. 71: 7490–501.

Suzuki, H., S. Takatsuka, H. Akashi, E. Yamamoto, M. Nojima, R. Maruyama, M. Kai, H.O. Yamano, Y. Sasaki, T. Tokino, Y. Shinomura, K. Imai, and M. Toyota. 2011. Genome-wide profiling of chromatin signatures reveals epigenetic regulation of MicroRNA genes in colorectal cancer. Cancer Res. 71: 5646–58.

Svoboda, M., L. Izakovicova Holla, R. Sefr, I. Vrtkova, I. Kocakova, B. Tichy, and J. Dvorak. 2008. Micro-RNAs miR125b and miR137 are frequently upregulated in response to capecitabine chemoradiotherapy of rectal cancer. Int J Oncol. 33: 541–7.

Takada, S. and H. Mano. 2007. Profiling of microRNA expression by mRAP. Nat Protoc. 2: 3136–45.

Tang, J.T., J.L. Wang, W. Du, J. Hong, S.L. Zhao, Y.C. Wang, H. Xiong, H.M. Chen, and J.Y. Fang. 2011. MicroRNA 345, a methylation-sensitive microRNA is involved in cell proliferation and invasion in human colorectal cancer. Carcinogenesis. 32: 1207–15.

To, K.K., Z. Zhan, T. Litman, and S.E. Bates. 2008. Regulation of ABCG2 expression at the 3' untranslated region of its mRNA through modulation of transcript stability and protein translation by a putative microRNA in the S1 colon cancer cell line. Mol Cell Biol. 28: 5147–61.

Tong, J.L., C.P. Zhang, F. Nie, X.T. Xu, M.M. Zhu, S.D. Xiao, and Z.H. Ran. 2011. MicroRNA 506 regulates expression of PPAR alpha in hydroxycamptothecin-resistant human colon cancer cells. FEBS Lett. 585: 3560–8.

Tsang, W.P. and T.T. Kwok. 2009. The miR-18a* microRNA functions as a potential tumor suppressor by targeting on K-Ras. Carcinogenesis. 30: 953–9.

Tsuchida, A., S. Ohno, W. Wu, N. Borjigin, K. Fujita, T. Aoki, S. Ueda, M. Takanashi, and M. Kuroda. 2011. miR-92 is a key oncogenic component of the miR-17-92 cluster in colon cancer. Cancer Sci. 102: 2264–71.

Tsui, N.B., E.K. Ng, and Y.M. Lo. 2002. Stability of endogenous and added RNA in blood specimens, serum, and plasma. Clin Chem. 48: 1647–53.

Turchinovich, A., L. Weiz, A. Langheinz, and B. Burwinkel. 2011. Characterization of extracellular circulating microRNA. Nucleic Acids Res. 39: 7223–33.

Valeri, N., P. Gasparini, C. Braconi, A. Paone, F. Lovat, M. Fabbri, K.M. Sumani, H. Alder, D. Amadori, T. Patel, G.J. Nuovo, R. Fishel, and C.M. Croce. 2010. MicroRNA-21 induces resistance to 5-fluorouracil by down-regulating human DNA MutS homolog 2 hMSH2. Proc Natl Acad Sci USA. 107: 21098–103.

Varallyay, E., J. Burgyan, and Z. Havelda. 2008. MicroRNA detection by northern blotting using locked nucleic acid probes. Nat Protoc. 3: 190–6.

Volinia, S., G.A. Calin, C.G. Liu, S. Ambs, A. Cimmino, F. Petrocca, R. Visone, M. Iorio, C. Roldo, M. Ferracin, R.L. Prueitt, N. Yanaihara, G. Lanza, A. Scarpa, A. Vecchione, M. Negrini, C.C. Harris, and C.M. Croce. 2006. A microRNA expression signature of human solid tumors defines cancer gene targets. Proc Natl Acad Sci USA. 103: 2257–61.

Wang, C.J., Z.G. Zhou, L. Wang, L. Yang, B. Zhou, J. Gu, H.Y. Chen, and X.F. Sun. 2009a. Clinicopathological significance of microRNA-31, -143 and -145 expression in colorectal cancer. Dis Markers. 26: 27–34.

Wang, P., F. Zou, X. Zhang, H. Li, A. Dulak, R.J. Tomko, Jr., J.S. Lazo, Z. Wang, L. Zhang, and J. Yu. 2009b. microRNA-21 negatively regulates Cdc25A and cell cycle progression in colon cancer cells. Cancer Res. 69: 8157–65.

Wellner, U., J. Schubert, U.C. Burk, O. Schmalhofer, F. Zhu, A. Sonntag, B. Waldvogel, C. Vannier, D. Darling, A. Zur Hausen, V.G. Brunton, J. Morton, O. Sansom, J. Schuler, M.P. Stemmler, C. Herzberger, U. Hopt, T. Keck, S. Brabletz, and T. Brabletz. 2009. The EMT-activator ZEB1 promotes tumorigenicity by repressing stemness-inhibiting microRNAs. Nat Cell Biol. 11: 1487–95.

Wu, J., G. Wu, L. Lv, Y.F. Ren, X.J. Zhang, Y.F. Xue, G. Li, X. Lu, Z. Sun, and K.F. Tang. 2012. MicroRNA-34a inhibits migration and invasion of colon cancer cells via targeting to Fra-1. Carcinogenesis. 33: 519–28.

Xi, Y., A. Formentini, M. Chien, D.B. Weir, J.J. Russo, J. Ju, M. Kornmann, and J. Ju. 2006. Prognostic values of microRNAs in colorectal cancer. Biomark Insights. 2: 113–121.

Xi, Y., G. Nakajima, E. Gavin, C.G. Morris, K. Kudo, K. Hayashi, and J. Ju. 2007. Systematic analysis of microRNA expression of RNA extracted from fresh frozen and formalin-fixed paraffin-embedded samples. Rna. 13: 1668–74.

Yamakuchi, M., M. Ferlito, and C.J. Lowenstein. 2008. miR-34a repression of SIRT1 regulates apoptosis. Proc Natl Acad Sci USA. 105: 13421–6.

Yamakuchi, M. and C.J. Lowenstein. 2009. MiR-34, SIRT1 and p53: the feedback loop. Cell Cycle. 8: 712–5.

Yamakuchi, M., C.D. Lotterman, C. Bao, R.H. Hruban, B. Karim, J.T. Mendell, D. Huso, and C.J. Lowenstein. 2010. P53-induced microRNA-107 inhibits HIF-1 and tumor angiogenesis. Proc Natl Acad Sci USA. 107: 6334–9.

Yamakuchi, M., S. Yagi, T. Ito, and C.J. Lowenstein. 2011. MicroRNA-22 regulates hypoxia signaling in colon cancer cells. PLoS One. 6: e20291.

Yu, X.F., J. Zou, Z.J. Bao, and J. Dong. 2011. miR-93 suppresses proliferation and colony formation of human colon cancer stem cells. World J Gastroenterol. 17: 4711–7.

Zhang, H., W. Li, F. Nan, F. Ren, H. Wang, Y. Xu, and F. Zhang. 2011a. MicroRNA expression profile of colon cancer stem-like cells in HT29 adenocarcinoma cell line. Biochem Biophys Res Commun. 404: 273–8.

Zhang, W., T. Winder, Y. Ning, A. Pohl, D. Yang, M. Kahn, G. Lurje, M.J. Labonte, P.M. Wilson, M.A. Gordon, S. Hu-Lieskovan, D.J. Mauro, C. Langer, E.K. Rowinsky, and H.J. Lenz. 2011b. A let-7 microRNA-binding site polymorphism in 3'-untranslated region of KRAS gene predicts response in wild-type KRAS patients with metastatic colorectal cancer treated with cetuximab monotherapy. Ann Oncol. 22: 104–9.

Zhou, J., Y. Zhou, B. Yin, W. Hao, L. Zhao, W. Ju, and C. Bai. 2010. 5-Fluorouracil and oxaliplatin modify the expression profiles of microRNAs in human colon cancer cells *in vitro*. Oncol Rep. 23: 121–8.

CHAPTER 9

MicroRNAs IN LUNG CANCER: FROM GENOMICS TO CLINICAL APPLICATIONS

Elena Arechaga-Ocampo,[1,a,*] Jose Diaz-Chavez,[2,b]
Cesar Lopez-Camarillo,[3] Eduardo Lopez-Urrutia,[4]
Claudia H. Gonzalez de la Rosa,[5] Carlos Perez-
Plasencia,[1,c,6] Yamilet Noriega Reyes[1,d] and
Luis A. Herrera[2]

ABSTRACT

Lung cancer is the neoplasia with the highest incidence and mortality in men and women worldwide. Lung cancer is classified into two large histologic subgroups: small cell lung carcinoma (SCLC) and non-small cell lung carcinoma (NSCLC) according to cellular origin, molecular changes, clinical-pathological features, and response to treatments. NSCLC constitutes 80% of the all cases of lung cancer; nevertheless the molecular mechanisms underlying the tumoral etiology still remain poorly understood. MicroRNAs (miRNAs) are evolutionarily conserved small noncoding RNAs that negatively regulate gene expression at the post-transcriptional level by repressing translation or decreasing mRNA stability in numerous biological processes. In cancer, miRNAs have differential spatial and temporal expression, which is related to several clinical, biological, molecular and genomic features of tumors. miRNA expression profiles, polymorphisms and epigenetic modification in

Authors' affiliations given at the end of the chapter.

NSCLC have been studied, and these studies have improved NSCLC diagnosis and classification, as well as provided prognostic information. In this review, we describe some of the better-characterized miRNAs in NSCLC and how they could improve lung cancer prognosis and therapy. We summarize the current understanding in expression of miRNAs and their involvement in process biology of NSCLC, as well as their potential as biomarkers for risk stratification, outcome prediction and classification of histologic subtypes, in addition to circulating biomarkers and miRNA-based NSCLC therapy.

1. INTRODUCTION

MicroRNAs (miRNAs) are single-stranded, 20–23 nucleotide–long RNA molecules that control gene expression in many cellular processes (Zhang 2009). These molecules bind to specific sequences in the 3' untranslated region (3'UTR) of mRNAs to reduce their stability and translation efficiency. miRNA genes can be expressed individually or within clusters, and can be found in introns of protein-coding genes as well as within repetitive regions and transposable elements (Huang et al. 2011). In oncogenesis, miRNAs have an important role in control of gene transcription, and based on their target, might be grouped in oncomiRNAs and miRNAs have been found to be deregulated in almost all human cancer, including lung cancer. miRNAs are involved in regulation of cellular processes linked to cancer as proliferation, differentiation, apoptosis, metastases and angiogenesis, however the exact molecular mechanisms leading to the malignant phenotype have not been unraveled yet (Garzon et al. 2006). The effects of miRNAs are mediated by modification in their abundance in tumor cells; these changes may be due to aberrant expression owing to gene mutations and deletions, genomic instability and chromosomal fragile sites that generate abnormal DNA copy numbers. In addition, miRNA gene expression in cancer cells can be regulated by abnormal epigenetic modifications such as methylation of their promoter regions (Lujambio et al. 2008). MiRNAs frequently target hundreds of mRNAs, including those of genes that mediate processes in tumorigenesis, such as cell cycle regulation, cell proliferation, apoptosis, differentiation, metabolism, stress response, inflammation and invasion (Hwang et al. 2006). The identification of novel molecular markers in cancer is a high priority in order to reduce morbidity and mortality, and provide new strategies for targeted cancer therapy. In lung cancer, patient prognosis still remains poor. Consequently, new molecular therapies that target important pathways depending of EGFR and RAS have been developed. The genes of these molecules are known to be commonly mutated in lung cancer. Thus EGFR and KRAS are considered biomarkers in lung cancer. There are many studies that strongly support the potential of miRNAs as biomarkers. Some miRNAs are known to be intimately involved in regulation of KRAS

(Johnson et al. 2005), as well as in the initiation, progression and prognosis of NSCLC. More recently, miRNAs have been detected in peripheral blood of lung cancer patients (Fanini et al. 2011), which also makes them attractive candidates as biomarkers for noninvasive and early lung cancer diagnosis. The potential of restoring levels of aberrantly under-expressed miRNAs with miRNA mimics, or inactivating over-expressed miRNAs with miRNA inhibitors has been explored and could be the next generation of therapeutic strategies.

2. LUNG CANCER

Lung cancer remains as one of the most aggressive cancer types with nearly 1.6 million new cases worldwide each year. There were an estimated 222,520 new cases and 157,300 deaths from lung cancer in the United States in 2010 (Jemal et al. 2010). The majority of patients with lung cancer present with advanced disease; for these patients the use of systemic chemotherapy has brought above modest improvements in overall survival and quality of life, though, the survival is still low, a median survival rate of 8 to 10 months (Jemal et al. 2010). Once recurred or metastasized, the disease is essentially incurable with survival rates at 5 years of less than 5%, and this has improved only marginally during the past 25 years (Jemal et al. 2010). Lung cancer is classified in two sub-groups as non-small cell lung cancer (NSCLC) and small cell lung cancer (SCLC), which constitute 85 and 15% of all lung cancer cases respectively. NSCLC comprises three major histological subtypes: adenocarcinoma, squamous cell carcinoma, and large cell carcinoma. This classification has important implications for the clinical management and prognosis of the disease (Silvestri et al. 2009). Cigarette smoke is the principal risk factor for the development of this neoplasia (Mao et al. 1997), however, the cases of lung cancer unrelated to smoking are growing, suggesting environmental or genetic determinants in disease initiation and progression (Sun et al. 2007).

Epidemiologic studies showing an association between family history and an increased risk of lung cancer have provided evidence of host susceptibility. Lung-cancer susceptibility and risk is associated with rare germ-line mutations in p53, retinoblastoma, epidermal growth factor receptor (EGFR) and other genes like ERCC1 (Sato et al. 2007). Other molecular and genetic studies have shown that some molecules contribute to sporadic tumors of NSCLC, actually they are useful as predictive biomarkers. EGFR regulates important tumorigenic processes, including proliferation, apoptosis, angiogenesis, and invasion. Along with its ligands, EGFR is frequently overexpressed during the development and progression of NSCLC. EGFR gene is amplified and over-expressed in 6% of NSCLC. However, activating mutations in the EGFR kinase domain occur early in the development of adenocarcinomas that are generally unrelated to smoking. Mutated EGFR is present in 10–15% of NSCLC tumors (Tang et al. 2005). Gene amplification and mutations in the kinase domain of C-erbB2 (HER-2/neu), a member of EGFR family, have been identified in patients with lung adenocarcinomas with a frequency of less than 5% and 5 to 10% respectively, and its overexpression is involved in ~25% of NSCLC cases (Shigematsu et al. 2005). EGFR and HER-2 kinase domain mutations have similar associations with female sex, non-smoking status, and Asian background in patients with adenocarcinoma (Pao et al. 2004).

The RAS–RAF–MEK pathway is involved in signaling downstream from EGFR leading the growth and tumor progression in NSCLC. Activating KRAS gene mutation occurs in ~30% of cases of NSCLC, mostly adenocarcinomas. KRAS mutations are localized in exon 12 (in 90% of patients) or exon 13, and they are the smoking-related G→T transversion and the nonsmoking-related G→A transition (Riely et al. 2008). KRAS mutations appear to be an early event in smoking-related lung adenocarcinoma, representing a poor prognosis in these patients. Other promising predictive markers in NSCLC are BRAF (Kobayashi et al. 2011) and the oncogenic fusion gene of EML4-ALK (Soda et al. 2007). BRAF, an effector molecule of RAS pathway, is mutated in about 2% of adenocarcinomas that do not show KRAS gene mutations. While EML4-ALK is present in 2% to 7% of NSCLC cases; essentially, this fusion gene is present in young patients with adenocarcinoma and no exposure to smoking (Brose et al. 2002). Some other molecules have been identified based on expression and genomic data, such as MYC and Cyclin D1, which are amplified and over-expressed in 2.5–10%, and 5% of NSCLC respectively, while BCL-2 overexpression is involved in ~25% of cases of NSCLC (Sato et al. 2007). Recent data have shown that gene promoter methylation is a common event in NSCLC, which contributes to oncogene over-expression or silencing of tumor suppressor genes. These epigenetic changes may be an early event in NSCLC, since the promoter region of the p16 gene is frequently methylated in smokers and premalignant lesions of lung cancer (Belinsky et al. 2006). Mutation status in EGFR and amplification in HER-2 genes correlated with good response to treatment with tyrosine kinase inhibitors (TKI e.g. gefitinib or erlotinib) (Sequist et al. 2007) and cytotoxic chemotherapy (Eberhard et al. 2005). HER-2 kinase domain mutations are associated with resistance to such TKI, but also with sensitivity to HER-2-targeted therapy (Wang et al. 2006), while KRAS mutations are correlated with poor response to treatment with TKI (Sequist et al. 2007). Alterations in PI3K-AKT-mTOR pathway have been described in NSCLC. AKT overexpression has been described in a subgroup of NSCLC tumors in conjunction with mutations or amplification of the PI3KCA gene. These genomic modifications are related to enhanced activity of PI3-K pathway, mainly in squamous cell carcinoma tumors (Rekhtman et al. 2012). On the other hand, tissues of smoking patients show higher levels of angiogenic factors such as VEGF. VEGF expression is increased in relation to tumoral grade, which in turn, correlates with increased microvessel density, development and worse prognosis of lung cancer. Tumoral angiogenesis and angiogenic factors are regulated by hypoxia-regulated pathways such as hypoxic inductor factor (HIF) 1α and 2α or through oncogenes as EGFR, KRAS and p53 (Giatromanolaki 2001). Currently a major focus of NSCLC research has been the identification of new biomarkers and molecule targets in order to have better diagnosis, prognosis and therapeutic treatments of lung cancer patients. Gene expression signature obtained through specific gene expression profile, offer a subset of expressed genes usually associated with a specific phenotype. This kind of analysis has brought about progress in the identification of markers, mutations, and genomic signatures specific expressed in NSCLC (Branica et al. 2012). In recent years, there has been great expectation of the potential of miRNAs as possible biomarkers and molecular targets in cancer. Lung cancer tissues have demonstrated unique spatial and temporal miRNA expression patterns; and some studies have identified specific miRNA expression signatures associated with clinical outcome of NSCLC patients (Wang et al. 2012). The current understanding in expression and predicted target genes of miRNAs as well as their promising potential as circulating biomarkers suggests that miRNAs might have a potential as therapeutic targets and agents in NSCLC.

3. BIOGENESIS OF miRNAs

MiRNAs are fundamental regulators of protein abundance in human cells; therefore, they have emerged as key regulators of almost every cellular biological process (Bartel et al. 2004). MiRNA biogenesis begins in the nucleus where RNA polymerase II transcribes miRNA genes to form primary miRNA transcripts (pri-miRNAs). Like other Pol II transcripts, pri-miRNAs are 5'-capped and 3'-polyadenilated transcripts (Zhang et al. 2010). The pri-miRNAs are cleaved by the RNase III enzyme Drosha/DGCR8 complex to form precursor miRNAs (pre-miRNAs), which are hairpin-shaped RNA molecules, 70–100 base pairs (bp) in lenght. Then, exportin 5 exports the pre-miRNAs from the nucleus into the cytoplasm. In the cytosol, the pre-miRNAs are cleaved by the RNase III enzyme Dicer/TRBP (transactivator RNA-binding protein) complex. The pre-miRNAs cleavage results in a small double-stranded RNA (dsRNA) duplex that contains both the mature miRNA, which targets mRNAs containing complementary sequences, and its complementary strand that will be degraded. The mature 20–25 nt long miRNA interacts with a member of the Ago (Argonaute) protein family to form a miRNA-induced silencing complex (miRISC). The miRNA guides such protein complexes to partially complementary target sites on mRNAs. Basically miRNAs function as of post-transcriptional regulators of gene expression and protein translation; it is estimated that they regulate up to 30% of all protein-coding genes (He et al. 2004). The mechanisms through which miRNAs regulate protein abundance in the cell rely upon their complementarity with protein-coding mRNAs sequence targets. MiRNAs bind to sequence-specific regulatory regions located in the 3´-UTR of mRNA targets although it has been shown that miRNAs can bind to 5´-UTR and even to the coding regions of mRNA targets. Binding of miRNA to mRNA sequences with perfect base-pairing homology induces the mRNA cleavage by Ago in the RISC, leading to inhibition of gene expression. Nevertheless, the imperfect binding to partially complementary sequences in 3'-UTR of mRNA targets, leads to repression of protein translation, this last mechanism is more commonly used by miRNAs in the cells (Pillai et al. 2007) (Fig. 1).

MiRNAs are frequently located in introns of protein-coding genes; therefore host gene promoters could control their expression, so there may be a strong correlation between mRNA and miRNA expression. However, recent reports indicate that miRNAs genes located whithin intronic regions

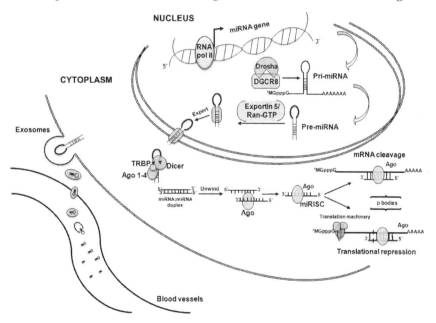

Figure 1. Schematic representation of miRNA biogenesis. MiRNA genes are transcribed in the nucleus, then, protein complex-induced nucleolytic modifications form a pre-miRNA which later is transported from the nucleus to the cytoplasm. In the cytoplasm, additional protein complexes break pre-miRNA and generate a mature miRNA, which targets sequences in mRNAs to regulate cellular proteins synthesis. MiRNAs can be released from the cell through exosomes into the bloodstream.

could have self-promoters (Golan et al. 2010). Other miRNAs are often arranged as clusters in the genome and they are transcribed as part of long non-coding RNAs to produce one pri-miRNA, which will be processed into several functional miRNAs (Bartel et al. 2004). MiRNA activity and abundance is regulated at various levels; regulation of pri-miRNA transcription, processing, target site binding and miRNA stability are among the most important (Obernosterer et al. 2006). Each cell type expresses a specific subset of miRNAs to ensure establishment and maintenance of cell type-specific mRNA profiles (Liang et al. 2007). Because of the involvement of miRNAs in the regulation of abundance of mRNAs and proteins that participate in cell mechanisms such as differentiation, cell growth, proliferation and apoptosis, it is not surprising that miRNAs have an important role in pathogenesis of cancer. 52.5% of miRNA genes are located in cancer-associated regions or fragile sites (prone to amplification, translocation or deletion) (Calin et al. 2004), besides, there are descriptions of transcriptional modifications and aberrant epigenetic events that impact in expression of miRNAs in tumors compared to normal tissues (Saito et al. 2006). These observations support the complex dual role of miRNAs, as either oncomiRNAs or anti-oncomiRNAs depending on whether they show up- or down-regulated expression in tumorigenic progression (Hwang et al. 2006). Not only do changes in expression of miRNAs in cancer have a biological implication, they also have clinical significance, which has been explored in recent years. In NSCLC cancer, there are many reports that show abnormal expression of miRNAs, that impact several modifications in the biological function of cancer cells. These observations have been the basis for the clinical significance of miRNAs in lung cancer (Wang et al. 2012).

4. GENOMIC AND EPIGENOMIC CHANGES OF miRNAs IN LUNG CANCER

4.1 Genetic Modifications of miRNAs

The genomic organization of miRNAs often dictates translational control. MiRNAs tend to be located in fragile chromosomal regions that are susceptible to translocations, microdeletions and amplifications. It has been recently reported that most known miRNAs are in regions of genomic aberration associated with cancer (Calin et al. 2004). They analyzed a panel of B-CLL samples with known deletions at 13q14 and a set of lung cancer cell lines. *Mir-16a* expression (located at 13q14) was low or absent in the majority of B-CLL cases. In contrast, both *mir-26a* (at 3p21) and *mir-99a* (at 21q11.2) were not expressed or were expressed at low levels in lung cancer cell lines, and this finding correlates with their location in regions of

LOH/HD in lung tumors cell lines. On the contrary, the expression of *mir-16a* was the same in the majority of cell lines, as they were in normal lung cells (Calin et al. 2004). In keeping with these observations, *let-7* and *mir-17-92* are linked to chromosomal deletions and gains respectively, indicating copy number variations as potential mechanisms for their deregulation in lung tumors (Calin et al. 2004, Hayashita et al. 2005, Rodriguez et al. 2004). Single nucleotide polimorphisms (SNPs) in pre-miRNAs could also alter miRNA processing, expression, and/or binding to target mRNA. It has been demonstrated that the *hsa-mir-196a2* rs11614913 variant homozygote is associated with poor survival and susceptibility to NSCLC as well as significantly increased mature *hsa-mir-196a* expression; moreover, it can directly influence the target binding of *hsa-mir-196a2-3p* (Hu et al. 2008, Tian et al. 2009). Deletions in miRNAs have been found in lung cancer; *mir-218* is deleted and down-regulated in lung squamous cell carcinoma and was found to be associated with a history of cigarette smoking (Davidson et al. 2010). Recently, a deletion of *mir-101* was also found in lung cancer, associated with reduced *mir-101* expression. Lung *in situ* carcinoma also harbored *mir-101* deletions, suggesting that genomic loss may be an early event in lung cancer development (Thu et al. 2011). The influence of copy number alterations at miRNA loci in the context of drug response has also been investigated recently. For example, *mir-662* was found to have a differential pattern of copy number alteration between sensitive and resistant cancer cell lines for most tested drugs. *miR-662* is located on chromosomal region 16p13.3 and was found to be more frequently gained in cell lines highly resistant to AZ628, erlotinib, geldanamycin, Gö-6976, HKI-272 (Neratinib), and MK-0457. All of these drugs are TKIs, except for geldanamycin, which is an antibody that targets HSP90 (Enfield et al. 2011).

4.2 Methylation of miRNA Genes

More than 100 species of known miRNAs are embedded within or near the CpG islands of the human genome and are potentially subject to control by epigenetic alterations such as DNA methylation and histone modification. Systematic assessments of miRNA expression and epigenetic modifications among cell lines and primary tumor specimens have revealed the existence of the epigenetic regulation of miRNAs in multiple tumor types (Kunej et al. 2011, Kozaki et al. 2008, Saito et al. 2006, Watanabe 2011). The miR-34 family consists of three miRNAs (miR-34a, miR-34b and miR-34c) that are derived from two transcripts (*mir-34a* on chromosome 1 and *mir-34b/c* on chromosome 11). MiR-34s have been shown to be direct targets of p53 (Corney et al. 2007, He et al. 2007, Bommer et al. 2007). Interestingly, *mir-34a* is most highly expressed in the brain, whereas *mir-34b/c* is most highly expressed in the lung with a low expression in the brain and no expression in any other tissues (Bommer et al. 2007), suggesting that mir-34b/c plays an important role in the p53 tumor suppressive pathway, at least in lung tissue. Lujambio et al. identified CpG island hypermethylation of *mir-148*, *mir-34b/c*, and the *mir-9* family, through expression microarray analysis on DNA-demethylating drug treatment in cancer cells. These results were confirmed on a group of primary tumor samples, including colon, lung, breast, head, and neck cancer, and melanoma. When miRNA hypermethylation was evaluated with respect to the existence or not of lymph node metastasis, the presence of *mir-34b/c*, *mir-148*, and *mir-9-3* CpG island hypermethylation in the primary tumor (lung, breast, melanoma) was significantly associated with those tumors that were positive for metastatic cancer cells in the corresponding lymph nodes, which highlights the importance of the *in vivo* role of miRNA epigenetic silencing in metastasis formation (Lujambio et al. 2008). *Mir-34b* hypermethylation was also recently observed in 41% of 99 primary NSCLC (Watanabe et al. 2011). In this study, the DNA methylation of *mir-34b* was also significantly associated with lymphatic invasion. These results suggest that miRNA methylation may be used in clinical practice as a marker to predict tumor prognosis and metastatic behavior. Recently, promoter hypermethylation of *mir-34b/c* was proposed as a potential prognostic factor for stage I NSCLC (Wang et al. 2011). It was also demonstrated that methylation of *miR-34b/c* is frequent in small-cell lung cancer (Tanaka et al. 2011). On the other hand, methylation of *mir-152*, *mir-9-3*, *mir-124-1*, *mir-124-2*, and *mir124-3* was analyzed in 96 NSCLC specimens using a combined bisulfite restriction analysis. Methylation of *mir-9-3*, *mir-124-2*, and *mir-124-3* was individually associated with an advanced T factor independent of age, sex, and smoking habit. Moreover, the methylation of multiple miRNAs loci has been associated with a poorer progression-free survival in a univariate analysis with a median observation period of 49.5 months (Kitano et al. 2011). The silencing of tumor-suppressive miRNAs, *mir-200b*, *mir-200c*, and *mir-205* was demonstrated in an *in vitro* premalignancy lung model consisting of 4 week-exposure of immortalized human bronchial epithelial cells to tobacco carcinogens which can induce a persistent, irreversible, and multifaceted dedifferentiation program marked by epithelial mesenchymal transition (EMT) and the emergence of stem cell-like properties (Tellez et al. 2011). The miR-200 family and miR-205 are key determinants of the epithelial phenotype by directly targeting ZEB1 and ZEB2, showing that miRNAs can indirectly regulate E-cadherin expression (Gregory et al. 2008, Park et al. 2008). So, the induction of EMT is epigenetically driven, initially by chromatin remodeling with ensuing promoter DNA methylation sustaining stable silencing of the miR-200b, miR-200c, and miR-205 implicated in this developmental program (Tellez et al. 2011).

4.3 miRNA Expression Profiles

Specific miRNA expression profiles can improve lung cancer classification and identify new pharmacological targets. MiRNA signatures have been linked to the prognosis of clinical subsets of lung cancers (Yu et al. 2008), which can help classify this malignancy by precisely defining the miRNA expression profiles that are characteristic of different histopathologic types of lung cancer (Landi et al. 2010). Expression profiles of miRNAs can provide prognostic information in lung cancer. Specific miRNA expression patterns can suggest survival outcomes for NSCLC patients. miRNA microarray analysis have identified statistical unique profiles, which could discriminate lung cancers from noncancerous lung tissues as well as molecular signatures that differ in tumor histology. The first evidence of deregulated miRNA expression in lung cancer came from the study by Volinia et al. who identified a group of miRNAs frequently aberrantly expressed in tumor tissues with respect to the normal tissue counterpart (Volinia et al. 2006). Yanaihara et al. compared miRNA patterns of expression in tumor versus adjacent uninvolved lung tissue in 104 cases of lung cancer. They identified 43 differentially expressed miRNAs between lung tumors and adjacent uninvolved lung tissue. In addition, five miRNAs (miR-155, -17-3p, -145, -21 and let-7a-2) predicted poor prognosis among patients with lung cancer (Yanaihara et al. 2006). Moreover, the same group identified six miRNAs (hsa-mir-205, -99b, -203, -202, -102, and -204-prec) that were expressed differently in adenocarcinoma and squamous cell carcinoma (Yanaihara et al. 2006). Recent findings on human serum containing stably expressed miRNA have revealed a great potential of serum miRNA signature as a disease fingerprint to predict survival. The first significant effort in this direction came from Chen et al. By deep sequencing of pooled sera from patients with and without lung cancer, 8 miRNAs were identified as differentially expressed in comparison populations. These miRNAs then were validated in a few clinical samples by RT-PCR (Chen et al. 2008). Another exploratory study looked at the potential for microarray profiling of serum-derived miRNAs (Lodes et al. 2009). In this work, the authors demonstrate the technical feasibility of performing such profiling and demonstrate the high accuracy of predicting cancer presence versus absence by cross-validation analysis (Lodes et al. 2009). Hu et al. analyzed miRNA expression profiles in sera of 303 patients with stage I to IIIA NSCLC. They detected miRNA level alterations between patients with shorter and longer survival time. Moreover, their results revealed that four miRNAs, including miR-486,

-30d, -1, and -499 could predict overall survival (Hu et al. 2010). Chen et al. identified, in a genome-wide serum miRNA expression study, a specific panel of 10 miRNAs that was able to distinguish NSCLC cases from controls with high sensitivity and specificity and that correlated with the stage of NSCLC. Furthermore, this 10 serum-miRNA profile could accurately classify serum samples collected up to 3 years prior to the clinical NSCLC diagnosis (Chen et al. 2012). By expression analysis of two serum miRNAs (hsa-miR-1254 and hsa-miR-574-5p), Foss et al. were able to discriminate early stage NSCLC samples from controls with a sensitivity of 82% and a specificity of 77% in a training cohort and with a sensitivity of 73% and a specificity of 71% in a validation cohort (Foss et al. 2011). Shen et al. recently identified a panel of four miRNAs, namely miR-21, -126, -210 and -486-5p, that distinguished NSCLC patients from healthy controls with 86.22% sensitivity and 96.55% specificity (Shen et al. 2011). Furthermore, Keller et al. showed in a multicenter study that different types of cancer or non-cancer diseases could be differentiated by blood-borne miRNA profiles (Keller et al. 2011). In addition, Leidinger et al. reported miRNA signatures that differentiated blood samples of lung cancer patients from blood samples of patients with non-malignant chronic obstructive pulmonary disease with 89.2% specificity, and 91.7% sensitivity (Leidinger et al. 2011). Likewise, Boeri et al. were able to predict lung cancer in plasma samples 1–2 years prior to diagnosis using CT. For the time being, however, the source of circulating miRNAs is elusive (Boeri et al. 2011). It has been suggested that they are released due to apoptosis or active exocytosis processes (Kosaka et al. 2010). This hypothesis is supported in a study by Rabinowits et al. that showed a similarity between the circulating exosomal miRNAs and the lung tumor-derived miRNA pattern.

MiRNAs are also present in other body fluids. Yu et al. showed that miRNAs were stably present in sputum. They were able to differentiate lung adenocarcinoma patients from healthy individuals by using a panel of four sputum miRNAs, namely miR-486, -21, -200b and -375, with high sensitivity (80.6%) and a specificity of 91.7% (Yu et al. 2010). Interestingly, two of these miRNAs, miR-21 and -486, show an overlap with the study on serum by Shen et al. (2011). The same group identified three sputum miRNAs, namely miR-205, -210 and -708, that distinguished squamous cell lung carcinoma patients from healthy individuals with 73% sensitivity and 96% specificity (Xing et al. 2010). On the other hand, few researches have studied the correlations between miRNA expression and radiotherapy sensitivity of lung cancer. Shin et al. explored the alteration of miRNA profiles by ionizing radiation in A549 human non-small cell lung cancer cells and identified 12 and 18 miRNAs in 20 Gy- and 40 Gy-exposed cells respectively, that exhibited more than 2-fold changes in their expression levels (Shin et al. 2009). Their results showed that miR-22 expression was modified in response to radiotherapy, which suggests that this miRNA could have implications in cell radio-response. At the moment, we know that miRNAs are a powerful tool for regulating the gene expression, in this context, it is logical to think that aberrant expression of miRNAs in lung cancer will result, also, in modifications of gene expression. According to their functions, a single miRNA could regulate several genes whereas several miRNAs could regulate one gene. For this reason, it has been complicated to specifically determine which genes or biological pathways could be modified as a consequence of aberrant miRNA expression. In Table 1 we summarize miRNA targets in lung cancer which have been validated in experimental models.

Table 1. Experimentally validated miRNAs in lung cancer.

miRNA	Expression in lung cancer	Function	Target gene	Reference
miR-1	Down	Represses proliferation and increases apoptosis	Met, Pim-1, FoxP1, HDAC4	Nasser et al. 2008
miR-7	Up	Represses apoptosis and increases proliferation	EGFR, Bcl-2	Webster et al. 2008, Xiong et al. 2011
Let-7	Down	Decreases proliferation and increases apoptosis	Cdk6, CDC25A, NRAS, KRAS, HMGA2	Lee et al. 2007, Kumar et al. 2008, Johnson et al. 2007, Esquela-Kerscher et al. 2008
miR-15/16	Down	Induces Rb-dependent cell cycle arrest and apoptosis	Bcl-2, CCND1, CCND2, CCNE1, WNT3A	Cimmino et al. 2005, Bandi et al. 2009
miR-17-92	Up	Induces angiogenesis, repress apoptosis and increase proliferation	Tsp1, CTGF, HIF1α□□PTEN, E2F1-3, BIM	Dews et al. 2006, Xiao et al. 2008, Sylvestre et al. 2007
miR-21	Up	Represses apoptosis, increases proliferation and induces invasion	PDCD4, PTEN, TPM1	Chan et al. 2005, Talotta et al. 2009, Zhang et al. 2010, Zhu et al. 2007
miR-29	Down	Demethylation, induces apoptosis decreases tumorigenicity, migration and invasion	DNMT3A, DNMT3B, Mcl-1, ID1	Fabbri et al. 2007, Mott et al. 2007, Rothschild et al. 2012

Table 1. contd....

Table 1. contd....

miRNA	Expression in lung cancer	Function	Target gene	Reference
miR-31	Up	Increases cell growth and tumorigenicity	LATS2, PPP2R2A	Liu et al. 2010
miR-34a-c	Down	Induces p53-dependent cellular stress responses	CDK4/6, Bcl-2, CCNE2, E2F3	He et al. 2007, Sun et al. 2008, Bommer et al. 2007
miR-93	Up	Increases proliferation	FUS1	Du et al. 2009
miR-98	Up	Increases proliferation and represses apoptosis	FUS1	Du et al. 2009
miR-101	Down	Inhibits cell proliferation and invasion	EZH2	Zhang et al. 2011
miR-126	Down	Reduces invasion, migration, adhesion and decreases angiogenesis	CRK, VEGF-A	Crawford et al. 2008, Liu et al. 2009
miR-128b	Down	Reduces EGFR-mediated proliferation	EGFR	Weiss et al. 2008
miR-130a	Down	Induces apoptosis and reduces migration	MET	Acunzo et al. 2012
miR-145	Down	Inhibits cell growth	c-myc	Chen et al. 2010
miR-197	Up	Increases proliferation	FUS1	Du et al. 2009
miR-183	Down	Inhibits migration and invasion	Ezrin	Wang et al. 2008
miR-200/429	Down	Increases E-cadherin expression, represses EMT and reduces metastasis	ZEB1/ZEB2, Flt1	Park et al. 2008, Gregory et al. 2008, Burk et al. 2008, Korpal et al. 2008, Gibbons et al. 2009, Roybal et al. 2011
miR-210	Down	Increases apoptosis	SDH	Puisségur et al. 2011
miR-221, miR-222	Up	Represses apoptosis, induces proliferation and restricts angiogenesis	P27 (kip1), c-kit, TIMP3	Garofalo et al. 2008, Poliseno et al. 2006, le Sage et al. 2007, Gillies et al. 2007, Galardi et al. 2007
miR-451	Down	Represses apoptosis and induces proliferation	RAB14	Wang et al. 2011
miR-494	Down	Represses cell proliferation and induces senescence	IGF2BP1	Ohdaira et al. 2012

5. CLINICAL SIGNIFICANCE OF miRNAs IN LUNG CANCER

Despite remarkable advances in the understanding of lung cancer over the past decades, early detection strategies, improvement in treatment strategies, better diagnosis and prognosis of NSCLC still remains deficient, which is reflected in the low treatment response rates and patient survival (Jemal et al. 2010). It is now clear that NSCLC is a highly heterogeneous cancer and that the molecular events that lead to NSCLC cancer development and progression may be dependent upon multiple biologic pathways. The advances in understanding of NSCLC biology have permitted its subdivision into various molecular subgroups; not only based on histological classification but also on the presence of molecular markers (Branica et al. 2012). Analysis of the genome and proteome of NSCLC has identified signatures for diagnosis and prognosis, as well as biological targets. Several studies in NSCLC cancer have demonstrated a deregulation of miRNA expression suggesting that they are important regulators in lung cancer pathogenesis through the modification of the expression of critical targets. In recent years, many research groups have identified individual targets and pathways of miRNAs relevant to NSCLC tumorigenesis. Studies that examine global changes in miRNA expression and the effects of individual miRNAs on lung cancer cell phenotype suggest that miRNAs are involved in NSCLC tumor development and progression and may potentially serve as biomarkers for diagnosis and prognosis (Wang et al. 2012).

5.1 miRNAs in Lung Cancer Diagnosis, Prognosis and Survival

Variability in lung cancer histology and prognosis means that better methods are needed for classifying lung tumors and stratifying patients for traditional cytotoxic therapies. A number of miRNAs are involved in lung cancer initiation, progression and prognosis. Further characterization of the highly complex miRNA regulatory networks will lead to the identification of new hallmarks of malignant growth, as well as new therapeutic targets and agents. As mentioned above, miRNA expression profiles have been useful for accurately distinguished adenocarcinomas from squamous cell carcinomas and identified the multi-step lung carcinogenesis. Changes in expression analysis of 365 miRNAs by RT-PCR have been reported during progression from hiperplasia to invasive squamous cell carcinoma (Mascaux et al. 2009). miR-15a, -32 and -34c decreased progressively whereas miR-199a and -139 were step-specific. On another novel approach to identifying miRNA signatures in lung cancer, investigators identified a 17 gene signature of genes targeted by miR-34b/34c/449 that accurately distinguished between NSCLC histological types, with a sensitivity of 90% for detecting lung cancer in the airway when combined with cytopathology (Liang 2008). Besides the identification of cancer type specific miRNA signatures, research is also aiming at the identification of specific miRNAs that are suited to differentiate between histological lung cancer subclasses. As treatment depends on the histological subtype, such miRNAs are likely to be useful for decision making in clinical treatment. Lebanony et al. were able to provide a highly accurate subclassification of NSCLC patients. They analyzed the expression of 141 miRNAS in adenocarcinomas versus squamous cell carcinomas and then determined that the relative levels between miR-205 and miR-21 can be used to distinguish the two NSCLC sub-types with 96 and 90% of sensitivity and specificity respectively. The authors propose that expression of miR-205 is a specific biomarker for squamous cell lung carcinoma, which may be useful for classification of NSCLC sub-types (Lebanony et al. 2009). Several studies have demonstrated the ability of miRNA expression for predicting the outcome of lung cancer patients, highlighting its possible diagnostic or prognostic value (Wang et al. 2012). The normal non-cancerous state of the lung is one of the tissues with the highest expression of the let-7 miRNA family (Johnson et al. 2007). Let-7 regulates the RAS pathway through repression of KRAS activity (Johnson et al. 2005), and activates mutations in KRAS that are commonly implicated in lung adenocarcinoma. Yu et al. found that expression of five miRNAs, let-7a, miR-221, -137, -372 and -182* which were analyzed by real time RT-PCR, had a predictive value for overall survival and relapse-free survival in 101 patients of stage I-III NSCLC. For instance, let-7 and miR-221 are related with a protective value, while, miR-137, miR-372, miR-182 and the miR-17-92 cluster were reported as a high-risk miRNAs. The miR-17-92 cluster is often over-expressed in lung cancer and has been found to promote tumor cell proliferation in these tumors (Yu et al. 2008). The poor survival time and the high relapse rates after surgery in lung cancer call for new methods to detect the disease at an early stage. Recently Võsa et al. provided evidence for miR-374 as a potential marker for early stage NSCLC (Võsa et al. 2011). As previously reviewed, Yanaihara et al. obtained expression profiles of miRNAs in an independent study of 143 resected lung tumors (Yanaihara et al. 2006). They proposed that under-expressed levels of let-7 correlated with prognosis of patients with adenocarcinoma independently of disease stage, consistent with its known tumor-suppressor role. It was reported that reduced let-7 expression was associated to significantly shorter survival in patients, regardless of disease stage (Takamizawa et al. 2004). Associated to this, Yanaihara et al. found that low expression of let-7a and high expression of the miR-155 precursor were associated with poor prognosis in NSCLC patients (Yanaihara et al. 2006). In 205 patients with stage I–IIIA of squamous cell carcinoma, Landi et al. identified a signature of down-regulated miRNAs including let-7e, miR-34a, miR-34c–5p, miR-25, and miR-191, which were associated with poor survival in male smokers (Landi et al. 2010), while Patnaik et al. demonstrated that a miRNA expression profile could predict recurrence of the disease in patients with NSCLC in stage I who underwent radical surgery (Patnaik et al. 2010). All of these reports strongly suggested that miRNA deregulation could be a good prognostic factor in NSCLC. Moreover, recently, different groups were able to identify miRNAs that differentiated NSCLC patients with brain metastases from patients without brain metastases (Arora et al. 2011, Nasser et al. 2011). Biomarkers that allow identification of NSCLC patients with increased risk for brain metastases will be of great value for decision-making in preventive radiation treatment.

5.2 miRNAs Associated to Therapy Response

Specific miRNA profiles could predict response to particular treatment, and the variable presence of individual miRNAs during the course of treatment could also be used for monitoring therapeutic response and assessing the degree of residual disease. Raponi et al. show that reduced expression of miR-146b predicted reduced overall survival in patients undergoing radical surgery for lung squamous cell carcinoma (Raponi et al. 2009). Yanaihara et al. also reported than low levels of another miRNA, miR-155, was associated with reduced overall survival after surgical resection (Yanaihara et al. 2006). While altered expression of let-7 was observed in response to radiotherapy, forced over-expression of let-7 in lung cancer cells can sensitize them to radiotherapy *in vitro* (Weidhaas et al. 2007). Wang et al found that 12 miRNAs expression were significantly different between radiotherapy sensitive and resistant patients (Wang et al. 2011). According to Shin's report, the expression of miR-22 was associated to radiotherapy treatment (Shin et al. 2009). Moreover, miRNAs have also shown to be predictive of response to cancer pharmacological therapy. Galluzzi et al. demonstrated that miR-630 was able to inhibit p53-regulated pro-apoptotic signaling pathways that are specifically induced by cisplatin and carboplatin (Galluzzi et al. 2010). Therapy response prediction based on miRNA expression profiles has been studied in NSCLC cell-lines resistant to apoptosis induced with TRAIL. This study reported that miR-221, -222, -100, -125 and -15b were down-regulated, while miR-9 and -96 showed up-regulation. It has been previously determined that miR-221 and miR-222 contribute to lung cancer resistance to TRAIL through silencing PTEN and TIMP3 tumor-suppressors (Garofalo et al. 2009). Recently, Voortman et al. conducted a large study on 639 patients treated with radically resected for NSCLC stage I–III, and then randomized to receive cisplatin-based adjuvant chemotherapy or follow-up in order to assess the prognostic and predictive value of miRNA expression (Voortman et al. 2010). There was no significant association among chemotherapy response and survival of patients, and expression of NSCLC-relevant miRNAs as miR-34a/b/c, -21, -29b, -155 and let-7a and their association with possible targets such as TP53, EGFR and KRAS mutation status and expression of p16. Results from Voortman's group suggested that just low expression of miR-21 could be a deleterious prognostic factor, but no single or combinatorial miRNA expression profile predicted response to adjuvant cisplatin-based chemotherapy. MiR-21 has been also found up-regulated by EGFR-activating mutations, especially in never-smokers (Seike et al. 2009). Therefore, miRNAs could affect the sensitivity to TKIs and other biological treatments that target EGFR and its downstream effectors. *In vitro* studies have described a possible role of miRNAs in response to anti-oncogenic treatment. MiR-7A has a role as a strong suppressor of the EGFR pathway; this miRNA is able to target EGFR, RAF1, AKT, and ERK, all of which are key players of this oncogenic network. Additional *in vitro* studies, have described that miR-128b can regulate EGFR in NSCLC cell lines. According to this, in a study with NSCLC tumors, it has been observed that 55% of cases shown loss of heterozygosity for miR-128b, which was correlated with enhanced clinical response to EGFR-targeted therapy with Gefitinib and survival following treatment (Weiss et al. 2008). Other miRNAS has been described as modulators of EGFR pathways. miR-145 inhibited lung cancer cell growth in patients with EGFR-activating mutations (Cho et al. 2009).

5.3 Circulating miRNAs

Currently, the identification and analysis of biomarkers in fluids as whole blood, serum, plasma or sputum is an attractive area in the study of lung cancer, because it represents a non-invasive method to establish an early and more accurate diagnosis, forecast of therapy response and to determine a prognosis of the disease. In serum from NSCLC patients has been identified a large amount of miRNAs. This finding suggested the potential of miRNAs as biomarkers in body fluids (Gilad et al. 2008). Chen et al. reported a large amount of stable miRNAs and different expression profiles in serum of patients with some kinds of tumors including lung cancer (Chen et al. 2008). Altered expression of 100 miRNAs was identified in serum of NSCLC patients compared to serum from healthy donors. Twenty-eight of them were absent while 63 new miRNAS were identified in the NSCLC patients. The same authors (Hu et al. 2010) showed over-expression of miR-23 and -225; then they report that levels of miR-486, -30d, -1 and -499 were associated with survival in a study with 303 NSCLC patients. In the present year, Chen et al. validated their previous reports. They used serum from 400 patients and 220 controls to evaluate miRNA expression using Taqman probe-based quantitative RT-PCR. The results showed that 10 miRNAs were differentially expressed between NSCLC patients and control serum samples. This miRNA panel, comprising miR-20a, -24, -25, -145, -152, -199a-5p, -221, -222, -223 and -320, was correlated with the stage of NSCLC in younger patients with current smoking habits, and most importantly, these results suggested that this miRNA panel expression could be a tool for early NSCLC detection (Chen et al. 2012). In other report, Yanaihara et al. identified a 12 miRNA diagnostic signature of NSCLC, including miR-17-3p, miR-21, miR-106a, miR-146, miR-155, miR-191, miR-192, miR-203, miR-205, miR-210, miR-212, and miR-214 (Yanaihara et al. 2006). These 12 miRNAs were evaluated in the bloodstream of 27 patients with lung adenocarcinoma and 9 healthy controls. Rabinowits et al. elegantly illustrate that circulating exosomal miRNA signatures reflect those of the primary lung tumor, accurately discriminating cancer cases from controls, and thus propose this type of analysis as a potential screening tool for the early detection of lung cancer (Rabinowits et al. 2009). By using microarrays, Liu et al. analyzed the miRNA expression of six paired lung cancer and normal tissues and identified three differentially expressed miRNAs, miR-21, -141, and -200c. High expression of miR-21 and -200c in the tumor and of miR-21 in serum were associated with poor survival in NSCLC patients (Liu et al. 2012). MiRNAs also have been quantified in sputum from NSCLC patients. High levels of miR-21 in 23 sputum samples from NSCLC patients in stage I–IV and 17 controls had a much better cancer sensitivity and specificity compared to its evaluation in sputum cytology (Yu et al. 2010). All of these results suggest that miRNAs as biomarkers in body fluids could be a revolution in clinical management of patients with NSCLC. In Table 2 we summarize all the miRNAs that have been implicated in any clinical factor in lung cancer.

5.4 miRNA-based Cancer Therapy

Anticancer therapy based in miRNAs is currently under investigation. As we know, one miRNA may have several targets, which implies that it could regulate several signaling pathways. Due to promiscuity of miRNAs in the modulation of gene expression, it has been difficult to implement powerful strategies against cancer cells using miRNAs as therapy targets or as pharmacological agents, since it could result in an undesired or unspecific effect on cancer cells. However, there are some advances in miRNAs-based lung cancer therapy. An approach to the direct manipulation of miRNAs for therapeutic effect could be through knockdown using

Table 2. Clinical implications of miRNAs in lung cancer.

miRNAs	Site of detection	Expression and diagnosis in lung cancer	References
hsa-mir-99b and hsa-mir-102, hsa-miR-126*, hsa-miR-145, hsa-miR-182, hsa-miR-183	Primary tumor	↑Adenocarcinoma	Yanaihara et al. 2006, Cho et al. 2009
hsa-miR-21, hsa-miR-205	Primary tumor	↑Lung cancer	Yanaihara et al. 2006, Seike et al. 2009, Cho et al. 2009
hsa-miR-126*	Primary tumor	↓Lung cancer	Yanaihara et al. 2006, Seike et al. 2009
hsa-miR-141, hsa-miR-210, hsa-miR-200b, hsa-miR-346	Primary tumor	↑Lung cancer (never smokers)	Seike et al. 2009, Cho et al. 2009, Volinia et al. 2006
hsa-miR-126, hsa-miR-30a, hsa-miR-30b, hsa-miR-30c, hsa-miR-30d, hsa-miR-486, hsa-miR-129, hsa-miR-451, hsa-miR-521, hsa-miR-138, hsa-miR-516a, hsa-miR-520	Primary tumor	↓Lung cancer (never smokers)	Seike et al. 2009
miR-21, miR-141, and miR-200c	Primary tumor	↑Lung cancer	Liu et al. 2012
hsa-miR-205	Primary tumor	↑Squamous cell lung carcinoma	Lebanony et al. 2009
hsa-let-7g, hsa-let-7b, hsa-let-7c, hsa-miR-29a, hsa-let-7f, hsa-let-7d, hsa-miR-98, hsa-let-7i, hsa-miR-26a, hsa-miR-30b, hsa-miR-146b-5p, hsa-miR-106b, hsa-let-7a, hsa-mir-663, hsa-miR-30d, hsa-miR-17, hsa-miR-498*, hsa-miR-26b, hsa-let-7e, hsa-mir-654-5p*, hsa-miR-181a, hsa-miR-103, hsa-miR-195, hsa-miR-191	Primary tumor	↑Adenocarcinoma vs. squamous cell lung carcinoma	Landi et al. 2010
hsa-miR-453*, hsa-miR-509-3p	Primary tumor	↓Adenocarcinoma vs. squamous cell lung carcinoma	Landi et al. 2010
hsa-miR-675, hsa-miR-93*, hsa-miR-1224-3p	Primary tumor	↑Lung cancer vs. COPD	Leidinger et al. 2011
hsa-miR-513b	Primary tumor	↓Lung cancer vs. COPD	Leidinger et al. 2011
miR-7, miR-21, miR-200b, miR-210, miR-219-1, miR-324	Primary tumor	↑Lung cancer	Boeri et al. 2011

miR-126, miR-451, miR-30a, and miR-486	Primary tumor	↑Lung cancer	Boeri et al. 2011
hsa-miR-9, hsa-miR-182, hsa-miR-200a*, hsa-miR-151, hsa-miR-205, hsa-miR-183, hsa-miR-130b*, hsa-miR-149, hsa-miR-193b, hsa-miR-339-5p, hsa-miR-196b, hsa-miR-224, hsa-miR-31, hsa-miR-196a, hsa-miR-423-3p, hsa-miR-708, hsa-miR-106b*, hsa-miR-210	Primary tumor	↑Lung cancer	Võsa et al. 2011
hsa-miR-1273, hsa-miR-206, hsa-miR-140-3p, hsa-miR-338-3p, hsa-miR-101, hsa-miR-144, hsa-miR-1285, hsa-miR-130a, hsa-miR-486-5p hsa-miR-24-2*, hsa-miR-144*, hsa-miR-30a	Primary tumor	↓Lung cancer	Võsa et al. 2011
miR-128b, miR-152, miR-125b, miR-205, miR-27a, miR-146a, miR-222, miR-23a, miR-24, miR-150	Serum	↑Lung cancer	Chen et al. 2008
miR-20a, miR-24, miR-25, miR-145, miR-152, miR-199a-5p, miR-221, miR-222, miR-223, miR-320	Serum	↑Lung cancer	Chen et al. 2012
hsa-miR-1254 and hsa-miR-574-5p	Serum	Early-stage NSCLC	Foss et al. 2011
miR-21, miR-126, miR-210, miR-486-5p	Serum	↑Stage I NSCLC (lung adenocarcinomas)	Shen et al. 2011
hsa-miR-140-3p, hsa-miR-22	Serum	↑Lung cancer	Keller et al. 2011
hsa-let7d, hsa-miR-106b, hsa-miR-98, hsa-miR-126	Serum	↓Lung cancer	Keller et al. 2011
miR-21, miR-375, miR-200b	sputum	↑Lung cancer	Yu et al. 2010
miR-486	sputum	↓Lung cancer	Yu et al. 2010
miR-205, miR-210, miR-708	sputum	↓Lung cancer	Xing et al. 2010
Prognosis			
miR-34b/c	Primary tumor	↓Predict recurrence and poor survival (stage I NSCLC)	Wang et al. 2011
miR-129-5p, miR-194*, miR-631, miR-200b*, miR-585, miR-623, miR-617, miR-622, miR-638, miRPlus_27560	Primary tumor	↓Predict recurrence (stage I NSCLC after resection)	Patnaik et al. 2010

Table 2. contd....

Table 2. contd....

miRNAs	Site of detection	Expression and diagnosis in lung cancer	References
miR-24, miR-141, miR-27b, miR-16, miR-21, miR-30c, miR-106a, miR-15b, miR-23b, miR-130a	Primary tumor	↑Predict recurrence (stage I NSCLC after resection)	Patnaik et al. 2010
miR-21	Primary tumor	↓ Poor survival (adjuvant chemotherapy after complete resection)	Voortman et al. 2010
miR-146b	Primary tumor	↑Poor survival	Raponi et al. 2009
hsa-miR-155, has-miR-213	Primary tumor	↑Poor survival	Yanaihara et al. 2006, Volinia et al. 2006
hsa-let-7a-2	Primary tumor	↓Poor survival	Yanaihara et al. 2006, Takamizawa et al. 2004, Yu et al. 2008
hsa-miR-221	Primary tumor	↓ Poor survival	Yu et al. 2008
hsa-miR-137, hsa-miR-372, and hsa-miR-182*	Primary tumor	↑Poor survival	Yu et al. 2008
miR-34b/c	Primary tumor	↓ Metastasis	Lujambio et al. 2008, Watanabe et al. 2011
hsa-miR-329-1-pre, hsa-miR-326-pre, hsa-miR-495-pre, hsa-miR-500*, hsa-miR-326, hsa-miR-370, hsa-miR-218, hsa-miR-330-3p, hsa-miR-122a, hsa-miR-325, hsa-miR-489-pre, hsa-miR-599, hsa-miR-328, hsa-miR-329-2-pre, hsa-miR-346, hsa-miR-650-Pre, hsa-miR-193-b-pre, hsa-miR-103	Primary tumor	↑Brain metastasis	Arora et al. 2011
hsa-miR-92a	Primary tumor	↓ Brain metastasis	Arora et al. 2011
miR-374a	Primary tumor	↓ Poor survival (early-stage NSCLC)	Võsa et al. 2011
miR-486, miR-30d	Serum	↑Poor survival	Hu et al. 2010
miR-1, miR-499	Serum	↓ Poor survival	Hu et al. 2010
miR-27b; miR-106a, miR-19b, miR-15b miR-16, mir-21	Serum	↑ Recurrence in stage I NSCLC	Boeri et al. 2011

		Therapy Response	
miR-221, miR-222	Serum	↑ Lung cancer aggressiveness	Boeri et al. 2011
miR-21	Serum	↑ Node metastasis and poor survival (advanced clinical stage of NSCLC)	Liu et al. 2012
miR-22	Primary tumor	Response to radiotherapy	Shin et al. 2009, Wang et al. 2011
miR-126, miR-let-7a, miR-495, miR-451, miR-128b	Primary tumor	↑ Response to radiotherapy	Wang et al 2011
miR-130a, miR-106b, miR-19b, miR-22, miR-15b, miR-17-5p and miR-21	Primary tumor	↓ Response to radiotherapy	Wang et al 2011
miR-181a	Primary tumor	↓ Resistance to CDDP	Galluzi et al. 2010
miR-630	Primary tumor	↑ Resistance to CDDP	Galluzi et al. 2010
miR-128b	Primary tumor	↓ Response to gefitinib	Weiss et al. 2008
miR-221, miR-222	Primary tumor	↑ TRAIL resistance	Garofalo et al. 2009
miR-662	Primary tumor	↑ Resistance to AZ628, Erlotinib, Geldanamycin, Gö- 6976, HKI-272 (Neratinib), and MK-0457	Enfield et al. 2011
hsa-miR-10b	Primary tumor	↑ Resistance to MG-132	Enfield et al. 2011
hsa -miR-193b	Primary tumor	↑ Resistance to AZ628, MK-0457	Enfield et al. 2011
hsa-miR-328	Primary tumor	↑ Resistance to Geldanamycin	Enfield et al. 2011
hsa-mir-628	Primary tumor	↑ Resistance to PF-2341066	Enfield et al. 2011

↑ Up-regulation, ↓ Down-regulation. COPD, chronic obstructive pulmonary disease; NSCLC, non-small cell lung cancer; CDDP, cis-diamminedichloroplatinum.

antisense oligonucleotides or through administration of synthetic miRNA analogues focused on enhancing traditional or standard cancer treatments. Several studies suggest that re-expression of specific down-regulated miRNAs in cancer cells could have a therapeutic benefit by stopping or even reversing tumor growth (Cho et al. 2012). One of most convincing evidences for miRNA replacement, as a strategy for cancer therapy is the manipulation of the tumor suppressor let-7 levels in lung cancer models (Barh et al. 2010). Loss of let-7 induces tumor formation and growth in various murine models of human lung cancer, this tumor-inducing effect appears to be carried out, at least partially, through the loss of regulation on oncogenes RAS and MYC (Johnson et al. 2005). Intratumoral injection of synthetic let-7 miRNA in a mouse model of lung cancer has been shown to reduce tumor burden (Kumar et al. 2008). Intranasal administration of let-7 in a KRAS-mutant mouse model of lung cancer effectively reduced tumor growth. Exogenous delivery of a synthetic let-7 mimic has been used to mediate remission of established NSCLC tumors in mice (Esquela-Kerscher et al. 2008). *In vitro* assays of radiation-induced cell death in lung cancer cells overexpressing members of the let-7 family of miRNAs, result in increased sensitivity to radiation therapy whereas decreasing let-7 levels induce a state of radioresistance (Weidhaas et al. 2007). Despite these encouraging results, the direct application of let-7 as a therapeutic agent for cancer is premature as yet, given that details of the immunogenic and cytotoxic effect of let-7 remain to be explored. Despite advances in the study of let-7 as a therapeutic agent in lung cancer, there are some others miRNAs that have been explored. For instance, miR-29 family members have been shown to reduce tumorigenic potential in a lung cancer model (Fabbri et al. 2007). Advances in understanding the involvement of miRNAs in lung cancer biology have allowed to establish the functions of miRNAs as let-7, which is useful as basis for implementing miRNA-based lung cancer therapy with promising results. All of these works suggest that the use of miRNAs as anti-cancer agents could be a promising strategy for new treatments of NSCLC and other malignances.

6. CONCLUSIONS

Lung cancer is one of the most aggressive tumor types in worldwide. Current research focuses on study of NSCLC in order to have better diagnostic methods and therapeutic strategies. The knowledge of genomic and epigenetic modifications in NSCLC have permitted the identification of some molecules than play an important role in the carcinogenesis process of lung cells, which implies that the known complexity of lung carcinogenesis is increasing. MiRNA functions have been associated to biological cell processes such as proliferation, DNA repair, cell death and angiogenesis. Genomic modifications and aberrant expression of miRNAs are phenomena that underlie the modified gene expression and protein translation observed in lung cancer cells. Because of their implication in lung carcinogenesis, miRNAs might represent good biomarkers for NSCLC, therefore they are emerging as intriguing and potentially powerful candidates in the arsenal to combat this kind of tumors. Analysis of miRNAs might represent a new strategy for better and more efficient diagnostic, prognostic and therapeutic approaches that decrease mortality and morbidity of lung cancer patients.

7. SUMMARY POINTS

- NSCLC is one of the tumors with the highest incidence and mortality in worldwide.
- MiRNAs are evolutionarily conserved small noncoding RNAs that negatively regulate gene expression at the post-transcriptional level.
- MiRNAs have differential spatial and temporal expression in cancer cells, which is related to several clinical, biological, molecular and genomic features of tumors.
- The study of expression profiles, polymorphisms and epigenetic modifications of miRNAs in NSCLC have permitted an understanding of the regulation of gene expression as a molecular mechanism underlying the tumoral etiology.
- MiRNAs tend to be located in fragile chromosomal regions that are susceptible to translocations, microdeletions and amplifications; moreover, they are enclosed within or near the CpG islands of the human genome and are potentially subject to control by DNA methylation.
- MiRNA expression may be used in clinical practice as a marker for lung tumor classification, metastatic behavior and survival outcome of NSCLC patients. Moreover, miRNAs themselves could represent new pharmacological targets.
- Several miRNAs as let-7, mir-34 and mir-29 have been validated as clinically significant in lung cancer.
- MiRNAs might represent good biomarkers for NSCLC, therefore they are emerging as intriguing and potentially powerful candidates in the arsenal to combat this kind of tumors.

ACKNOWLEDGEMENTS

Authors gratefully acknowledge the financial support from the National Council of Science and Technology (CONACyT), Mexico (grants 112454, 115306, 115552 and 115591), and The Institute of Science and Technology (ICyT-DF), Mexico (grant PIUTE10-147).

ABBREVIATIONS

Ago	:	Argonaute
bp	:	Base pairs
dsRNA	:	Double strand RNA
COPD	:	Chronic obstructive pulmonary disease
CDDP	:	Cis-diamminedichloroplatinum
EGFR	:	Epidermal growth factor receptor
EMT	:	Epithelial mesenchymal transition
HER-2	:	Human Epidermal Receptor 2
HIF	:	Hypoxic inductor factor
miRNAs	:	MicroRNAs
miRISC	:	MiRNA-induced silencing complex
nt	:	Nucleotide
NSCLC	:	Non-small cell lung cancer
OncomiRNAs	:	Oncogenic miRNAs
pre-miRNAs	:	Precursor miRNAs
pri-miRNAs	:	Primary miRNA transcripts
SCLC	:	Small cell lung carcinoma
SNP	:	Single nucleotide polimorphism
TRBP	:	Transactivator RNA-binding protein
TKI	:	Tyrosine kinase inhibitor
UTR	:	Untranslated region

REFERENCES

Acunzo, M., R. Visone, G. Romano, A. Veronese, F. Lovat, D. Palmieri et al. 2012. miR-130a targets MET and induces TRAIL-sensitivity in NSCLC by downregulating miR-221 and 222. Oncogene. 31: 634–642.

Arora, S., A.R. Ranade, N.L. Tran, S. Nasser, S. Sridhar, R.L. Korn et al. 2011. MicroRNA-328 is associated with (non-small) cell lung cancer (NSCLC) brain metastasis and mediates NSCLC migration. Int J Cancer. 129: 2621–2631.

Bandi, N., S. Zbinden, M. Gugger, M. Arnold, V. Kocher, L. Hasan et al. 2009. miR-15a and miR-16 are implicated in cell cycle regulation in a Rb-dependent manner and are frequently deleted or down-regulated in non-small cell lung cancer. Cancer Res. 69: 5553–5559.

Barh, D., R. Malhotra, B. Ravi, and P. Sindhurani. 2010. MicroRNA let-7: an emerging next-generation cancer therapeutic. Curr Oncol. 1: 70–80.

Bartel, D.P. 2004. MicroRNAs: genomics, biogenesis, mechanism, and function. Cell. 116: 281–297.

Belinsky, S.A., K.C. Liechty, F.D. Gentry, H.J. Wolf, J. Rogers, K. Vu et al. 2006. Promoter hypermethylation of multiple gene in sputum precedes lung cancer incidence in a high-risk cohort. Cancer Res. 66: 3338–3344.

Boeri, M., C. Verri, D. Conte, L. Roz, P. Modena, F. Facchinetti et al. 2011. MicroRNA signatures in tissues and plasma predict development and prognosis of computed tomography detected lung cancer. Proc Natl Acad Sci USA. 108: 3713–3718.

Bommer, G.T., I. Gerin, Y. Feng, A.J. Kaczorowski, R. Kuick, R.E. Love et al. 2007. p53-mediated activation of miRNA34 candidate tumor-suppressor genes. Curr Biol. 17: 1298–1307.

Brose, M.S., P. Volpe, M. Feldman, M. Kumar, I. Rishi, R. Gerrero et al. 2002. BRAF and RAS mutations in human lung cancer and melanoma. Cancer Res. 62: 6997–7000.

Burk, U., J. Schubert, U. Wellner, O. Schmalhofer, E. Vincan, S. Spaderna et al. 2008. A reciprocal repression between ZEB1 and members of the miR-200 family promotes EMT and invasion in cancer cells. EMBO Rep. 9: 582–589.

Calin, G.A., C. Sevignani, C.D. Dumitru, T. Hyslop, E. Noch, S. Yendamuri, M. Shimizu et al. 2004. Human microRNA genes are frequently located at fragile sites and genomic regions involved in cancers. Proc Natl Acad Sci USA. 101: 2999–3004.

Chan, J.A., A.M. Krichevsky, and K.S. Kosik. 2005. MicroRNA-21 is an antiapoptotic factor in human glioblastoma cells. Cancer Res. 65: 6029–6033.

Chen, X., Y. Ba, L. Ma, X. Cai, Y. Yin, K. Wang et al. 2008. Characterization of microRNAs in serum: a novel class of biomarkers for diagnosis of cancer and other diseases. Cell Res. 18: 997–1006.

Chen, X., Z. Hu, W. Wang, Y. Ba, L. Ma, C. Zhang et al. 2012. Identification of ten serum microRNAs from a genome-wide serum microRNA expression profile as novel noninvasive biomarkers for non-small cell lung cancer diagnosis. Int J Cancer. 130: 1620–1628.

Chen, Z., H. Zeng, Y. Guo, P. Liu, H. Pan, A. Deng et al. 2010. miRNA-145 inhibits non-small cell lung cancer cell proliferation by targeting c-Myc. J Exp Clin Cancer Res. 29: 151–161.

Cho, W.C., A.S. Chow, and J.S. Au. 2009. Restoration of tumour suppressor hsa-miR-145 inhibits cancer cell growth in lung adenocarcinoma patients with epidermal growth factor receptor mutation. Eur J Cancer. 45: 2197–2206.

Cho, W.C. 2012. Exploiting the therapeutic potential of microRNAs in human cancer. Expert Opin Ther Targets. 16: 345–350.

Cimmino, A., G.A. Calin, M. Fabbri, M.V. Iorio, M. Ferracin, M. Shimizu et al. 2005. miR-15 and miR-16 induce apoptosis by targeting BCL2. Proc Natl Acad Sci USA. 102: 13944–13949.

Corney, D.C., A. Flesken-Nikitin, A.K. Godwin, W. Wang, and A.Y. Nikitin. 2007. MicroRNA-34b and MicroRNA-34c are targets of p53 and cooperate in control of cell proliferation and adhesion-independent growth. Cancer Res. 67: 8433–8438.

Crawford, M., E. Brawner, K. Batte, L. Yu, M.G. Hunter, G.A. Otterson et al. 2008. MicroRNA-126 inhibits invasion in non-small cell lung carcinoma cell lines. Biochem Biophys Res Commun. 373: 607–162.

Davidson, M.R., J.E. Larsen, I.A. Yang, N.K. Hayward, B.E. Clarke, E.E. Duhig et al. 2010. MicroRNA-218 is deleted and downregulated in lung squamous cell carcinoma. PLoS One. 5: e12560.

Dews, M., A. Homayouni, D. Yu, D. Murphy, C. Sevignani, E. Wentzel et al. 2006. Augmentation of tumor angiogenesis by a Myc-activated microRNA cluster. Nat Genet. 38: 1060–1065.

Du, L., J.J. Schageman, M.C. Subauste, B. Saber, S.M. Hammond, L. Prudkin et al. 2009. miR-93, miR-98, and miR-197 regulate expression of tumor suppressor gene FUS1. Mol Cancer Res. 7: 1234–1243.

Eberhard, D.A., B.E. Johnson, L.C. Amler, A.D. Goddard, S.L. Heldens, R.S. Herbst et al. 2005. Mutations in the epidermal growth factor receptor and in KRAS are predictive and prognostic indicators in patients with non-small-cell lung cancer treated with chemotherapy alone and in combination with erlotinib. J Clin Oncol. 23: 5900–5909.

Enfield, K.S., G.L. Stewart, L.A. Pikor, C.E. Alvarez, S. Lam, W.L. Lam et al. 2011. MicroRNA gene dosage alterations and drug response in lung cancer. J Biomed Biotechnol. 2011: 474632–474647.

Esquela-Kerscher, A., P. Trang, J.F. Wiggins, L. Patrawala, A. Cheng, L. Ford et al. 2008. The let-7 microRNA reduces tumor growth in mouse models of lung cancer. Cell Cycle 7: 759–764.

Fabbri, M., R. Garzon, A. Cimmino, Z. Liu, N. Zanesi, E. Callegari et al. 2007. MicroRNA-29 family reverts aberrant methylation in lung cancer by targeting DNA methyltransferases 3A and 3B. Proc Natl Acad Sci USA. 104: 15805–15810.

Fanini, F., I. Vannini, D. Amadori, and M. Fabbri. 2011. Clinical implications of microRNAs in lung cancer. Semin. Oncol 6: 776–780.

Foss, K.M., C. Sima, D. Ugolini, M. Neri, K. E. Allen, and G.J. Weiss. 2011. miR-1254 and miR-574-5p: serum-based microRNA biomarkers for early stage non-small cell lung cancer. J Thorac Oncol. 6: 482–488.

Galardi, S., N. Mercatelli, E. Giorda, S. Massalini, G.V. Frajese, S.A. Ciafre et al. 2007. miR-221 and miR-222 expression affects the proliferation potential of human prostate carcinoma cell lines by targeting p27Kip1. J Biol Chem. 282: 23716–23724.

Galluzzi, L., E. Morselli, I. Vitale, O. Kepp, L. Senovilla, A. Criollo et al. 2010. miR-181a and miR-630 regulate cisplatin-induced cancer cell death. Cancer Res. 70: 1793–1803.

Garofalo, M., C. Quintavalle, G. Di Leva, C. Zanca, G. Romano, C. Taccioli et al. 2008. MicroRNA signatures of TRAIL resistance in human non-small cell lung cancer. Oncogene. 27: 3845–3855.

Garofalo, M., G. Di Leva, G. Romano, G. Nuovo, S.S. Suh, A. Ngankeu et al. 2009. miR-221&222 regulate TRAIL resistance and enhance tumorigenicity through PTEN and TIMP3 downregulation. Cancer Cell. 16: 498–509.

Garzon, R., M. Fabbri, A. Cimmino, G.A. Calin, and C.M. Croce. 2006. MicroRNA expression and function in cancer. Trends Mol Med. 12: 580–587.

Giatromanolaki, A. 2001. Prognostic role of angiogenesis in non-small cell lung cancer. Anticancer Res. 6B: 4373–4382.

Gibbons, D.L., W. Lin, C.J. Creighton, Z.H. Rizvi, P.A. Gregory, G.J. Goodall et al. 2009. Contextual extracellular cues promote tumor cell EMT and metastasis by regulating miR-200 family expression. Genes Dev. 23: 2140–2151.

Gilad, S., E. Meiri, Y. Yogev, S. Benjamin, D. Lebanony, N. Yerushalmi et al. 2008. Serum miRNAs are promising novel biomarkers. PLoS One. 3: e3148.

Gillies, J.K. and I.A. Lorimer. 2007. Regulation of p27Kip1 by miRNA 221/222 in glioblastoma. Cell Cycle 6: 2005–2009.

Golan, D., C. Levy, B. Friedman, and N. Shomron. 2010. Biased hosting of intronic microRNA genes. Bioinformatics 26: 992–995.

Gregory, P.A., A.G. Bert, E.L. Paterson, S.C. Barry, A. Tsykin, G. Farshid et al. 2008. The miR-200 family and miR-205 regulate epithelial to mesenchymal transition by targeting ZEB1 and SIP1. Nat Cell Biol. 10: 593–601.

Hayashita, Y., H. Osada, Y. Tatematsu, H. Yamada, K. Yanagisawa, S. Tomida et al. 2005. A polycistronic microRNA cluster, miR-17-92, is overexpressed in human lung cancers and enhances cell proliferation. Cancer Res. 65: 9628–9632.

He, L. and G.J. Hannon. 2004. MicroRNAs: small RNAs with a big role in gene regulation. Nat Rev Genet. 5: 522–531.

He, L., X. He, L.P. Lim, E. de Stanchina, Z. Xuan, Y. Liang et al. 2007. A microRNA component of the p53 tumour suppressor network. Nature. 447: 1130–1134.

Hu, Z., J. Chen, T. Tian, X. Zhou, H. Gu, L. Xu et al. 2008. Genetic variants of miRNA sequences and non-small cell lung cancer survival. J Clin Invest. 18: 2600–2608.

Hu, Z., X. Chen, Y. Zhao, T. Tian, G. Jin, Y. Shu et al. 2010. Serum microRNA signatures identified in a genome-wide serum microRNA expression profiling predict survival of non-small-cell lung cancer. J Clin Oncol. 28: 1721–1726.

Huang, Y., Z.J. Shen, Q. Zou, S.P. Wang, S.M. Tang, and G.Z. Zhang. 2011. Biological functions of microRNAs: a review. J Physiol Biochem. 67: 129–139.

Hwang, H.W. and J.T. Mendell. 2006. MicroRNAs in cell proliferation, cell death, and tumorigenesis. Br J Cancer 94: 776–778.

Jemal, A., R. Siegel, J. Xu, and E. Ward. 2010. Cancer statistics, 2010. Cancer J Clin. 60: 277–300.

Johnson, C.D., A. Esquela-Kerscher, G. Stefani, M. Byrom, K. Kelnar, D. Ovcharenko et al. 2007. The let-7 microRNA represses cell proliferation pathways in human cells. Cancer Res. 67: 7713–7722.

Johnson, S.M., H. Grosshans, J. Shingara, M. Byrom, R. Jarvis, A. Cheng et al. 2005. RAS is regulated by the let-7 miRNA family. Cell. 120: 635–647.

Keller, A., P. Leidinger, A. Bauer, A. ElSharawy, J. Haas, C. Backes et al. 2011. Toward the blood-borne miRNome of human diseases. Nat Methods. 8: 841–843.

Kitano, K., K. Watanabe, N. Emoto, H. Kage, E. Hamano, T. Nagase et al. 2011. CpG island methylation of microRNAs is associated with tumor size and recurrence of non-small-cell lung cancer. Cancer Sci. 102: 2126–2131.

Kobayashi, M., M. Sonobe, T. Takahashi, A. Yoshizawa, M. Ishikawa, R. Kikuchi et al. 2011. Clinical significance of BRAF gene mutations in patients with non-small cell lung cancer. Anticancer Res. 12: 4619–4623.

Korpal, M., E.S. Lee, G. Hu, and Y. Kang. 2008. The miR-200 family inhibits epithelial-mesenchymal transition and cancer cell migration by direct targeting of E-cadherin transcriptional repressors ZEB1 and ZEB2. J Biol Chem. 283: 14910–14914.

Kosaka, N., H. Iguchi, and T. Ochiya. 2010. Circulating microRNA in body fluid: a new potential biomarker for cancer diagnosis and prognosis. Cancer Sci. 101: 2087–2092.

Kozaki, K., I. Imoto, S. Mogi, K. Omura, and J. Inazawa. 2008. Exploration of tumor suppressive microRNAs silenced by DNA hypermethylation in oral cancer. Cancer Res. 68: 2094–2105.

Kumar, M.S., S.J. Erkeland, R.E. Pester, C.Y. Chen, M.S. Ebert, P.A. Sharp et al. 2008. Suppression of non-small cell lung tumor development by the let-7 microRNA family. Proc Natl Acad Sci USA. 105: 3903–5908.

Kunej, T., I. Godnic, J. Ferdin, S. Horvat, P. Dovc, and G.A. Calin. 2011. Epigenetic regulation of microRNAs in cancer: an integrated review of literature. Mutat Res. 717: 77–84.

Landi, M.T., Y. Zhao, M. Rotunno, J. Koshiol, H. Liu, A.W. Bergen et al. 2010. MicroRNA expression differentiates histology and predicts survival of lung cancer. Clin Cancer Res. 16: 430–441.

le Sage, C., R. Nagel, D.A. Egan, M. Schrier, E. Mesman, A. Mangiola et al. 2007. Regulation of the p27(Kip1) tumor suppressor by miR-221 and miR-222 promotes cancer cell proliferation. EMBO J. 26: 3699–3708.

Lebanony, D., H. Benjamin, S. Gilad, M. Ezagouri, A. Dov, K. Ashkenazi et al. 2009. Diagnostic assay based on hsa-miR-205 expression distinguishes squamous from non-squamous non-small-cell lung carcinoma. J Clin Oncol. 27: 2030–2037.

Lee, Y.S. and A. Dutta. 2007. The tumor suppressor microRNA let-7 represses the HMGA2 oncogene. Genes Dev. 21: 1025–1030.

Leidinger, P., A. Keller, A. Borries, H. Huwer, M. Rohling, J. Huebers et al. 2011. Specific peripheral miRNA profiles for distinguishing lung cancer from COPD. Lung Cancer. 74: 41–47.

Liang, Y., D. Ridzon, I. Wong, and C. Chen. 2007. Characterization of miRNA expression profiles in normal human tissues. BMC Genomics. 8: 166–172.

Liang, Y. 2008. An expression meta-analysis of predicted microRNA targets identifies a diagnostic signature for lung cancer. BMC Med Genomics. 1: 61–77.

Liu, B., X.C. Peng, X.L. Zheng, J. Wang, and Y.W. Qin. 2009. MiR-126 restoration down-regulate VEGF and inhibit the growth of lung cancer cell lines *in vitro* and *in vivo*. Lung Cancer. 66: 169–175.

Liu, X., L.F. Sempere, H. Ouyang, V.A. Memoli, A.S. Andrew, Y. Luo et al. 2010. MicroRNA-31 functions as an oncogenic microRNA in mouse and human lung cancer cells by repressing specific tumor suppressors. J Clin Invest. 120: 1298–1309.

Liu, X.G., W.Y. Zhu, Y.Y. Huang, L.N. Ma, S.Q. Zhou, Y.K. Wang et al. 2012. High expression of serum miR-21 and tumor miR-200c associated with poor prognosis in patients with lung cancer. Med Oncol. 29: 618–626.

Lodes, M.J., M. Caraballo, D. Suciu, S. Munro, A. Kumar, and B. Anderson. 2009. Detection of cancer with serum miRNAs on an oligonucleotide microarray. PLoS One. 4: e6229.

Lujambio, A., G.A. Calin, A. Villanueva, S. Ropero, M. Sanchez-Cespedes, D. Blanco et al. 2008. A microRNA DNA methylation signature for human cancer metastasis. Proc Natl Acad Sci USA. 105: 13556–13561.

Mao, L., J.S. Lee, J.M. Kurie, Y.H. Fan, S.M. Lippman, J.J. Lee et al. 1997. Clonal genetic alterations in the lungs of current and former smokers. J Natl Cancer. 89: 857–862.

Mascaux, C., J.F. Laes, G. Anthoine, A. Haller, V. Ninane, A. Burny, and J.P. Sculier. 2009. Evolution of microRNA expression during human bronchial squamous carcinogenesis. Eur Respir J. 3: 352–359.

Mott, J.L., S. Kobayashi, S.F. Bronk, and G.J. Gores. 2007. mir-29 regulates Mcl-1 protein expression and apoptosis. Oncogene. 26: 6133–6140.

Nasser, M.W., J. Datta, G. Nuovo, H. Kutay, T. Motiwala, S. Majumder et al. 2008. Down-regulation of micro-RNA-1 (miR-1) in lung cancer. Suppression of tumorigenic property of lung cancer cells and their sensitization to doxorubicin-induced apoptosis by miR-1. J Biol Chem. 283: 33394–33405.

Nasser, S., A.R. Ranade, S. Sridhar, L. Haney, R.L. Korn, M.B. Gotway et al. 2011. Biomarkers associated with metastasis of lung cancer to brain predict patient survival. Int J Data Min Bioinform. 5: 287–307.

Obernosterer, G., P.J. Leuschner, M. Alenius, and J. Martinez. 2006. Post-transcriptional regulation of miRNA expression. RNA. 12: 1161–1167.

Ohdaira, H., M. Sekiguchi, K. Miyata, and K. Yoshida. 2012. MicroRNA-494 suppresses cell proliferation and induces senescence in A549 lung cancer cells. Cell Prolif. 45: 32–38.

Pao, W., V. Miller, M. Zakowski, J. Doherty, K. Politi, I. Sarkaria et al. 2004. EGF receptor gene mutations are common in lung cancers from "never smokers" and are associated with sensitivity of tumors to gefitinib and erlotinib. Proc Natl Acad Sci USA. 101: 13306–13311.

Park, S.M., A.B. Gaur, E. Lengyel, and M.E. Peter. 2008. The miR-200 family determines the epithelial phenotype of cancer cells by targeting the Ecadherin repressors ZEB1 and ZEB2. Genes Dev. 22: 894–907.

Patnaik, S.K., E. Kannisto, S. Knudsen, and S. Yendamuri. 2010. Evaluation of microRNA expression profiles that may predict recurrence of localized stage I non-small cell lung cancer after surgical resection. Cancer Res. 70: 36–45.

Pillai, R.S., S.N. Bhattacharyya, and W. Filipowicz. 2007. Repression of protein synthesis by miRNAs: how many mechanisms? Trends Cell Biol. 17: 118–126.

Poliseno, L., A. Tuccoli, L. Mariani, M. Evangelista, L. Citti, K.Woods et al. 2006. MicroRNAs modulate the angiogenic properties of HUVECs. Blood. 108: 3068–3071.

Puisségur, M.P., N.M. Mazure, T. Bertero, L. Pradelli, S. Grosso, K. Robbe-Sermesant et al. 2011. miR-210 is overexpressed in late stages of lung cancer and mediates mitochondrial alterations associated with modulation of HIF-1 activity. Cell Death Differ. 18: 465–478.

Rabinowits, G., C. Gerçel-Taylor, J.M. Day, D.D. Taylor, and G.H. Kloecker. 2009. Exosomal microRNA: a diagnostic marker for lung cancer. Clin Lung Cancer. 10: 42–46.

Raponi, M., L. Dossey, T. Jatkoe, X. Wu, G. Chen, H. Fan et al. 2009. MicroRNA classifiers for predicting prognosis of squamous cell lung cancer. Cancer Res. 69: 5776–5783.

Rekhtman, N., P.K. Paik, M.E. Arcila, L.J. Tafe, G.R. Oxnard, A.L. Moreira et al. 2012. Clarifying the spectrum of driver oncogene mutations in biomarker-verified squamous carcinoma of lung: Lack of EGFR/KRAS and Presence of PIK3CA/AKT1 Mutations. Clin Cancer Res. 18: 1167–1176.

Riely, G.J., M.G. Kris, D. Rosenbaum, J. Marks, A. Li, D.A. Chitale et al. 2008. Frequency and distinctive spectrum of *kras* mutations in never smokers with lung adenocarcinoma. Clin Cancer Res. 14: 5731–5734.

Rodriguez, A., S. Griffiths-Jones, J.L. Ashurst, and A. Bradley. 2004. Identification of mammalian microRNA host genes and transcription units. Genome Res. 14: 1902–1910.

Rothschild, S.I., M.P. Tschan, E.A. Federzoni, R. Jaggi, M.F. Fey, M. Gugger et al. 2012. MicroRNA-29b is involved in the Src-ID1 signaling pathway and is dysregulated in human lung adenocarcinoma. Oncogene. doi: 10.1038/onc.2011.578.

Roybal, J.D., Y. Zang, Y.H. Ahn, Y. Yang, D.L. Gibbons, B.N. Baird et al. 2011. miR-200 Inhibits lung adenocarcinoma cell invasion and metastasis by targeting Flt1/VEGFR1. Mol Cancer Res. 9: 25–35.

Ruan, K., X. Fang, and G. Ouyang. 2009. MiRNAs: novel regulators in the hallmarks of human cancer. Cancer Lett. 285: 116–126.

Saito, Y. and P.A. Jones. 2006. Epigenetic activation of tumor suppressor microRNAs in human cancer cells. Cell Cycle. 5: 2220–2222.

Sato, M., D.S. Shames, A.F. Gazdar, and J.D. Minna. 2007. A translational view of the molecular pathogenesis of lung cancer. J Thorac Oncol. 2: 327–343.

Seike, M., A. Goto, T. Okano, E.D. Bowman, A.J. Schetter, I. Horikawa et al. 2009. MiR-21 is an EGFR-regulated anti-apoptotic factor in lung cancer in never-smokers. Proc Natl Acad Sci USA. 106: 12085–12090.

Sequist, L.V., D.W. Bell, T.J. Lynch, and D.A. Haber. 2007. Molecular predictors of response to epidermal growth factor receptor antagonists in non-small-cell lung cancer. J Clin Oncol. 25: 587–595.

Shen, J., N.W. Todd, H. Zhang, L. Yu, X. Ling, Y. Mei et al. 2011. Plasma microRNAs as potential biomarkers for non-small-cell lung cancer. Lab Invest. 91: 579–587.

Shigematsu, H., T. Takahashi, M. Nomura, K. Majmudar, M. Suzuki, H. Lee et al. 2005. Somatic mutations of the HER2 kinase domain in lung adenocarcinomas. Cancer Res. 65: 1642–1646.

Shin, S., H.J. Cha, E.M. Lee, S.J. Lee, S.K. Seo, H.O. Jin et al. 2009. Alteration of miRNA profiles by ionizing radiation in A549 human non-small cell lung cancer cells. Int J Oncol. 35: 81–86.

Silvestri, G.A., A.J. Alberg, and J. Ravenel. 2009. The changing epidemiology of lung cancer with a focus on screening. Br Med J. 339: 451–454.

Soda, M., Y.L. Choi, M. Enomoto, S. Takada, Y. Yamashita, S. Ishikawa et al. 2007. Identification of the transforming EML4- ALK fusion gene in non-small-cell lung cancer. Nature. 448: 561–566.

Sun, F., H. Fu, Q. Liu, Y. Tie, J. Zhu, R. Xing et al. 2008. Downregulation of CCND1 and CDK6 by miR-34a induces cell cycle arrest. FEBS Lett. 582: 1564–1568.

Sun, S., J.H. Schiller, and A.F. Gazdar. 2007. Lung cancer in never smokers-a different disease. Nat Rev Cancer. 7: 778–790.

Sylvestre, Y., V. De Guire, E. Querido, U.K. Mukhopadhyay, V. Bourdeau, F. Major et al. 2007. An E2F/miR-20a autoregulatory feedback loop. J Biol Chem. 282: 2135–2143.

Takamizawa, J., H. Konishi, K. Yanagisawa, S. Tomida, H. Osada, H. Endoh et al. 2004. Reduced expression of the let-7 microRNAs in human lung cancers in association with shortened postoperative survival. Cancer Res. 64: 3753–3756.

Talotta, F., A. Cimmino, M.R. Matarazzo, L. Casalino, G. De Vita, M. D'Esposito et al. 2009. An autoregulatory loop mediated by miR-21 and PDCD4 controls the AP-1 activity in RAS transformation. Oncogene. 28: 73–84.

Tanaka, N., S. Toyooka, J. Soh, T. Kubo, H. Yamamoto, Y. Maki et al. 2011. Frequent methylation and oncogenic role of microRNA-34b/c in small-cell lung cancer. Lung Cancer Epub ahead of print.

Tang, X., H. Shigematsu, B.N. Bekele, J.A. Roth, J.D. Minna, W.K. Hong et al. 2005. EGFR tyrosine kinase domain mutations are detected in histologically normal respiratory epithelium in lung cancer patients. Cancer Res. 65: 7568–7572.

Tellez, C.S., D.E. Juri, K. Do, A.M. Bernauer, C.L. Thomas, L.A. Damiani et al. 2011. EMT and stem cell-like properties associated with miR-205 and miR-200 epigenetic silencing are early manifestations during carcinogen-induced transformation of human lung epithelial cells. Cancer Res. 71: 3087–3097.

Thu, K.L., R. Chari, W.W. Lockwood, S. Lam, and W.L. Lam. 2011. miR-101 DNA copy loss is a prominent subtype specific event in lung cancer. J Thorac Oncol. 6: 1594–1598.

Tian, T., Y. Shu, J. Chen, Z. Hu, L. Xu, G. Jin et al. 2009. A functional genetic variant in microRNA-196a2 is associated with increased susceptibility of lung cancer in Chinese. Cancer Epidemiol Biomarkers Prev. 18: 1183–1187.

Volinia, S., G.A. Calin, C.G. Liu, S. Ambs, A. Cimmino, F. Petrocca et al. 2006. A microRNA expression signature of human solid tumors defines cancer gene targets. Proc Natl Acad Sci USA. 103: 2257–2261.

Voortman, J., A. Goto, J. Mendiboure, J.J. Sohn, A.J. Schetter, M. Saito et al. 2010. MicroRNA expression and clinical outcomes in patients treated with adjuvant chemotherapy after complete resection of non-small cell lung carcinoma. Cancer Res. 70: 8288–8298.

Võsa, U., T. Vooder, R. Kolde, K. Fischer, K. Välk, N. Tõnisson et al. 2011. Identification of miR-374a as a prognostic marker for survival in patients with early-stage non-small cell lung cancer. Genes Chromosomes Cancer. 50: 812–822.

Vrabec-Branica, B. and S. Gajovic. 2012. Molecular biomarkers of lung carcinoma. Front Biosci (Elite Ed). 4: 865–875.

Wang, G., W. Mao, and S. Zheng. 2008. MicroRNA-183 regulates Ezrin expression in lung cancer cells. FEBS Lett. 582: 3663–3668.

Wang, Q., S. Wang, H. Wang, P. Li, and Z. Ma. 2012. MicroRNAs: novel biomarkers for lung cancer diagnosis, prediction and treatment. Exp Biol Med. doi:10.1258/ebm.2011.011192.

Wang, R., Z.X. Wang, J.S. Yang, X. Pan, W. De, and L.B. Chen. 2011. MicroRNA-451 functions as a tumor suppressor in human non-small cell lung cancer by targeting Ras-related protein 14 (RAB14). Oncogene. 30: 2644–2658.

Wang, S.E., A. Narasanna, M. Perez-Torres, B. Xiang, F.Y. Wu, S. Yang et al. 2006. HER2 kinase domain mutation results in constitutive phosphorylation and activation of HER2 and EGFR and resistance to EGFR tyrosine kinase inhibitors. Cancer Cell. 10: 25–38.

Wang, X.C., L.Q. Du, L.L. Tian, H.L. Wu, X.Y. Jiang, H. Zhang et al. 2011. Expression and function of miRNA in postoperative radiotherapy sensitive and resistant patients of non-small cell lung cancer. Lung Cancer. 72: 92–99.

Wang, Z., Z. Chen, Y. Gao, N. Li, B. Li, F. Tan et al. 2011. DNA hypermethylation of microRNA-34b/c has prognostic value for stage non-small cell lung cancer. Cancer Biol Ther. 11: 490–496.

Watanabe, K., N. Emoto, E. Hamano, M. Sunohara, M. Kawakami, H. Kage et al. 2011. Genome structure-based screening identified epigenetically silenced microRNA associated with invasiveness in non-small-cell lung cancer. Int J Cancer. doi: 10.1002/ijc.26254.

Webster, R.J., K.M. Giles, K.J. Price, P.M. Zhang, J.S. Mattick, and P.J. Leedman. 2009. Regulation of epidermal growth factor receptor signaling in human cancer cells by microRNA-7. J Biol Chem. 284: 5731–5741.

Weidhaas, J.B., I. Babar, S.M. Nallur, P. Trang, S. Roush, M. Boehm et al. 2007. MicroRNAs as potential agents to alter resistance to cytotoxic anticancer therapy. Cancer Res. 67: 11111–11116.

Weiss, G.J., L.T. Bemis, E. Nakajima, M. Sugita, D.K. Birks, W.A. Robinson et al. 2008. EGFR regulation by microRNA in lung cancer: correlation with clinical response and survival to gefitinib and EGFR expression in cell lines. Ann Oncol. 19: 1053–1059.

Xiao, C., L. Srinivasan, D.P. Calado, H.C. Patterson, B. Zhang, J. Wang et al. 2008. Lymphoproliferative disease and autoimmunity in mice with increased miR-17-92 expression in lymphocytes. Nat Immunol. 9: 405–414.

Xing, L., N.W. Todd, L. Yu, H. Fang, and F. Jiang. 2010. Early detection of squamous cell lung cancer in sputum by a panel of microRNA markers. Mod Pathol. 23: 1157–1164.

Xiong, S., Y. Zheng, P. Jiang, R. Liu, X. Liu, and Y. Chu. 2011. MicroRNA-7 inhibits the growth of human non-small cell lung cancer A549 cells through targeting BCL-2. Int J Biol Sci. 7: 805–814.

Yanaihara, N., N. Caplen, E. Bowman, M. Seike, K. Kumamoto, M. Yi et al. 2006. Unique microRNA molecular profiles in lung cancer diagnosis and prognosis. Cancer Cell. 9: 189–198.

Yu, L., N.W. Todd, L. Xing, Y. Xie, H. Zhang, Z. Liu et al. 2010. Early detection of lung adenocarcinoma in sputum by a panel of microRNA markers. Int J Cancer. 127: 2870–2878.

Yu, S.L., H.Y. Chen, G.C. Chang, C.Y. Chen, H.W. Chen, S. Singh et al. 2008. MicroRNA signature predicts survival and relapse in lung cancer. Cancer Cell. 13: 48–57.

Zhang, C. 2009. Novel functions for small RNA molecules. Curr Opin Mol Ther. 11: 641–651.

Zhang, J.G., J.J. Wang, F. Zhao, Q. Liu, K. Jiang, and G.H. Yang. 2010. MicroRNA-21 (miR-21) represses tumor suppressor PTEN and promotes growth and invasion in non-small cell lung cancer (NSCLC). Clin Chim Acta. 411: 846–852.

Zhang, J.G., J.F. Guo, D.L. Liu, Q. Liu, and J.J. Wang. 2011. MicroRNA-101 exerts tumor-suppressive functions in non-small cell lung cancer through directly targeting enhancer of zeste homolog 2. J Thorac Oncol. 6: 671–678.

Zhang, X. and Y. Zeng. 2010. Regulation of mammalian miRNA expression. J Cardiovasc Trans Res. 3: 197–203.

Zhu, S., M.L. Si, H. Wu, and Y.Y. Mo. 2007. MicroRNA-21 targets the tumor suppressor gene tropomyosin 1 (TPM1). J Biol Chem. 282: 14328–14336.

[1]Oncogenomics Lab, National Institute of Cancerology, Mexico.
[2]Cancer Biomedical Research Unit, National Institute of Cancerology, Mexico.
[3]Genomics Science Program, Oncogenomics and Cancer Proteomics Lab, Autonomous University of Mexico City, Mexico.
Email: cesar.lopez@uacm.edu.mx
[4]Molecular Biochemistry Lab, UBIPRO, FES-I, National Autonomous University of Mexico, Mexico.
Email: e_urrutia@me.com
[5]Department of Natural Science. Metropolitan Autonomous University-C, Mexico.
Email: cgonzalez@correo.cua.uam.mx
[6]Massive Sequencing Unite, National Institute of Cancerology-Genomics Lab FES-I, UBIMED, National Autonomous University of Mexico, Mexico.
[a]Email: earechagao@incan.edu.mx
[b]Email: josediaz030178@hotmail.com
[c]Email: carlos.pplas@campus.iztacala.unam.mx
[d]Email: yamilet_noriega@hotmail.com
[e]Email: herreram@biomedicas.unam.mx
*Corresponding author

CHAPTER 10

MicroRNAs IN MELANOMA

Federica Felicetti,[a,*] Maria Cristina Errico,[b]
Gianfranco Mattia,[c] Nadia Felli[d] and Alessandra Carè[e]

ABSTRACT

MicroRNAs (miRNAs) are small non coding RNAs regulating gene expression of target mRNAs at post-transcriptional level by pairing to specific sequences in the 3'UTR of target genes. Many studies demonstrated miRNAs involvement in all the main biological processes, including tumor development. The first publications on the role of distinct miRNAs in melanoma date back only four years ago and since then a large number of results were collected. Among many others, Robinson and his group (2008) linked the expression of the single miRNA-137 with the expression of MITF, a 'master regulator' of melanogenesis. Moreover, Felicetti and co-workers have shown that miRNA-221/-222 control melanoma progression through multiple oncogenic mechanisms by downregulating c-KIT receptor and p27Kip1. Recently, through the analyses of gene expression profiles obtained by the technique of "microarray", it was possible to discriminate

Department of Hematology, Oncology, and Molecular Medicine, Istituto Superiore Sanità, Viale Regina Elena, 299-00161 Rome, Italy.
[a]Email: federica.felicetti@iss.it
[b]Email: mariacristina.errico@iss.it
[c]Email: gianfranco.mattia@iss.it
[d]Email: nadia.felli@iss.it
[e]Email: alessandra.care@iss.it
*Corresponding author

subgroups of patients phenotypically indistinguishable identifying a number of miRNAs that could be validated as diagnostic markers for melanoma. In the case of uveal melanoma the expression profiles have already demonstrated this miRNA discriminating capability. Circulating miRNAs have also been detected in both plasma and serum of healthy and diseased subjects. These extracellular RNAs appear to be involved in the intercellular communication between cancer cells and their surrounding microenvironment, representing useful blood biomarkers. In conclusion it could be assumed that, in a not too distant future, miRNAs can be the basis for new therapeutic strategies. The identification of new molecular targets is particularly relevant in melanoma whose treatment, despite many efforts, still represents a challenge.

1. INTRODUCTION

Cutaneous melanoma is an aggressive neoplasia whose incidence is steadily increasing being the number of melanoma cases worldwide rising faster than any other cancer. Screening and early detection are still the best prognostic factors leading to 99% of favorable outcome if the primary lesion is detected very early (more than 90% survival in stage I melanomas). At the initial phases of the disease surgical excision mostly represents a definitive solution, but no effective therapies have been determined for advanced stages and the overall 5-year survival of metastatic melanomas remains lower than 10%. This fact gets even more aggravating, considering metastasis to distinct organs as a very early event in the progression of this disease. It is then obvious how important might be to study in depth the molecular oncogenic pathways implicated in transformation and progression in order to identify new representative markers for diagnosis, prognosis and eventually therapeutic treatments. Several melanoma biomarkers, such as the mitotic and the Ki-67 marker expression indexes, have been evaluated for their prognostic utility with promising early results (Vereecken et al. 2007); however, to date, none has been proven to be clinically useful in large-scale studies (Larson et al. 2009). Although several molecular abnormalities have been associated with melanoma progression, as the loss of AP-2 transcription factor (Huang et al. 1998) or the high mutation rate of the B-RAF oncogene (Dhomen and Marais 2007), the mechanisms underlying the differential gene expression are still largely unknown and the conventional histological classification remains the best prognostic factor (Clark et al. 1984). The Clark model describes the histological changes that accompany the progression from normal melanocytes to malignant melanoma (Fig. 1).

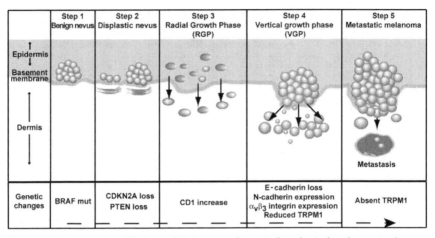

	Step 1 Benign nevus	Step 2 Displastic nevus	Step 3 Radial Growth Phase (RGP)	Step 4 Vertical growth phase (VGP)	Step 5 Metastatic melanoma
Genetic changes	BRAF mut	CDKN2A loss PTEN loss	CD1 increase	E-cadherin loss N-cadherin expression $\alpha_v\beta_3$ integrin expression Reduced TRPM1	Absent TRPM1

Figure 1. Representative 'Clark model' showing the main histological and genetic changes from normal melanocytes to malignant melanoma.

In this model, five distinct steps of melanoma development and progression are distinguished: a mature melanocyte acquires mutations that lead from benign (Step 1) to dysplastic nevi (Step 2). The subsequent radial growth phase (RGP) primary melanoma (Step 3) is the first recognizable malignant stage in which cells do not possess metastatic potential, but are already locally invasive. RGP is followed by the vertical growth phase (VGP) (Step 4), in which melanoma cells infiltrate and invade the dermis showing metastatic potential. This process finally results in metastases to distant organs by an overgrowth of disseminated tumor cells at these sites (Step 5) (Gray-Schopfer et al. 2007). The main clinical and histopathological prognostic factors that are currently in use for melanoma include tumor depth (i.e. Breslow thickness), diameter, ulceration, anatomic site (i.e. acral, mucosal, cutaneous) and sentinel lymph-node status (Balch et al. 2001). The identification of other molecules as putative prognostic or diagnostic markers and, eventually, as new targets to aim at in new therapeutic approaches is therefore particularly relevant.

The recent identification of microRNA (miRNA) family and the growing evidence of their altered expression in cancer, where they are capable of acting as oncogenes as well as tumor suppressor genes, provided a new level of molecular regulation. The miRNA-based mechanisms appear of particular relevance considering that each miRNA targets multiple mRNAs and that in theory each single mRNA might be regulated by several miRNAs according to complex combinatorial circuitries.

Gene expression profiling has already demonstrated the existence of specific patterns of miRNAs and the consequent chance of utilizing specific miRNAs as diagnostic and/or prognostic markers. MiRNA signatures will allow the identification of a continuously increasing number of miRNA specifically involved in major cellular processes (e.g. proliferation and differentiation, apoptosis, and angiogenesis) and whose deregulated expression is causally related to melanoma development/progression.

The understanding of these phenomena will not only shed light in the regulation of these molecular mechanisms, but it will also represent another step in the possibility of considering miRNAs as a novel candidate for therapeutic approaches.

Growing evidence is showing that miRNAs are not strictly cellular, but are secreted through the release of small vesicles called "exosomes" and are therefore present extracellularly in the peripheral blood and in cell-culture media. The presence of exosomes in patients' serum suggests their potential use as source of prognostic and/or diagnostic markers; moreover it has been suggested that exosome-associated miRNAs might have a role in intercellular communication, although concrete evidence for this has been lacking (Pegtel et al. 2010).

2. LET-7 FAMILY

The first two discovered microRNAs, lin-4 and let-7, were originally identified in the nematode *Caenorhabditis Elegans* where they act as regulators of developmental timing and cell proliferation. Let-7 was subsequently found as the first known human miRNA playing a tumor suppressor function being able to inhibit cellular proliferation and promote differentiation. The let-7 family of microRNAs is highly conserved across species in sequence and function, and it is frequently lost in many types of cancer (Boyerinas et al. 2010).

One of the first publications on the role of tumor suppressor of let-7 in melanoma has been reported by Kunz and his group (Schultz et al. 2008). They reported that the expression of five members of the let-7 family (let-7a, -7b, -7d, -7e, and -7g) is significantly reduced in primary melanoma compared to benign nevi (similar results have been described in the literature for lung and colon cancer). They also demonstrated that overexpression of let-7b in melanoma cells represses cyclins (D1, D3 and A) and cyclin-dependent kinase 4 (CDK4), as well as the ectopic expression of let-7b decreases the anchorage-independent growth ability of melanoma cells and inhibits cell-cycle progression. In the same year, another study by Müller and Bosserhoff confirmed that let-7a was lost in melanoma and malignant cell lines when compared with primary melanocytes and the authors identified a really important direct target for this microRNA, the integrin-$\beta3$ (ITG$\beta3$). This protein is known to be highly related to melanoma progression and its high expression enhances migratory and invasive potentials of melanoma cells. A recent report by Fu and his collaborators showed that let-7b can suppress the expression of Basigin (Bsg), also known as extracellular matrix metalloproteinases inducer (EMMPRIN), in B16-F10 mouse melanoma cell line (Fu et al. 2010). This protein is highly expressed on the surface of tumor cells. The upregulation of let-7b in mouse melanoma cells causes its downregulation and indirectly suppresses another metalloproteinase, MMP9, with consequent reduction of proliferation, invasion and migration as well as lung metastatization.

Avery interesting work published by Fisher and his group has been related to the analysis of miRNAtranscriptional regulation (Ozsolak et al. 2008). In particular they identified a group of miRNAs regulated by MITF (Microphthalmia-associated transcription factor) in melanoma cells using nucleosome mapping and linker sequence analyses. These miRNAs include, among others, some members of the let-7 family (let-7a-1, -7d, -7f-1, and -7i).

Given the crucial roles of the let-7 family, additional studies should be conducted in melanoma cells to provide a better understanding of the transcriptional regulation of these microRNAs and, in turn, the regulation of their target genes in cancer, with the final proposal of identifying novel strategies diagnosis and treatment of this devastating disease.

3. MicroRNA-221/-222

Recent evidence indicates that microRNAs play a crucial role in the initiation and progression of human cancer. Specifically several groups have focused their attention on miRNA-221 and miRNA-222 demonstrating their capability to act as oncogenes.

MiRNA-221 and miRNA-222 are clustered on the X chromosome and transcribed in a common precursor of 2.1 kb RNA, suggestive of a coordinate functional role and their upregulation has been reported in many types of cancers in comparison with their normal counterparts (Sun et al. 2009a,b, Ciafre et al. 2005, Visone et al. 2007, Zhang et al. 2010, Garofalo et al. 2009, Miller et al. 2008, Medina et al. 2008, Felicetti et al. 2008). Specifically, Felicetti and co-workers reported miRNA-221/-222 increasingly expressed during the multistep process from normal melanocytes to advanced metastatic melanomas (Felicetti et al. 2008). To demonstrate the functional role of miRNA-221/-222, they used a lentiviral vector system to over-express both microRNAs in a moderately aggressive melanoma cell line, selected on the basis of its low but detectable levels of miRNA-221/-222 and of its ability to produce melanin, a function often lost in more advanced melanomas. As a direct effect, the ectopic miRNA-221 and miRNA-222 expression resulted in a significant increase in the proliferative growth rate, in the invasive and chemotactic capabilities, as well as in the anchorage-independent growth. They also confirmed the induction of a more tumorigenic phenotype by miRNA-221/-222 in an *in vivo* model: tumor volumes of miRNA-221/-222-expressing melanoma cells were increased when compared with controls. Conversely, the silencing of miRNA-221/-222 in metastatic cells by chemically modified oligomers (antagomirs), inhibited the main functional properties associated with advanced melanomas *in vitro* and *in vivo* (Felicetti et al. 2008). These findings suggest the inhibition of miRNA-221/-222 as an attractive approach for translation into the clinical setting, particularly important in advanced melanoma still lacking effective treatments. Accordingly, other studies reported the feasibility and safety of prolonged administration of a locked nucleic acid (LNA)–modified phosphorothioate oligonucleotide that antagonizes the function of a specific miRNA in a highly relevant disease model (Lanford et al. 2010).

MiRNA-221/-222 exert their function by repressing p27Kip1 and c-KIT receptor, thus promoting both melanoma cells proliferation and differentiation blockade (Felicetti et al. 2008, Igoucheva et al. 2009). p27Kip1 is a cell-cycle regulatory protein that interacts with cyclin-CDK2 and -CDK4, inhibiting cell cycle progression at the G1/S checkpoint. Regulation of p27Kip1 by miRNA-221 and miRNA-222 represents an additional oncogenic mechanism underlying the abnormal cell cycle rate of advanced melanoma and of many other tumors (le Sage et al. 2007, Medina et al. 2008, Felicetti et al. 2008). The miRNA-221/-222 binding to the 3′UTR of p27Kip1 is favored by Pumilio (PUM)-1, a RNA-binding protein, that in cycling cells induces a local conformational change in the p27Kip1 transcript making it accessible to miRNA regulation (Kedde et al. 2010). In addition Fornari and co-workers have shown that the down-modulation of another member of CDKIs, p57, was associated with miRNA-221 upregulation and increased aggressiveness in human hepatocarcinoma (Fornari et al. 2008), thus further substantiating the biological significance of miRNA-221 and -222 as cell cycle regulators. The miRNA-221/-222 contribution to melanogenesis is exerted by repressing the c-KIT tyrosine kinase receptor (Felicetti et al. 2008). The SCF/c-KIT signaling pathway is essential during melanocytes migration from the neural crest to the skin. c-Kit receptor for the stem cell factor (SCF) plays an important role in melanogenesis, cell growth, migration, and survival (Alexeev et al. 2006). While ubiquitously expressed in mature melanocytes, c-KIT is downregulated in up to 70% of metastases, allowing melanoma cells to escape SCF/c-KIT–triggered apoptosis (Willmore-Payne et al. 2005). The transduction signal generated by SCF/c-KIT interaction induces the activation of MITF protein, known to regulate a broad repertoire of genes whose functions in melanocytes range from development, differentiation, survival, cell-cycle regulation and pigment production. In particular, MITF controls the melanin production directly regulating the main melanogenic enzymes such as tyrosinase (TYR) and tyrosinase-related protein 1 (TYRP1). These notions indicates that whereas miRNA-221 and -222 suppress the expression of c-KIT, there is a subsequent inhibition of c-KIT/MAPK signaling cascade resulting in a moderate downregulation of MITF, TYR and TRP-1. In accordance, miRNA-221/-222 overexpression in a primary melanoma cell line caused reduced differentiation and melanogenesis, as hallmarks of an oncogenic progression (Felicetti et al. 2008). The regulation of MITF expression in melanoma cells appears extremely complex as, according to a proposed model, MITF high levels result in cell cycle arrest and differentiation, intermediate levels promote proliferation and tumorigenesis, whereas low amounts lead to cell cycle arrest and apoptosis (Wellbrock and Marais 2005, Goding et al. 2006, Gray-Shopfer et al. 2007, Hoek and Goding 2010). Therefore microRNAs, as miRNA-221/-222, miRNa-137 or miRNA-148/152, should tightly adjust MITF at the favoured levels. Consistent with this regulation, there is less MITF expressed in melanoma cells than in normal melanocytes (Haflidadóttir et al. 2010, Felicetti et al. 2008).

Recent reports highlighted miRNA-221/-222 capability to directly target the tumor suppressor PTEN in NSCLC (Garofalo et al. 2009) and the proapoptotic protein PUMA in glioblastoma (Zhang et al. 2010). Given that PUMA expression is dramatically reduced in metastatic compared to primary melanomas (Karst et al. 2005) and the loss of PTEN function seems to be responsible for many of the phenotypic features of melanoma (Aguissa-Touré and Li 2011), it would be important to study if both PUMA and PTEN are also downregulated by miRNA-221/-222 in melanoma, adding a new small piece in the elucidation of the complex mechanism of regulation.

Regarding the machinery involved in the transcriptional regulation of miRNA-221/-222 expression, Felicetti and coworkers identified the promyelocytic leukemia zinc finger (PLZF), previously reported to be expressed in normal melanocytes and down-modulated in melanomas (Felicetti et al. 2004), as a repressor of miRNA-221 and miRNA-222 via direct binding to their putative regulatory regions. Authors demonstrated that the lack of PLZF in melanomas unblocks miRNA-221 and miRNA-222, which are increasingly expressed along with the disease progression and, by inhibiting c-KIT and p27 translation, favors the induction of a malignant phenotype (Felicetti et al. 2008). Furthermore, Ozsolak and coworkers reported a number of miRNAs, including miRNA-221/-222, to be regulated by MITF in melanoma cells by using nucleosome mapping and linker sequence analyses (Ozsolak et al. 2008).

Finally, Howell and co-workers reviewed positive and negative feedback loops involving miRNA-221/-222 transcriptional regulators and targets (Howell et al. 2010). Specifically, they described the miRNA-221/-222 involvement in melanoma progression as the result of both p27Kip1 and PTEN inhibition. As a consequence of p27Kip1 reduction, cyclin-dependent kinase 2 (CDK2) mediates PLZF phosphorylation, thus causing its ubiquitination and degradation. In addition PTEN reduction promotes the binding of the astrocyte elevated gene 1 (AEG-1) to PLZF, preventing it from its transcriptional repressive action.

3.1 Regulation of microRNA Transcription: The Example of miRNA-221-222

Little is known on microRNA transcriptional regulation. The majority of miRNAs are transcribed by RNA Pol II and many characteristics of miRNA gene promoters, such as the relative frequencies of CpG islands, TATA box, TFIIB recognition, initiator elements, are similar to the promoters of proteincoding genes. Although, their individual regulatory machinery has not been finely studied yet, the DNA-binding factors that regulate miRNA transcription largely overlap with those that control protein-coding genes.

Among diverse miRNAs, particular attention has been paid on the transcriptional regulation of miRNA-221/-222 in different types of cancers, including melanoma, where different groups evidenced both positive and negative regulators of transcription (Garofalo et al. 2009).

These regulatory binding sites have been identified in the putative promoter region up-stream to pre- miRNA-222 between –400 and –50 bp (Felicetti et al. 2008, Di Leva et al. 2010, Mattia et al. 2011). Felicetti and colleagues highlighted the first negative regulation system of miRNA-221/-222 transcription played by the transcription factor in melanoma cell lines. PLZF is a tumor suppressor gene, barely or not detectable in melanomas, but expressed in normal melanocytes. A clear down-regulation of both miRNAs, paralleled by a more differentiated melanocyte-like phenotype, was observed when PLZF was restored in melanoma cells. Two putative consensus binding sequences for PLZF are located upstream to pre-miRNA-222 and a third site is localized in the intragenic region between the two miRNA sequences. Promoter luciferase and ChIP assays confirmed the capability of PLZF to bind the consensus sequences upstream to miRNA-221/222, negatively regulating their expression. Thus, in advanced melanomas the lack of PLZF allows miRNA-221/-222 up-modulation and, in turn, the activation of two oncogenic pathways involved in melanoma progression, through p27Kip1 and c-KIT deregulation (Felicetti et al. 2008).

A more complex function played by ETS-1(v-ets erythroblastosis virus E26 oncogene homolog 1) transcription factor on miRNA-221/222 regulation has been demonstrated in melanoma (Mattia et al. 2011). ETS-1 transcription factor is the founding member of the ETS gene superfamily, encoding a class of phosphoproteins characterized by a conserved domain that recognizes and binds to a GGAA/T DNA core sequence. ETS-1 is involved in an array of cellular functions and, dimerizing with different partners, it can play either positive or negative functions (He 2007, Nakayama 1999). The complexity of ETS-1 action has been well demonstrated by the authors that discovered the presence of either negative or positive regulation on miRNA-221/222 transcription, in early and advanced melanoma cells, respectively. ETS-1 post-translational modifications, rather than its total protein content, seem to be functionally relevant to melanoma. In fact, taking in mind the activating phosphorylation at Thr38, in low grade malignant cells the significant amounts of barely or not phosphorylated ETS-1 represses miRNA-221/-222 transcription. Conversely, in metastatic melanomas the persistent activation of the MAPK-ERK1/2 cascade increases the fraction of Thr-38 phosphorylated ETS-1 inducing miRNA-221/222 transcription and possibly tumor malignancy. Bioinformatic analyses showed the presence of a canonical seed for miRNA-221/222 in the 3'UTR of ETS-1. Interestingly, ETS-1 resulted directly targeted by miRNA-222, but not miRNA-221, thus indicating the capabilities of these two miRNAs also of uncoupled functions demonstrating the existence of their uncoupled functionalities in addition to the common ones (Mattia et al. 2011).

The AP-1 transcription complex, characterized by c-Jun and c-Fos heterodimerized proteins, is predicted to bind and transcriptionally activate miRNA-221/-222 promoter. In metastatic melanomas, the ERK constitutive signaling activates c-JUN (Lopez-Bergami et al. 2007) and P-T38-ETS-1 (Yang et al. 1996), which in turn cooperates towards miRNA-221/-222 activation (Garofalo et al. 2009, Mattia et al. 2011).

Recently, a report by Galardi and co-authors showed that the ectopic modulation of NF-κB modifies miRNA-221/-222 expression in prostate carcinoma and glioblastoma cell lines (Galardi et al. 2011). The identification of two separate distal regions upstream of miRNA-221/-222 promoter bound by the NF-κB subunit p65 and driving efficient transcription have been demonstrated. In this distal enhancer region it has been defined a second binding site for c-Jun that cooperates with p65 fully accounting for the observed up-regulation of miRNA-221/222. Thus, this study finds out an additional mechanism through which NF-κB and c-Jun, two transcription factors deeply involved in cancer onset and progression, contribute to oncogenesis by promoting miRNA-221/222 transcription. Whether or not this distal enhancer regulatory region is also involved in miRNA-221/-222 transcriptional regulation in melanoma has to be investigated.

4. MicroRNA-214

As already reported, several microRNAs such as miRNA-221/-222, miRNA-137, miRNA-182, let7 and miRNA-34a have been described to be involved in the progression of melanoma through the regulation of their respective target genes including c-KIT, p27Kip1, MITF, FOXO3, ITGB3, CCND1 and MET (Mueller and Bosserhoff 2009). Nevertheless the exact mechanisms by which these microRNAs can pursue their function have not yet been identified. A recent work published by Penna in the EMBO journal has led to the identification of a single microRNA, miRNA-214, that can predict the metastatic development of melanoma cells. This microRNA, is able to control the metastatic dissemination of this tumor, thus increasing the capacity of migration, invasion, extravasation and survival of the cells (Penna et al. 2011). Penna et al. have utilized the poorly metastatic A375 parental (A375P) cell line and its highly metastatic variants selected by repeated cycling in nude mice. In this system they demonstrated that miRNA-214 is over-expressed in the metastatic variants. Moreover this modulation is more pronounced than that observed for the other miRNAs previously found involved in melanoma (miRNA-137, miRNA-221/-222, and miRNA-34a). Among the downstream target genes for miRNA-214, the most important are AP-2γ transcription factor and the integrin-α3 (ITGA3). Moreover the expression of AP-2αtranscription factor was also found to be regulated by miRNA-214 although it is unclear whether this is a direct or mediated effect. The loss of AP-2α is one of the molecular abnormalities historically associated with melanoma progression, being able to regulate a cascade of genes known to play a role in melanoma metastasis (c-KIT, MCAM-MUC18, HER-2, VEGF, MMP-2 and the thrombin receptor PAR-1). It was also demonstrated that in melanoma PAR-1isable to regulate the transcriptional activation and secretion of several proangiogenic factors (such as IL-8, PDGF and VEGF) as well as to control the expression of the tumor suppressor gene Maspin. This study appears to be really important demonstrating the main role of miRNA-214 in tumor progression such to be considered as a target for therapeutic intervention.

5. MicroRNA-211

Most of miRNA studies focused on oncomirs, over-expressed in melanoma and downregulating their relative targets. Thus far very little is known regarding miRNAs systematically repressed in these neoplastic cells. Realizing this gap in knowledge, Mazar and co-authors (Mazar et al. 2010) examined the expression levels of human miRNAs in a panel of melanoma cell lines and clinical melanoma samples. They reported that miRNA-211 levels are consistently reduced in non-pigmented melanoma cells compared to its levels in melanocytes. The expression levels of several potential miRNA-211 target mRNAs are elevated in melanoma cells; particularly they observed an increased expression of its target transcript potassium large conductance calcium-activated channel, subfamily M, alpha member 1 (KCNMA1). In fact miRNA-211 directly targets the KCNMA1 transcript whose expression seems to be partially responsible for the high capacity of cell proliferation and invasiveness of melanoma cell lines. In the same paper the authors also demonstrate that MITF expression, a master regulator of melanocyte development (see also miRNA-221/222 and miRNA-137 sections), is important for the coordinate expression of miRNA-211 and of the transient receptor potential cation channel, subfamily M, member (TRPM1). TRPM1 gene is a suppressor of melanoma metastasis (although the molecular basis of this mechanism is not understood) which encodes multiple polypeptide isoforms including melastatin-1. Interestingly miRNA-211 gene is included in TRPM1 sixth intron. Miller and colleagues (Miller et al. 2004) have already demonstrated the presence of a MITF-binding motif (GCTCACATGT) in the promoter of TRPM1, where MITF acts as a transcriptional regulator. Mazar demonstrated that MITF can also transcriptionally coregulate miRNA-211 expression via the TRPM1 promoter.

Another study has found that the tumor suppressive activity in melanoma cells was not mediated by melastatin, but instead by miRNA-211 (Levy et al. 2010). The authors reported that increasing expression of miRNA-211, but not melastatin, reduced migration and invasion of malignant and that highly invasive human melanomas are characterized by low levels of both melastatin and miRNA-211.

They also identified insulin-like growth factor 2 receptor (IGF2R), transforming growth factor-b receptor 2 (TGFBR2) and nuclear factor of activated T-cells 5 (NFAT5) as central node genes "directly or indirectly" reduced by miRNA-211 and in turn leading to reduced invasion of malignant melanomas. Whereas NFAT5 and IGF2R have not been previously linked to melanoma metastasis, increased TGFBR2 expression is linked to TGFB which has been previously shown to downregulate MITF, reducing pigmentation, cell adhesion and invasion of blood vessels.

Moreover BRN2 (also known as POU3F2, POU-domain transcription factor) was identified as a direct functional target of miRNA-211. In agreement with earlier reports, the authors proposed that melanoma tumor cells lacking MITF expression show significantly lower levels of miRNA-211 and TRPM1, allowing elevated expression of BRN2 linked with a more invasive phenotype (Boyle et al. 2011). These and other studies have shown that miRNA-211 expression is abundant in melanocytes and markedly decline in melanoma progression (Xu et al. 2012). On this basis, the recent advances in microRNA-based therapeutics might include miRNA-211 among the novel therapeutic targets for melanoma treatment.

6. miRNA-137

In a recent work Bemis reported that the chromosomal location of miRNA-137 is commonly indicated to harbor a gene contributing to melanoma susceptibility (Bemis et al. 2008). Human miRNAs are located within either exonic or intronic regions of coding or non-coding transcription units. miRNA-137 is placed in a large non-coding transcript with a variable nucleotide tandem repeat (VNTR). The alteration of these 15 bp sequences in primary miRNA-137 interferes with the maturation process, leading to predicted changes in the folding structure of pre-miRNA. MiRNA-137 has been shown to target the MITF "the master regulator" of cell growth, maturation, apoptosis and pigmentation of melanocytes. The role of MITF in melanoma development is complex: high levels of MITF have been demonstrated to exert an anti-proliferative activity whereas very low levels or no expression induce apoptotic pathways. In addition MITF amplification has been found in a significant number of malignant melanoma (20%) and this genetic alteration correlates with decreased overall patient survival. Most likely, MITF plays both cancer-promoting and cancer-inhibiting roles depending on the expression level and/or activity, and proper regulation of MITF is required for cell growth and survival of tumor cells. It is likely that melanoma cells have developed a way to maintain MITF at the appropriate level for tumorigenesis (Russo et al. 2009). In this context, miRNA-137 plays an important role in melanoma biology as regulator of MITF expression level. In melanoma cells lines the presence of alterations in pre- miRNA-137 sequence prevents the normal interaction with MITF 3′UTR. These findings, confirmed by Haflidadóttir and colleagues, indicate that miRNA-137 plays a key functional role in melanoma transformation, in some way maintaining MITF level in the range compatible with tumor progression (Haflidadóttir et al. 2010). Moreover, as recently shown, MITF mRNA isoform containing the miRNA-137 site in its 3′UTR is present in normal melanocytes, but not in melanoma cells, indicating that miRNA-137 repression escape is important for neoplastic transformation (Goswami et al. 2010).

As a support of the important role of miRNA-137 in melanoma biology, Deng recently provided evidences of miRNA-137 dependent inhibition of the carboxi-terminal binding protein I (CtBPI). CtBP1 is an evolutionarily conserved transcriptional corepressor that plays an important role in tumorigenesis and tumor progression, mediating repression of several tumor suppressor genes (Deng et al. 2011).

Expression studies showed that miRNA-137 is down-modulated in metastatic melanoma cells, unblocking the translation of key functional proteins. In this regard, the authors focused on the CtBp1 protein, based on two observations: (*i*) the bioinformatic analysis and the luciferase assay indicated that miRNA-137 interacts with the 3' UTR of CtBp1 mRNA; and (*ii*) miRNA-137 level is inversely related to that of CtBp1 protein and its downstream genes, such as E-cadherin and Bax in the analysed melanoma cell lines.

The tumor suppressor function of miRNA-137 in uveal melanoma has been recently demonstrated by Chan and colleagues. The expression level of miRNA-137 in uveal melanoma cells is lower than in normal melanocytes, and the restored expression of miRNA-137 causes impaired proliferative potential, coupled with down-modulation of MITF protein. Moreover, miRNA-137 transduction in uveal melanoma cells downregulates c-Met and CDK6 protein level, important oncogenic tyrosine kinase receptor and cell cycle related protein respectively (Chan et al. 2011). Furthermore, tumor cells transduced with miRNA-137 have showed a significant reduction of phosphorylated-retinoblastoma protein (p-Rb), a known target of CDK6.

All together these results suggest that miRNA-137 plays a key role in melanoma cells and its dysregulation may contribute to their oncogenic and metastatic potential.

7. MicroRNA-34 FAMILY

In mammalians, the miRNA-34 family comprises three processed miRNAs that are encoded by two different genes: miRNA-34a is encoded by its own transcript, whereas miRNA-34b and miRNA-34c share a common primary transcript (Hermeking et al. 2010). The importance of studying miRNA-34 family members in cancer is directly linked to p53 functions, the most important tumor suppressor gene, frequently silenced or inactivated in cancer, and directly connected to increased metastatic potential. Different types of DNA double-strand breaks promote accumulation and increased transcriptional activity of p53, whose level directly mediates either cell cycle arrest or apoptosis in consideration of the presence of light or severe DNA damage (Hollstein et al. 1991, Vogelstein et al. 2000, Oren 2003). Reports from several laboratories have shown that members of the miRNA-34 family are directly regulated by p53. In addition the up-regulation of miRNA-34a/b/c induced cell-cycle arrest and apoptosis, suggesting these miRNAs as effectors of p53 functions (Chang et al. 2007, Raver-Shapira et al. 2007, Tarasov et al. 2007).

Examples among different target genes of miRNA-34 family members include CDK4/6, Cyclin E2, MET and Bcl-2 (He et al. 2007). Their induction allows p53 to regulate tumor suppressive mechanisms through cell cycle arrest, senescence or apoptosis. On the contrary, their inactivation may be a selective advantage for cancer cells. The most studied miRNA-34a resides on the chromosomal locus 1p36 and its homozygous deletion or epigenetic inactivation are present in neuroblastoma and in melanoma, respectively (Welch et al. 2007 and Cozzolino et al. 2011).

In addition, inspection of the genomic region upstream to miRNA-34a, revealed a prominent CpG island, including the p53 binding site. Bisulfite sequencing confirmed a heavily methylation in prostate as well as in pancreatic, breast and malignant melanoma cell lines. Accordingly, in the IGR-39 melanoma cell line expressing wild-type p53, CpG methylation of miRNA-34a regulatory region did not allow p53 induction of miRNA-34a even after etoposide treatment (Lodygin et al. 2008). This effect may provide a selective advantage for cancer cells which are exposed to stimuli that lead to activation of p53. CpG methylation of miRNA-34a/b/c was found in malignant melanoma, but not in normal melanocytes, in good connection with the metastatic potential and the induction of epithelial-mesenchimal transition (EMT) (Siemens et al. 2011). Since p53 inhibition of EMT has been described as a mode of tumor suppression which presumably prevents metastasis (Hwang et al. 2011), the frequent inactivation of p53 and/or miRNA-34a/b/c found in cancer and in melanoma, may shift the equilibrium of these reciprocal regulations towards the mesenchymal state, thereby blocking the cells in a metastatic state (Siemens et al. 2011).

It is also important to mention that another member of the p53 family, p63, correlating with the growth ability and regenerative capability of keratinocytes directly depends on miRNA-34a and -34c repression (Antonini et al. 2010). MiRNA-34a has also been indicated as a potential tumor suppressor in uveal melanoma where it is prevalently silenced. Its enforced expression led to a significant decrease in cell growth and migration with reduction of c-Met mRNA expression and therefore, down-regulation of phosphorylated Akt and cell cycle related proteins (Yan et al. 2009).

More recently, it has been demonstrated that malignant cells express ligands for the natural killer cell immune-receptor NKG2D, which sensitizes to early recognition and elimination by cytotoxic lymphocytes and provides an innate barrier against tumor development. Tumor cells can escape from NKG2D immune surveillance by an enhanced proteolytic shedding of the NKG2D ligand ULBP2, also causative for the increased levels of soluble ligands in sera of cancer patients (Groh et al. 2002). Recently, an elevated level of soluble ULBP2 in sera from melanoma patients was identified and classified as a strong independent predictor of poor prognosis. Experimental proofs indicate that miRNA-34a and miRNA-34c control ULBP2 expression and that level of miRNA-34a inversely correlated with expression of ULBP2 surface molecules. Accordingly, treatment of melanoma cells with miRNA inhibitors leads to up-regulation of ULBP2, whereas miRNA-34 mimics lead to down-regulation of ULBP2, diminishing tumor cell recognition by NK cells. Thus, the loss of miRNA-34 members found in melanoma cells might assist tumor immune surveillance by enhancing cell associated ULBP2 (Heinemann et al. 2011).

Finally, different experimental data indicate that miRNA-34 family expression might have diagnostic or prognostic potential and can be possibly used as a predictor of therapy response in different tumor types. The group around Scotlandi defined a signature of five miRNAs including, besides miRNA-34a, also miRNA-23a, miRNA-92a, miRNA-490-3p, and miRNA-130b, as an independent predictor of risk to disease progression and survival in patient of Ewing's sarcoma. Results were particularly robust for miRNA-34a, which appeared associated to either event-free or overall survival: patients with the highest expression of miRNA-34a did not experience adverse events in five years, whereas patients with the lowest expression recurred within two years (Nakatami et al. 2012). Further studies are required to possible extend the diagnostic, prognostic and therapeutic functions of miRNA-34 members in melanoma.

8. CIRCULATING miRNAs

Circulating miRNAs have been detected in both plasma and serum of healthy and diseased subjects, where they represent potential noninvasive molecular markers. In order to actually utilize miRNAs as accurate predictors in cancer, a number of points should be evaluated more in depth. Although it is realistic to directly and consistently amplify miRNAs from serum avoiding RNA extraction, difficult quantification and the lack of representative internal controls still represent a drawback.

As ribonucleases are highly present in body fluids, the presence of circulating miRNAs suggests that they are mostly included in lipid or lipoprotein complexes, such as apoptotic bodies, microvesicles or exosomes resulting, therefore, protected (Skog et al. 2008). These extracellular RNAs appear to be involved in the intercellular communication between cancer cells and their surrounding microenvironment, providing a useful tool for studying the genetic changes relative to tumor progression simply analyzing serum samples, presumably regardless of the cancer type or site of dissemination (Kosaka et al. 2010).

Many kinds of circulating miRNAs have been associated with various types of cancers, but at this time only the levels of serum miRNA-221 has been reported as a possible tumor marker of metastatic melanoma (Kanemaru et al. 2011). Specifically, Kanemaru and co-workers reported that serum miRNA-221 levels were significantly higher in metastatic melanoma patients than either in melanoma *in situ* or in normal subjects.

On the contrary, the authors did not observe a difference of miRNA-221 between healthy controls and patients with melanoma *in situ*. Very interestingly, the authors found a correlation between miRNA-221 and tumor thickness indicating miRNA-221 as a marker of cancer prognosis. A small number of patients have been analyzed yet and further studies with larger sample numbers will be required.

Melanoma exosomes/microvesicles have already been shown to exert paracrine functions possibly to prepare the neighboring microenvironment to metastatic dissemination. Results demonstrated that exosomes can act on endothelial cells by inducing a proangiogenic program and on sentinel lymph nodes by regulating inflammatory cytokines (Hood et al. 2011). In the last years, literature is indeed flourishing with examples proving the role of tumor exosomes in the transfer of growth factors and cognate receptors to homologous or heterologous target cells (Camussi et al. 2010). In particular, in melanoma cells functional studies have confirmed the competence of these microvesicular particles to convey the metastatic assets of an advanced donor melanoma into a less aggressive recipient cell line. Furthermore Yuan et al. have shown that microvesicles (MVs) derived from embryonic stem cells may also transfer their subset of microRNA in an acceptor cell population of fibroblasts consequently altering the expression levels of microRNA target mRNAs in surrounding cells (Yuan et al. 2009).

Finally, based on their peculiar lipid composition, exosomes will possibly represent new and useful vehicles for therapeutic drug, including microRNA deliveries.

9. MicroRNAs EXPRESSION PROFILING

Microarray technology enabled us to evaluate simultaneously gene expression of coding and non coding RNAs. In these studies microRNA expression profiles resulted really useful in distinguishing tumors from normal samples, also specifying different subtypes of tumors. Moreover microRNAs turned out to be representative of the original tumor for metastases of unknown origin or for poorly differentiated neoplasias. In addition microRNA profiles could reveal initial alterations associated with cancer, thus being quite useful for early detection or changes predictive of recurrence and metastasis. Thus, the potential clinical utility of microRNA may potentially serve for diagnosis, prognosis, treatment response and clinical outcome.

One of the first publications on miRNA expression profiling in tumor cells that included malignant melanoma samples was published by Lu and coauthors in 2005. In this paper the authors examined the expression of 217 mammalian miRNAs in as much as 334 samples. The miRNA profiles thus obtained were found to be surprisingly informative, because they reflect the developmental lineage and differentiation state of the tumors, whereas messenger RNA profiles were highly inaccurate when applied to the same samples. These results have spotlighted the importance of the potential of miRNA profiling in cancer diagnosis (Lu et al. 2005).

Numerous published works have shown that it was possible to identify miRNAs selectively modulated in the genetic program of the neoplastic cell with respect to the normal counterpart, by using the analysis of miRNAs expression profiles (miRNoma) acquired by the technique of "microarray". A great number of published data started from this type of analysis, for stopping on selected differentially expressed microRNA in order to analyse their functional roles and identify their direct or indirect molecular targets.

In the case of uveal melanoma expression profiles obtained have already demonstrated the ability of diagnostic and prognostic microRNAs, because they allow discriminating subgroups of patients phenotypically indistinguishable. In fact, six microRNAs (miRNA-146b, -214, 199th-*,-199a, -143, let-7b) are sufficient to distinguish with 100% specificity the subgroup of patients at high risk of metastasis from those at low risk (Worley et al. 2008).

Global miRNA expression profiles were determined also on melanoma lymph node metastases from clinical samples compared with melanoma cell lines and melanocytes. Interestingly, the abnormally expressed miRNAs were linked with clinical characteristics, patient survival and mutational status of BRAF (Homo sapiens v-raf murine sarcoma viral oncogene homolog B1) and NRAS (neuroblastoma RAS viral (v-ras) oncogene homolog). For example, miRNA-193a, miRNA-338 and miRNA-565 are downregulated in presence of a BRAF mutation. Furthermore, low miRNA-191 and high of miRNA-193b expressions are associated with poor melanoma-specific survival (Caramuta et al. 2010).

It is finally important that microRNAs have also been shown as useful indicators of patient subgroups for selected treatments. It is evident how adequately characterize the molecular profiles represents a key element for a correct treatment of patients, which allows to exclude those that do not meet the required criteria, avoiding expensive and unnecessary treatment with serious side effects if not highly motivated.

10. CONCLUSIONS

A series of accumulating evidences is showing that modulation of miRNAs represents a novel attractive strategy. MiRNAs are strongly involved in cellular behavior through their role in regulating hundreds of targets involved in different networks (Baek et al. 2008 and Selbach et al. 2008). Therefore therapeutic modulation of a single miRNA may affect many pathways simultaneously to achieve clinical benefit. On the other hand, the cascade of effects, possibly deriving from targeting single miRNAs, should be stringently evaluated in proper preclinical models to rule out any potential toxicity. In the future an important goal should be represented by an effective specific delivery.

To date most *in vivo* studies targeting miRNAs have aimed to inhibit the so called "oncomir" by using antisense reagents, such as antagomirs, LNA or other oligomers (Elmen et al. 2008 and Krutzfeldt et al. 2005). Even if this is a promising approach, the therapeutic delivery of tumor suppressor miRNAs appears more useful. As reported for miRNA-26 (Kota et al. 2009), miRNA replacement therapies for those miRNAs that are highly expressed in normal cells, but lost in cancer, may have unique advantages. Indeed high physiological levels of miRNAs should make normal cells tolerant to an enforced reexpression. In addition, as each miRNA is theoretically able to target multiple transcripts, the probability of acquired resistance should be really lowered in that many simultaneous mutations would be required.

11. SUMMARY POINTS

- Cutaneous melanoma is an aggressive neoplasia whose incidence is rising faster than any other cancer.
- Although several molecular abnormalities have been associated with melanoma progression, the mechanisms underlying the differential gene expression are still largely unknown. The recent identification of microRNA family and the growing evidence of their altered expression in cancer provided a new level of molecular regulation.
- Some studies have shown that overexpression of several microRNAs, such as miRNA-221/-222 or miRNA-214, are involved in melanoma progression through multiple oncogenic mechanisms. Conversely other miRNAs, asmiRNA-34, miRNA-211 and let-7 family, playing a tumor suppressor role, are systematically repressed in melanoma cells. In addition miRNA-221/-222 and miRNA-137 have been reported as direct regulators of MITF, a well known 'master gene' of the melanocytic lineage (Fig. 2). Circulating miRNAs have also been detected in both plasma and serum of healthy and diseased subjects. These extracellular RNAs appear to be involved in the intercellular communications, representing useful blood biomarkers.
- Through the analyses of gene expression profiles obtained by the technique of "microarray", it possible to discriminate a number of miRNAs as diagnostic markers for melanoma.

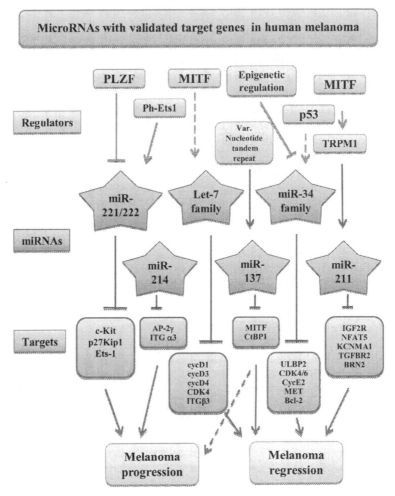

Figure 2. Representative picture showing regulatory factors, miRNAs and validated target genes in human melanoma.

- In a not too distant future, the modulation of selected miRNAs can be the basis for novel attractive therapeutic strategies.

ACKNOWLEDGMENTS

We thank L. Bottero for editorial assistance and G. Loreto for graphics.

ABBREVIATIONS

3'UTR	:	3' untranslated region
AP-2	:	activating enhancer binding protein 2
Bcl-2	:	B-cell CLL/lymphoma 2
BRAF	:	Homo sapiens v-raf murine sarcoma viral oncogene homolog B1
BRN2	:	POU-domain transcription factor class 3 homeobox 2 (POU3F2)
CCND1	:	cyclin D1
CDK4	:	cyclin-dependent kinase
c-Fos	:	FBJ murine osteosarcoma viral oncogene homolog
c-Jun	:	jun proto-oncogene (AP-1)
c-Kit	:	v-kit Hardy-Zuckerman 4 feline sarcoma viral oncogene homolog
CtBP1	:	carboxi-terminal binding protein I
Cyc	:	cyclins
ETS-1	:	v-ets erythroblastosis virus E26 oncogene homolog 1
FOXO3	:	forkhead box O3
HER-2	:	v-erb-b2 erythroblastic leukemia viral oncogene homolog 2
IGF2R	:	insulin-like growth factor 2 receptor
IL-8	:	interleukin 8
ITG	:	integrin
KCNMA1	:	potassium large conductance calcium-activated channel, subfamily M, alpha member 1
LNA	:	locked nucleic acid
MCAM-MUC18:		melanoma cell adhesion molecule
MET	:	met proto-oncogene (hepatocyte growth factor receptor)
miRNA	:	microRNA
MITF	:	Microphthalmia-associated transcription factor
MMP	:	matrix metalloproteinase
MVs	:	microvesicles
NFAT5	:	nuclear factor of activated T-cells 5
NF-κB	:	nuclear factor of kappa light polypeptide gene enhancer in B-cells 1
NKG2D	:	killer cell lectin-like receptor subfamily K, member 1
NRAS	:	neuroblastoma RAS viral (v-ras) oncogene homolog
NSCLC	:	non small cell lung cancer
p27Kip1	:	cyclin-dependent kinase inhibitor 1B
PAR-1	:	Protease-activated receptor 1

PDGF	:	platelet-derived growth factor
PLZF	:	promyelocytic leukemia zinc finger
PTEN	:	phosphatase and tensin homolog
PUMA	:	p53 up-regulated modulator of apoptosis
SCF	:	stem cell factor (c-Kit ligand)
TGFBR2	:	transforming growth factor-b receptor 2
TRPM1	:	transient receptor potential cation channel, subfamily M, member
TYR	:	tyrosinase
TYRP1	:	tyrosinase-related protein 1
ULBP2	:	UL16 binding protein 2
VEGF	:	vascular endothelial growth factor

REFERENCES

Aguissa-Touré, A.H. and G. Li. 2011. Genetic alterations of PTEN in human melanoma. Cell. Mol. Life Sci. EPUB ahead of print

Alexeev, V. and K. Yoon. 2006. Distinctive role of the c-kit receptor tyrosine kinase signaling in mammalian melanocytes. J Invest Dermatol. 126: 1102–1110.

Antonini, D., M.T. Russo, L. De Rosa, M. Gorrese, L. Del Vecchio, and C. Missero. 2010. Transcriptional repression of miR-34 family contributes to p63-mediated cell cycle progression in epidermal cells. J Invest Dermatol. 130: 1249–1257.

Baek, D., J. Villén, C. Shin, F.D. Camargo, S.P. Gygi, and D.P. Bartel. 2008. The impact of microRNAs on protein output. Nature. 455: 64–71.

Balch, C.M., A.C. Buzaid, S.J. Soong, M.B. Atkins, N. Cascinelli, D.G. Coit, I.D. Fleming, J.E Gershenwald, A. Jr. Houghton, J.M. Kirkwood, K.M. McMasters, M.F. Mihm, D.L. Morton, D.S. Reintgen, M.I. Ross, A. Sober, J.A. Thompson, and J.F. Thompson. 2001. Final version of the American Joint Committee on Cancer staging system for cutaneous melanoma. J Clin Oncol. 19: 3635–3648.

Bemis, L.T., R. Chen, C.M. Amato, E.H. Classen, S.E. Robinson, D.G. Coffey, P.F. Erickson, Y.G. Shellman, and W.A. Robinson. 2008. MicroRNA-137 targets microphthalmia-associated transcription factor in melanoma cell lines.Cancer Res. 68: 1362–1368.

Bommer, G.T., I. Gerin, Y. Feng, A.J. Kaczorowski, R. Kuick, R.E. Love, Y. Zhai, T.J. Giordano, Z.S. Quin, B.B. Moore, O.A. MacDougald, K.R. Cho, and E.R. Fearon. 2007. p53-mediated activation of miRNA34 candidate tumor-suppressor genes. Curr Biol. 17: 1298–1307.

Boyerinas, B., S.M. Park, A. Hau, A.E. Murmann, and M.E. Peter. 2010. The role of let-7 in cell differentiation and cancer. Endocr Relat Cancer. 17: F19–36.

Boyle, G.M., S.L. Woods, V.F. Bonazzi, M.S. Stark, E. Hacker, L.G. Aoude, K. Dutton-Register, A.L. Cook, R.A. Sturm, and N.K. Hayward. 2011. Melanoma cell invasiveness is regulated by miR-211 suppression of the BRN2 transcription factor. Pigment Cell Melanoma Res. 24: 525–537.

Camussi, G., M.C. Deregibus, S. Bruno, V. Cantaluppi, and L. Biancone. 2010. Exosomes/microvesicles as a mechanism of cell-to-cell communication. Kidney Int. 78: 838–848.

Caramuta, S., S. Egyha'zi, M. Rodolfo, D. Witten, J. Hansson, C. Larsson, and W.-O. Lui. 2010. MicroRNA Expression Profiles Associated with Mutational Status and Survival in Malignant Melanoma. Journal of Investigative Dermatology. 130: 2062–2070.

Chan, E., R. Patel, S. Nallur, E. Ratner, A. Bacchiocchi, K. Hoyt, S. Szpakowski, S. Godshalk, S. Ariyan, M. Sznol, R. Halaban, M. Krauthammer, D. Tuck, F.J. Slack, and J.B.

Weidhaas. 2011. MicroRNA signatures differenziate melanoma subtypes. Cell Cycle. 10: 1845–1852.

Chang, F., L.S. Steelman, J.T. Lee, J.G. Shelton, P.M. Navolanic, W.L. Blalock, R.A. Franklin, and J.A. McCubrey. 2003. Signal transduction mediated by the Ras/Raf/MEK/ERK pathway from cytokine receptors to transcription factors: potential targeting for therapeutic intervention. Leukemia. 17: 1263–1293.

Chang, T.C., E.A. Wentzel, O.A. Kent, K. Ramachandran, M. Mullendore, K.H. Lee, G. Feldmann, M. Yamakuchi, M. Ferlito, C.J. Lowenstein, D.E. Arking, M.A. Beer, A. Maitra, and J.T. Mendell. 2007. Transactivation of miR-34a by p53 broadly influences gene expression and promotes apoptosis. Mol Cell. 26: 745–752.

Ciafre, S.A., S. Galardi, A. Mangiola, M. Ferracin, C.G. Liu, G. Sabatino, M. Negrini, G. Maira, C.M. Croce, and M.G. Farace. 2005. Extensivemodulation of a set of microRNAs in primary glioblastoma. Biochem Biophys Res Commun. 334: 1351–1358.

Clark, W.H., D.E. Elder, D. Guerry 4th, M.N. Epstein, M.H. Greene, and M. Van Horn. 1984. A study of tumor progression: the precursor lesion of superficial spreading and nodular melanoma. Hum Pathol. 15: 1147–1165.

Cozzolino, A.M., L. Pedace, M. Castori, P. De Simone, N. Preziosi, I. Sperduti, C. Panetta, V. Mogini, C. De Bernardo, A. Morrone, C. Catricalà, and P. Grammatico. 2011. Analysis of the miR-34a locus in 62 patients with familial cutaneous melanoma negative for CDKN2A/CDK4 screening. Familial Cancer EPUB ahead of print.

Deng, Y., H. Deng, F. Bi, J. Liu, L.T. Bemis, D. Norris, X.J. Wang, and Q. Zhang. 2011. MicroRNA-137 targets carboxyl-terminal binding protein 1 in melanoma cell lines. Int J Biol Sci. 7: 133–137.

Dhomen, N. and R. Marais. 2007. New insight into BRAF mutation in cancer. Curr Opin Genet Dev. 17: 31–39.

Di Leva, G., P. Gasparini, C. Piovan, A. Ngankeu, M. Garofalo, C. Taccioli, M.V. Iorio, M. Li, S. Volinia, H. Alder, T. Nakamura, G. Nuovo, Y. Liu, K.P. Nephew, and C.M. Croce. 2010. MicroRNA cluster 221-222 and estrogen receptor alpha interactions in breast cancer. J Natl Cancer Inst. 102: 706–721.

Elmén, J., M. Lindow, A. Silahtaroglu, M. Bak, M. Christensen, A. Lind-Thomsen, M. Hedtjärn, J.B. Hansen, H.F. Hansen, E.M. Straarup, K. McCullagh, P. Kearney, and S. Kauppinen. 2008. Antagonism of microRNA-122 in mice by systemically administered LNA-antimiR leads to up-regulation of a large set of predicted target mRNAs in the liver. Nucleic Acids Res. 36: 1153–1162.

Felicetti, F., L. Bottero, N. Felli, G. Mattia, C. Labbaye, E. Alvino, C. Peschle, M.P. Colombo, and A. Carè. 2004. Role of PLZF in melanoma progression. Oncogene. 3: 4567–4576.

Felicetti, F., C.M. Errico, L. Bottero, P. Segnalini, A. Stoppacciaro, M. Biffoni, N. Felli, G. Mattia, M. Petrini, M.P. Colombo, C. Peschle, and A. Carè. 2008. The Promyelocytic Zinc Finger-MicroRNA-221/-222 Pathway Controls Melanoma Progression through Multiple Oncongenic Mechanisms. Cancer Res. 68: 2745–2754.

Felicetti, F., I. Parolini, L. Bottero, K. Fecchi, M.C. Errico, C. Raggi, M. Biffoni, F. Spadaro, M.P. Lisanti, M. Sargiacomo, and A. Carè. 2009. Caveolin-1 tumor-promoting role in human melanoma. Int J Cancer. 125: 1514–1522.

Fornari, F., L. Gramantieri, M. Ferracin, A. Veronese, S. Sabbioni, G.A. Calin, G.L. Grazi, C. Giovannini, C.M. Croce, L. Bolondi, and M. Negrini. 2008. miR-221 controls CDKN1C/p57 and CDKN1B/p27 expression in human hepatocellular carcinoma. Oncogene. 27: 5651–5661.

Fu, T.Y., C.C. Chang, C.T. Lin, C.H. Lai, S.Y. Peng, Y.J. Ko, and P.C. Tang. 2011. Let-7b-mediated suppression of basigin expression and metastasis in mouse melanoma cells. Exp Cell Res. 317: 445–451.

Galardi, S., N. Mercatelli, M.G. Farace, and S.A. Ciafrè. 2011. NF-kB and c-Jun induce the expression of the oncogenic miR-221 and miR-222 in prostate carcinoma and glioblastoma cells. Nucleic Acids Res. 39: 3892–3902.

Garofalo, M., G. Di Leva, G. Romano, G. Nuovo, S.S. Suh, A. Ngankeu, C. Taccioli, F. Pichiorri, H. Alder, P. Secchiero, P. Gasparini, A. Gonelli, S. Costinean, M. Acunzo, G. Condorelli, and C.M. Croce. 2009. A miR-221&222 regulate TRAIL resistance and enhance tumorigenicity through PTEN and TIMP3 downregulation. Cancer Cell. 16: 498–509.

Garofalo, M., C. Quintavalle, G. Romano, C.M. Croce, and G. Condorelli. 2012. miR221/222 in cancer: their role in tumor progression and response to therapy. Curr Mol Med. 12: 27–33.

Goding, C. and F.L. Meyskens Jr. 2006. Microphthalmic associated transcription factor integrates melanocyte biology and melanoma progression. Clinical Cancer Research. 12: 1069–1073.

Goldberg, M., M. Treier, J. Ghysdael, and D. Bohmann. 1994. Repression of AP-1-stimulated transcription by c-Ets-1. J Biol Chem. 269: 16566–16573.

Goswami, S., R.S. Tarapore, J.J. Teslaa, Y. Grinblat, V. Setaluri, and V.S. Spiegelman. 2010. MicroRNA-340-mediated degradation of microphthalmia-associated transcription factor mRNA is inhibited by the coding region determinant-binding protein. J Biol Chem. 85: 20532–20540.

Gray-Schopfer, V., C. Wellbrock, and R. Marais. 2007. Melanoma biology and new targeted therapy. Nature. 445: 851–857.

Groh, V., J. Wu, C. Yee, and T. Spies. 2002. Tumour-derived soluble MIC ligands impair expression of NKG2D and T-cell activation. Nature. 419: 734–738.

Haflidadóttir, B.S., K. Bergsteinsdóttir, C. Praetorius, and E. Steingrímsson. 2010. miR-148 regulates Mitf in melanoma cells. PLoS One. 5: e11574.

He, J., Y. Pan, J. Hu, C. Albarracin, Y. Wu, and J.L. Dai. 2007. Profile of Ets gene expression in human breast carcinoma. Cancer Biol Ther. 6: 76–82.

He, L., X. He, L.P. Lim, E. de Stanchina, Z. Xuan, Y. Liang, W. Xue, L. Zender, J. Magnus, D. Ridzon, A.L. Jackson, P.S. Linsley, C. Chen, S.W. Lowe, M.A. Cleary, and G.J. Hannon. 2007. A microRNA component of the p53 tumour suppressor network. Nature. 447: 1130–1134.

Heinemann, A., F. Zhao, S. Pechlivanis, J. Eberle, A. Steinle, S. Diederichs, D. Schadendorf, and A. Paschen. 2011. Tumor Suppressive MicroRNAs miR-34a/c Control Cancer Cell Expression of ULBP2, a Stress-Induced Ligand of the Natural Killer Cell Receptor NKG2D. Cancer Research. 18: 460–471.

Hermeking, H. 2010. The miR-34 family in cancer and apoptosis. Cell Death and Differentiation. 17: 193–199.

Hoek, K.S. and C.R. Goding. 2010. Cancer stem cells versus phenotype-switching in melanoma. Pigment Cell and Melanoma Research. 23: 746–759.

Hollstein, M., D. Sidransky, B. Vogelstein, and C.C. Harris. 1991. p53 mutations in human cancers. Science. 253: 49–53.

Howell, P.M. Jr., X. Li, A.I. Riker, and Y. Xi. 2010. MicroRNA in Melanoma. Ochsner J. 10: 83–92.

Huang, S., D. Jean, M. Luca, M.A Tainsky, and M. Bar-Eli. 1998. Loss of AP-2 results in downregulation of c-KIT and enhancement of melanoma tumorigenicity and metastasis. EMBO J. 17: 4358–4369.

Hwang, C.I., A. Matoso, D.C. Corney, A. Flesken-Nikitin, S. Körner, W. Wang, C. Boccaccio, S.S. Thorgeirsson, P.M. Comoglio, H. Hermeking, and A.Y. Nikitin. 2011. Wild-type p53 controls cell motility and invasion by dual regulation of MET expression. PNAS. 23: 14240–14245.

Igoucheva, O. and V. Alexeev. 2009. MicroRNA-dependent regulation of cKit in cutaneous melanoma. Biochem Biophys Res Commun. 379: 790–794.

Kanemaru, H., S. Fukushima, J. Yamashita, N. Honda, R. Oyama, A. Kakimoto, S. Masuguchi, T. Ishihara, Y. Inoue, M. Jinnin, and H. Ihn. 2011. The circulating microRNA-221 level in patients with malignant melanoma as a new tumor marker. J Dermatol Sci. 61: 187–193.

Karst, A.M., D.L. Dai, M. Martinka, and G. Li. 2005. PUMA expression is significantly reduced in human cutaneous melanomas. Oncogene. 24: 1111–1116.

Kedde, M., M. van Kouwenhove, W. Zwart, J.A. Oude Vrielink, R. Elkon, and R. Agami. 2010. A Pumilio-induced RNA structure switch in p27-3′ UTR controls miR-221 and miR-222 accessibility. Nat Cell Biol. 12: 1014–1020.

Kosaka, N., H. Iguchi, and T. Ochiya. 2010. Circulating microRNA in body fluid: a new potential biomarker for cancer diagnosis and prognosis. Cancer Sci. 101: 2087–2092.

Kota, J., R.R. Chivukula, K.A. O'Donnell, E.A. Wentzel, C.L. Montgomery, H.W. Hwang, T.C. Chang, P. Vivekanandan, M. Torbenson, K.R. Clark, J.R. Mendell, and J.T. Mendell. 2009. Therapeutic microRNA delivery suppresses tumorigenesis in a murine liver cancer model. Cell. 137: 1005–1017.

Krützfeldt, J., N. Rajewsky, R. Braich, K.G. Rajeev, T. Tuschl, M. Manoharan, and M. Stoffel. 2005. Silencing of microRNAs *in vivo* with 'antagomirs'. Nature. 438: 685–689.

Lanford, R.E., E.S. Hildebrandt-Eriksen, A. Petri, R. Persson, M. Lindow, M.E. Munk, S. Kauppinen, and H. Ørum. 2010. Therapeutic silencing of microRNA-122 in primates with chronic hepatitis C virus infection. Science. 327: 198–201.

Larson, A.R., E. Konat, and R.M. Alani. 2009. Melanoma biomarkers: current status and vision for the future. Nat Clin Pract Oncol. 6: 105–117.

le Sage, C., R. Nagel, D.A. Egan, M. Schrier, E. Mesman, A. Mangiola, C. Anile, G. Maira, N. Mercatelli, S.A. Ciafrè, M.G. Farace, and R. Agami. 2007. Regulation of the p27Kip1 tumor suppressor by miR-221 and miR-222 promotes cancer cell proliferation. EMBO J. 26: 3699–3708.

Levy, C., M. Khaled, D. Iliopoulos, M.M. Janas, S. Schubert, S. Pinner, P.H. Chen, S. Li, A.L. Fletcher, S. Yokoyama, K.L. Scott, L.A. Garraway, J.S. Song, S.R. Granter, S.J. Turley, D.E. Fisher, and C.D. Novina. 2010. Intronic miR-211 assumes the tumor suppressive function of its host gene in melanoma. Mol Cell. 40: 841–849.

Lodygin, D., V. Tarasov, A. Epanchintsev, C. Berking, T. Knyazeva, H. Körner, P. Knyazev, J. Diebold, and H. Hermeking. 2008. Inactivation of miR-34a by aberrant CpG methylation in multiple types of cancer. Cell Cycle. 16: 2591–2600.

Lopez-Bergami, P., C. Huang, J.S. Goydos, D. Yip, M. Bar-Eli, M. Herlyn, K.S. Smalley, A. Mahale, A. Eroshkin, S. Aaronson, and Z. Ronai. 2007. Rewired ERK-JNK signaling pathways in melanoma. Cancer Cell. 11: 447–460.

Lu, J., G. Getz, E.A. Miska, E. Alvarez-Saavedra, J. Lamb, D. Peck, A. Sweet-Cordero, B.L. Ebert, R.H. Mak, A.A. Ferrando, J.R. Downing, T. Jacks, H.R. Horvitz, and T.R. Golub. 2005. MicroRNA expression profiles classify human cancers. Nature. 435: 834–838.

Mattia, G., M.C. Errico, F. Felicetti, M. Petrini, L. Bottero, L. Tomasello, P. Romania, A. Boe, P. Segnalini, A. Di Virgilio, M.P. Colombo, and A. Carè. 2011. Constitutive activation of the ETS-1-miR-222 circuitry in metastatic melanoma. Pigment Cell Melanoma Res. 24: 953–965.

Mazar, J., K. DeYoung, D. Khaitan, E. Meister, A. Almodovar, J. Goydos, A. Ray, and R.J. Perera. 2010. The Regulation of miRNA-211 Expression and Its Role in Melanoma Cell Invasiveness. PLoS One. 5: e13779.

Medina, R., S.K. Zaidi, C.G. Liu, J.L. Stein, A.J. van Wijnen, C.M. Croce, and G.S. Stein. 2008. MicroRNAs 221 and 222 bypass quiescence and compromise cell survival. Cancer Res. 68: 2773–8.

Miller, A.J., J. Du, S. Rowan, C.L. Hershey, H.R. Widlund, and D.E. Fisher. 2004. Transcriptional regulation of the melanoma prognostic marker melastatin (TRPM1) by MITF in melanocytes and melanoma. Cancer Res. 64: 509–516.

Miller, T.E., K. Ghoshal, B. Ramaswamy, S. Roy, J. Datta, C.L. Shapiro, S. Jacob, and S. Majumder. 2008. MicroRNA-221 / 222 confers tamoxifen resistance in breast cancer by targeting p27Kip1. J Biol Chem. 283: 29897–29903.

Mueller, D.W. and A.K. Bosserhoff. 2009. Role of miRNAs in the progression of malignant melanoma. Br J Cancer. 101: 551–556.

Müller, D.W. and A.-K. Bosserhoff. 2008. Integrin β_3 expression is regulated by let-7a miRNA in malignant melanomaRegulation of integrin β3 in melanoma. Oncogene. 27: 6698–6706.

Nakatani, F., M. Ferracin, M.C. Manara, S. Ventura, V. Del Monaco, S. Ferrari, M. Alberghini, A. Grilli, S. Knuutila, K.L. Schaefer, G. Mattia, M. Negrini, P. Picci, M. Serra, and K. Scotlandi. 2012. miR-34a predicts survival of Ewing's sarcoma patients and directly influences cell chemo-sensitivity and malignancy. J of Phatology. 226: 796–805.

Nakayama, T., M. Ito, A. Ohtsuru, S. Naito, M. Nakashima, and I. Sekine. 1999. Expression of the ets-1 proto-oncogene in human thyroid tumor. Mod Pathol. 12: 61–68.

Oren, M. 2003. Decision making by p53: life, death and cancer. Cell Death Differ. 10: 431–442.

Ozsolak, F., L.L. Poling, Z. Wang, H. Liu, X.S. Liu, R.G. Roeder, X. Zhang, J.S. Song, and D.E. Fisher. 2008. Chromatin structure analyses identify miRNA promoters. Genes Dev. 22: 3172–3183.

Pegtel, D.M., K. Cosmopoulos, D.A. Thorley-Lawson, M.A. van Eijndhoven, E.S. Hopmans, J.L. Lindenberg, T.D. de Gruijl, T. Würdinger, and J.M. Middeldorp. 2010. Functional delivery of viral miRNAs via exosomes. Proc Natl Acad Sci USA. 107(14): 6328–33.

Penna, E., F. Orso, D. Cimino, E. Tenaglia, A. Lembo, E. Quaglino, L. Poliseno, A. Haimovic, S. Osella-Abate, C. De Pittà, E. Pinatel, M.B. Stadler, P. Provero, M.G. Bernengo, I. Osman, and D. Taverna. 2011. microRNA-214 contributes to melanoma tumour progression through suppression of TFAP2C. EMBO J. 30(10): 1990–2007.

Raver-Shapira, N., E. Marciano, E. Meiri, Y. Spector, N. Rosenfeld, N. Moskovits, Z. Bentwich, and M. Oren. 2007. Transcriptional activation of miR-34a contributes to p53-mediated apoptosis. Mol Cell. 26: 731–743.

Russo, A.E., E. Torrisi, Y. Bevelacqua, R. Perrotta, M. Libra, J.A. McCubrey, D.A. Spandidos, F. Stivala, and G. Malaponte. 2009. Melanoma: molecular pathogenesis and emerging target therapies. Int J Oncol. 34(6): 1481–9.

Schultz, J., P. Lorenz, G. Gross, S. Ibrahim, and M. Kunz. 2008. MicroRNA let-7b targets important cell cycle molecules in malignant melanoma cells and interferes with anchorage-independent growth. Cell Res. 18(5): 549–57.

Selbach, M., B. Schwanhäusser, N. Thierfelder, Z. Fang, R. Khanin, and N. Rajewsky. 2008. Widespread changes in protein synthesis induced by microRNAs. Nature. 455(7209): 58–63.

Siemens, H., R. Jackstadt, S. Hünten, M. Kaller, A. Menssen, U. Götz, and H. Hermeking. 2011. miR-34 and SNAIL form a double-negative feedback loop to regulate epithelial-mesenchymal transitionset al. Cell Cycle. 10: 4256–71.

Skog, J., T. Würdinger, S. van Rijn, D.H. Meijer, L. Gainche, M. Sena-Esteves, W.T. Jr Curry, B.S. Carter, A.M. Krichevsky, and X.O. Breakefield. 2008. Glioblastoma microvesicles transport RNA and proteins that promote tumour growth and provide diagnostic biomarkers. Nat Cell Biol. 10(12): 1470–6.

Sun, T., Q. Wang, S. Balk, M. Brown, G.S. Lee, and P. Kantoff. 2009b. The role of microRNA-221and microRNA-222 in androgen-independent prostate cancer cell lines. Cancer Res. 69: 3356–63.16.

Sun, T., M. Yang, P. Kantoff, and G.S. Lee. 2009a. Role of microRNA-221/-222 in cancer development and progression. Cell Cycle. 8: 2315–6.

Tarasov, V., P. Jung, B. Verdoodt, D. Lodygin, A. Epanchintsev, A. Menssen, G. Meister, and H. Hermeking. 2007. Differential regulation of microRNAs by p53 revealed by massively parallel sequencing: miR-34a is a p53 target that induces apoptosis and G1-arrest. Cell Cycle. 6: 1586–1593.

Vereecken, P., M. Laporte, and M. Heenen. 2007. Significance of cell kinetic parameters in the prognosis of malignant melanoma: a review. J Cutan Pathol. 34(2): 139–45.

Visone, R., L. Russo, P. Pallante, I. De Martino, A. Ferraro, V. Leone, E. Borbone, F. Petrocca, H. Alder, C.M. Croce, and A. Fusco. 2007. MicroRNAs (miR)-221 and miR-222, both overexpressed in human thyroidpapillary carcinomas, regulate p27Kip1 protein levels and cell cycle. Endocr Relat Cancer. 14: 791–8.

Vogelstein, B., D. Lane and A.J. Levine. 2000. Surfing the p53 network. Nature. 408: 307–310.

Welch, C., Y. Chen, and R.L. Stallings. 2007. MicroRNA-34a functions as a potential tumor suppressor by inducing apoptosis in neuroblastoma cells. Oncogene. 26: 5017–22.

Wellbrock, C. and R. Marais. 2005. Elevated expression of MITF counteracts B-RAF-stimulated melanocyte and melanoma cell proliferation. Journal of Cell Biology. 170: 703–708.

Willmore-Payne, C., J.A. Holden, S. Tripp, and L.J. Layfield. 2005. Human.malignant.melanoma: detection of BRAF- and c-kit-activating mutations by high-resolution amplicon melting analysis. Hum Pathol. 36(5): 486–93.

Worley, L.A., M.D. Long, M.D. Onken, and J.W. Harbour. 2008. Micro-RNAs associated with metastasis in uveal melanoma identified by multiplexed microarray profiling. Melanoma Res. 18(3): 184–90.

Xu, Y., T. Brenn, E.R. Brown, V. Doherty, and D.W Melton. 2012. Differential expression of microRNAs during melanoma progression: miR-200c, miR-205 and miR-211 are downregulated in melanoma and act as tumour suppressors. Br J Cancer. 106(3): 553–61.

Yan, D., X. Zhou, X. Chen, D.N. Hu, X.D. Dong, J. Wang, F. Lu, L. Tu, and J. Qu. 2009. MicroRNA-34a Inhibits Uveal Melanoma Cell Proliferation and Migration through Downregulation of c-Met. Invest Ophthalmol Vis Sci. 50: 1559–1565.

Yang, B.S., C.A. Hauser, G. Henkel, M.S. Colman, C. Van Beveren, K.J. Stacey, D.A. Hume, R.A. Maki, and M.C. Ostrowski. 1996. Ras-mediated phosphorylation of a conserved threonine residue enhances the transactivation activities of c-Ets1 and c-Ets2. Mol Cell Biol. 16: 538–547.

Yuan, A., E.L. Farber, A.L. Rapoport, D. Tejada, R. Deniskin, N.B. Akhmedov, and D.B. Farber. 2009. Transfer of microRNAs by embryonic stem cell microvesicles. PLoS One. 4: e4722.

Zhang, C., J. Zhang, A. Zhang, Y. Wang, L. Han, Y. You, P. Pu, and C. Kang. 2010. PUMA is a novel target of miR-221/222 in human epithelial cancers. Int J Oncol. 37(6): 1621–6.

Zhang, C.Z., J.X. Zhang, A.L. Zhang, Z.D. Shi, L. Han, Z.F. Jia, W.D. Yang, G.X. Wang, T. Jiang, Y.P. You, P.Y. Pu, J.Q. Cheng, and C.S. Kang. 2010. MiR-221 and miR-222 target PUMA to induce cell survival in glioblastoma. Mol Cancer. 9: 229.

Zhang, J., L. Han, Y. Ge, X. Zhou, A. Zhang, C. Zhang, Y. Zhong, Y. You, P. Pu, and C. Kang. 2010. miR-221/222 promote malignant progression of glioma through activation of the Akt pathway. Int J Oncol. 36: 913–20.

CHAPTER 11

MicroRNAs IN
MEDULLOBLASTOMA

Tarek Shalaby,[a] Martin Baumgartner,[b]
Giulio Fiaschetti[c] and Michael A. Grotzer[d,*]

ABSTRACT

Medulloblastoma (MB), the most frequent malignant brain tumor of childhood, is an invasive embryonal tumor of the cerebellum with an inherent tendency to metastasize via the subarachnoidal space. Up to 60–70% of children with MB can be cured using surgery, radiotherapy and chemotherapy, but significant tumor- and treatment-related side effects have to be considered. Mechanisms driving the formation of MB point towards impaired cerebellar development as an important factor. MicroRNAs (miRNAs) have recently emerged as key regulators of gene expression during development and their expression is frequently de-regulated in human disease states including cancer. It is becoming clear that miRNAs are essential regulators of key pathways implicated in MB pathogenesis. This chapter will focus on miRNAs targeting the expression of proteins involved in developmental pathways such as hedgehog signaling as well as proteins that are known to impact

Oncology Department, University Children's Hospital of Zurich, Switzerland.
[a]Email: tarek.shalaby@kispi.uzh.ch
[b]Email: Martin.Baumgartner@kispi.uzh.ch
[c]Email: Giulio.Fiaschetti@kispi.uzh.ch
[d]Email: Michael.Grotzer@kispi.uzh.ch
*Corresponding author

biological activity and clinical features of MB such as ErbB2 and MYC. Finally, we will discuss the potential of miRNAs as a new MB therapeutic target.

1. INTRODUCTION

MB is a cerebellar tumor that accounts for approximately one fifth of all pediatric brain tumors. Despite aggressive treatment regimes the overall mortality rate remains relatively high (Rood et al. 2004, Rossi et al. 2008, Saran 2009). Improvements in therapy can be expected from the clarification of the molecular mechanisms contributing to tumor pathogenesis, which is essential for the development of new options for targeted therapy. Advances in cancer biology research have increasingly linked pediatric neoplasms with disordered mechanisms of normal development, supporting the model of embryonal tumorigenesis (Grimmer and Weiss 2006). Likewise the origins of MB are intrinsically linked to the development of the cerebellum: experimental evidences in mouse models proved that the tumor arises from neural stem cell precursors in the external germinal layer of the cerebellum during neurogenesis. However, there is still a debate ongoing about the exact cellular origin in humans (Gilbertson and Ellison 2008, Grimmer and Weiss 2006, Rodini et al. 2010, Ward et al. 2009, Yang et al. 2008). Pathways, such as sonic hedgehog (SHH), NOTCH and MYC, regulating the normal cerebellar development, play a crucial role in the MB tumorigenesis (Marino 2005). These pathways are considered to be the master regulators of cerebellum development and several genes as well as miRNAs belonging to these pathways are frequently deregulated in human MBs (Zollo 2011). Hence dissecting the molecular alterations associated with cerebellum development should significantly improve our understanding of the biological networks driving MB tumorigenesis and may allow future targeted therapies to become available.

MicroRNAs (miRNAs) are small (about 21 to 23 nucleotides in length) evolutionarily conserved non-coding RNAs that act as expression regulators of genes involved in fundamental cellular processes, such as development, differentiation, proliferation, survival and death (Bartel 2004). Over the past few years, specific miRNAs have been reported to be implicated in various human cancers including leukemia, breast, colon, testicular, lung, thyroid, head and neck cancers as well as brain tumors (Calin and Croce 2009, Corcoran et al. 2011, De Krijger et al. 2011, He et al. 2005, King et al. 2011, Lin et al. 2010, Lv et al. 2010, Murray et al. 2011, Si et al. 2007, Tran et al. 2007, Voorhoeve et al. 2006, Zhu et al. 2007, 2008). miRNAs were linked to tumorigenesis due to the fact that 50% of the miRNAs genes are located at genomic sites associated with cancer-specific chromosomal rearrangements as well as due to the proximity of their genes to chromosomal breakpoints

(Calin et al. 2004, Palmero et al. 2011). It has been found that miRNAs have the ability, in the context of human cancer, to behave like oncogenes or tumor suppressors, a finding which has important implications for cancer etiology and diagnosis (Calin and Croce 2006, Esquela-Kerscher and Slack 2006, Medina and Slack 2008). Both losses and gains of miRNA functions through deletions, amplifications or mutations in miRNA loci or epigenetic silencing and the deregulation of transcription factors that target specific miRNAs, can contribute to cancer development (Croce 2009). The frequent aberrant expression and functional implication of miRNAs in human cancers have lifted these small cellular components to the ranks of preferred drug targets that operates by a new mechanism of action (Bader et al. 2011). Treatment with miRNA-blocking drugs could be particularly effective in treating tumors that developed due to an excess of specific microRNAs. Although the role of miRNA function in brain tumor development is still partially understood, it is becoming clear that miRNAs are essential regulators of many of the key pathways implicated in MB pathogenesis and progression (Ferretti et al. 2008). This chapter will focus on the current understanding of the role that miRNAs play in MB development and progression and will discuss their potential as new therapeutic targets.

2. NORMAL AND ONCOGENIC ROLES OF miRNAs IN THE DEVELOPMENT OF THE NERVOUS SYSTEM

2.1 Role of miRNAs in Embryonal Development

The importance of miRNA pathways for proper embryonic cell function has been confirmed in stem cell systems (Hatfield et al. 2005, Jin and Xie 2007, Kanellopoulou et al. 2005, Murchison et al. 2005, Park et al. 2007, Wu 2011). Numerous publications demonstrate the importance of the developmental effects and the defects related to the miRNA-mediated regulation of gene expression during early embryogenesis in different systems, including *C. elegans, drosophila,* zebrafish and mice (Cao et al. 2006, 2007, Houbaviy et al. 2003, Lakshmipathy et al. 2010, Makeyev et al. 2007, Slack 2010, Suh et al. 2004, Visvanathan et al. 2007, Yekta et al. 2004). In humans, experimental data shows that miRNAs control the fate of human embryonal stem cells (ESCs) during early development (Bar et al. 2008, Gangaraju and Lin 2009, Goff et al. 2009, Lakshmipathy et al. 2010, Lee et al. 2006, Mineno et al. 2006, Morin et al. 2008, Viswanathan et al. 2008). However, which of these individual miRNAs are important for the proper function of particular stem cell types is not well understood (Hatfield and Ruohola-Baker 2008, Suh et al. 2004). Suh et al. (Suh et al. 2004) showed that human (ESCs) express a unique set of microRNAs (miR-200c, 368, 154*, 371, 372, 373*, 373) that are rapidly downregulated during differentiation and may serve as molecular markers for the early embryonic stage and for undifferentiated stem cells. Interestingly the same group identified a subset of miRNAs that are expressed in ESCs as well as in embryonal cancer stem cells (miR-200c, 368, 154*, 371, 372, 373*, 373) [reviewed in (Navarro and Monzo 2010)].

2.2 miRNAs Associated with Neural Differentiation

Increasing evidence suggests that embryogenesis and tumorigenesis display common characteristics. Both processes involve temporally and spatially precisely coordinated proliferation, differentiation and cell migration, which, if dysregulated, can contribute to tumor formation. Based on these parallels, researchers begun to examine the role of miRNAs in neural precursor development and cell fate determination in order to determine whether brain developmental pathways are linked to brain tumor etiology. The first evidence for a functional role of miRNAs in the nervous system came from genetic studies conducted in invertebrates that indicated the role of two miRNAs, lsy-6 and miR-273, in neural cell fate specification in *C. elegans* (Erdos et al. 1968, Johnston et al. 2005). Cell fate specification during organogenesis by miRNAs is not restricted to worms because miR-9a and miR-7 play a major role in cell fate specification in developing sensory organs in *Drosophila* (Li and Carthew 2005). Recently, the role of miRNAs in the development of mammalian brains was demonstrated in mouse mutants lacking the activity of Dicer, the type III RNAse that is necessary for miRNA biogenesis, which die at embryonic day 7 [reviewed in (Coolen and Bally-Cuif 2009)]. Using zebrafish as a model, researchers discovered that specific miRNAs such as miR-92b, miR-124, miR-222, miR-218a, miR-26 and miR-29, miR-9 are expressed differentially between neural precursors and stem cells (Giraldez et al. 2005, Kapsimali et al. 2007). In differentiating ESCs of mammals, particular miRNAs such as let-7a, miR-218, and miR-125, miR-200a, miR-200b, miR-429, miR138 and miR-141 were found to be specifically expressed or enriched in brain tissues while others such as miR-124 and miR-9, were found to be increased sharply in abundance during the transition from neuronal precursors to neurons (Cheng et al. 2009, Choi et al. 2008, Conaco et al. 2006, Kim et al. 2007, Landgraf et al. 2007, Makeyev et al. 2007, Miska et al. 2004, Packer et al. 2008, Sempere et al. 2004, Visvanathan et al. 2007, Yu et al. 2008, Zhao et al. 2009). Studying the expression pattern of miRNAs during mammalian brain development, Kim et al. (Kim et al. 2004) identified 8 miRNAs (miR-103, -124a, -128, -323, -326, -329, -344, -192-2) to be temporally expressed in the rat cortex and cerebellum during neural differentiation and development [reviewed in (Fernandez et al. 2009)]. In another study, using human embryonic cells that differentiate into neurons upon retinoic acid treatment, Sempere et al. showed that miR-9/9, -103-1, -124a, -124b, -128, -135, -156, -218 were co-ordinately induced during the neural differentiation process (Sempere et al. 2004). Some of these miRNAs are the same as the candidates Kim and colleagues showed to be developmentally regulated in rat brain corticogenesis, indicating important roles for these miRNAs in neuronal cell fate specification and differentiation. Remarkably some of these miRNAs were reported to be also active in neuron-developmental disorders (Zhao et al. 2008).

2.3 Oncogenic Roles of miRNAs in Brain Development

The notion that cancer is fuelled by self-renewing stem cells is gaining prominence, and so is the idea that miRNAs can direct cell fate. Yu et al. recently brought the two fields together by showing that a single miRNA molecule (miRNA/let-7) can regulate stemness in breast cancer cells (Yu et al. 2007). The important role of stem cells both in tissue and cancer development has driven much of the research into neural cancer stem cell biology. Recently specific miRNA genes were identified to be responsible for ESCs proliferation and differentiation as well as contributing to the initiation and progression of cancer stem cells (Croce and Calin 2005, DeSano and Xu 2009, Hatfield and Ruohola-Baker 2008, Silber et al. 2008, Yu et al. 2007). Silber et al. reported that miR-124 and miR-137 induce differentiation of neural stem cells and glioblastoma stem cells (Silber et al. 2008). DeSano et al. (DeSano and Xu 2009) reported that miR-34 deficiency is involved in the self-renewal and survival of cancer stem cells and in cancer cells lacking functional p53, restoration of miR-34 was able to re-establish the tumor suppressing signaling pathway (DeSano and Xu 2009). Remarkably, specific miRNAs, such as miR-302, were recently found to be capable of reprogramming the cancer cells back into a pluriopotent embryonic stem cell-like state, which then could be induced into tissue-specific mature cells (Lin et al. 2008). Together these reports suggest that miRNA expression is vital for normal as well as abnormal ECS development that can lead to cancer stem cell initiation [reviewed in (DeSano and Xu 2009)]. Cancer stem cells have been identified, isolated and characterized recently in embryonal neural malignancies including MB (Fan and Eberhart 2008, Mantamadiotis and Taraviras 2011, Singh et al. 2003, Singh et al. 2004, Yang et al. 2008). With the knowledge that MB harbor cancer initiating cells with stem cell properties, there is now a great effort by scientists to understand the involvement of aberrantly expressed miRNAs in embryonal cancer stem cells and to elucidate the mechanisms which distinguish these cells from normal stem cells. Functional studies on miRNAs within the cancer stem cells of MB will be crucial for elucidating the mechanisms behind oncogenesis in these deadly malignancies. The gained insight might reveal novel therapeutic targets, whose inhibition should aim at eliminating the cancer-initiating cells and thus greatly reduce the risk of cancer relapse.

3. MOLECULAR BASIS OF MEDULLOBLASTOMA DEVELOPMENT

Embryonal tumors have long been suspected to be caused by faulty activation of developmental signaling pathways. Studies on embryonic cancer stem cells revealed that tumors in their gene expression repertoire harbor hallmarks of cells during embryonal development (Strizzi et al. 2009). Thus, to better understand MB development, scientists examined the relationship between development of cerebellum and MB formation (Raffel 2004). Human cerebellum develops over a long period of time, extending from the early embryonic period until the first postnatal years. This prolonged development time makes the cerebellum vulnerable to a broad spectrum of developmental disorders. The human cerebellum derives from two germinal matrices: ventricular zone (VZ)-derived cells eventually differentiate into the Purkinje cells and the deep cerebellar nuclei forming the early cerebellar anlage. The secondary germinal matrix, termed the rhombic lip, gives rise to granule cell precursors (GCPs) that migrate to form the external granule layer (EGL). The postnatal EGL undergoes rapid clonal expansion generating post-mitotic neurons within the internal granular cell layer (IGL) that differentiate and migrate through the molecular layer along Bergmann glia to form the IGL. The EGL eventually disappears as all cell division ceases and all postmitotic neurons move to the IGL (Raffel 2004, Ten Donkelaar et al. 2003). GCPs in the outer EGL display an upregulation in genes that mark the granule neuron lineage and genes involved in cell proliferation such as Ptch, Smo, Gli1 and Gli2, while early post-mitotic granule neurons, located in the inner EGL, display increases in neuron-specific class III β tubulin levels, which mark cell cycle exit and early neuronal differentiation. Hence, the balance between proliferation and cell cycle exit is crucial to ensure proper cerebellar pattern formation [reviewed in (Srivastava and Nalbantoglu 2010)]. Based on the close similarity of microscopic appearance of MB tumor cells and GCPs, and their location on the surface of the cerebellum, it has been proposed that a subset of MBs arises from GCPs that give rise to granule neurons (Gibson et al. 2010, Gilbertson 2011, Katsetos et al. 2003, Marino 2005, Raffel 2004). In support of this view, immunohistochemical studies have demonstrated that MB cells express markers commonly associated with GCPs, such as p75NTR, TrkC, Zic1 and Math1 (Buhren et al. 2000, Pomeroy et al. 1997, Salsano et al. 2004, Yokota et al. 1996). Furthermore comparative studies on the molecular control of GCP neurogenesis and MB formation have identified genes and their regulators within signaling pathways, as well as noncoding RNAs, that have crucial roles in both normal cerebellar development and MB pathogenesis (Hatten and Roussel 2011). These include SHH, NOTCH, ErbB and MYC pathways.

4. miRNAs ABERRANTLY EXPRESSED IN MEDULLOBLASTOMA

Despite the intense investigation on the involvement of miRNAs in a variety of cancer types, knowledge of their roles in MB pathogenesis has only recently started to build up. Pierson et al. (Pierson et al. 2008) were the first to report the involvement of miRNA in MB. The authors found that the expression of miR-124a—a brain enriched miRNA preferentially expressed in mature neurons and in external granule cells of cerebellum that are thought to be cells of origins of MBs (Cheng et al. 2009, Conaco et al. 2006, Makeyev et al. 2007)—is significantly decreased in MB cells compared to normal cerebellum. Similar decreases were also observed in other brain tumors (Fernandez et al. 2009, Silber et al. 2008). Interestingly restoration of miR-124 function inhibits MB cell proliferation, suggesting that it may act as a suppressor of tumor growth. In a computational-based approach to predict miR-124 target candidates, Pierson et al. discovered and validated two targets of miR-124: cyclin dependent kinase 6 (CDK6) that was identified as an adverse prognostic marker in MB and solute carrier family 16, member 1(SLC16A1) that may represent a novel therapeutic target for treatment of malignant MBs (Mendrzyk et al. 2005, Northcott et al. 2009). In a distinct, but parallel, study Ferretti et al. (Ferretti et al. 2008) used high-throughput screening to examine miRNA expression profiles in 34 patients with MB. Their investigation revealed a set of 78 miRNAs with altered expression in MB, compared to normal adult and fetal cerebellar cells. Amazingly the authors identified a unique set of four upregulated miRNAs (let-7g, miR-19a, miR-106b and miR-191) that are capable of distinguishing anaplastic variants from classic and/or desmoplastic MBs. Although further verification is required to validate the data, these findings imply that miRNAs may have a wide clinical applicability as potential diagnostic and prognostic biomarkers. Moreover, the authors found a number of developmentally regulated miRNAs that maintained levels of expression in MB similar to those observed in fetal cerebellum, supporting the notion that MBs exhibit some embryonic features. For example in MB samples they observed impaired expression of several neuronal miRNAs known to be expressed during neuronal development such as miR-9 and miR-125a, miR-128a, 128b and miR-181b suggesting that some of these miRNAs might be involved in MB tumorigenesis (Pang et al. 2009). Notably all of these impaired miRNAs were identified before in glioblastomas (Ciafre et al. 2005), suggesting a common role of these miRNAs in the etiology of malignant brain tumors. The authors selected miR-9 and miR-125a, as two neuronal candidates that were downregulated in MB, for functional studies. Ectopic overexpression of miR-9 and miR-125a decreased proliferation, augmented apoptosis, and promoted arrest of MB tumor growth suggesting a role for both miRNAs as tumor suppressors (Turner et al. 2010). These miRNAs mediated their effects through targeting the pro-proliferative truncated isoform of the neurotrophin receptor TrkC, which is commonly upregulated in MB cells. Such findings are in line with those observed in neuroblastomas (Laneve et al. 2007, Pang et al. 2009). Consistently, other miRNAs such as miR-216, miR-135b, miR-217, miR-592, miR-340 were found to be overexpressed in pediatric brain tumors including MB, whereas miR-92b, miR-23a, miR-27a, miR-146b, miR-22) were found to be underexpressed when compared to normal brain tissues (Birks et al. 2011).

Together these reports clearly point to a critical role for miRNAs in MB carcinogenesis. Such discoveries not only provide new insights into the molecular pathogenesis of MB tumors, but also raise hopes for translating miRNA expression and functions in diagnosis, prognostication and therapy.

5. miRNAs TARGETING THE DEVELOPMENTAL PATHWAYS INVOLVED IN MEDULLOBLASTOMA

5.1 Sonic Hedgehog/Patched Signaling

SHH, a soluble extracellular factor secreted by Purkinje cells underneath the EGL, is the ligand for Patched (Ptch1), a transmembrane protein, and negative regulator of the signaling pathway (Wechsler-Reya and Scott 1999). The effector molecule, smoothened (Smo), is a G-protein coupled receptor that is essential for SHH signaling. In the classic model of SHH signaling, Smo is associated with Ptch1 at the cell membrane, where Ptch1 suppresses Smo activity in the absence of SHH. When SHH binds the Ptch1 receptor, a conformational change of the Ptch1-Smo complex relieves the inhibition of Smo, allowing for Smo downstream signaling that activates the zinc finger protein transcription factors Gli1 and Gli2, and inactivates the transcriptional repressor Gli3 (Saran 2009). Together, these proteins regulate the transcriptional program in the cell nucleus (Wechsler-Reya and Scott 1999).

The SHH pathway is pivotal to the development of most mammalian organs and its aberration has been implicated in birth defects and a multitude of tumor types (Machold et al. 2003, Riobo and Manning 2007, Wang et al. 2007). SHH pathway is a master regulator of the development of cerebellar GCPs which proliferate in response to Purkinje cell-produced SHH (Raffel 2004) [reviewed in (Ruiz et al. 2007)]. Constitutive activation of the SHH signaling alters the development of GCPs that remain overgrown, thus becoming susceptible to malignant transformation into MB (Pietsch et al. 1997, Pomeroy and Sturla 2003, Raffel et al. 1997, Ruiz i Altaba et al. 2002, Taylor et al. 2002). Furthermore, studies on the inactivation of Gli alleles in Ptch1 heterozygous mouse models for MB suggest that Gli1 functions in the transformation of GCPs and highlight the key role of SHH pathway activation in MB (Hatten and Roussel 2011, Ho and Scott 2002, Kimura et al. 2005, Thompson et al. 2006, Wechsler-Reya and Scott 1999).

To investigate miRNAs targeting the SHH developmental signaling that is involved in some MBs formation (Goodrich et al. 1997, Hallahan et al. 2004, Hatten 1999), Uziel et al. characterized the miRNA expression pattern of proliferating GNPs from P6 wild-type during mouse cerebellar development and in MBs derived from two mouse models with: Ink4c−/−, Ptch +/− and Ink4c−/−, p53−/− genetic background. The group found that miR-17/92 cluster was specifically upregulated in MB tumors with the activated SHH/PTCH signal pathway (Uziel et al. 2009). A comparison of P6 GPCs wild-type and MB tumor-prone mice to postmitotic differentiated neurons revealed 26 miRNAs that showed increased expression and 24 miRNAs that showed decreased expression. Among the 26 overexpressed miRNAs, nine were encoded by the miR-17/92 cluster or its paralogs. Moreover, enforced expression of miR-17/92 induced MBs with complete penetrance after orthotopic transplant into the brains of recipient immunocompromised animals. These results provided the first evidence that the SHH/PTCH signaling pathway and miR-17/92 functionally interact and contribute to MB development, however the exact mechanism by which SHH and miR-17/92 collaborate during cerebellar development and MB formation is unclear. In agreement with Uziel's results, Northcott and colleagues reported that profiling the expression of 427 mature miRNAs in a series of 90 primary human MBs revealed that components of the miR-17/92 polycistron are the most highly up-regulated miRNAs in MB. Expression of miR-17/92 was highest in the subgroup of MB associated with activation of SHH signaling pathway compared with other subgroups of MB. Interestingly MBs in which miR-17/92 was up-regulated also had elevated levels of MYC/MYCN expression. Consistent with its regulation by SHH, the authors observed that SHH treatment of primary cerebellar granule neuron precursors (CGNP), proposed cells of origin for the SHH-associated MBs, resulted in increased miR-17/92 expression. Moreover, ectopic miR-17/92 expression in CGNPs increased cell proliferation and also enabled them to proliferate in the absence of SHH. The group concluded that miR-17/92 is a positive effector of SHH-mediated proliferation and that aberrant expression/amplification of this miRNA confers a growth advantage to MBs (Northcott et al. 2009).

In a distinct but parallel study using Gli1 status to discriminate SHH-driven MB from non-SHH cases, Ferretti et al. (Ferretti et al. 2008) stratified a series of 31 primary MBs into two classes (Gli1-high and Gli1-low) and carried out miRNA profiling on a subset of 250 miRNAs. Three candidate miRNAs exhibiting reduced expression in Gli1-high tumors, miR-125b, miR-324-5p and miR-326, were tested in a functional study because of their predicted capacity to repress Smo and Gli1 transcripts. All three candidates downregulated Smo levels when overexpressed in DAOY MB cells and inhibited cell proliferation and anchorage-independent growth in soft agar colony formation assay. Additionally, expression of these candidate miRNAs correlated positively with CGNP differentiation state *in vitro* and their exogenous expression promoted neurite outgrowth in the same cell type. These results indicate a mechanism for the regulation of the SHH signaling pathway in MBs that depends on miR-125b, miR-324-5p and miR-326 function (Medina et al. 2008) and suggest that deregulation of specific miRNAs is responsible for abnormal activation of SHH signaling and MB formation [reviewed in (Pang et al. 2009)]. However, it is important to note that miR-17/92 was not reported as differentially expressed between Gli1-high and Gli1-low MB in this study (Fernandez et al. 2009).

In a recent study, using miRNA expression profiling the group of Bai (Bai et al. 2011) demonstrated that the retinal miR-183-96-182 cluster on chromosome 7q32 is highly overexpressed in non-sonic hedgehog MBs and miR-182 is significantly overexpressed in metastatic MB as compared to non-metastatic tumors. The group reported that overexpression of miR-182 in non-SHH-MB *in vitro* as well as *in vivo* increases cell invasion and xenografts overexpressing miR-182 spread to the leptomeninges, while knockdown of the same miRNA decreases cell migration.

5.2 The NOTCH Signaling Pathway

The evolutionarily conserved NOTCH pathway is involved in many developmental processes and participates in cell fate determination during embryonic and postnatal life. To date, four NOTCH receptors (NOTCH-1-4) have been identified in mammals, as well as several ligands such as the Delta-like (Dll1, Dll3 and Dll4), and the Jagged (Jag1 and Jag2) ligands. The most characterized effectors of the NOTCH signaling are the transcription factors Hairy and enhancer of split (HES) and Hairy/enhancer-of-split related with YRPW motif protein 1 (HEY1). Binding of a ligand on one cell to the NOTCH receptor on a neighboring cell triggers proteolytic cleavage of the receptor, which intracellularly releases the soluble domain of NOTCH (NICD). Once translocated into the nucleus, NICD promotes the transcriptional activation of NOTCH-responsive genes, the HES and HEY family of transcription factors.

It is well established that the NOTCH signaling pathway regulates the differentiation of granule neuron precursor cells (Marino 2005, Solecki et al. 2001). Like the SHH pathway, the NOTCH pathway is involved in both cerebellar development and in MB pathogenesis (Fogarty et al. 2005, Marino 2005). NOTCH2 expression is increased in about 15% of MBs (Srivastava and Nalbantoglu 2010) and increased expression of HES1 was found to be associated with poor prognosis in MB patients (Fan et al. 2004, Hallahan et al. 2004, Ingram et al. 2008). The NOTCH pathway has been linked specifically to a unique fraction of MB tumor cells that harbor precursor stem-cell markers (Eberhart 2007, Fan et al. 2006). Studies using human MB cell lines showed that NOTCH signaling is required for maintaining subpopulations of progenitor-like cells potentially capable of re-populating tumors after initial therapy, and that inhibition of the pathway can limit tumor cell growth (Fan et al. 2006, Hallahan et al. 2004). One inhibitory mechanism involves miR-199b-5p, which binds the 3' UTR sequence of HES1 and thereby represses the main effector of the NOTCH signaling (HES1) in MB cell lines (Garzia et al. 2009). Consistently, down-regulation of HES1 expression by miR-199b-5p negatively regulates proliferation and growth of MB cells. Moreover, the overexpression of miR-199b-5p in MB cells diminishes the stem-cell-like subpopulation and impairs the engrafting of MB cells in the cerebellum of nude mice. Importantly, miR-199b-5p downregulation was found significantly associated with metastasis in 61 MB samples, whereas high levels of miR-199b expression correlated with better overall survival. Thus, the clear association of miR-199b with disease outcome highlights the potential clinical relevance of this miRNA. Another microRNA (miR-34a) targets the NOTCH pathway further upstream by targeting of the NOTCH ligand delta-like 1, which results in decreased CD15+/CD133+ tumor-propagating cells and supports neural differentiation in MB (de Antonellis et al. 2011).

5.3 The ErbB-2 Signaling Pathway

ErbB2 (V-erb-b2 erythroblastic leukemia viral oncogene homolog 2, alias HER-2/NEU), is a proto-oncogene that encodes for a transmembrane glycoprotein with extensive homology to the human epidermal growth factor receptor (EGFR) (Britsch 2007). Unlike EGFR, however, ErbB2 is an orphan receptor, devoid of an extracellular ligand binding domain and therefore not able to bind growth factors. It tightly binds to ligand-bound growth factor receptors of the EGFR receptor family. This hetero-dimerization stabilizes the ligand binding and thus the signal transmission. ErbB2 protein expression is regulated by both transcriptional and post-transcriptional mechanisms, and is involved in the development of numerous types of human cancer such as ovarian, breast, renal and prostate cancer as well as MB (Gilbertson et al. 1998). Changes in expression of ErbB2 have been demonstrated to impact biological activity and clinical features of MB (Gilbertson et al. 1997). The abnormal (high) expression of ErbB2 protein correlates with more aggressive clinicopathological features, drug resistance and sensitivity to specific chemotherapy and specific hormonal therapy regimens in MB (Gilbertson et al. 1997, Herms et al. 1997, Pollack et al. 2002). ErbB2 expression in combination with high ErbB4 results in a poor prognostic impact in MB and was found to be associated with a high mitotic index and the presence of metastases (Bal et al. 2006, Gilbertson et al. 1998, Hernan et al. 2003, Srivastava and Nalbantoglu 2010). Regarding natural miRNAs acting on ErbB2, Scott et al. reported recently a coordinate suppression of ErbB2 by enforced expression of micro-RNA miR-125a or miR-125b (Scott et al. 2007), while Adachi and his co-workers have found MiR-205 to be negatively regulated by ErbB2 overexpression in breast cancer (Adachi et al. 2011). In MBs, Ferretti et al. identified a miRNA signature in a group of MB overexpressing ErbB2. After subdividing 34 MBs into either 13 overexpressing and 21 not overexpressing ErbB2 the authors identified a group of six miRNAs (miR-10b, miR-135a, miR-135b, miR-125b, miR-153, miR-199b) that exhibited significantly different expression level in tumors overexpressing ErbB2 that is known to correlate with poor survival among MB patients, providing prospects for these miRNAs to be further considered as additional prognosis biomarkers (Ferretti et al. 2009). Additionally in their study the amount of expression change of miR-153 correlated with disease risk where lower expression levels of miR-153 were observed in high versus average risk group [reviewed in (Ajeawung et al. 2010)]. Taken together, these data highlight the important role of miRNAs associated with the ErbB2 oncogene and illustrate the significant contribution of miRNA control to a pathway that is critical for MB pathogenesis.

5.4 The MYC Signaling Pathway

Amplification and overexpression of the proto-oncogene MYC have been associated with poor prognosis for MB patients in multiple studies (Das et al. 2009, de Haas et al. 2008, Ebinger et al. 2006, Grotzer et al. 2001, 2007, Herms et al. 2000, Lamont et al. 2004, Pfister et al. 2009, Raabe and Eberhart 2010, Rutkowski et al. 2007, Scheurlen et al. 1998, Takei et al. 2009, von Hoff et al. 2010). It is well established that hyperactivity of the MYC oncogenic transcription factor dramatically reprograms gene expression, often facilitating cellular proliferation and tumorigenesis. These effects are elicited by MYC through the coordination, the activation and the repression of an extensive network of protein-coding genes as well as noncoding RNAs including miRNAs (Bui and Mendell 2010). The first demonstration that miRNAs could indeed function as critical components of MYC oncogenic pathways came with the discovery that c-MYC directly activates transcription of the polycistronic miR-17/92 cluster (Dang et al. 2005, O'Donnell et al. 2005). Induction of these miRNAs occurs through the canonical mechanism whereby MYC binds directly to an E-box within the first intron of the gene encoding the miR-17/92 primary transcript. In brain tumors, members of the miR-17/92 gene cluster family, act as proto-oncogenes, with roles in enhancing proliferation and angiogenesis (Nicoloso and Calin 2008). Evidence now exists, which convincingly implicates the miR-17/92 cluster as a *bona fide* oncogene implicated in MB pathogenesis (Northcott et al. 2009, Uziel et al. 2009). This, in turn, leads in SHH-dependent MBs to oncogenic activation of c-Myc and to the modulation of E2F1 [reviewed in (Ajeawung et al. 2010)]. Thus, within the MYC target genes' network, miRNAs play an important role and contribute markedly to the phenotypic outcome controlled by MYC pathway that is critical for both normal and cancerogenic development. The transcriptional activation of the miR-17-92 cluster by MYC represents only one mechanism of a much broader MYC-regulated miRNA network. For example, MYC can also repress or block the maturation of specific miRNAs including many with documented tumor suppressor activity including let-7 family members, miR-15a/16-1, miR-26a, miR-29 family members, and miR-34a (Bui and Mendell 2010). Therefore the mechanisms through which MYC represses miRNAs are of particular interest, since targeted reversion of these effects could have important therapeutic implications.

Given the ability of MYC to regulate diverse and important cell functions, it is not surprising that the transcription of MYC is delicately regulated at transcriptional, post-transcriptional, translational, and post-translational levels through a variety of mechanisms (Secombe et al. 2004). Although a number of protein regulators have been shown to be involved in these regulations, recent studies in different types of cancer also divulged several miRNA regulators of MYC, such as miR-24, miR-22, miR-145 or miR-185 (Lal et al. 2009, Liao and Lu 2011, Sachdeva et al. 2009, Xiong et al. 2010). Other miRNAs also affect expression of MYC regulators, which may indirectly affect MYC expression (Bueno and Malumbres 2011). However, the mechanism through which miRNAs control c-MYC activity in MBs was unclear until recently. Lv et al. 2011a screened 48 MBs for mutation, deletion and amplification of a selection of nine miRNA genes was conducted on the basis of both the presence of sequences targeting the 3'-untranslated region for the MYC mRNA and of chromosomal loci reported to be deleted or amplified in MBs (Lv et al. 2011b). The authors observed an association between deletion of the miR-512-2 gene and overexpression of MYC. Their *in vitro* study confirmed that miR-512-2 could target MYC expression in MB tumor cells. Consistently, knockdown of miR-512-2 resulted in significant MYC upregulation, whereas forced overexpression of miR-512-2 resulted in the downregulation of both mRNA and MYC protein. In the literature the amplification of the MYC gene has been observed only in a small fraction of MBs (5–10%) (Aldosari et al. 2002, Grotzer et al. 2000), whereas MYC overexpression appears to be much more common (30–50%) (Eberhart et al. 2004) and does not correlate with amplification of the MYC gene (Grotzer et al. 2000) suggesting that an alternative mechanism is involved. Therefore Lv et al. results suggest that alterations in the miRNA genes that target MYC may be an alternative mechanism leading to overexpression of MYC in MBs. In order to verify whether miRNA signatures can identify MB tumor subsets with distinct molecular features, Ferretti et al. investigated the association of miRNA expression patterns with MBs expressing high MYC protein. By subdividing 34 MB samples into either 18 overexpressing and 16 not overexpressing c-MYC the authors found that c-MYC overexpressing tumors exhibited a small group of miRNAs (miR-181b, miR-128a, miR-128b) that were differentially expressed compared to MBs not overexpressing c-MYC (Ferretti et al. 2008). Although based on a limited number of tumor samples, these findings suggest that MB subsets with high c-MYC have distinct miRNA signatures and more generally highlight the molecular heterogeneity and complexity of regulation of signaling pathways modulated by miRNAs. Taken together, the sum of these studies clearly indicates that the de-regulated expression of miRNAs plays a critical role in MB pathogenesis and tumor behavior, by targeting specific gene products involved in MB tumorigenesis.

6. miRNAs INVOLVED IN MEDULLOBLASTOMA METASTASIS

In addition to their role in promoting the development of primary tumors, miRNAs have also been implicated in tumor progression, and metastatic dissemination of tumor cells. Several cell biological processes, including those controlling adhesion, migration, and invasion, are involved in the dissemination of primary tumor cells to distant sites. Alterations in miRNA function were found to be involved in the regulation of these processes and can influence metastatic potential (Ma and Weinberg 2008, Ventura and Jacks 2009). For example, miR-10b, miR-125a, miR-141, miR-200a,b,c, miR-429, miR-661, mir-9, miR-10a, miR-17-92, miR, miR-27a,b, miR-29a, miR-30d, miR-107, miR-130a, miR-143, miR-151, miR-155, miR-182, miR-210, miR-224, miR-373, miR-378, miR-486, miR-520c, miR-520h, miR-532 have been reported to promote metastasis in different human cancers (Aigner 2011). The normal function of some of these miRNAs is to promote the proper translocation of cells during embryonal development, whereas they also have been found highly expressed in aggressive human cancers (Ma et al. 2007). Among several pro-metastatic miRNAs, miR-21 is up-regulated in a variety of solid tumors, including glioblastoma, breast, lung, colon, prostate, pancreas and stomach cancers (White et al. 2011) and has been causally linked to cell migration of several cancer cell lines (Asangani et al. 2008). We recently reported the aberrant expression of miR-21 in MB cells and showed that miR-21 regulates the expression of multiple target proteins associated with tumor dissemination, many of which are implicated in MB biology (Grunder et al. 2011). Importantly, repression of microRNA-21 decreased MB cell migration *in vitro*, possibly due to diminished repression of the tumor suppressor PDCD4. Thus, miR-21 appears to be a pro-metastatic miRNA in MB. The development of pharmaceutical strategies directed against miR-21 could improve the curability of selected MBs.

In summary, the altered miRNA expression signatures in MB highlight the functional significance of these novel oncogenic molecules involved in MB tumor formation and progression. Thus, a better understanding of miRNA function is essential for unraveling mechanisms responsible for the development of human MB and may allow the identification of targets for novel therapeutic strategies.

7. miRNAs AS POTENTIAL CANCER THERAPEUTICS

The rationale for using miRNAs as anticancer drugs is based on two major findings: 1) miRNA expression is deregulated in cancer and 2) the cancer phenotype can be changed by targeting miRNA expression. Depending on expression levels and function of miRNAs in the diseased tissue, two approaches for developing miRNA-based therapies can be envisaged: antagonists (inhibition of oncogenic miRNAs) and mimetics (replacement of tumor suppressive miRNAs). Hypothetically, normal cells should tolerate such targeting of miRNA pathways that are deregulated in the tumor, because these pathways are already activated or repressed by the endogenous miRNA. Moreover and in contrast to cancer cells, normal cells do not display oncogenic addiction—in the miRNA scenario to a single miRNA regulation—and are consequently less vulnerable to events that target one specific molecule (Gandellini et al. 2011). Strategies that have been used to modulate miRNA activity for cancer therapeutic purposes are summarized below.

7.1 miRNA Antagonists

This application involves strategies that aim to inhibit oncogenic miRNAs that promote increase of function in human diseases. The most common strategy to ablate miRNA function is achieved by single-stranded oligonucleotides with miRNA complementary sequences (antisense). The backbones of these are chemically modified in order to increase the affinity towards the endogenous miRNA or, alternatively, to trigger the degradation of the endogenous miRNA. Examples of miRNA antagonists are anti-miRNA oligonucleotides (AMOs), locked-nucleic acids (LNA), miRNA sponges, miRNA masking and small molecule inhibitors specific for certain miRNAs (Bader et al. 2011, Gandellini et al. 2011). The most widely used AMO are 2¢-O-methyl AMOs, 2¢-O-methoxyethyl AMOs, and (LNA) AMOs (Krutzfeldt et al. 2005, Meister et al. 2004, Stenvang et al. 2008, Weiler et al. 2006, Wu et al. 2011). miR-21 inhibition *in vitro* provides a good example of an anti-miR application in cancer cells. Knockdown of miR-21 suppressed cell growth and activated apoptosis in cultured glioblastoma and breast cancer cells (Chan et al. 2005, Si et al. 2007), reduced invasion in metastatic breast cancer, lung metastasis (Zhu et al. 2008), colon cancer (Asangani et al. 2008), glioblastoma (Gabriely et al. 2008) and MB culture cells (Grunder et al. 2011). *In vivo* studies also showed that treatment with LNA–anti-miR-21 oligonucleotides disrupted glioma growth (Moriyama et al. 2009). Other studies have shown that intravenous administration of AMOs against miR-16, miR-122, miR-192 and miR-194 in animals offer efficient and sustained silencing of corresponding miRNAs (Bhardwaj et al. 2010). As an alternative to chemically modified antisense oligos, vector-encoded RNA molecules, known as 'miRNA sponges' have recently been designed to contain multiple tandem binding sites to a miRNA of interest (Ebert et al. 2007, Ebert and Sharp 2010, Lu et al. 2009). miRNA sponges function by immobilizing the target miRNAs and thus blocking their regulatory function in the cell. miRNA sponges are transcripts under the control of strong promoters (RNA polymerase II) and contain multiple tandem binding sites to a miRNA of interest. miRNA sponges are able to inhibit miRNA function as strongly as AMOs. In addition miRNA sponges come with the advantage of being stably integrated into the genome, thus opening up the possibility of creating stable cell lines and transgenic animals that are functionally deficient for a specific miRNA family [reviewed in (Gandellini et al. 2011, Gentner et al. 2009, Johnnidis et al. 2008)]. However, the risks of insertional mutagenesis in target cells, as well as challenges in the delivery of these non-small-molecules make sponges not ideal for therapeutic applications. "miRNA masking" is another alternative strategy which refers a sequence with perfect complementarity to the binding site of an endogenous miRNA in the target gene. This miRNA masking can form duplex with the target mRNA with higher affinity, therefore blocking the access of endogenous miRNA to its binding site without the potential side effects of mRNA degradation by AMOs (Xiao et al. 2007).

The recent development of small molecules having the ability to modulate the function of miRNAs, renders miRNA pathways susceptible to small molecule inhibitors. Small molecule inhibitors of the miRNA pathway have several advantages over traditional nucleic acid-based tools to manipulate miRNA function, since they are more easily delivered into humans or animals, they are more stable intracellularly, and they are less expensive to manufacture. Based on these promising developments, it can be expected that several new small molecule inhibitors of miRNA function will be discovered in the near future (Xiao et al. 2007).

7.2 miRNA Replacement Therapy

Since the expression of protein-coding tumor suppressors can often inhibit tumor growth, it has been proposed that restoring tumor suppressive miRNAs may also have an anti tumor effect. The strategy of restoring tumor suppressive miRNAs lost in the cancer cells to re-establish their normal functions is called miRNA replacement therapy. In this regard miRNA mimics are either single or double-stranded miRNAs that have the same sequence as the depleted, naturally occurring miRNA. As a consequence, miRNA replacement therapy is expected to have limited off-target effects. The simplicity, specificity and potency of mimicking miRNA precursors have made this an exciting approach to target genes for both research and therapeutic purposes (Bader et al. 2010).

Several technologies have proven effective in delivering therapeutic miRNAs to tumor tissues *in vivo*. These include adenoviral or lentiviral vector-based systems that were originally developed for gene therapy. A viral approach has been used to reintroduce Let-7 expression in an orthotopic lung cancer mouse model (Takamizawa et al. 2004). The reintroduction of Let-7 expression significantly reduced lung tumor formation *in vivo* (Trang et al. 2010). In line with this is the reintroduction of miR-26a in a liver cancer mouse model, which resulted in the inhibition of cancer cell proliferation, induction of tumor-specific apoptosis, and prevented disease progression (Kota et al. 2009). Another example that shows the value of miRNA replacement is provided by miR-34a (Wiggins et al. 2011). In this case, therapeutic delivery of miR-34a led to an accumulation of miR-34a in the tumor tissue, suppression of known miR-34a target genes, and, most importantly, inhibition of lung tumor growth (Bader et al. 2011). Similar to the replacement therapy approach but differing in miRNA delivery, unmodified miR-145 and miR-33a have been delivered by using polyethylenimine (PEI) in a mouse model of colon carcinoma, a method that proved efficacious (Ibrahim et al. 2011). Likewise transient transfection with synthetic miR-16 significantly reduced cell proliferation in prostate cancer cells *in vitro* and reduced metastasis *in vivo* (Takeshita et al. 2010). Another example for the success of miRNA replacement therapy is demonstrated for miR-15 and miR-16 (Cimmino et al. 2005) that are often deleted in CLL patients (Bonci et al. 2008, Calin et al. 2002, Cimmino et al. 2005) [reviewed in (Li et al. 2009)].

Together, this evidence showing that miRNAs can function as tumor suppressors and that the synthetic versions of these miRNAs robustly interfere with tumor growth in animal models reflects the importance of miRNAs in cancer and strongly supports the development of miRNA mimics as ideal candidates for therapeutic intervention.

7.3 Challenges of miRNA-based Therapies

Biological instability in body fluids and tissues, the difficulties of specific delivery and the insufficient uptake for effective target inhibition challenge the development of miRNA-based therapies similarly to small interfering RNA therapies. Chemical modifications in oligonucleotides have been investigated to overcome these obstacles and include morpholinos, peptide nucleic acids, cholesterol conjugation and phosphorothioate backbone modifications. However, the drawbacks of improved delivery to tissues after chemical modifications are impaired biological activity and increased toxicity. To overcome these problems, novel miRNA formulations, including nanoparticles and polymers or viral transduction approaches are currently developed. Moreover, effort is put into the engineering of effective systems that deliver synthetic miRNAs selectively to the diseased tissues in order to limit potentially toxic off-target effects [reviewed in (Garzon et al. 2010)]. Finally homeostasis of the brain is dependent on the blood-brain barrier (BBB). This intricate barrier, which is comprised of special brain endothelial cells, dictates the exchange of essential nutrients and limits the entrance of molecules and immune cells into the central nervous system. This barrier therefore forms yet another important hurdle for effective miRNAs treatment of malignant brain cancer such as MB. Hence effective brain delivery for miRNAs may require the design of particular therapeutic approaches that overcome this physiological obstruction such as the use of nanoparticles, immunoliposomes, peptide vectors or carrier-mediated transport through the BBB [reviewed in (Deeken and Loscher 2007).

8. CONCLUSIONS

Within the last decade, researchers have made substantial progress in understanding MB development through the discovery of deregulations in highly conserved developmental signaling pathways. However, effective targeted therapies remain still elusive and the development of novel therapeutic strategies remains an urgent goal. The discovery of the integral involvement of miRNAs in MB pathogenesis provides new hope for accomplishing this task. Supported by solid evidence for a critical role in cancer and strengthened by a unique mechanism of action, miRNAs are likely to yield a new class of targeted therapeutics. Nevertheless the challenge remains to translate our increasing understanding of miRNA biology in cancer into readily available medicine. From a translational point of view, delivery, target specificity, pharmacokinetics of miRNAs and long-term safety of miRNA drugs will be of major concern and should be addressed thoroughly. From a basic science point of view, great effort is still required to decipher miRNA networks, to understand upstream regulators and to achieve accurateness in correlating miRNA expression with functional consequences. However, the rapid expansion of this field and the extent of the perspectives are sources of hope for exciting future advances and for an efficient translation basic knowledge into clinical applications.

ACKNOWLEDGMENT

Grant Support: Swiss Research Foundation Child and Cancer.

ABBREVIATIONS

AMOs	:	Anti-miRNA oligonucleotides
BBB	:	Blood-brain barrier
CLL	:	Chronic lymphatic leukaemia
CSF	:	Cerebrospinal fluid
CT	:	Computed tomography
EGFR	:	Epidermal growth factor receptor
EGL	:	External granule layer
ESCs	:	Embryonal stem cells
GCPs	:	Granule cell precursors
IGL	:	Internal granular cell layer
LNA	:	Locked-nucleic acids

REFERENCES

Adachi, R., S. Horiuchi, Y. Sakurazawa, T. Hasegawa, K. Sato, and T. Sakamaki. 2011. ErbB2 down-regulates microRNA-205 in breast cancer. Biochem Biophys Res Commun. 411: 804–8.

Aigner, A. 2011. MicroRNAs (miRNAs) in cancer invasion and metastasis: therapeutic approaches based on metastasis-related miRNAs. J Mol Med (Berl). 89: 445–57.

Ajeawung, N.F., B. Li, and D. Kamnasaran. 2010. Translational applications of microRNA genes in medulloblastomas. Clin Invest Med. 33: E223–33.

Aldosari, N., S.H. Bigner, P.C. Burger, L. Becker, J.L. Kepner, H.S. Friedman, and R.E. Mclendon. 2002. MYCC and MYCN oncogene amplification in medulloblastoma. A fluorescence *in situ* hybridization study on paraffin sections from the Children's Oncology Group. Arch Pathol Lab Med. 126: 540–545.

Asangani, I.A., S.A. Rasheed, D.A. Nikolova, J.H. Leupold, N.H. Colburn, S. Post, and H. Allgayer. 2008. MicroRNA-21 (miR-21) post-transcriptionally downregulates tumor suppressor Pdcd4 and stimulates invasion, intravasation and metastasis in colorectal cancer. Oncogene 27: 2128–36.

Bader, A.G., D. Brown, and M. Winkler. 2010. The promise of microRNA replacement therapy. Cancer Res. 70: 7027–30.

Bader, A.G., D. Brown, J. Stoudemire, and P. Lammers. 2011. Developing therapeutic microRNAs for cancer. Gene Ther. Dec. 18(12): 1121–6.

Bai, A.H., T. Milde, M. Remke, C.G. Rolli, T. Hielscher, Y.J. Cho, M. Kool, P.A. Northcott, M. Jugold, A.V. Bazhin, S.B. Eichmuller, A.E. Kulozik, A. Pscherer, A. Benner, M.D. Taylor, S.L. Pomeroy, R. Kemkemer, O. Witt, A. Korshunov, P. Lichter, and S.M. Pfister. 2011. MicroRNA-182 promotes leptomeningeal spread of non-sonic hedgehog-medulloblastoma. Acta Neuropathol. Dec 2.

Bal, M.M., B. Das Radotra, R. Srinivasan, and S.C. Sharma. 2006. Expression of c-erbB-4 in medulloblastoma and its correlation with prognosis. Histopathology. 49: 92–3.

Bar, M., S.K. Wyman, B.R. Fritz, J. Qi, K.S. Garg, R.K. Parkin, E.M. Kroh, A. Bendoraite, P.S. Mitchell, A.M. Nelson, W.L. Ruzzo, C. Ware, J.P. Radich, R. Gentleman, H. Ruohola-Baker, and M. Tewari. 2008. MicroRNA discovery and profiling in human embryonic stem cells by deep sequencing of small RNA libraries. Stem Cells. 26: 2496–505.

Bartel, D.P. 2004. MicroRNAs: genomics, biogenesis, mechanism, and function. Cell. 116: 281–97.

Bhardwaj, A., S. Singh, and A.P. Singh. 2010. MicroRNA-based cancer therapeutics: big hope from small RNAs. Mol Cell Pharmacol. 2: 213–219.

Birks, D.K., V.N. Barton, A.M. Donson, M.H. Handler, R. Vibhakar, and N.K. Foreman. 2011. Survey of MicroRNA expression in pediatric brain tumors. Pediatr Blood Cancer. 56: 211–6.

Bonci, D., V. Coppola, M. Musumeci, A. Addario, R. Giuffrida, L. Memeo, L. D'urso, A. Pagliuca, M. Biffoni, C. Labbaye, M. Bartucci, G. Muto, C. Peschle, and R. De Maria. 2008. The miR-15a-miR-16-1 cluster controls prostate cancer by targeting multiple oncogenic activities. Nat Med. 14: 1271–7.

Britsch, S. 2007. The neuregulin-I/ErbB signaling system in development and disease. Adv Anat Embryol Cell Biol. 190: 1–65.

Bueno, M.J. and M. Malumbres. 2011. MicroRNAs and the cell cycle. Biochim Biophys Acta. 1812: 592–601.

Buhren, J., A.H. Christoph, R. Buslei, S. Albrecht, O.D. Wiestler, and T. Pietsch. 2000. Expression of the neurotrophin receptor p75NTR in medulloblastomas is correlated with distinct histological and clinical features: evidence for a medulloblastoma subtype derived from the external granule cell layer. J Neuropathol Exp Neurol. 59: 229–240.

Bui, T.V. and J.T. Mendell. 2010. Myc: Maestro of MicroRNAs. Genes Cancer. 1: 568–575.

Calin, G.A., C.D. Dumitru, M. Shimizu, R. Bichi, S. Zupo, E. Noch, H. Aldler, S. Rattan, M. Keating, K. Rai, L. Rassenti, T. Kipps, M. Negrini, F. Bullrich, and C.M. Croce. 2002. Frequent deletions and down-regulation of micro- RNA genes miR-15 and miR-16 at 13q14 in chronic lymphocytic leukemia. Proc Natl Acad Sci USA. 99: 15524–9.

Calin, G.A., C. Sevignani, C.D. Dumitru, T. Hyslop, E. Noch, S. Yendamuri, M. Shimizu, S. Rattan, F. Bullrich, M. Negrini, and C.M. Croce. 2004. Human microRNA genes are frequently located at fragile sites and genomic regions involved in cancers. Proc Natl Acad Sci USA. 101: 2999–3004.

Calin, G.A. and C.M. Croce. 2006. MicroRNA signatures in human cancers. Nat Rev Cancer. 6: 857–66.

Calin, G.A. and C.M. Croce. 2009. Chronic lymphocytic leukemia: interplay between noncoding RNAs and protein-coding genes. Blood. 114: 4761–70.

Cao, X., G. Yeo, A.R. Muotri, T. Kuwabara, and F.H. Gage. 2006. Noncoding RNAs in the mammalian central nervous system. Annu Rev Neurosci. 29: 77–103.

Cao, X., S.L. Pfaff, and F.H. Gage. 2007. A functional study of miR-124 in the developing neural tube. Genes Dev. 21: 531–6.

Chan, J.A., A.M. Krichevsky, and K.S. Kosik. 2005. MicroRNA-21 is an antiapoptotic factor in human glioblastoma cells. Cancer Res. 65: 6029–33.

Cheng, L.C., E. Pastrana, M. Tavazoie, and F. Doetsch. 2009. miR-124 regulates adult neurogenesis in the subventricular zone stem cell niche. Nat Neurosci. 12: 399–408.

Choi, P.S., L. Zakhary, W.Y. Choi, S. Caron, E. Alvarez-Saavedra, E.A. Miska, M. Mcmanus, B. Harfe, A.J. Giraldez, H.R. Horvitz, A.F. Schier, and C. Dulac. 2008. Members of the miRNA-200 family regulate olfactory neurogenesis. Neuron. 57: 41–55.

Ciafre, S.A., S. Galardi, A. Mangiola, M. Ferracin, C.G. Liu, G. Sabatino, M. Negrini, G. Maira, C.M. Croce, and M.G. Farace. 2005. Extensive modulation of a set of microRNAs in primary glioblastoma. Biochem Biophys Res Commun. 334: 1351–8.

Cimmino, A., G.A. Calin, M. Fabbri, M.V. Iorio, M. Ferracin, M. Shimizu, S.E. Wojcik, R.I. Aqeilan, S. Zupo, M. Dono, L. Rassenti, H. Alder, S. Volinia, C.G. Liu, T.J. Kipps, M. Negrini, and C.M. Croce. 2005. miR-15 and miR-16 induce apoptosis by targeting BCL2. Proc Natl Acad Sci USA. 102: 13944–9.

Conaco, C., S. Otto, J.J. Han, and G. Mandel. 2006. Reciprocal actions of REST and a microRNA promote neuronal identity. Proc. Natl Acad Sci USA. 103: 2422–7.

Coolen, M. and L. Bally-Cuif. 2009. MicroRNAs in brain development and physiology. Curr Opin Neurobiol. 19: 461–70.

Corcoran, C., A.M. Friel, M.J. Duffy, J. Crown, and L. O'driscoll. 2011. Intracellular and extracellular microRNAs in breast cancer. Clin Chem. 57: 18–32.

Croce, C.M. and G.A. Calin. 2005. miRNAs, cancer, and stem cell division. Cell. 122: 6–7.

Croce, C.M. 2009. Causes and consequences of microRNA dysregulation in cancer. Nat Rev Genet. 10: 704–14.

Dang, C.V., K.A. O'donnell, and T. Juopperi. 2005. The great MYC escape in tumorigenesis. Cancer Cell. 8: 177–8.

Das, P., T. Puri, V. Suri, M.C. Sharma, B.S. Sharma, and C. Sarkar. 2009. Medulloblastomas: a correlative study of MIB-1 proliferation index along with expression of c-Myc, ERBB2, and anti-apoptotic proteins along with histological typing and clinical outcome. Childs Nerv Syst. 25: 825–35.

De Antonellis, P., C. Medaglia, E. Cusanelli, I. Andolfo, L. Liguori, G. De Vita, M. Carotenuto, A. Bello, F. Formiggini, A. Galeone, G. De Rosa, A. Virgilio, I. Scognamiglio, M. Sciro, G. Basso, J.H. Schulte, G. Cinalli, A. Iolascon, and M. Zollo. 2011. MiR-34a targeting of NOTCH ligand delta-like 1 impairs CD15+/CD133+ tumor-propagating cells and supports neural differentiation in medulloblastoma. PLoS One. 6: e24584.

De Haas, T., N. Hasselt, D. Troost, H. Caron, M. Popovic, L. Zadravec-Zaletel, W. Grajkowska, M. Perek, M.C. Osterheld, D. Ellison, F. Baas, R. Versteeg, and M. Kool. 2008. Molecular risk stratification of medulloblastoma patients based on immunohistochemical analysis of MYC, LDHB, and CCNB1 expression. Clin Cancer Res. 14: 4154–60.

De Krijger, I., L.J. Mekenkamp, C.J. Punt, and I.D. Nagtegaal. 2011. MicroRNAs in colorectal cancer metastasis. J Pathol. 224: 438–47.

Deeken, J.F. and W. Loscher. 2007. The blood-brain barrier and cancer: transporters, treatment, and Trojan horses. Clin Cancer Res. 13: 1663–74.

Desano, J.T. and L. Xu. 2009. MicroRNA regulation of cancer stem cells and therapeutic implications. Aaps J. 11: 682–92.

Eberhart, C.G., J. Kratz, Y. Wang, K. Summers, D. Stearns, K. Cohen, C.V. Dang, and P.C. Burger. 2004. Histopathological and molecular prognostic markers in medulloblastoma: c-Myc, N-myc, TrkC, and anaplasia. J Neuropathol Exp Neurol. 63: 441–9.

Eberhart, C.G. 2007. In search of the medulloblast: neural stem cells and embryonal brain tumors. Neurosurg. Clin N Am. 18: 59–69, viii–ix.

Ebert, M.S., J.R. Neilson, and P.A. Sharp. 2007. MicroRNA sponges: competitive inhibitors of small RNAs in mammalian cells. Nat Methods. 4: 721–6.

Ebert, M.S. and P.A. Sharp. 2010. MicroRNA sponges: progress and possibilities. RNA. 16: 2043–50.

Ebinger, M., L. Senf, and W. Scheurlen. 2006. Risk stratification in medulloblastoma: screening for molecular markers. Klin Padiatr. 218: 139–42.

Erdos, E.G., L.L. Tague, and I. Miwa. 1968. Kallikrein in granules of a salivary gland. Proc West Pharmacol Soc. 11: 26–8.

Esquela-Kerscher, A. and F.J. Slack. 2006. Oncomirs—microRNAs with a role in cancer. Nat Rev Cancer. 6: 259–69.

Fan, X., I. Mikolaenko, I. Elhassan, X. Ni, Y. Wang, D. Ball, D.J. Brat, A. Perry, and C.G. Eberhart. 2004. NOTCH1 and NOTCH2 have opposite effects on embryonal brain tumor growth. Cancer Res. 64: 7787–93.

Fan, X., W. Matsui, L. Khaki, D. Stearns, J. Chun, Y.M. Li, and C.G. Eberhart. 2006. NOTCH pathway inhibition depletes stem-like cells and blocks engraftment in embryonal brain tumors. Cancer Res. 66: 7445–52.

Fan, X. and C.G. Eberhart. 2008. Medulloblastoma stem cells. J Clin Oncol. 26: 2821–7.

Fernandez, L.A., P.A. Northcott, M.D. Taylor, and A.M. Kenney. 2009. Normal and oncogenic roles for microRNAs in the developing brain. Cell Cycle. 8: 4049–54.

Ferretti, E., E. De Smaele, E. Miele, P. Laneve, A. Po, M. Pelloni, A. Paganelli, L. Di Marcotullio, E. Caffarelli, I. Screpanti, I. Bozzoni, and A. Gulino. 2008. Concerted microRNA control of Hedgehog signalling in cerebellar neuronal progenitor and tumour cells. EMBO J. 27: 2616–27.

Ferretti, E., E. De Smaele, A. Po, L. Di Marcotullio, E. Tosi, M.S. Espinola, C. Di Rocco, R. Riccardi, F. Giangaspero, A. Farcomeni, I. Nofroni, P. Laneve, U. Gioia, E. Caffarelli, I. Bozzoni, I. Screpanti, and A. Gulino. 2009. MicroRNA profiling in human medulloblastoma. Int J Cancer. 124: 568–77.

Fogarty, M.P., J.D. Kessler, and R.J. Wechsler-Reya. 2005. Morphing into cancer: the role of developmental signaling pathways in brain tumor formation. J Neurobiol. 64: 458–75.

Gabriely, G., T. Wurdinger, S. Kesari, C.C. Esau, J. Burchard, P.S. Linsley, and A.M. Krichevsky. 2008. MicroRNA 21 promotes glioma invasion by targeting matrix metalloproteinase regulators. Mol Cell Biol. 28: 5369–80.

Gandellini, P., V. Profumo, M. Folini, and N. Zaffaroni. 2011. MicroRNAs as new therapeutic targets and tools in cancer. Expert Opin Ther Targets. 15: 265–79.

Gangaraju, V.K. and H. Lin. 2009. MicroRNAs: key regulators of stem cells. Nat Rev Mol Cell Biol. 10: 116–25.

Garzia, L., I. Andolfo, E. Cusanelli, N. Marino, G. Petrosino, D. De Martino, V. Esposito, A. Galeone, L. Navas, S. Esposito, S. Gargiulo, S. Fattet, V. Donofrio, G. Cinalli, A. Brunetti, L.D. Vecchio, P.A. Northcott, O. Delattre, M.D. Taylor, A. Iolascon, and M. Zollo. 2009. MicroRNA-199b-5p impairs cancer stem cells through negative regulation of HES1 in medulloblastoma. PLoS One. 4: e4998.

Garzon, R., G. Marcucci and C.M. Croce. 2010. Targeting microRNAs in cancer: rationale, strategies and challenges. Nat Rev Drug Discov. 9: 775–89.

Gentner, B., G. Schira, A. Giustacchini, M. Amendola, B.D. Brown, M. Ponzoni, and L. Naldini. 2009. Stable knockdown of microRNA *in vivo* by lentiviral vectors. Nat Methods. 6: 63–6.

Gibson, P., Y. Tong, G. Robinson, M.C. Thompson, D.S. Currle, C. Eden, T.A. Kranenburg, T. Hogg, H. Poppleton, J. Martin, D. Finkelstein, S. Pounds, A. Weiss, Z. Patay, M. Scoggins, R. Ogg, Y. Pei, Z.J. Yang, S. Brun, Y. Lee, F. Zindy, J.C. Lindsey, M.M. Taketo, F.A. Boop, R.A. Sanford, A. Gajjar, S.C. Clifford, M.F. Roussel, P.J. Mckinnon, D.H. Gutmann, D.W. Ellison, R. Wechsler-Reya, and R.J. Gilbertson. 2010. Subtypes of medulloblastoma have distinct developmental origins. Nature. 468: 1095–9.

Gilbertson, R.J., R.H. Perry, P.J. Kelly, A.D. Pearson, and J. Lunec. 1997. Prognostic significance of HER2 and HER4 coexpression in childhood medulloblastoma. Cancer Res. 57: 3272–80.

Gilbertson, R.J., S.C. Clifford, W. Macmeekin, W. Meekin, C. Wright, R.H. Perry, P. Kelly, A.D. Pearson, and J. Lunec. 1998. Expression of the ErbB-neuregulin signaling network during human cerebellar development: implications for the biology of medulloblastoma. Cancer Res. 58: 3932–41.

Gilbertson, R.J. and D.W. Ellison. 2008. The origins of medulloblastoma subtypes. Annu Rev Pathol. 3: 341–65.

Gilbertson, R.J. 2011. Mapping cancer origins. Cell. 145: 25–9.

Giraldez, A.J., R.M. Cinalli, M.E. Glasner, A.J. Enright, J.M. Thomson, S. Baskerville, S.M. Hammond, D.P. Bartel, and A.F. Schier. 2005. MicroRNAs regulate brain morphogenesis in zebrafish. Science. 308: 833–8.

Goff, L.A., J. Davila, M.R. Swerdel, J.C. Moore, R.I. Cohen, H. Wu, Y.E. Sun, and R.P. Hart. 2009. Ago2 immunoprecipitation identifies predicted microRNAs in human embryonic stem cells and neural precursors. PLoS One. 4: e7192.

Goodrich, L.V., L. Milenkovic, K.M. Higgins, and M.P. Scott. 1997. Altered neural cell fates and medulloblastoma in mouse patched mutants. Science. 277: 1109–13.

Grimmer, M.R. and W.A. Weiss. 2006. Childhood tumors of the nervous system as disorders of normal development. Curr Opin Pediatr. 18: 634–8.

Grotzer, M.A., A.J. Janss, K. Fung, J.A. Biegel, L.N. Sutton, L.B. Rorke, H. Zhao, A. Cnaan, P.C. Phillips, V.M. Lee, and J.Q. Trojanowski. 2000. TrkC expression predicts good clinical outcome in primitive neuroectodermal brain tumors. J Clin Oncol. 18: 1027–35.

Grotzer, M.A., M.D. Hogarty, A.J. Janss, X. Liu, H. Zhao, A. Eggert, L.N. Sutton, L.B. Rorke, G.M. Brodeur, and P.C. Phillips. 2001. c-Myc messenger RNA expression predicts survival outcome in childhood primitive neuroectodermal tumor/medulloblastoma. Clin Cancer Res. 7: 2425–33.

Grotzer, M.A., K. Von Hoff, A.O. Von Bueren, T. Shalaby, W. Hartmann, M. Warmuth-Metz, A. Emser, R.D. Kortmann, J. Kuehl, T. Pietsch, and S. Rutkowski. 2007. Which clinical and biological tumor markers proved predictive in the prospective multicenter trial HIT'91--implications for investigating childhood medulloblastoma. Klin Padiatr. 219: 312–7.

Grunder, E., R. D'ambrosio, G. Fiaschetti, L. Abela, A. Arcaro, T. Zuzak, H. Ohgaki, S.Q. Lv, T. Shalaby, and M. Grotzer. 2011. MicroRNA-21 suppression impedes medulloblastoma cell migration. Eur J Cancer. 47: 2479–90.

Hallahan, A.R., J.I. Pritchard, S. Hansen, M. Benson, J. Stoeck, B.A. Hatton, T.L. Russell, R.G. Ellenbogen, I.D. Bernstein, P.A. Beachy, and J.M. Olson. 2004. The SmoA1 mouse model reveals that NOTCH signaling is critical for the growth and survival of sonic hedgehog-induced medulloblastomas. Cancer Res. 64: 7794–800.

Hatfield, S.D., H.R. Shcherbata, K.A. Fischer, K. Nakahara, R.W. Carthew, and H. Ruohola-Baker. 2005. Stem cell division is regulated by the microRNA pathway. Nature. 435: 974–8.

Hatfield, S. and H. Ruohola-Baker. 2008. microRNA and stem cell function. Cell Tissue Res. 331: 57–66.

Hatten, M.E. 1999. Expansion of CNS precursor pools: a new role for Sonic Hedgehog. Neuron. 22: 2–3.

Hatten, M.E. and M.F. Roussel. 2011. Development and cancer of the cerebellum. Trends Neurosci. 34: 134–42.

He, H., K. Jazdzewski, W. Li, S. Liyanarachchi, R. Nagy, S. Volinia, G.A. Calin, C.G. Liu, K. Franssila, S. Suster, R.T. Kloos, C.M. Croce, and A. De La Chapelle. 2005. The role of microRNA genes in papillary thyroid carcinoma. Proc Natl Acad Sci USA. 102: 19075–80.

Herms, J., I. Neidt, B. Luscher, A. Sommer, P. Schurmann, T. Schroder, M. Bergmann, B. Wilken, S. Probst-Cousin, P. Hernaiz-Driever, J. Behnke, F. Hanefeld, T. Pietsch, and H.A. Kretzschmar. 2000. C-MYC expression in medulloblastoma and its prognostic value. Int J Cancer. 89: 395–402.

Herms, J.W., J. Behnke, M. Bergmann, H.J. Christen, R. Kolb, M. Wilkening, E. Markakis, F. Hanefeld, and H.A. Kretzschmar. 1997. Potential prognostic value of C-erbB-2 expression in medulloblastomas in very young children. J Pediatr Hematol Oncol. 19: 510–5.

Hernan, R., R. Fasheh, C. Calabrese, A.J. Frank, K.H. Maclean, D. Allard, R. Barraclough, and R.J. Gilbertson. 2003. ERBB2 up-regulates S100A4 and several other prometastatic genes in medulloblastoma. Cancer Res. 63: 140–8.

Ho, K.S. and M.P. Scott. 2002. Sonic hedgehog in the nervous system: functions, modifications and mechanisms. Curr Opin Neurobiol. 12: 57–63.

Houbaviy, H.B., M.F. Murray, and P.A. Sharp. 2003. Embryonic stem cell-specific MicroRNAs. Dev Cell. 5: 351–8.

Ibrahim, A.F., U. Weirauch, M. Thomas, A. Grunweller, R.K. Hartmann, and A. Aigner. 2011. MicroRNA replacement therapy for miR-145 and miR-33a is efficacious in a model of colon carcinoma. Cancer Res. 71: 5214–24.

Ingram, W.J., K.I. Mccue, T.H. Tran, A.R. Hallahan, and B.J. Wainwright. 2008. Sonic Hedgehog regulates Hes1 through a novel mechanism that is independent of canonical NOTCH pathway signalling. Oncogene. 27: 1489–500.

Jin, Z. and T. Xie. 2007. Dcr-1 maintains *Drosophila* ovarian stem cells. Curr Biol. 17: 539–44.

Johnnidis, J.B., M.H. Harris, R.T. Wheeler, S. Stehling-Sun, M.H. Lam, O. Kirak, T.R. Brummelkamp, M.D. Fleming, and F.D. Camargo. 2008. Regulation of progenitor cell proliferation and granulocyte function by microRNA-223. Nature. 451: 1125–9.

Johnston, R.J., Jr., S. Chang, J.F. Etchberger, C.O. Ortiz, and O. Hobert. 2005. MicroRNAs acting in a double-negative feedback loop to control a neuronal cell fate decision. Proc Natl Acad Sci USA. 102: 12449–54.

Kanellopoulou, C., S.A. Muljo, A.L. Kung, S. Ganesan, R. Drapkin, T. Jenuwein, D.M. Livingston, and K. Rajewsky. 2005. Dicer-deficient mouse embryonic stem cells are defective in differentiation and centromeric silencing. Genes Dev. 19: 489–501.

Kapsimali, M., W.P. Kloosterman, E. De Bruijn, F. Rosa, R.H. Plasterk, and S.W. Wilson. 2007. MicroRNAs show a wide diversity of expression profiles in the developing and mature central nervous system. Genome Biol. 8: R173.

Katsetos, C.D., L. Del Valle, A. Legido, J.P. De Chadarevian, E. Perentes, and S.J. Mork. 2003. On the neuronal/neuroblastic nature of medulloblastomas: a tribute to Pio del Rio Hortega and Moises Polak. Acta Neuropathol. 105: 1–13.

Kim, J., A. Krichevsky, Y. Grad, G.D. Hayes, K.S. Kosik, G.M. Church, and G. Ruvkun. 2004. Identification of many microRNAs that copurify with polyribosomes in mammalian neurons. Proc Natl Acad Sci USA. 101: 360–5.

Kim, J., K. Inoue, J. Ishii, W.B. Vanti, S.V. Voronov, E. Murchison, G. Hannon, and A. Abeliovich. 2007. A MicroRNA feedback circuit in midbrain dopamine neurons. Science. 317: 1220–4.

Kimura, H., D. Stephen, A. Joyner, and T. Curran. 2005. Gli1 is important for medulloblastoma formation in Ptc1+/– mice. Oncogene. 24: 4026–36.

King, C.E., L. Wang, R. Winograd, B.B. Madison, P.S. Mongroo, C.N. Johnstone, and A.K. Rustgi. 2011. LIN28B fosters colon cancer migration, invasion and transformation through let-7-dependent and -independent mechanisms. Oncogene. 30: 4185–93.

Kota, J., R.R. Chivukula, K.A. O'donnell, E.A. Wentzel, C.L. Montgomery, H.W. Hwang, T.C. Chang, P. Vivekanandan, M. Torbenson, K.R. Clark, J.R. Mendell, and J.T. Mendell. 2009. Therapeutic microRNA delivery suppresses tumorigenesis in a murine liver cancer model. Cell. 137: 1005–17.

Krutzfeldt, J., N. Rajewsky, R. Braich, K.G. Rajeev, T. Tuschl, M. Manoharan, and M. Stoffel. 2005. Silencing of microRNAs *in vivo* with 'antagomirs'. Nature. 438: 685–9.

Lakshmipathy, U., J. Davila, and R.P. Hart. 2010. miRNA in pluripotent stem cells. Regen Med. 5: 545–55.

Lal, A., F. Navarro, C.A. Maher, L.E. Maliszewski, N. Yan, E. O'day, D. Chowdhury, D.M. Dykxhoorn, P. Tsai, O. Hofmann, K.G. Becker, M. Gorospe, W. Hide, and J. Lieberman. 2009. miR-24 inhibits cell proliferation by targeting E2F2, MYC, and other cell-cycle genes via binding to "seedless" 3'UTR microRNA recognition elements. Mol Cell. 35: 610–25.

Lamont, J.M., C.S. Mcmanamy, A.D. Pearson, S.C. Clifford, and D.W. Ellison. 2004. Combined histopathological and molecular cytogenetic stratification of medulloblastoma patients. Clin Cancer Res. 10: 5482–93.

Landgraf, P., M. Rusu, R. Sheridan, A. Sewer, N. Iovino, A. Aravin, S. Pfeffer, A. Rice, A.O. Kamphorst, M. Landthaler, C. Lin, N.D. Socci, L. Hermida, V. Fulci, S. Chiaretti, R. Foa, J. Schliwka, U. Fuchs, A. Novosel, R.U. Muller, B. Schermer, U. Bissels, J. Inman, Q. Phan, M. Chien, D.B. Weir, R. Choksi, G. De Vita, D. Frezzetti, H.I. Trompeter, V. Hornung, G. Teng, G. Hartmann, M. Palkovits, R. Di Lauro, P. Wernet, G. Macino, C.E. Rogler, J.W. Nagle, J. Ju, F.N. Papavasiliou, T. Benzing, P. Lichter, W. Tam, M.J. Brownstein, A. Bosio, A. Borkhardt, J.J. Russo, C. Sander, M. Zavolan, and T. Tuschl. 2007. A mammalian microRNA expression atlas based on small RNA library sequencing. Cell. 129: 1401–14.

Laneve, P., L. Di Marcotullio, U. Gioia, M.E. Fiori, E. Ferretti, A. Gulino, I. Bozzoni, and E. Caffarelli. 2007. The interplay between microRNAs and the neurotrophin receptor tropomyosin-related kinase C controls proliferation of human neuroblastoma cells. Proc Natl Acad Sci USA. 104: 7957–62.

Lee, C.T., T. Risom, and W.M. Strauss. 2006. MicroRNAs in mammalian development. Birth Defects Res C Embryo Today. 78: 129–39.

Li, C., Y. Feng, G. Coukos, and L. Zhang. 2009. Therapeutic microRNA strategies in human cancer. Aaps J. 11: 747–57.

Li, X. and R.W. Carthew. 2005. A microRNA mediates EGF receptor signaling and promotes photoreceptor differentiation in the *Drosophila* eye. Cell. 123: 1267–77.

Liao, J.M. and H. Lu. 2011. Autoregulatory suppression of c-Myc by miR-185-3p. J Biol Chem. 286: 33901–9.

Lin, P.Y., S.L. Yu, and P.C. Yang. 2010. MicroRNA in lung cancer. Br J Cancer. 103: 1144–8.

Lin, S.L., D.C. Chang, S. Chang-Lin, C.H. Lin, D.T. Wu, D.T. Chen, and S.Y. Ying. 2008. Mir-302 reprograms human skin cancer cells into a pluripotent ES-cell-like state. RNA. 14: 2115–24.

Lu, Y., J. Xiao, H. Lin, Y. Bai, X. Luo, Z. Wang, and B. Yang. 2009. A single anti-microRNA antisense oligodeoxyribonucleotide (AMO) targeting multiple microRNAs offers an improved approach for microRNA interference. Nucleic Acids Res. 37: e24.

Lv, M., X. Zhang, H. Jia, D. Li, B. Zhang, H. Zhang, M. Hong, T. Jiang, Q. Jiang, J. Lu, X. Huang, and B. Huang. 2011a. An oncogenic role of miR-142-3p in human T-cell acute lymphoblastic leukemia (T-ALL) by targeting glucocorticoid receptor-alpha and cAMP/PKA pathways. Leukemia. Oct 7. doi: 10.1038/leu.2011.273.

Lv, S.Q., Y.H. Kim, F. Giulio, T. Shalaby, S. Nobusawa, H. Yang, Z. Zhou, M. Grotzer, and H. Ohgaki. 2011b. Genetic Alterations in MicroRNAs in Medulloblastomas. Brain Pathol. Jul 27. doi: 10.1111/j.1750-3639.2011.00523.x.

Ma, L., J. Teruya-Feldstein, and R.A. Weinberg. 2007. Tumour invasion and metastasis initiated by microRNA-10b in breast cancer. Nature. 449: 682–8.

Ma, L. and R.A. Weinberg. 2008. Micromanagers of malignancy: role of microRNAs in regulating metastasis. Trends Genet. 24: 448–56.

Machold, R., S. Hayashi, M. Rutlin, M.D. Muzumdar, S. Nery, J.G. Corbin, A. Gritli-Linde, T. Dellovade, J.A. Porter, L.L. Rubin, H. Dudek, A.P. Mcmahon, and G. Fishell. 2003. Sonic hedgehog is required for progenitor cell maintenance in telencephalic stem cell niches. Neuron. 39: 937–50.

Makeyev, E.V., J. Zhang, M.A. Carrasco, and T. Maniatis. 2007. The MicroRNA miR-124 promotes neuronal differentiation by triggering brain-specific alternative pre-mRNA splicing. Mol Cell. 27: 435–48.

Mantamadiotis, T. and S. Taraviras. 2011. Self-renewal mechanisms in neural cancer stem cells. Front Biosci. 16: 598–607.

Marino, S. 2005. Medulloblastoma: developmental mechanisms out of control. Trends Mol Med. 11: 17–22.

Medina, P.P. and F.J. Slack. 2008. microRNAs and cancer: an overview. Cell Cycle. 7: 2485–92.

Medina, R., S.K. Zaidi, C.G. Liu, J.L. Stein, A.J. Van Wijnen, C.M. Croce, and G.S. Stein. 2008. MicroRNAs 221 and 222 bypass quiescence and compromise cell survival. Cancer Res. 68: 2773–80.

Meister, G., M. Landthaler, Y. Dorsett, and T. Tuschl. 2004. Sequence-specific inhibition of microRNA- and siRNA-induced RNA silencing. RNA. 10: 544–50.

Mendrzyk, F., B. Radlwimmer, S. Joos, F. Kokocinski, A. Benner, D.E. Stange, K. Neben, H. Fiegler, N.P. Carter, G. Reifenberger, A. Korshunov, and P. Lichter. 2005. Genomic and protein expression profiling identifies CDK6 as novel independent prognostic marker in medulloblastoma. J Clin Oncol. 23: 8853–62.

Mineno, J., S. Okamoto, T. Ando, M. Sato, H. Chono, H. Izu, M. Takayama, K. Asada, O. Mirochnitchenko, M. Inouye and I. Kato. 2006. The expression profile of microRNAs in mouse embryos. Nucleic Acids Res. 34: 1765–71.

Miska, E.A., E. Alvarez-Saavedra, M. Townsend, A. Yoshii, N. Sestan, P. Rakic, M. Constantine-Paton, and H.R. Horvitz. 2004. Microarray analysis of microRNA expression in the developing mammalian brain. Genome Biol. 5: R68.

Morin, R.D., M.D. O'connor, M. Griffith, F. Kuchenbauer, A. Delaney, A.L. Prabhu, Y. Zhao, H. Mcdonald, T. Zeng, M. Hirst, C.J. Eaves, and M.A. Marra. 2008. Application of massively parallel sequencing to microRNA profiling and discovery in human embryonic stem cells. Genome Res. 18: 610–21.

Moriyama, T., K. Ohuchida, K. Mizumoto, J. Yu, N. Sato, T. Nabae, S. Takahata, H. Toma, E. Nagai, and M. Tanaka. 2009. MicroRNA-21 modulates biological functions of pancreatic cancer cells including their proliferation, invasion, and chemoresistance. Mol Cancer Ther. 8: 1067–74.

Murchison, E.P., J.F. Partridge, O.H. Tam, S. Cheloufi, and G.J. Hannon. 2005. Characterization of Dicer-deficient murine embryonic stem cells. Proc Natl Acad Sci USA. 102: 12135–40.

Murray, M.Y., S.A. Rushworth, and D.J. Macewan. 2011. MicroRNAs as a new therapeutic target towards leukaemia signalling. Cell. Signal. 2: 363–8.

Navarro, A. and M. Monzo. 2010. MicroRNAs in human embryonic and cancer stem cells. Yonsei Med J. 51: 622–32.

Nicoloso, M.S. and G.A. Calin. 2008. MicroRNA involvement in brain tumors: from bench to bedside. Brain Pathol. 18: 122–9.

Northcott, P.A., L.A. Fernandez, J.P. Hagan, D.W. Ellison, W. Grajkowska, Y. Gillespie, R. Grundy, T. Van Meter, J.T. Rutka, C.M. Croce, A.M. Kenney, and M.D. Taylor. 2009. The miR-17/92 polycistron is up-regulated in sonic hedgehog-driven medulloblastomas and induced by N-myc in sonic hedgehog-treated cerebellar neural precursors. Cancer Res. 69: 3249–55.

O'donnell, K.A., E.A. Wentzel, K.I. Zeller, C.V. Dang, and J.T. Mendell. 2005. c-Myc-regulated microRNAs modulate E2F1 expression. Nature. 435: 839–43.

Packer, A.N., Y. Xing, S.Q. Harper, L. Jones, and B.L. Davidson. 2008. The bifunctional microRNA miR-9/miR-9* regulates REST and CoREST and is downregulated in Huntington's disease. J Neurosci. 28: 14341–6.

Palmero, E.I., S.G. De Campos, M. Campos, N.C. De Souza, I.D. Guerreiro, A.L. Carvalho, and M.M. Marques. 2011. Mechanisms and role of microRNA deregulation in cancer onset and progression. Genet Mol Biol. 34: 363–70.

Pang, J.C., W.K. Kwok, Z. Chen, and H.K. Ng. 2009. Oncogenic role of microRNAs in brain tumors. Acta Neuropathol. 117: 599–611.

Park, J.K., X. Liu, T.J. Strauss, D.M. Mckearin, and Q. Liu. 2007. The miRNA pathway intrinsically controls self-renewal of *Drosophila* germline stem cells. Curr Biol. 17: 533–8.

Pfister, S., M. Remke, A. Benner, F. Mendrzyk, G. Toedt, J. Felsberg, A. Wittmann, F. Devens, N.U. Gerber, S. Joos, A. Kulozik, G. Reifenberger, S. Rutkowski, O.D. Wiestler, B. Radlwimmer, W. Scheurlen, P. Lichter, and A. Korshunov. 2009. Outcome prediction in pediatric medulloblastoma based on DNA copy-number aberrations of chromosomes 6q and 17q and the MYC and MYCN loci. J Clin Oncol. 27: 1627–36.

Pierson, J., B. Hostager, R. Fan, and R. Vibhakar. 2008. Regulation of cyclin dependent kinase 6 by microRNA 124 in medulloblastoma. J Neurooncol. 90: 1–7.

Pietsch, T., A. Waha, A. Koch, J. Kraus, S. Albrecht, J. Tonn, N. Sorensen, F. Berthold, B. Henk, N. Schmandt, H.K. Wolf, A. Von Deimling, B. Wainwright, G. Chenevix-Trench, O.D. Wiestler, and C. Wicking. 1997. Medulloblastomas of the desmoplastic variant carry mutations of the human homologue of *Drosophila* patched. Cancer Res. 57: 2085–8.

Pollack, I.F., J. Biegel, A. Yates, R. Hamilton, and S. Finkelstein. 2002. Risk assignment in childhood brain tumors: the emerging role of molecular and biologic classification. Curr Oncol Rep. 4: 114–22.

Pomeroy, S.L., M.E. Sutton, L.C. Goumnerova, and R.A. Segal. 1997. Neurotrophins in cerebellar granule cell development and medulloblastoma. J Neurooncol. 35: 347–52.

Pomeroy, S.L. and L.M. Sturla. 2003. Molecular biology of medulloblastoma therapy. Pediatr Neurosurg. 39: 299–304.

Raabe, E.H. and C.G. Eberhart. 2010. High-risk medulloblastoma: does c-myc amplification overrule histopathology? Pediatr Blood Cancer. 54: 344–5.

Raffel, C., R.B. Jenkins, L. Frederick, D. Hebrink, B. Alderete, D.W. Fults, and C.D. James. 1997. Sporadic medulloblastomas contain PTCH mutations. Cancer Res. 57: 842–845.

Raffel, C. 2004. Medulloblastoma: molecular genetics and animal models. Neoplasia. 6: 310–22.

Riobo, N.A. and D.R. Manning. 2007. Pathways of signal transduction employed by vertebrate Hedgehogs. Biochem J. 403: 369–79.

Rodini, C.O., D.E. Suzuki, A.M. Nakahata, M.C. Pereira, L. Janjoppi, S.R. Toledo, and O.K. Okamoto. 2010. Aberrant signaling pathways in medulloblastomas: a stem cell connection. Arq Neuropsiquiatr. 68: 947–52.

Rood, B.R., T.J. Macdonald, and R.J. Packer. 2004. Current treatment of medulloblastoma: recent advances and future challenges. Semin Oncol. 31: 666–75.

Rossi, A., V. Caracciolo, G. Russo, K. Reiss, and A. Giordano. 2008. Medulloblastoma: from molecular pathology to therapy. Clin Cancer Res. 14: 971–6.

Ruiz I. Altaba, A., P. Sanchez, and N. Dahmane. 2002. Gli and hedgehog in cancer: tumours, embryos and stem cells. Nat Rev Cancer. 2: 361–72.

Ruiz I. Altaba, A., C. Mas, and B. Stecca. 2007. The Gli code: an information nexus regulating cell fate, stemness and cancer. Trends Cell Biol. 17: 438–47.

Rutkowski, S., A. Von Bueren, K. Von Hoff, W. Hartmann, T. Shalaby, F. Deinlein, M. Warmuth-Metz, N. Soerensen, A. Emser, U. Bode, U. Mittler, C. Urban, M. Benesch, R.D. Kortmann, P.G. Schlegel, J. Kuehl, T. Pietsch, and M. Grotzer. 2007. Prognostic relevance of clinical and biological risk factors in childhood medulloblastoma: results of patients treated in the prospective multicenter trial HIT'91. Clin Cancer Res. 13: 2651–7.

Sachdeva, M., S. Zhu, F. Wu, H. Wu, V. Walia, S. Kumar, R. Elble, K. Watabe, and Y.Y. Mo. 2009. p53 represses c-Myc through induction of the tumor suppressor miR-145. Proc Natl Acad Sci USA. 106: 3207–12.

Salsano, E., B. Pollo, M. Eoli, M.T. Giordana, and G. Finocchiaro. 2004. Expression of MATH1, a marker of cerebellar granule cell progenitors, identifies different medulloblastoma sub-types. Neurosci Lett. 370: 180–5.

Saran, A. 2009. Medulloblastoma: role of developmental pathways, DNA repair signaling, and other players. Curr Mol Med. 9: 1046–57.

Scheurlen, W.G., G.C. Schwabe, S. Joos, J. Mollenhauer, N. Sorensen, and J. Kuhl. 1998. Molecular analysis of childhood primitive neuroectodermal tumors defines markers associated with poor outcome. J Clin Oncol. 16: 2478–85.

Scott, G.K., A. Goga, D. Bhaumik, C.E. Berger, C.S. Sullivan, and C.C. Benz. 2007. Coordinate suppression of ERBB2 and ERBB3 by enforced expression of micro-RNA miR-125a or miR-125b. J Biol Chem. 282: 1479–86.

Secombe, J., S.B. Pierce, and R.N. Eisenman. 2004. Myc: a weapon of mass destruction. Cell 117: 153–6.

Sempere, L.F., S. Freemantle, I. Pitha-Rowe, E. Moss, E. Dmitrovsky, and V. Ambros. 2004. Expression profiling of mammalian microRNAs uncovers a subset of brain-expressed microRNAs with possible roles in murine and human neuronal differentiation. Genome Biol. 5: R13.

Si, M.L., S. Zhu, H. Wu, Z. Lu, F. Wu, and Y.Y. Mo. 2007. miR-21-mediated tumor growth. Oncogene. 26: 2799–803.

Silber, J., D.A. Lim, C. Petritsch, A.I. Persson, A.K. Maunakea, M. Yu, S.R. Vandenberg, D.G. Ginzinger, C.D. James, J.F. Costello, G. Bergers, W.A. Weiss, A. Alvarez-Buylla, and J.G. Hodgson. 2008. miR-124 and miR-137 inhibit proliferation of glioblastoma multiforme cells and induce differentiation of brain tumor stem cells. BMC Med. 6: 14.

Singh, S.K., I.D. Clarke, M. Terasaki, V.E. Bonn, C. Hawkins, J. Squire, and P.B. Dirks. 2003. Identification of a cancer stem cell in human brain tumors. Cancer Res. 63: 5821–8.

Singh, S.K., C. Hawkins, I.D. Clarke, J.A. Squire, J. Bayani, T. Hide, R.M. Henkelman, M.D. Cusimano, and P.B. Dirks. 2004. Identification of human brain tumour initiating cells. Nature. 432: 396–401.

Slack, F.J. 2010. Stem cells: Big roles for small RNAs. Nature. 463: 616.

Solecki, D.J., X.L. Liu, T. Tomoda, Y. Fang, and M.E. Hatten. 2001. Activated NOTCH2 signaling inhibits differentiation of cerebellar granule neuron precursors by maintaining proliferation. Neuron. 31: 557–68.

Srivastava, V.K. and J. Nalbantoglu. 2010. The cellular and developmental biology of medulloblastoma: current perspectives on experimental therapeutics. Cancer Biol Ther. 9: 843–52.

Stenvang, J., A.N. Silahtaroglu, M. Lindow, J. Elmen, and S. Kauppinen. 2008. The utility of LNA in microRNA-based cancer diagnostics and therapeutics. Semin Cancer Biol. 18: 89–102.

Strizzi, L., K.M. Hardy, E.A. Seftor, F.F. Costa, D.A. Kirschmann, R.E. Seftor, L.M. Postovit, and M.J. Hendrix. 2009. Development and cancer: at the crossroads of Nodal and NOTCH signaling. Cancer Res. 69: 7131–4.

Suh, M.R., Y. Lee, J.Y. Kim, S.K. Kim, S.H. Moon, J.Y. Lee, K.Y. Cha, H.M. Chung, H.S. Yoon, S.Y. Moon, V.N. Kim, and K.S. Kim. 2004. Human embryonic stem cells express a unique set of microRNAs. Dev Biol. 270: 488–98.

Takamizawa, J., H. Konishi, K. Yanagisawa, S. Tomida, H. Osada, T. Endoh, T. Harano, Y. Yatabe, M. Nagino, Y. Nimura, T. Mitsudomi, and T. Takahashi. 2004. Reduced expression of the let-7 microRNAs in human lung cancers in association with shortened postoperative survival. Cancer Res. 64: 3753–6.

Takei, H., Y. Nguyen, V. Mehta, M. Chintagumpala, R.C. Dauser, and A.M. Adesina. 2009. Low-level copy gain versus amplification of myc oncogenes in medulloblastoma: utility in predicting prognosis and survival. Laboratory investigation. J Neurosurg Pediatr. 3: 61–5.

Takeshita, F., L. Patrawala, M. Osaki, R.U. Takahashi, Y. Yamamoto, N. Kosaka, M. Kawamata, K. Kelnar, A.G. Bader, D. Brown, and T. Ochiya. 2010. Systemic delivery of synthetic microRNA-16 inhibits the growth of metastatic prostate tumors via downregulation of multiple cell-cycle genes. Mol Ther. 18: 181–7.

Taylor, M.D., L. Liu, C. Raffel, C.C. Hui, T.G. Mainprize, X. Zhang, R. Agatep, S. Chiappa, L. Gao, A. Lowrance, A. Hao, A.M. Goldstein, T. Stavrou, S.W. Scherer, W.T. Dura, B. Wainwright, J.A. Squire, J.T. Rutka, and D. Hogg. 2002. Mutations in SUFU predispose to medulloblastoma. Nat Genet. 31: 306–10.

Ten Donkelaar, H.J., M. Lammens, P. Wesseling, H.O. Thijssen, and W.O. Renier. 2003. Development and developmental disorders of the human cerebellum. J Neurol. 250: 1025–36.

Thompson, M.C., C. Fuller, T.L. Hogg, J. Dalton, D. Finkelstein, C.C. Lau, M. Chintagumpala, A. Adesina, D.M. Ashley, S.J. Kellie, M.D. Taylor, T. Curran, A. Gajjar, and R.J. Gilbertson. 2006. Genomics identifies medulloblastoma subgroups that are enriched for specific genetic alterations. J Clin Oncol. 24: 1924–31.

Tran, N., T. Mclean, X. Zhang, C.J. Zhao, J.M. Thomson, C. O'brien, and B. Rose. 2007. MicroRNA expression profiles in head and neck cancer cell lines. Biochem Biophys Res Commun. 358: 12–7.

Trang, P., P.P. Medina, J.F. Wiggins, L. Ruffino, K. Kelnar, M. Omotola, R. Homer, D. Brown, A.G. Bader, J.B. Weidhaas, and F.J. Slack. 2010. Regression of murine lung tumors by the let-7 microRNA. Oncogene. 29: 1580–7.

Turner, J.D., R. Williamson, K.K. Almefty, P. Nakaji, R. Porter, V. Tse, and M.Y. Kalani. 2010. The many roles of microRNAs in brain tumor biology. Neurosurg Focus. 28: E3.

Uziel, T., F.V. Karginov, S. Xie, J.S. Parker, Y.D. Wang, A. Gajjar, L. He, D. Ellison, R.J. Gilbertson, G. Hannon, and M.F. Roussel. 2009. The miR-17~92 cluster collaborates with the Sonic Hedgehog pathway in medulloblastoma. Proc Natl Acad Sci USA. 106: 2812–7.

Ventura, A. and T. Jacks. 2009. MicroRNAs and cancer: short RNAs go a long way. Cell. 136: 586–91.

Visvanathan, J., S. Lee, B. Lee, J.W. Lee, and S.K. Lee. 2007. The microRNA miR-124 antagonizes the anti-neural REST/SCP1 pathway during embryonic CNS development. Genes Dev. 21: 744–9.

Viswanathan, S.R., G.Q. Daley, and R.I. Gregory. 2008. Selective blockade of microRNA processing by Lin28. Science. 320: 97–100.

Von Hoff, K., W. Hartmann, A.O. Von Bueren, N.U. Gerber, M.A. Grotzer, T. Pietsch, and S. Rutkowski. 2010. Large cell/anaplastic medulloblastoma: outcome according to myc status, histopathological, and clinical risk factors. Pediatr Blood Cancer. 54: 369–76.

Voorhoeve, P.M., C. Le Sage, M. Schrier, A.J. Gillis, H. Stoop, R. Nagel, Y.P. Liu, J. Van Duijse, J. Drost, A. Griekspoor, E. Zlotorynski, N. Yabuta, G. De Vita, H. Nojima, L.H. Looijenga, and R. Agami. 2006. A genetic screen implicates miRNA-372 and miRNA-373 as oncogenes in testicular germ cell tumors. Cell. 124: 1169–81.

Wang, L., Z.G. Zhang, S.R. Gregg, R.L. Zhang, Z. Jiao, Y. Letourneau, X. Liu, Y. Feng, J. Gerwien, L. Torup, M. Leist, C.T. Noguchi, Z.Y. Chen, and M. Chopp. 2007. The Sonic hedgehog pathway mediates carbamylated erythropoietin-enhanced proliferation and differentiation of adult neural progenitor cells. J Biol Chem. 282: 32462–70.

Ward, R.J., L. Lee, K. Graham, T. Satkunendran, K. Yoshikawa, E. Ling, L. Harper, R. Austin, E. Nieuwenhuis, I.D. Clarke, C.C. Hui, and P.B. Dirks. 2009. Multipotent CD15+ cancer stem cells in patched-1-deficient mouse medulloblastoma. Cancer Res. 69: 4682–90.

Wechsler-Reya, R.J. and M.P. Scott. 1999. Control of neuronal precursor proliferation in the cerebellum by Sonic Hedgehog. Neuron. 22: 103–14.

Weiler, J., J. Hunziker, and J. Hall. 2006. Anti-miRNA oligonucleotides (AMOs): ammunition to target miRNAs implicated in human disease? Gene Ther. 13: 496–502.

White, N.M., E. Fatoohi, M. Metias, K. Jung, C. Stephan, and G.M. Yousef. 2011. Metastamirs: a stepping stone towards improved cancer management. Nat Rev Clin Oncol. 8: 75–84.

Wiggins, J.F., L. Ruffino, K. Kelnar, M. Omotola, L. Patrawala, D. Brown, and A.G. Bader. 2010. Development of a lung cancer therapeutic based on the tumor suppressor microRNA-34. Cancer Res. 70: 5923–30.

Wu, W. 2011. Modulation of microRNAs for potential cancer therapeutics. Methods Mol Biol. 676: 59–70.

Xiao, J., B. Yang, H. Lin, Y. Lu, X. Luo, and Z. Wang. 2007. Novel approaches for gene-specific interference via manipulating actions of microRNAs: examination on the pacemaker channel genes HCN2 and HCN4. J Cell Physiol. 212: 285–92.

Xiong, J., Q. Du, and Z. Liang. 2010. Tumor-suppressive microRNA-22 inhibits the transcription of E-box-containing c-Myc target genes by silencing c-Myc binding protein. Oncogene 29: 4980–8.

Yang, Z.J., T. Ellis, S.L. Markant, T.A. Read, J.D. Kessler, M. Bourboulas, U. Schuller, R. Machold, G. Fishell, D.H. Rowitch, B.J. Wainwright, and R.J. Wechsler-Reya. 2008. Medulloblastoma can be initiated by deletion of Patched in lineage-restricted progenitors or stem cells. Cancer Cell. 14: 135–45.

Yekta, S., I.H. Shih, and D.P. Bartel. 2004. MicroRNA-directed cleavage of HOXB8 mRNA. Science. 304: 594–6.

Yokota, J., J. Adachi, A. Sasaki, and T. Kohno. 1996. Genetic defects in cancer cells and cancer prone families. Eur J Cancer Prev. 5 Suppl. 2: 33–6.

Yu, F., H. Yao, P. Zhu, X. Zhang, Q. Pan, C. Gong, Y. Huang, X. Hu, F. Su, J. Lieberman, and E. Song. 2007. let-7 regulates self renewal and tumorigenicity of breast cancer cells. Cell. 131: 1109–23.

Yu, J.Y., K.H. Chung, M. Deo, R.C. Thompson, and D.L. Turner. 2008. MicroRNA miR-124 regulates neurite outgrowth during neuronal differentiation. Exp Cell Res. 314: 2618–33.

Zhao, C., W. Deng, and F.H. Gage. 2008. Mechanisms and functional implications of adult neurogenesis. Cell. 132: 645–60.

Zhao, C., G. Sun, S. Li, and Y. Shi. 2009. A feedback regulatory loop involving microRNA-9 and nuclear receptor TLX in neural stem cell fate determination. Nat Struct Mol Biol. 16: 365–71.

Zhu, R., Y. Ji, L. Xiao, and A. Matin. 2007. Testicular germ cell tumor susceptibility genes from the consomic 129.MOLF-Chr19 mouse strain. Mamm Genome. 18: 584–95.

Zhu, S., H. Wu, F. Wu, D. Nie, S. Sheng, and Y.Y. Mo. 2008. MicroRNA-21 targets tumor suppressor genes in invasion and metastasis. Cell Res. 18: 350–9.

Zhu, W., X. Liu, J. He, D. Chen, Y. Hunag, and Y.K. Zhang. 2010. Overexpression of members of the microRNA-183 family is a risk factor for lung cancer: A case control study. BMC Cancer. 11: 393.

Zollo, M. Andolfo, I. De Antonellis, P. 2011. Cancer stem cells theories and practice (ISBN 978-953-307-225-8) Review: MicroRNAs and cancer stem cells in medulloblastoma. Available from: http://www.intechopen.com/books/cancer-stem-cells-theories-and-practice/micrornas-and-cancer-stem-cells-in-medulloblastoma.

CHAPTER 12

EXTRACELLULAR MicroRNAs AS POTENTIAL BIOMARKERS AND THERAPEUTIC TOOLS IN CANCER

Muriel Thirion[a] and Takahiro Ochiya[b],*

ABSTRACT

Though circulating extracellular nucleic acids are not a recent discovery, the studies made over the past several years, especially on microRNAs (miRNAs), have demonstrated their importance in cancer. These nucleic acids, essential in cellular homeostasis and intercellular communication, have been detected in plasma, serum, urine, and other body fluids from healthy subjects as well as in diseased patients. These developments point to the role that miRNAs may play as novel diagnostic and prognostic markers. It has also opened the path to new therapeutic strategies such as immunotherapeutic nanovesicles and RNA interference (RNAi) delivery systems. In this review, we will summarize the present knowledges of the secretory mechanism, biological function and use of miRNAs as potential biomarkers and therapeutic tools in disease treatment.

Division of Molecular and Cellular Medicine, National Cancer Center Research Institute, 5-1-1 Tsukiji, Chuo-ku, Tokyo, 104-0045, Japan.
[a]Email: muriel.th@hotmail.com
[b]Email: tochiya@ncc.go.jp
*Corresponding author

1. INTRODUCTION

Prior to the discovery of the double-helical structure of DNA by Watson and Crick, Mandel and Métais reported the presence of extracellular nucleic acids in the blood plasma (Mandel and Métais 1948). Due to lack of interest and proper molecular biological techniques to study circulating nucleic acids, this noteworthy work was unfortunately buried for almost 20 years until Tan and colleagues demonstrated the presence of circulating DNA in patients with systemic lupus erythematosus (Tan et al. 1966). The real breakthrough took place in 1994 when two groups detected the presence of mutated tumor-associated oncogenes in the plasma of patients with cancer (Sorenson et al. 1994, Vasioukhin et al. 1994). Since then, many studies demonstrated the presence of several extracellular nucleic acids (DNA, mRNA, miRNA) at high concentration in the blood of diseased patients (Shinozaki et al. 2007, Lawrie et al. 2008, Mitchell et al. 2008). These data were remarkable because they opened the path to new alternative approaches to the usual cancer screening tests that are invasive and inefficient in the early stages' detection.

Among the extracellular nucleic acids, miRNAs are of particular interest. Indeed, these small regulatory RNA molecules can modulate the expression of numerous specific mRNA targets and therefore play key roles in a variety of physiological and pathological processes (Bartel 2004). The expression pattern of miRNAs seems to be tissue-specific and altered expression has been reported in various cancers. In this review, we will thus focus on the potential usefulness of extracellular miRNAs as biomarkers and therapeutic tools in cancer.

2. ORIGIN OF EXTRACELLULAR miRNAs

In addition to being of small size, relatively abundant and tissue specific, circulating miRNAs are also very stable. However, RNases are present in large amounts in the blood. To assess the resistance mechanism of these extracellular miRNAs, Mitchell and colleagues designed synthetic miRNA with no homology to human sequences and added them to the plasma (Mitchell et al. 2008). These miRNAs were rapidly degraded unlike the endogenous plasma miRNAs, demonstrating that the latters should exit in a form resistant to RNase activity. At the same time, the circulating miRNAs were demonstrated to remain stable after being subjected to harsh conditions such as boiling, low/high pH or freeze-thaw cycles, conditions under which most RNA would be degraded (Chen et al. 2008). El-Hefnawy and coworkers showed that extracellular RNA was destroyed in presence of detergents, suggesting a protection by inclusion within lipid or lipoprotein vesicles or by attachment to proteins (El-Hefnawy et al. 2004).

2.1 Extracellular miRNAs Carried in Secreted Vesicles

Recent studies have revealed a novel genetic exchange between cells using miRNAs incorporated in and carried by extracellular vesicles (EV). These EV are called microvesicles, exosomes or others regarding to the size, density and secretion mechanisms. Their composition, origin and properties will be discussed later in this chapter. Presently the focus is on the actual knowledge about their origin (Fig. 1). The classical secretory pathway allows the cell to release soluble proteins and requires carrier and secretory vesicles that contain intraluminal components. By contrast, the secreted membranes vesicles contain cytoplasmic components. They can form at the plasma membrane by direct budding into the extracellular environment and produce large size (100–1000 nm), irregular shape microvesicles or smaller size round membrane particles. Alternatively, there is another type of particles called exosome-like vesicles. These are thought to be released from multivesicular bodies (MVB) but the mechanism is not clear yet. Finally, secreted vesicles can form inside internal compartments that subsequently fusion with the plasma membrane to release small size (30–100 nm) cup shape exosomes. A

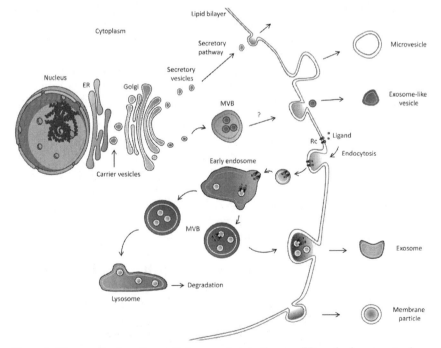

Figure 1. Biogenesis of various cellular secretory pathways. ER, endoplasmic reticulum; MVB, multivesicular bodies; Rc, receptor.

recent paper from our laboratory demonstrated that miRNAs are released through a ceramide-dependent secretory machinery and that the secretory miRNAs are transferable and functional in the recipient cells (Kosaka et al. 2010) (Fig. 2). In agreement with the finding by Kosaka and colleagues, some other reports showed the inhibition of neutral sphingomyelinase-2 (nSMase2), the enzyme that regulates the ceramide-dependent pathway, impairs the cellular export of miRNAs (Mittelbrunn et al. 2011, Kogure et al. 2011). Nevertheless, these data do not exclude other mechanisms for the secretion of miRNAs from the cells.

One of the first papers to report the existence of miRNA in small particles was from Valadi and coworkers (Valadi et al. 2007). The authors showed that vesicles released from human and murine mast cell lines contained mRNA and miRNA. Later, Hunter and colleagues demonstrated the presence of miRNAs in the blood's EV. These identified miRNAs had been previously shown to regulate the cell differentiation of blood cells, metabolic pathways and immune function (Hunter et al. 2008). To date, a large number of studies report extracellular particles as an important

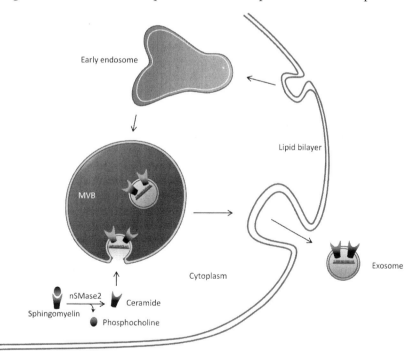

Figure 2. A model of secretory mechanism of microRNA involving exosomes. miRNA are brought to the multivesicular bodies (MVB), packaged into the exosomes and released from the cells, stimulated by the surge of cellular ceramide. nSMase2, neutral sphingomyelinase 2.

source of miRNA in the circulation and suggest a regulated miRNA-sorted mechanism. However, the exact mechanism of sorting and incorporation of miRNA to the vesicles and their secretion still needs to gain better understanding.

2.2 Extracellular miRNAs Associated with Protein Complexes

While the role of secreted vesicle as miRNA carrier is becoming increasingly recognized, some reports showed that an important part of the circulating miRNA is associated with RNA-binding proteins such as argonaute 2 protein (Ago2) and nucleophosmin 1 (NPM1).

Ago2 is the key effector protein of miRNA-mediated silencing. By using size-exclusion chromatography and immunoprecipitation techniques, Arroyo and coworkers demonstrated that circulating miRNA cofractionated with Ago2 complexes rather than with vesicles (Arroyo et al. 2011).

NPM1 is a nucleolar RNA-binding protein that is involved in ribosome biogenesis and transport. Wang and colleagues found that NPM1 was released outside the cell, independently from the vesicular fraction, and that this extracellular NPM1 bound miRNA and protected it from RNAse activity (Wang et al. 2010a).

It is still unclear whether these miRNA-protein complexes could target specifically and be functional in intercellular communication or rather be by-products of normal or dying cells. However, Wang and colleagues have shown that intracellular levels of ATP affected the exportation of miRNA, implying this process to be largely energy dependent (Wang et al. 2010a). This points to a possible role of the miRNA-protein complexes in cell–cell communication, as it has already been reported for miRNA-exosomes. This aspect is critical in determining the usefulness of circulating miRNAs as biomarkers and particularly in the context of therapeutic targets that will be discussed in the point 4 below.

2.3 Extracellular miRNAs Associated with Lipoprotein Complexes

Finally, extracellular miRNA was also found in association with lipoproteins. Vickers and coworkers demonstrated that high-density lipoprotein (HDL) could load plasma miRNA and deliver it to cells, leading to altered gene expression of these recipient cells (Vickers et al. 2011) (Fig. 3). To this end, this group first separated the exosome, low-density lipoprotein (LDL) and

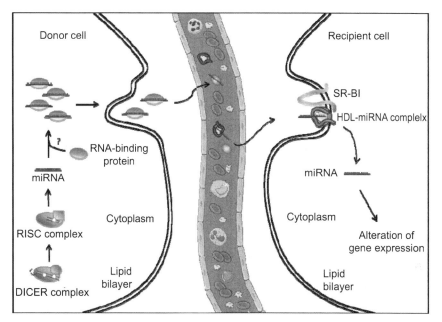

Figure 3. A model of miRNA intercellular communication mechanism involving high-density lipoprotein (HDL). Exported miRNA are carried and delivered to the recipient cells by HDL. The delivery process into the cell is dependent on the scavenger receptor class B type I protein (SR-BI). RISC, RNA-induced silencing complex.

HDL fractions. LDL and HDL have similar physico-chemical properties but differ in size, composition and biogenesis. The authors found the presence of miRNA in all the three separated fractions. Nevertheless, the miRNA-profile was different in HDL fraction compared to exosome and LDL fractions, with miR-223 highly expressed in HDL.

Vickers and colleagues demonstrated the capacity of HDL to load miRNA molecules *in vivo* by using wild-type and apolipoprotein E-null mutant mice. The authors also showed *in vitro* that native HDL could deliver functional miRNA to recipient cells, mediated by scavenger receptor class B type I (SR-BI). But, in contrast to miRNA-loading process into exosomes, inhibition of the nSMase2 increased the amount of miR-223 exported to HDL, suggesting likely common pathways but distinct mechanisms.

This was the first study to demonstrate a new role of HDL apart from its involvement in cholesterol dynamics. These findings open a perspective of use of HDL-miRNA complex as biomarker, therapeutic tool for disorders related to lipid metabolism or as delivery tool for other diseases.

3. EXTRACELLULAR miRNAs AS POTENTIAL BIOMARKERS IN CANCER

The origin of extracellular miRNAs has been explored here above and the different transporters identified to date are summarized in Fig. 4. Until now, we described only the miRNAs secreted in the bloodstream. However, recent studies have also demonstrated the presence of miRNAs in most of the other body fluids (urine, saliva, etc.), increasing the potential of the circulating miRNAs as non-invasive bioclinical tools. Through various techniques and samples, numerous groups have shown that cancers can have specific altered miRNA profiles in those body fluids (Table 1).

Lawrie and colleagues were the first to demonstrate the presence of miRNAs (miR-21, miR-155 and miR-210) at higher levels in the serum of patients suffering from diffuse large B-cell lymphoma (DLBCL) (Lawrie et al. 2008) than in control subjects. The authors also found that miR-21, highly expressed in DLBCL, was correlated with relapse-free survival. Mitchell and coworkers investigated the miRNAs that are usually expressed in prostate cancer cells and discovered that, in serum samples, the expression of miR-141 could differentiate patients from healthy controls (Mitchell et al. 2008). In this study, the group also demonstrated that miRNAs originating from human prostate cancer xenografts entered the circulation, were detected in the plasma and could distinguish xenografted mice from controls. In the breast cancer, blood circulating miR-195 and let-7a were found to be increased in patients, compared to the healthy controls, and decreased following tumor resection (Heneghan et al. 2010). Moreover, this group correlated the expression of let-7a and miR-21 to clinicopathological

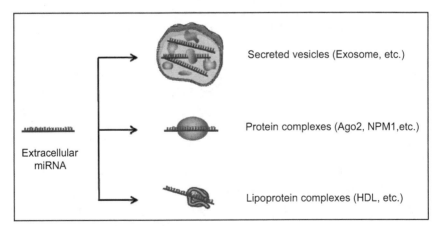

Figure 4. Transporters of extracellular miRNAs. Extracellular miRNAs are carried in secreted vesicles, protein complexes, and lipoprotein complexes. Ago2, argonaute 2 protein; NPM1, nucleophosmin 1; HDL, high-density lipoprotein.

Table 1. Circulating miRNA as cancer biomarkers.

Type of cancer	Body fluid	Potential miRNA biomarker	Justification	Usefulness	References
Bladder cancer	Urina	miR-126 and miR-182	Increased	Diagnosis	Hanke et al. 2010
		miR-96 and miR-183	Increased, Correlated with tumor grade, Decreased after surgical treatment	Diagnosis Prognosis	Yamada et al. 2011
Breast cancer	Serum/plasma	miR-195 and let-7a, let-7a and miR-21	Increased, Decreased after surgical treatment, Correlated with nodal status and ER status	Diagnosis Prognosis	Heneghan et al. 2010
		miR-126, miR-199a, miR-335, miR-21, miR-106a and miR-155	Increased or decreased, Correlated with ER/PR status and tumor grade	Diagnosis Prognosis	Wang et al. 2010b
Colorectal cancer	Serum/plasma	miR-92	Increased, Decreased after surgical treatment	Diagnosis	Ng et al. 2009
		miR-29a and miR-92a	Increased	Diagnosis	Huang et al. 2010
		miR-141	Increased	Prognosis	Cheng et al. 2011
Diffuse large B cell lymphoma	Serum/plasma	miR-21, miR-155 and miR-210 miR-21	Increased Correlated with disease-free survival	Diagnosis Prognosis	Lawrie et al. 2008
Esophageal squamous cell carcinoma	Serum/plasma	miR-10a, miR-22, miR-100, miR-148b, miR-223, miR-133a, and miR-127-3p	Increased	Diagnosis	Zhang et al. 2010
		miR-31	Increased, correlated with tumor stage	Diagnosis Prognosis	Zhang et al. 2011
Gastric cancer	Serum/plasma	miR-17-5p, miR-21, miR-106a, miR-106b and let-7a	Increased or decreased, Decreased after surgical treatment	Diagnosis	Tsujiura et al. 2010

Table 1. contd....

Table 1. contd....

Type of cancer	Body fluid	Potential miRNA biomarker	Justification	Usefulness	References
Gastric cancer	Serum/plasma	miR-1, miR-20a, miR-27a, miR-34 and miR-423-5p	Increased, Correlated with tumor stage	Diagnosis Prognosis	Liu et al. 2011a
		miR-378	Increased	Diagnosis	Liu et al. 2011b
Glioblastoma	Serum/plasma	miR-21	Increased, Found in EV	Diagnosis	Skog et al. 2008
Hepatocellular carcinoma (HCC)	Serum/plasma	miR-500	Increased, Decreased after surgical treatment	Diagnosis	Yamamoto et al. 2009
		miR-92a	Decreased, Increased after surgical treatment,	Diagnosis	Shigoka et al. 2010
		miR-25, miR-375, and let-7f	Correlated with HBV-positive HCC and -infection	Diagnosis	Li et al. 2010a
		miR-122, miR-192, miR-21, miR-223, miR-26a, miR-27a and miR-801	Could discriminate HCC group from other liver diseases' group and healthy group in a large cohort	Diagnosis	Zhou et al. 2011
Leukemia	Serum/plasma	miR-92a; ratio miR-92a/miR-638	Decreased, Ratio well correlated with AL diagnosis	Diagnosis	Tanaka et al. 2009
		miR-150*, miR-195, miR-222, MiR-29a;	Could discriminate CLL group from other hematologic malignancies' group and healthy group	Diagnosis	Moussay et al. 2011
		miR-20a	Correlated with CLL-ZAP-70 expression status	Prognosis	
Lung cancer	Serum/plasma	miR-25 and miR-223	Increased	Diagnosis	Chen et al. 2008
		miR-20a, -24, -25, -145, -152, -199a-5p, -221, -222, -223 and -320	Increased, Correlated with early diagnosis	Diagnosis	Chen et al. 2011
		miRNA in exosomes	Correlated with tumor-derived miRNA patterns	Diagnosis	Rabinowits et al. 2009
		let-7f, miR-20b and miR-30e-3p	Decreased, Found in EV, Correlated with tumor stage, disease-free survival and overall survival	Prognosis	Silva et al. 2011

Lung cancer	Serum/plasma	miR-1, miR-30d, miR-486 and miR-499	Correlated with overall survival	Prognosis	Hu et al. 2010
		miR-21, miR-210 and miR-486-5p	Increased or decreased	Diagnosis	Shen et al. 2011
		miR-21	Increased, Correlated with metastasis, Correlated with poor diagnosis in association with tumor miR-200c	Diagnosis, Prognosis	Wang et al. 2011, Wei et al. 2011 and Liu et al. 2011c
		miR-96, miR-182 and miR-183	Increased, Correlated with overall survival	Diagnosis Prognosis	Zhu et al. 2011
		miR-126 and miR-183	Correlated with tumor grade	Prognosis	Lin et al. 2012
		miR-155, miR-182, miR-197	Increased, Correlated with chemotherapy responsiveness	Diagnosis Prognosis	Zheng et al. 2011
		miR-10b, miR-141, miR-155 and miR-34a	Increased	Diagnosis	Roth et al. 2011
		miR-1254 and miR-574-5p	Increased	Diagnosis	Foss et al. 2011
	Effusion	miR-24, miR-26a and miR-30d, miR-152	Potentially correlated with malignant effusion and with chemotherapy responsiveness	Diagnosis Prognosis	Xie et al. 2010
Melanoma	Serum/plasma	16 miRNA	Increased and decreased	Diagnosis	Leidinger et al. 2010
		miR-221	Increased, Decreased after surgical treatment	Diagnosis Prognosis	Kanemaru et al. 2011
Oral cancer	Saliva	miR-125a and miR-200a	Decreased, Ago2 detected in the saliva	Diagnosis	Park et al. 2009
		miR-375	Decreased	Diagnosis	Wiklund et al. 2011
	Serum/plasma	miR-31	Increased, Decreased after surgical treatment, Also increased in saliva	Diagnosis	Liu et al. 2010d
		miR-184	Increased, Decreased after surgical treatment	Diagnosis	Wong et al. 2008

Table 1. contd....

Table 1. contd....

Type of cancer	Body fluid	Potential miRNA biomarker	Justification	Usefulness	References
Ovarian cancer	Serum/plasma	miR-21, miR-141, miR-200a, miR-200b, miR-200c, miR-203, miR-205 and miR-214	Increased, Found in EV and correlated with tumor-derived miRNA patterns, Correlated with tumor grade	Diagnosis Prognosis	Taylor and Gercel-Taylor 2008
		miR-21, miR-29a, miR-92, miR-93 and miR-126; miR-99b, miR-127, miR-155	Increased or decreased	Diagnosis	Resnick et al. 2009
Pancreatic cancer (PC)	Serum/plasma	miR-210	Increased	Diagnosis	Ho et al. 2010
		miR-200a and miR-200b	Increased	Diagnosis	Li et al. 2010b
		miR-20a, miR-21, miR-24, miR-25, miR-99a, miR-185, miR-191	Increased, Correlated with tumor grade, Correlated with overall survival, Could discriminate PC group from chronic pancreatitis group in a large cohort	Diagnosis Prognosis	Liu et al. 2011e
		miR-16, miR-196a	Increased, Improved diagnosis in combination with CA19-9 assay	Diagnosis	Liu et al. 2011f
		miR-18a	Increased, Decreased after surgical treatment and increased at recurrence in contrast to CA19-9	Diagnosis	Morimura et al. 2011
		miR-21, let-7 family, miR-146a	Increased or decreased	Diagnosis	Ali et al. 2010
Prostate cancer	Serum/plasma	miR-141	Increased	Diagnosis	Mitchell et al. 2008
		miR-30c, -26b, -451, -223, -24, -874, -1274a, -1207-5p, -93 and -106a	Increased or decreased, Correlated with risk of disease progression	Diagnosis Prognosis	Moltzahn et al. 2011
		miR-375 and miR-141	Increased, Correlated with metastasis	Diagnosis	Brase et al. 2011
Rhabdomyosarcoma	Serum/plasma	miR-206	Increased	Diagnosis	Miyachi et al. 2010

Abbreviations: AL, acute leukemia (comprising myeloid and lymphoblastic); CA19-9, carbohydrate antigen 19-9; CLL, chronic lymphocytic leukemia; ER, estrogen receptor; EV, extracellular vesicles; HBV, hepatitis B virus; HCC, hepatocellular carcinoma; PC, pancreatic cancer; PR, progesterone receptor; ZAP-70, zeta-chain-associated protein kinase 70 variables (nodal and estrogen receptor status). Interestingly, miR-195 has been previously correlated to estrogen receptor status in breast cancer biopsies (Mattie et al. 2006). This group therefore provided miRNAs with potential use in diagnosis, prognosis and surgical monitoring. Wang and colleagues increased the panel by finding miRNAs correlated with the progesterone receptor and the tumor grade (Wang et al. 2010b). Some studies even showed a link between miRNA-profile and the chemotherapy responsiveness. So, Zheng and coworkers demonstrated that the plasma levels of miR-155 and miR-197 were significantly decreased in the patients with lung cancer in the late phase of chemotherapy compared to the early phase (Zheng et al. 2011). The authors also found that the combination of three plasma miRNAs (miR-155, miR-197 and miR-182) yielded about 81% sensitivity and 87% specificity in discriminating lung cancer patients from controls. Similar approaches using blood samples were done on other cancers such as colorectal cancer (Ng et al. 2009), esophageal squamous cell carcinoma (Zhang et al. 2011), gastric cancer (Tsujiura et al. 2010 and Liu et al. 2011a), lung cancer (Hu et al. 2010) or bladder cancer using urine samples in this latter case (Yamada et al. 2011).

3.1 The Need for Specific Cancer Signature

In order to improve the diagnostic approach, several groups found a specific combination of miRNAs to distinguish a specific cancer from other diseases or malignancies. In the case of hepatocellular carcinoma (HCC), Zhou and colleagues constructed a receiver operating characteristic (ROC) curve on a large cohort of nearly 1000 subjects and used the area under the ROC curve (AUC) for diagnostic evaluation (Zhou et al. 2011). The group identified a 7 miRNA-panel that provided a high diagnostic accuracy of HCC and one that could differentiate HCC group from healthy (AUC = 0.941), chronic hepatitis B virus (HBV, AUC = 0.842) and cirrhosis group (AUC = 0.884), respectively. Similarly, Moussay and coworkers differentiated chronic lymphocytic leukemia (CLL) group from other hematologic malignancies and healthy groups (Moussay et al. 2011).

Another way to improve the detection of cancer is to combine existing routine assays with circulating miRNA profiles. In the case of pancreatic cancer, one of the routine diagnostic assays is the detection of the carbohydrate antigen 19-9 (CA19-9), an antigen released from the cell surface of pancreatic tumor cells. Liu and colleagues demonstrated that detection of plasma miR-16 and miR-196a, in combination with CA19-9 assay, improved the diagnosis, especially in early tumor screening (Liu et al. 2011f).

3.2 Extracellular miRNAs Carried or Not by EV?

Most of the current studies focused only on the detection of miRNAs in the body fluid and determined the altered expression and the potential use as biomarker, without distinction about the form of these circulating miRNAs. Nevertheless, some questioned this point. Park and colleagues were not only the first ones to investigate and demonstrate the presence of miRNAs in the saliva. The authors were also the first ones to hypothesize the association of extracellular miRNAs to RISC complex component Ago2 and to provide indirect evidence by performing Ago2-immunoblot analysis on the saliva samples (Park et al. 2009).

On the other hand, several reports demonstrated the presence of circulating EV that carry miRNA and that are related to tumor cells. Taylor and Gercel-Taylor were the first to demonstrate the presence of tumor-derived exosomes in the case of ovarian cancer (Taylor and Gercel-Taylor 2008). The authors showed that the levels of circulating exosomes increased in patients compared to healthy subjects and that the increase was parallel with the tumor grade. This group also found correlation between tumor-derived and circulating exosome-derived miRNA patterns. Rabinowits and colleagues reached similar conclusions for lung cancer (Rabinowits et al. 2009). Working on a wider cohort of patients, Silva and colleagues designed experiments that allowed identifying three miRNAs of potential interest (let-7f, miR-30e-3p and miR-20b) (Silva et al. 2011). The levels of these three miRNAs were lower in plasma vesicles of lung cancer patients. The expression levels of let-7f and miR-30e-3p differentiate two groups of patients for grade of disease. Moreover, miR-30-e-3p and let-7f correlated with disease-free survival and overall survival, respectively.

4. EXTRACELLULAR miRNAs AS THERAPEUTIC TARGETS AND THE EMERGENCE OF COMBINED EV AND RNAi AS NEW THERAPEUTIC TOOLS IN CANCER

The discovery of small non-coding RNA in cells and their crucial regulatory roles was a major breakthrough in the understanding of cell function. It was also at the origin of the short interfering RNA (siRNA) as new therapeutic agents, especially in the case of cancer. But the existence of extracellular miRNAs, in many body fluids, with functional impact on the intercellular communication and carried by EV, brought along tools potentially more powerful; it not only allows identification of new therapeutic targets to fight diseases, but it also provides more appropriate delivery systems.

4.1 siRNAs as Potential Therapeutic Agents

Similarly to miRNA, RNAi implicates sequence-specific gene silencing by Watson-Crick base pairing using small RNA called, in this case, siRNA (for a review see Carthew and Sontheimer 2009). MiRNA and siRNA have common pathways and partners including Dicer enzymes and Ago proteins. Nevertheless, their are two main differences between them. Firstly, miRNAs are endogenous whereas siRNAs are exogenous in origin, derived from transgenes incorporated in the cells. Secondly, the processing appears to occur from stem-loop precursors with incomplete double-stranded design for miRNA whereas it occurs from long, complete complementary double-stranded RNA for siRNA. Thanks to genetic engineering, synthetic siRNAs were thus developed and used to suppress specific gene expression, especially for the genes whose expression is altered in the case of cancer, viral infections or other diseases. Elbashir and colleagues were the first to show the siRNA-mediated gene silencing in mammalian cells (Elbashir et al. 2001). Many other studies followed the work of this group. For example, Scherr and colleagues designed siRNA to target the bcr-abl oncogene, responsible of chronic myeloid leukemia (CML) and bcr-abl-positive acute lymphoblastic leukemia (ALL) (Scherr et al. 2003). The authors found a reduction of bcr-abl mRNA up to 87% in bcr-abl-positive cell lines and in primary cells from CML patients. The reduction of mRNA was specific and affected the protein expression level of the encoded proteins as well. Moreover, by targeting the same bcr-abl mRNA, Wohlbold and coworkers increased the sensitivity to the selective tyrosine kinase inhibitor imatinib in leukemic cells expressing the imatinib-resistant form of Bcr-Abl protein (Wohlbold et al. 2003). Nieth and colleagues also succeeded to reverse a resistance-phenotype in tumor cells (Nieth et al. 2003). This group worked on the 'classical' multi-drug resistance (MDR) that is mediated by the adenosine triphosphate binding cassette (ABC)-transporter P-glycoprotein (MDR1/P-gp). The authors used siRNA duplexes to test the resistance to daunorubicin in both human pancreatic and gastric carcinoma cells. The specific inhibition decreased the resistance up to 89% and 58% in pancreatic and gastric cell lines, respectively. The *in vivo* experiment was then performed using MDR1/P-gp-xenograft mice and showed complete reversal of the MDR phenotype (Stein et al. 2008). The authors used the technology of jet-injection delivery to incorporate non-viral vectors expressing the short hairpin RNA (shRNA), a precursor of siRNA, directly into the tumor.

Despite the promising results of siRNA and the recent advances in delivery systems, one of the major issues remain that is the existence of clinically suitable, safe, effective and specific delivery vehicles.

4.2 EV as Potential Delivery Systems

As stated previously, EV are natural nuclease-resistant delivery carriers implicated in intercellular communication. Among these, exosomes are the best characterized to date. They are complex nano-sized particles composed of proteins, lipids, carbohydrates and nucleic acids (Théry et al. 2009). Exosomes can contain proteins involved in various cellular processes: T-cell stimulation (MHC class I and II), adhesion (integrins, tetraspanins), signalling pathways (syntenin, Gα), membrane transport and trafficking (ATPase channels, Rho GDI, annexins), cytoskeleton (actin, myosin, tubulin), MVB formation (Alix), chaperones (heat shock proteins), metabolism (phosphoglycerate kinase 1, α-enolase, ADP ribosylation factor) or transcription (histones). The lipid composition is usually rich in cholesterol, sphingomyelin and ceramide, which suggests, in correlation with the proteins identified in exosomes (flotillin 1), the existence of lipid rafts. Lipid rafts are microdomains in the lipid bilayer cellular membrane that are detergent-resistant and work as signalling and sorting platforms (Staubach and Hanish 2011). An on-line database named Exocarta was created to pool all the research findings about exosomes (www.exocarta.org).

Although physiopathological role of exosomes requires more investigation to improve our knowledge and comprehension, evidences point out the intimate relation between their functions and their cellular origins, origins that implicate specific exosome composition. Exosomes have been demonstrated to be secreted by many cells including reticulocytes (first description of exosomes) (Pan and Johnstone 1983), B cells (Raposo et al. 1996), T cells (Blanchard et al. 2002), mast cells (Skodos et al. 2001), platelets (Heijnen et al. 1999), intestinal epithelial cells (Van Niel et al. 2003), dendritic cells (Pêche et al. 2003) and tumor cells (Taylor and Gercel-Taylor 2008). The adhesion molecules on the exosome surface play an important role in the cell targeting. For example, the exosomes produced by the B cells present MHC class II that stimulate CD4+ T cells *in vitro* (Raposo et al. 1996). Exosomes have also been demonstrated to inhibit or promote the immune responses. The observed inhibition was correlated with tumor-derived exosomes and included induction of T-cell apoptosis (Huber et al. 2005) or reduction of the cytotoxicity of natural killer cells (Ashiru et al. 2010). Some pro-immune activities were also identified for tumor-derived exosomes but mainly restricted to stress-induced conditions (Gastpar et al. 2005). The promotion of immune responses was also observed in macrophages. When infected by various pathogens (*Mycobacterium* or *Toxoplasma*), these cells release exosomes containing pathogen-derived inflammatory molecules that induce the secretion of pro-inflammatory cytokines by the recipient macrophages (Bhatnagar and Schorey 2007).

Exosomes thus have a strong potential as suitable multi-functional vesicles that can carry and transfer safely and efficiently materials able to affect other cells.

4.3 Building on the Natural System: A Combined Approach of EV and RNAi as Therapeutic Tools

Skog and colleagues were among the first to suggest the use of EV as carriers to deliver nucleic acids. This group found that glioblastoma derived-EV contain mRNA, miRNA and angiogenic proteins (Skog et al. 2008). The authors reported that glioblastoma derived-EV could enter human brain microvascular endothelial cells (HBMVEC) in *in vitro* system and translate a reporter mRNA carried by these EV. Skog and coworkers also demonstrated that the tubule length of HBMVEC doubled within 16 h in presence of the EV, supporting a role of the latter in initiating angiogenesis in brain endothelial cells. The authors therefore concluded that tumor vesicles act as multicomponent delivery vehicle for mRNA, miRNA and proteins, to communicate genetic information and signalling proteins to the neighboring cells. On the other hand, Yang and colleagues showed that macrophages regulate the invasiveness of breast cancer cells through EV-mediated delivery of oncogenic miRNA (Yang et al. 2011). The authors used a transwell co-culture system of interleukin 4-activated macrophages and breast cancer cells and they tracked the signal of fluorescently-labeled exogenous miRNA originating from macrophages. This group were therefore able to verify the transport of miRNA without direct cell–cell contact from the macrophages to the breast cancer cells. In the EV, Yang and coworkers detected the presence of miR-223, a miRNA specific of the interleukin 4-activated macrophages that has been previously reported to be implicated in the progression of renal (Gottardo et al. 2007) and HCC cancers (Xu et al. 2011). By treating macrophages with a miR-223 anti-sense oligonucleotide, the authors could reduce the observed effects.

The group of Alvarez-Erviti was the first to develop and test *in vivo* delivery of siRNA to the mouse brain by systemic injection of targeted exosomes (Alvarez-Erviti et al. 2011). The authors initially selected dendritic cells with specific characteristics from the bone marrow of mice, cultured them and purified their exosomes. In order to target the purified EV to the desired tissues (muscle and brain), targeting peptides were fused to exosomal membrane protein. The group subsequently proceeded to the incorporation of specific siRNA into the modified vesicles before re-administrating them intravenously to the mice. In contrast with the mice to which naked siRNA was injected, exosome-encapsulated siRNA were resistant to nonspecific uptake by spleen, liver or kidney. Moreover,

Alvarez-Erviti and colleagues found the silencing of the targeted mRNA in several brain regions, demonstrating the efficiency of targeted exosomes. The authors also showed that multiple injections of these 'self' modified particles did not affect the delivery efficacy. Altogether, the conditions required of delivery systems for clinical applications were met. Finally, this group demonstrated for the first time the possibility to deliver therapeutic agents *in vivo* across the blood-brain barrier. This point is of great importance to address the actual issues in the treatment of neuronal diseases.

5. CONCLUSIONS

This review focused on the extracellular miRNAs, from their discovery to their potential clinical applications in cancer. Many studies highlighted the usefulness of these miRNAs as biomarkers in three different ways (Table 1): (i) differentiate normal from diseased states, (ii) help to make a prognosis by differentiating the tumor grade, and (iii) monitor the response to therapy. Nevertheless, the published data is still inconsistent, which is probably due to several reasons. Firstly, different detection and normalization methods (i.e. microarray *vs* single-gene PCR) have been used. Secondly, the origin of the tumor samples could vary (i.e. breast cancer, which can originate from different cells). Thirdly, great variability in the size of the cohort has been observed. Standardization is therefore required to obtain reliable non-invasive body fluid-based routine detection tests. Another improvement to be made is on the specificity. Indeed, some miRNAs such as miR-21 have been identified in numerous cancers. Specific miRNA-panel signature and/ or combination with already existing tests should thus be found to allow the correlation with specific cancer.

The discovery of intracellular and extracellular miRNAs also led to the development of RNAi strategies. However, these strategies need a safe and efficient carrier in order to deliver the therapeutic oligonucleotides into the targeted tissues or cells. Several small size cargo-loading vehicles such as liposomes, polymers or virus-like particles have already been designed for *in vivo* targeting but they still encounter some issues (de Wolf et al. 2007, Azuma et al. 2010, Guo et al. 2009). The EV are naturally occurring vesicles that are very attractive tools in clinical use because they retain the advantages of the existing nanoscale drug delivery systems while adding *in vivo* safety and low immunogenicity. The first proof-of-concept was provided recently by Alvarez-Erviti and colleagues (see above). The optimization of EV as a drug delivery system requires a better understanding of their biology (production, functions, targeting mechanisms, etc.). However due to their ability to cross impermeable biological barriers such as the blood-brain barrier and the possibility to use patient-derived cells as a source of tailored biocompatible therapeutic carriers, EV are seen to have enormous potential in the therapeutic research field. The next decade should see the start of human clinical applications of extracellular miRNAs as diagnostic, prognostic and therapeutic tools.

6. SUMMARY POINTS

- Altered expression levels of miRNA have been correlated with cancer. Recently, miRNAs have also been detected in many body fluids including plasma, serum, and urine.
- Due to their stability and their expression levels that differentiate diseased patients from healthy subjects, these circulating miRNAs have revealed great potential as novel diagnostic and prognostic markers.
- A substantial amount of extracellular miRNAs is carried by secreted extracellular vesicles and participates to intercellular genetic material exchange.
- Using EV as vehicle of drug delivery and RNAi to counteract the altered expression levels of targeted miRNA creates a new combined strategy that holds promise to overcome impediments in the field of therapeutics for cancer and other diseases.

ACKNOWLEDGMENTS

This work was supported in part by a grant-in-aid for the Third-Term Comprehensive 10-Year Strategy for Cancer Control of Japan; Project for Development of Innovative Research on Cancer Therapeutics (P-Direct); Scientific Research on Priority Areas Cancer from the Japanese Ministry of Education, Culture, Sports, Science, and Technology; and the Program for Promotion of Fundamental Studies in Health Sciences of the National Institute of Biomedical Innovation of Japan. The authors would like to thank Servier Medical Art for their image bank used to create the illustrations.

ABBREVIATIONS

ABC	:	ATP binding cassette
ADP	:	adenosine diphosphate
Ago	:	argonaute protein
ALL	:	acute lymphoblastic leukemia
ATP	:	adenosine triphosphate
AUC	:	area under the ROC curve
CA19-9	:	carbohydrate antigen 19-9
CD4	:	cluster of differentiation 4

326 MicroRNAs in Cancer

CLL	:	chronic lymphocytic leukemia
CML	:	chronic myeloid leukemia
DLBCL	:	diffuse large B-cell lymphoma
EV	:	extracellular vesicles
HBMVEC	:	human brain microvascular endothelial cells
HBV	:	hepatitis B Virus
HCC	:	hepatocellular carcinoma
HDL	:	high-density lipoprotein
LDL	:	low-density lipoprotein
MDR	:	multi-drug resistance
MDR1/P-gp	:	ABC-transporter P-glycoprotein
MHC	:	major histocompatibility complex
miRNA	:	microRNA
MVB	:	multivesicular bodies
NPM1	:	nucleophosmin 1
nSMase2	:	neutral sphingomyelinase-2
RhoGDI	:	Rho guanosine 5'-diphosphate-dissociation inhibitor
RNAi	:	RNA interference
ROC	:	receiver operating characteristic
SR-BI	:	scavenger receptor class B type I
shRNA	:	short hairpin RNA
siRNA	:	short interfering RNA

REFERENCES

Ali, S., K. Almhanna, W. Chen, P.A. Philip, and F.H. Sarkar. 2010. Differentially expressed miRNAs in the plasma may provide a molecular signature for aggressive pancreatic cancer. Am J Transl Res. 3(1): 28–47.

Alvarez-Erviti, L., Y. Seow, H. Yin, C. Betts, S. Lakhal, and M.J. Wood. 2011. Delivery of siRNA to the mouse brain by systemic injection of targeted exosomes. Nat Biotechnol. 29(4): 341–345.

Arroyo, J.D., J.R. Chevillet, E.M. Kroh, I.K. Ruf, C.C. Pritchard, D.F. Gibson, P.S. Mitchell, C.F. Bennett, E.L. Pogosova-Agadjanyan, D.L. Stirewalt, J.F. Tait, and M. Tewari. 2011. Argonaute2 complexes carry a population of circulating microRNAs independent of vesicles in human plasma. Proc Natl Acad Sci USA. 108(12): 5003–5008.

Ashiru, O., P. Boutet, L. Fernández-Messina, S. Agüera-González, J.N. Skepper, M. Valés-Gómez, and H.T. Reyburn. 2010. Natural killer cell cytotoxicity is suppressed by exposure to the human NKG2D ligand MICA*008 that is shed by tumor cells in exosomes. Cancer Res. 70(2): 481–489.

Azuma, K., K. Nakashiro, T. Sasaki, H. Goda, J. Onodera, N. Tanji, M. Yokoyama, and H. Hamakawa. 2010. Anti-tumor effect of small interfering RNA targeting the androgen receptor in human androgen-independent prostate cancer cells. Biochem Biophys Res Commun. 391(1): 1075–1079.

Bartel, D.P. 2004. MicroRNAs: genomics, biogenesis, mechanism, and function. Cell. 116: 281–297.

Bhatnagar, S. and J.S. Schorey. 2007. Exosomes released from infected macrophages contain Mycobacterium avium glycopeptidolipids and are proinflammatory. J Biol Chem. 282(35): 25779–25789.

Blanchard, N., D. Lankar, F. Faure, A. Regnault, C. Dumont, G. Raposo, and C. Hivroz. 2002. TCR activation of human T cells induces the production of exosomes bearing the TCR/ CD3/zeta complex. J Immunol. 168(7): 3235–3241.

Brase, J.C., M. Johannes, T. Schlomm, M. Fälth, A. Haese, T. Steuber, T. Beissbarth, R. Kuner, and H. Sültmann. 2011. Circulating miRNAs are correlated with tumor progression in prostate cancer. Int J Cancer. 128(3): 608–616.

Carthew, R.W. and E.J. Sontheimer. 2009. Origins and Mechanisms of miRNAs and siRNAs. Cell. 136(4): 642–655.

Chen, X., Y. Ba, L. Ma, X. Cai, Y. Yin, K. Wang, J. Guo, Y. Zhang, J. Chen, X. Guo, Q. Li, X. Li, W. Wang, Y. Zhang, J. Wang, X. Jiang, Y. Xiang, C. Xu, P. Zheng, J. Zhang, R. Li, H. Zhang, X. Shang, T. Gong, G. Ning, J. Wang, K. Zen, J. Zhang, and C.Y. Zhang. 2008. Characterization of microRNAs in serum: a novel class of biomarkers for diagnosis of cancer and other diseases. Cell Res. 18: 997–1006.

Chen, X., Z. Hu, W. Wang, Y. Ba, L. Ma, C. Zhang, C. Wang, Z. Ren, Y. Zhao, S. Wu, R. Zhuang, Y. Zhang, H. Hu, C. Liu, L. Xu, J. Wang, H. Shen, J. Zhang, K. Zen, and C.Y. Zhang. 2011. Identification of ten serum microRNAs from a genome-wide serum microRNA expression profile as novel noninvasive biomarkers for non-small cell lung cancer diagnosis. Int J Cancer. doi: 10.1002/ijc.26177.

Cheng, H., L. Zhang, D.E. Cogdell, H. Zheng, A.J. Schetter, M. Nykter, C.C. Harris, K. Chen, S.R. Hamilton, and W. Zhang. 2011. Circulating plasma MiR-141 is a novel biomarker for metastatic colon cancer and predicts poor prognosis. PLoS One. 6(3): e17745.

de Wolf, H.K., C.J. Snel, F.J. Verbaan, R.M. Schiffelers, W.E. Hennink, and G. Storm. 2007. Effect of cationic carriers on the pharmacokinetics and tumor localization of nucleic acids after intravenous administration. Int J Pharm. 331(2): 167–175.

Elbashir, S.M., J. Harborth, W. Lendeckel, A. Yalcin, K. Weber, and T. Tuschl. 2001. Duplexes of 21-nucleotide RNAs mediate RNA interference in cultured mammalian cells. Nature. 411(6836): 494–498.

El-Hefnawy, T., S. Raja, L. Kelly, W.L. Bigbee, J.M. Kirkwood, J.D. Luketich, and T.E. Godfrey. 2004. Characterization of amplifiable, circulating RNA in plasma and its potential as a tool for cancer diagnostics. Clin Chem. 50: 564–573.

Foss, K.M., C. Sima, D. Ugolini, M. Neri, K.E. Allen, and G.J. Weiss. 2011. miR-1254 and miR-574-5p: serum-based microRNA biomarkers for early-stage non-small cell lung cancer. J Thorac Oncol. 6(3): 482–488.

Gastpar, R., M. Gehrmann, M.A. Bausero, A. Asea, C. Gross, J.A. Schroeder, and G. Multhoff. 2005. Heat shock protein 70 surface-positive tumor exosomes stimulate migratory and cytolytic activity of natural killer cells. Cancer Res. 65(12): 5238–5247.

Gottardo, F., C.G. Liu, M. Ferracin, G.A. Calin, M. Fassan, P. Bassi, C. Sevignani, D. Byrne, M. Negrini, F. Pagano, L.G. Gomella, C.M. Croce, and R. Baffa. 2007. Micro-RNA profiling in kidney and bladder cancers. Urol Oncol. 25(5): 387–392.

Guo, B., Y. Zhang, G. Luo, L. Li, and J. Zhang. 2009. Lentivirus-mediated small interfering RNA targeting VEGF-C inhibited tumor lymphangiogenesis and growth in breast carcinoma. Anat Rec (Hoboken). 292(5): 633–639.

Hanke, M., K. Hoefig, H. Merz, A.C. Feller, I. Kausch, D. Jocham, J.M. Warnecke, and G. Sczakiel. 2010. A robust methodology to study urine microRNA as tumor marker: microRNA-126 and microRNA-182 are related to urinary bladder cancer. Urol Oncol. 28(6): 655–661.

Heijnen, H.F., A.E. Schiel, R. Fijnheer, H.J. Geuze, and J.J. Sixma. 1999. Activated platelets release two types of membrane vesicles: microvesicles by surface shedding and exosomes derived from exocytosis of multivesicular bodies and alpha-granules. Blood. 94(11): 3791–3799.

Heneghan, H.M., N. Miller, A.J. Lowery, K.J. Sweeney, J. Newell, and M.J. Kerin. 2010. Circulating microRNAs as novel minimally invasive biomarkers for breast cancer. Ann Surg. 251(3): 499–505.

Ho, A.S., X. Huang, H. Cao, C. Christman-Skieller, K. Bennewith, Q.T. Le, and A.C. Koong. 2010. Circulating miR-210 as a Novel Hypoxia Marker in Pancreatic Cancer. Transl Oncol. 3(2): 109–113.

Hu, Z., X. Chen, Y. Zhao, T. Tian, G. Jin, Y. Shu, Y. Chen, L. Xu, K. Zen, C. Zhang, and H. Shen. 2010. Serum microRNA signatures identified in a genome-wide serum microRNA expression profiling predict survival of non-small-cell lung cancer. J Clin Oncol. 28(10): 1721–1726.

Huang, Z., D. Huang, S. Ni, Z. Peng, W. Sheng, and X. Du. 2010. Plasma microRNAs are promising novel biomarkers for early detection of colorectal cancer. Int J Cancer. 127(1): 118–126.

Huber, V., S. Fais, M. Iero, L. Lugini, P. Canese, P. Squarcina, A. Zaccheddu, M. Colone, G. Arancia, M. Gentile, E. Seregni, R. Valenti, G. Ballabio, F. Belli, E. Leo, G. Parmiani, and L. Rivoltini. 2005. Human colorectal cancer cells induce T-cell death through release of proapoptotic microvesicles: role in immune escape. Gastroenterology. 128(7): 1796–1804.

Hunter, M.P., N. Ismail, X. Zhang, B.D. Aguda, E.J. Lee, L. Yu, T. Xiao, J. Schafer, M.L. Lee, T.D. Schmittgen, S.P. Nana-Sinkam, D. Jarjoura, and C.B. Marsh. 2008. Detection of microRNA expression in human peripheral blood microvesicles. PLoS One. 3(11): e3694.

Kanemaru, H., S. Fukushima, J. Yamashita, N. Honda, R. Oyama, A. Kakimoto, S. Masuguchi, T. Ishihara, Y. Inoue, M. Jinnin, and H. Ihn. 2011. The circulating microRNA-221 level in patients with malignant melanoma as a new tumor marker. J Dermatol Sci. 61(3): 187–193.

Kogure, T., W.L. Lin, I.K. Yan, C. Braconi, and T. Patel. 2011. Intercellular nanovesicle-mediated microRNA transfer: a mechanism of environmental modulation of hepatocellular cancer cell growth. Hepatology. 54(4): 1237–1248.

Kosaka, N., H. Iguchi, Y. Yoshioka, F. Takeshita, Y. Matsuki, and T. Ochiya. 2010. Secretory mechanisms and intercellular transfer of microRNAs in living cells. J Biol Chem. 285: 17442–17452.

Lawrie, C.H., S. Gal, H.M. Dunlop, B. Pushkaran, A.P. Liggins, K. Pulford, A.H. Banham, F. Pezzella, J. Boultwood, J.S. Wainscoat, C.S. Hatton, and A.L. Harris. 2008. Detection of elevated levels of tumour-associated microRNAs in serum of patients with diffuse large B-cell lymphoma. Br J Haematol. 141: 672–675.

Leidinger, P., A. Keller, A. Borries, J. Reichrath, K. Rass, S.U. Jager, H.P. Lenhof, and E. Meese. 2010. High-throughput miRNA profiling of human melanoma blood samples. BMC Cancer. 10: 262.

Li, A., N. Omura, S.M. Hong, A. Vincent, K. Walter, M. Griffith, M. Borges, and M. Goggins. 2010b. Pancreatic cancers epigenetically silence SIP1 and hypomethylate and overexpress miR-200a/200b in association with elevated circulating miR-200a and miR-200b levels. Cancer Res. 70(13): 5226–5237.

Li, L.M., Z.B. Hu, Z.X. Zhou, X. Chen, F.Y. Liu, J.F. Zhang, H.B. Shen, C.Y. Zhang, and K. Zen. 2010a. Serum microRNA profiles serve as novel biomarkers for HBV infection and diagnosis of HBV-positive hepatocarcinoma. Cancer Res. 70(23): 9798–9807.

Lin, Q., W. Mao, Y. Shu, F. Lin, S. Liu, H. Shen, W. Gao, S. Li, and D. Shen. 2012. A cluster of specified microRNAs in peripheral blood as biomarkers for metastatic non-small-cell lung cancer by stem-loop RT-PCR. J Cancer Res Clin Oncol. 138(1): 85–93.

Liu, C.J., S.Y. Kao, H.F. Tu, M.M. Tsai, K.W. Chang, and S.C. Lin. 2010d. Increase of microRNA miR-31 level in plasma could be a potential marker of oral cancer. Oral Dis. 16(4): 360–364.

Liu, H., L. Zhu, B. Liu, L. Yang, X. Meng, W. Zhang, Y. Ma, and H. Xiao. 2011b. Genome-wide microRNA profiles identify miR-378 as a serum biomarker for early detection of gastric cancer. Cancer Lett. 316(2): 196–203.

Liu, J., J. Gao, Y. Du, Z. Li, Y. Ren, J. Gu, X. Wang, Y. Gong, W. Wang, and X. Kong. 2011f. Combination of plasma microRNAs with serum CA19-9 for early detection of pancreatic cancer. Int J Cancer. doi: 10.1002/ijc.26422.

Liu, R., C. Zhang, Z. Hu, G. Li, C. Wang, C. Yang, D. Huang, X. Chen, H. Zhang, R. Zhuang, T. Deng, H. Liu, J. Yin, S. Wang, K. Zen, Y. Ba, and C.Y. Zhang. 2011a. A five-microRNA signature identified from genome-wide serum microRNA expression profiling serves as a fingerprint for gastric cancer diagnosis. Eur J Cancer. 47(5): 784–791.

Liu, R., X. Chen, Y. Du, W. Yao, L. Shen, C. Wang, Z. Hu, R. Zhuang, G. Ning, C. Zhang, Y. Yuan, Z. Li, K. Zen, Y. Ba, and C.Y. Zhang. 2011e. Serum microRNA expression profile as a biomarker in the diagnosis and prognosis of pancreatic cancer. Clin Chem. doi:10.1373/clinchem.2011.172767.

Liu, X.G., W.Y. Zhu, Y.Y. Huang, L.N. Ma, S.Q. Zhou, Y.K. Wang, F. Zeng, J.H. Zhou, and Y.K. Zhang. 2011c. High expression of serum miR-21 and tumor miR-200c associated with poor prognosis in patients with lung cancer. Med Oncol. doi: 10.1007/s12032-011-9923-y.

Mandel, P. and P. Métais. 1948. Les acides nucléiques du plasma sanguin chez l'homme. CR Acad Sci Paris. 142: 241–243.

Mattie, M.D., C.C. Benz, J. Bowers, K. Sensinger, L. Wong, G.K. Scott, V. Fedele, D. Ginzinger, R. Getts, and C. Haqq. 2006. Optimized high-throughput microRNA expression profiling provides novel biomarker assessment of clinical prostate and breast cancer biopsies. Mol Cancer. 5: 24.

Mitchell, P.S., R.K. Parkin, E.M. Kroh, B.R. Fritz, S.K. Wyman, E.L. Pogosova-Agadjanyan, A. Peterson, J. Noteboom, K.C. O'Briant, A. Allen, D.W. Lin, N. Urban, C.W. Drescher, B.S. Knudsen, D.L. Stirewalt, R. Gentleman, R.L. Vessella, P.S. Nelson, D.B. Martin, and M. Tewari. 2008. Circulating microRNAs as stable blood-based markers for cancer detection. Proc Natl Acad Sci USA. 105: 10513–10518.

Mittelbrunn, M., C. Gutiérrez-Vázquez, C. Villarroya-Beltri, S. González, F. Sánchez-Cabo, M.A. González, A. Bernad, and F. Sánchez-Madrid. 2011. Unidirectional transfer of microRNA-loaded exosomes from T cells to antigen-presenting cells. Nat Commun. 2: 282.

Miyachi, M., K. Tsuchiya, H. Yoshida, S. Yagyu, K. Kikuchi, A. Misawa, T. Iehara, and H. Hosoi. 2010. Circulating muscle-specific microRNA, miR-206, as a potential diagnostic marker for rhabdomyosarcoma. Biochem Biophys Res Commun. 400(1): 89–93.

Moltzahn, F., A.B. Olshen, L. Baehner, A. Peek, L. Fong, H. Stöppler, J. Simko, J.F. Hilton, P. Carroll, and R. Blelloch. 2011. Microfluidic-based multiplex qRT-PCR identifies diagnostic and prognostic microRNA signatures in the sera of prostate cancer patients. Cancer Res. 71(2): 550–560.

Morimura, R., S. Komatsu, D. Ichikawa, H. Takeshita, M. Tsujiura, H. Nagata, H. Konishi, A. Shiozaki, H. Ikoma, K. Okamoto, T. Ochiai, H. Taniguchi, and E. Otsuji. 2011. Novel diagnostic value of circulating miR-18a in plasma of patients with pancreatic cancer. Br J Cancer. 105(11): 1733–1740.

Moussay, E., K.Wang, J.H. Cho, K. van Moer, S. Pierson, J. Paggetti, P.V. Nazarov, V. Palissot, L.E. Hood, G. Berchem, and D.J. Galas. 2011. MicroRNA as biomarkers and regulators in B-cell chronic lymphocytic leukemia. Proc Natl Acad Sci USA. 2108(16): 6573–6578.

Ng, E.K., W.W. Chong, H. Jin, E.K. Lam, V.Y. Shin, J. Yu, T.C. Poon, S.S. Ng, and J.J. Sung. 2009. Differential expression of microRNAs in plasma of patients with colorectal cancer: a potential marker for colorectal cancer screening. Gut. 58(10): 1375–1381.

Nieth, C., A. Priebsch, A. Stege, and H. Lage. 2003. Modulation of the classical multidrug resistance (MDR) phenotype by RNA interference (RNAi). FEBS Lett. 545(2-3): 144–150.

Pan, B.T. and R.M. Johnstone. 1983. Fate of the transferrin receptor during maturation of sheep reticulocytes *in vitro*: selective externalization of the receptor. Cell. 33(3): 967–978.

Park, N.J., H. Zhou, D. Elashoff, B.S. Henson, D.A. Kastratovic, E. Abemayor, and D.T. Wong. 2009. Salivary microRNA: discovery, characterization, and clinical utility for oral cancer detection. Clin Cancer Res. 15(17): 5473–5477.

Pêche, H., M. Heslan, C. Usal, S. Amigorena, and M.C. Cuturi. 2003. Presentation of donor major histocompatibility complex antigens by bone marrow dendritic cell-derived exosomes modulates allograft rejection. Transplantation. 76(10): 1503–1510.

Rabinowits, G., C. Gerçel-Taylor, J.M. Day, D.D. Taylor, and G.H. Kloecker. 2009. Exosomal microRNA: a diagnostic marker for lung cancer. Clin Lung Cancer. 10(1): 42–46.

Raposo, G., H.W. Nijman, W. Stoorvogel, R. Liejendekker, C.V. Harding, C.J. Melief, and H.J. Geuze. 1996. B lymphocytes secrete antigen-presenting vesicles. J Exp Med. 183(3): 1161–1172.

Resnick, K.E., H. Alder, J.P. Hagan, D.L. Richardson, C.M. Croce, and D.E. Cohn. 2009. The detection of differentially expressed microRNAs from the serum of ovarian cancer patients using a novel real-time PCR platform. Gynecol Oncol. 112(1): 55–59.

Roth, C., S. Kasimir-Bauer, K. Pantel, and H. Schwarzenbach. 2011. Screening for circulating nucleic acids and caspase activity in the peripheral blood as potential diagnostic tools in lung cancer. Mol Oncol. 5(3): 281–291.

Scherr, M., K. Battmer, T. Winkler, O. Heidenreich, A. Ganser, and M. Eder. 2003. Specific inhibition of bcr-abl gene expression by small interfering RNA. Blood. 101(4): 1566–1569.

Shen, J., Z. Liu, N.W. Todd, H. Zhang, J. Liao, L. Yu, M.A. Guarnera, R. Li, L. Cai, M. Zhan, and F. Jiang. 2011. Diagnosis of lung cancer in individuals with solitary pulmonary nodules by plasma microRNA biomarkers. BMC Cancer. 11: 374.

Shigoka, M., A. Tsuchida, T. Matsudo, Y. Nagakawa, H. Saito, Y. Suzuki, T. Aoki, Y. Murakami, H. Toyoda, T. Kumada, R. Bartenschlager, N. Kato, M. Ikeda, T. Takashina, M. Tanaka, R. Suzuki, K. Oikawa, M. Takanashi, and M. Kuroda. 2010. Deregulation of miR-92a expression is implicated in hepatocellular carcinoma development. Pathol Int. 60(5): 351–357.

Shinozaki, M., S.J. O'Day, M. Kitago, F. Amersi, C. Kuo, J. Kim, H.J. Wang, and D.S. Hoon. 2007. Utility of circulating B-RAF DNA mutation in serum for monitoring melanoma patients receiving biochemotherapy. Clin Cancer Res. 13: 2068–2074.

Silva, J., V. García, A. Zaballos, M. Provencio, L. Lombardía, L. Almonacid, J.M. García, G. Domínguez, C. Peña, R. Diaz, M. Herrera, A. Varela, and F. Bonilla. 2011. Vesicle-related microRNAs in plasma of nonsmall cell lung cancer patients and correlation with survival. Eur Respir J. 37(3): 617–623.

Skog, J., T. Würdinger, S. van Rijn, D.H. Meijer, L. Gainche, M. Sena-Esteves, W.T. Jr. Curry, B.S. Carter, A.M. Krichevsky, and X.O. Breakefield. 2008. Glioblastoma microvesicles transport RNA and proteins that promote tumour growth and provide diagnostic biomarkers. Nat Cell Biol. 10(12): 1470–1476.

Skokos, D., S. Le Panse, I. Villa, J.C. Rousselle, R. Peronet, B. David, A. Namane, and S. Mécheri. 2001. Mast cell-dependent B and T lymphocyte activation is mediated by the secretion of immunologically active exosomes. J Immunol. 166(2): 868–876.

Sorenson, G.D., D.M. Pribish, F.H. Valone, V.A. Memoli, D.J. Bzik, and S.L. Yao. 1994. Soluble normal and mutated DNA sequences from single copy genes in human blood. Cancer Epidemiol Biomarkers Prev. 3: 67–71.

Staubach, S. and F.G. Hanisch. 2011. Lipid rafts: signaling and sorting platforms of cells and their roles in cancer. Expert Rev Proteomics. 8(2): 263–277.

Stein, U., W. Walther, A. Stege, A. Kaszubiak, I. Fichtner, and H. Lage. 2008. Complete *in vivo* reversal of the multidrug resistance phenotype by jet-injection of anti-MDR1 short hairpin RNA-encoding plasmid DNA. Mol Ther. 16(1):178–186.

Tan, E.M., P.H. Schur, R.I. Carr, and H.G. Kunkel. 1966. Deoxyribonucleic acid (DNA) and antibodies to DNA in the serum of patients with systemic lupus erythematosus. J Clin Invest. 45: 1732–1740.

Tanaka, M., K. Oikawa, M. Takanashi, M. Kudo, J. Ohyashiki, K. Ohyashiki, and M. Kuroda. 2009. Down-regulation of miR-92 in human plasma is a novel marker for acute leukemia patients. PLoS One. 4(5): e5532.

Taylor, D.D. and C. Gercel-Taylor. 2008. MicroRNA signatures of tumor-derived exosomes as diagnostic biomarkers of ovarian cancer. Gynecol Oncol. 110: 13–21.

Théry, C., M. Ostrowski, and E. Segura. 2009. Membrane vesicles as conveyors of immune responses. Nat Rev Immunol. 9(8): 581–593.

Tsujiura, M., D. Ichikawa, S. Komatsu, A. Shiozaki, H. Takeshita, T. Kosuga, H. Konishi, R. Morimura, K. Deguchi, H. Fujiwara, K. Okamoto, and E. Otsuji. 2010. Circulating microRNAs in plasma of patients with gastric cancers. Br J Cancer. 102(7): 1174–1179.

Valadi, H., K. Ekstrom, A. Bossios, M. Sjostrand, J.J. Lee, and J.O. Lotvall. 2007. Exosome mediated transfer of mRNAs and microRNAs is a novel mechanism of genetic exchange between cells. Nat Cell Biol. 9: 654–659.

Van Niel, G., J. Mallegol, C. Bevilacqua, C. Candalh, S. Brugière, E. Tomaskovic-Crook, J.K. Heath, N. Cerf-Bensussan, and M. Heyman. 2003. Intestinal epithelial exosomes carry MHC class II/peptides able to inform the immune system in mice. Gut. 52(12): 1690–1697.

Vasioukhin, V., P. Anker, P. Maurice, J. Lyautey, C. Lederrey, and M. Stroun. 1994. Point mutations of the N-ras gene in the blood plasma DNA of patients with myelodysplastic syndrome or acute myelogenous leukaemia. Br J Haematol. 86: 774–779.

Vickers, K.C., B.T. Palmisano, B.M. Shoucri, R.D. Shamburek, and A.T. Remaley. 2011. MicroRNAs are transported in plasma and delivered to recipient cells by high-density lipoproteins. Nat Cell Biol. 13(4): 423–433.

Wang, F., Z. Zheng, J. Guo, and X. Ding. 2010b. Correlation and quantitation of microRNA aberrant expression in tissues and sera from patients with breast tumor. Gynecol Oncol. 119(3): 586–593.

Wang, K., S. Zhang, J. Weber, D. Baxter, and D.J. Galas. 2010a. Export of microRNAs and microRNA-protective protein by mammalian cells. Nucleic Acids Res. 38(20): 7248–7259.

Wang, Z.X., H.B. Bian, J.R. Wang, Z.X. Cheng, K.M. Wang, and W. De. 2011. Prognostic significance of serum miRNA-21 expression in human non-small cell lung cancer. J Surg Oncol. 104(7): 847–851.

Wei, J., W. Gao, C.J. Zhu, Y.Q. Liu, Z. Mei, T. Cheng, and Y.Q. Shu. 2011. Identification of plasma microRNA-21 as a biomarker for early detection and chemosensitivity of non-small cell lung cancer. Chin J Cancer. 30(6): 407–414.

Wiklund, E.D., S. Gao, T. Hulf, T. Sibbritt, S. Nair, D.E. Costea, S.B. Villadsen, V. Bakholdt, J.B. Bramsen, J.A. Sørensen, A. Krogdahl, S.J. Clark, and J. Kjems. 2011. MicroRNA alterations and associated aberrant DNA methylation patterns across multiple sample types in oral squamous cell carcinoma. PLoS One. 6(11): e27840.

Wohlbold, L., H. van der Kuip, C. Miething, H.P. Vornlocher, C. Knabbe, J. Duyster, and W.E. Aulitzky. 2003. Inhibition of bcr-abl gene expression by small interfering RNA sensitizes for imatinib mesylate (STI571). Blood. 102(6): 2236–2239.

Wong, T.S., X.B. Liu, B.Y. Wong, R.W. Ng, A.P. Yuen, and W.I. Wei. 2008. Mature miR-184 as potential oncogenic microRNA of squamous cell carcinoma of tongue. Clin Cancer Res. 14(9): 2588–2592.

Xie, L., X. Chen, L. Wang, X. Qian, T. Wang, J. Wei, L. Yu, Y. Ding, C. Zhang, and B. Liu. 2010. Cell-free miRNAs may indicate diagnosis and docetaxel sensitivity of tumor cells in malignant effusions. BMC Cancer. 10: 591.

Xu, J., C. Wu, X. Che, L. Wang, D. Yu, T. Zhang, L. Huang, H. Li, W. Tan, C. Wang, and D. Lin. 2011. Circulating microRNAs, miR-21, miR-122, and miR-223, in patients with hepatocellular carcinoma or chronic hepatitis. Mol Carcinog. 50(2): 136–142.

Yamada, Y., H. Enokida, S. Kojima, K. Kawakami, T. Chiyomaru, S. Tatarano, H. Yoshino, K. Kawahara, K. Nishiyama, N. Seki, and M. Nakagawa. 2011. MiR-96 and miR-183 detection in urine serve as potential tumor markers of urothelial carcinoma: correlation with stage and grade, and comparison with urinary cytology. Cancer Sci. 102(3): 522–529.

Yamamoto, Y., N. Kosaka, M. Tanaka, F. Koizumi, Y. Kanai, T. Mizutani, Y. Murakami, M. Kuroda, A. Miyajima, T. Kato, and T. Ochiya. 2009. MicroRNA-500 as a potential diagnostic marker for hepatocellular carcinoma. Biomarkers. 14(7): 529–538.

Yang, M., J. Chen, F. Su, B. Yu, F. Su, L. Lin, Y. Liu, J.D. Huang, and E. Song. 2011. Microvesicles secreted by macrophages shuttle invasion-potentiating microRNAs into breast cancer cells. Mol Cancer. 10: 117.

Zhang, C., C. Wang, X. Chen, C. Yang, K. Li, J. Wang, J. Dai, Z. Hu, X. Zhou, L. Chen, Y. Zhang, Y. Li, H. Qiu, J. Xing, Z. Liang, B. Ren, C. Yang, K. Zen, and C.Y. Zhang. 2010. Expression profile of microRNAs in serum: a fingerprint for esophageal squamous cell carcinoma. Clin Chem. 56(12): 1871–1879.

Zhang, T., Q. Wang, D. Zhao, Y. Cui, B. Cao, L. Guo, and S.H. Lu. 2011. The oncogenetic role of microRNA-31 as a potential biomarker in oesophageal squamous cell carcinoma. Clin Sci (Lond). 121(10): 437–447.

Zheng, D., S. Haddadin, Y. Wang, L.Q. Gu, M.C. Perry, C.E. Freter, and M.X. Wang. 2011. Plasma microRNAs as novel biomarkers for early detection of lung cancer. Int J Clin Exp Pathol. 4(6): 575–586.

Zhou, J., L. Yu, X. Gao, J. Hu, J. Wang, Z. Dai, J.F. Wang, Z. Zhang, S. Lu, X. Huang, Z. Wang, S. Qiu, X. Wang, G. Yang, H. Sun, Z. Tang, Y. Wu, H. Zhu, and J. Fan. 2011. Plasma microRNA panel to diagnose hepatitis B virus-related hepatocellular carcinoma. J Clin Oncol. 29(36): 4781–4788.

Zhu, W., X. Liu, J. He, D. Chen, Y. Hunag, and Y.K. Zhang. 2011. Overexpression of members of the microRNA-183 family is a risk factor for lung cancer: a case control study. BMC Cancer. 11: 393.

CHAPTER 13

MicroRNAs AS POTENTIAL THERAPEUTIC TARGETS IN CANCER

Ailbhe M. McDermott,[a],* Helen M. Heneghan,[b]
Nicola Miller[c] and Michael J. Kerin[d]

ABSTRACT

Aberrant mi(cro)RNA expression is associated with most pathological disease processes, including carcinogenesis. The knowledge that miRNAs have dual roles, as oncogenes or tumour suppressors, unveils their remarkable potential in novel cancer therapeutics. Additionally, the ability of miRNA expression profiles to classify tumours according to clinico-pathological variables highlights their potential as cancer biomarkers. This may contribute to improved patient selection for adjuvant therapies and help monitor response to treatment. MiRNAs may be exploited as therapeutic agents in various ways: firstly through antisense-mediated inhibition of over-expressed miRNAs, secondly through 'replacement' of under-expressed miRNAs with either miRNA mimetics or viral vector-encoded miRNAs, and thirdly by modulating miRNA expression to augment a patient's response to

Surgery, School of Medicine, Clinical Science Institute, National University of Ireland, Galway, Ireland.
[a]Email: ailbhemcdermott@gmail.com
[b]Email: helenheneghan@hotmail.com
[c]Email: nicola.miller@nuigalway.ie
[d]Email: michael.kerin@nuigalway.ie
*Corresponding author

existing treatments. This chapter will outline the specific mechanisms and applications for each of these therapeutic strategies. We will also outline the progress achieved to date in this field, and the challenges that remain to be addressed before miRNA-based therapies become the next generation of cancer treatments.

1. INTRODUCTION

Mi(cro)RNAs are a class of small, non-coding RNA fragments that play important roles in most biological processes by regulating gene expression. They were first discovered almost 20 years ago, and have since become the focus of much scientific and translational research. MiRNAs are known to play functional roles in both the normal and pathological state. Aberrant miRNA expression has been described in several pathological processes, including carcinogenesis (Calin and Croce 2006). Indeed, investigation into their altered expression in cancer unveiled their dual roles, as oncogenes and tumour suppressor genes. MiRNAs have been identified as potential novel therapeutic targets, or agents, to employ in cancer treatment strategies.

MiRNA expression profiles can classify tumours by type or clinico-pathological characteristics, which has potential utility in cancer diagnostic, prognostic and predictive settings. This would permit more appropriate treatment selection, allow close surveillance of response to treatment and even spare those patients who have early disease or will not respond to systemic treatment, from the toxicities associated with adjuvant chemotherapy.

MiRNAs may be exploited as potential therapeutic targets in cancer by several approaches. Firstly, depleted miRNAs may be replaced by either miRNA mimetics or viral-vector encoded miRNAs; secondly, over-expressed miRNAs can be inhibited by miRNA masking, miRNA sponges or anti-miRNA oligonucleotides (AMOs) and thirdly, patient response to current treatment modalities can be enhanced by altering miRNA expression (Heneghan et al. 2010).

2. miRNA FUNCTION AND MECHANISMS OF ACTION

MiRNAs have been demonstrated to play key roles in almost every aspect of the cell cycle. A mature miRNA strand is incorporated into a miRNA-associated RNA-induced silencing complex (miRISC) (Lowery et al. 2008). It is in this arrangement that a specific miRNA interacts with its target mRNA and exhibits its function at a post-transcriptional level. Understanding the mechanism of miRNA sequence-specific modulation is critical in realizing their application as therapeutic targets in the cancer setting. The 'seed sequence', the important region on a miRNA for target recognition, is located on the five prime tail of the miRNA extending from bases two to eight (Bartel 2004). There are two potential mechanisms of action for each miRNA, determined by the degree of base-pair complementarity between the seed sequence of the miRNA and its mRNA target. If there is perfect complementarity, the RNA-mediated interface pathway is activated and the mRNA is cleaved. However, more commonly, there is imperfect base pair matching which results in repression of protein translation. The miRNA seed sequence binds with partial complementarity to sequences frequently located in the three prime untranstranslated region (UTR) of the mRNA target. It is worth noting however, that miRNAs are also capable of binding to the five prime UTR and the coding region of mRNAs (Jackson and Standart 2007). The mechanism of action responsible for translational repression as a consequence of imperfect base pairing has been proposed by Filipowicz et al. (Filipowicz et al. 2008). The efficacy of translation is decreased as a result of 'bulges' which are created in the central region (less commonly the three prime UTR) due to miss-matching of the base pairs. These miss-matched bulges can have an effect on cleavage of mRNA that is Argonaute (AGO)-mediated. To date, over 1,500 human miRNAs have been described. Each individual miRNA has the ability to bind with several mRNA targets, and exert a multitude of effects through imperfect base pairing. In this way, it is estimated that miRNAs govern in excess of 30% of protein coding genes (Miranda et al. 2006). This highlights the potential influence of miRNAs on practically every biological pathway. These potent regulatory molecules are involved in differentiation, apoptosis, proliferation, and cell-fate determination (Mann et al. 2010).

3. THE ROLE OF miRNAs IN CANCER

MiRNAs function at the initiation, promotion and metastatic stages of many malignancies. They are intricately involved in tumourigenesis and so, the recent ability to manipulate miRNAs has become an appealing anti-cancer treatment strategy. In the cancer state, there is a large body of evidence demonstrating how miRNAs function as oncogenes and tumour suppressor genes. Interestingly, a single miRNA can act dually, as both an oncogene and a tumour suppressor gene, depending on the specific cancer type. For example, *miR-125b* is known to have varying expression levels in different tumours; down-regulation is observed in anaplastic thyroid and serous ovarian carcinoma, indicating a potential tumour suppressor role (Nam et al. 2008, Visone et al. 2007) while oncogenic properties (over-expression) are noted in prostate cancer (Ozen et al. 2008). This is an important concept, as it highlights how each cancer must be considered a distinct disease entity when developing cancer-specific miRNA-based therapeutic strategies.

4. miRNA IDENTIFICATION AND PROFILING IN CANCER

Cancer comprises a heterogeneous group of complex diseases, several of which share fundamental pathological processes. Our improved understanding of miRNA function has allowed us to better understand the complexity of human malignancies, particularly the pathways involved in tumour initiation and progression. Indeed miRNA expression profiling of a range of human tissues, both normal and pathological, enhances our understanding of the developmental stages of several cancers. Specific profiles of dysregulated miRNAs can be created for almost every type of tumour. The use of miRNA expression profiles has several advantages over their predecessors, mRNA profiles, predominantly in permitting the classification of poorly differentiated tumours (Lu et al. 2005, Volinia et al. 2006). Unique miRNA signatures highlight the potential of miRNAs as cancer-specific diagnostic biomarkers, and also unveil the potential therapeutic strategy of restoring the miRNA expression profile to normal as an anti-cancer treatment approach. There are a range of different techniques available for high-throughput miRNA expression profiling. Here we describe some of the most common high-throughput miRNA profiling techniques.

4.1 Oligonucleotide miRNA-Microarray

Oligonucleotide based miRNA-microarray analysis was first described in 2004 (Liu et al. 2004). It the most commonly used method for genome-wide assessment of miRNA expression in human cancers, particularly in large studies. The microarray chip is printed with the sense strand, the miRNA gene specific oligonucleotide probe on the microarray hybridizes with labelled cDNA of the miRNA targets. The cDNA is synthesized by reverse transcription. The results are obtained after staining with Streptavidin-Alexa 647 conjugates and detected by laser scanning. This method is easily standardized and can be applied reliably to a large sample size. It is also cheaper to perform than some of the more novel approaches, permitting simultaneous analysis of over 300 miRNAs. However, drawbacks include a lack of specificity in detecting structurally similar miRNAs.

4.2 Bead-based Flow Cytometric Technology

Bead-based flow cytometric technology is a highly specific method of miRNA expression profiling. The technique was first described by Lu et al. in 2005 (Lu et al. 2005). Polystyrene beads are coated with antisense oligonucleotide probes which hybridize with biotin-labelled PCR amplicon dsDNA (target). Staining for streptavidin–phycoerythrin is performed before results are obtained from signal detection using flow cytometry. Bead-based technologies are technically more challenging to perform, allowing only small numbers of miRNAs to be studied at a time. In addition it has been suggested that bias may be introduced to this technique; the small RNA sample must be enriched prior to commencing by fractionation. Bias could also be introduced during PCR. However, this method permits higher miRNA specificity than oligonucleotide-based microarrays.

4.3 Tag-based Sequencing Methods

Other technologies in this field include the tag-based sequencing methods; miRNA serial analysis of gene expression (miRAGE or SAGE) (Cummins et al. 2006) and RNA-primed array-based Klenow enzyme (RAKE) assay (Nelson et al. 2004). SAGE involves the serial analysis of miRNA expression by small RNA purification, tagging and cloning. However its widespread use is restricted due to these laborious and expensive steps. The RAKE assay includes on-slide application of the Klenow fragment of DNA polymerase I. This assay is sensitive and specific; additionally it permits the discovery of novel miRNAs. However it is also labour intensive and requires a large amount of RNA.

4.4 Deep Sequencing Technology

Recent advances in scientific technology permit genome wide sequence analysis; deep sequence technology is one such large-scale platform. The application of this technology permits the simultaneous sequencing of millions of sequences on a single sample. Deep sequencing is free from many of the limitations confronted by other methods; there is no cross-hybridization and no prior sequence knowledge is required. Deep sequencing can be considered 'unbiased' in this regard and is an excellent choice for diseases in which molecular profiling has previously been unfruitful. The major difficulty associated with this method of miRNA expression profiling is the large amount of data that confronts the researcher. Advances in bioinformatic tools have permitted informative results to be deduced from the data produced during such experiments (Friedlander et al. 2008, Hackenberg et al. 2011, Wang et al. 2009, Yang et al. 2009). Indeed, some researchers consider this complex data analysis and large numbers of generated sequences to be a disadvantage associated with deep-sequencing. In addition, this method is expensive to perform.

4.5 The Future of miRNA Expression Profiling Techniques

The techniques discussed above can only be applied to samples or specimens that have been archived, as cell lysis or fixation is required. However, it is well known that miRNA expression patterns are likely to be transient, and reflect the host or disease micro-environment at that particular time. It could be that a miRNA profile of a specific cancer state represents a stage in the natural history of tumour development, constantly evolving with tumour progression. The next step in miRNA expression profiling would be to devise profiling techniques which could be applied *in vivo*. Molecular imaging is one such technique which permits the evaluation of the dynamic functioning of miRNAs within living cells. This remarkable advance supports repeated quantitative imaging of tumour and stem cells. Molecular imaging is superior in this regard as it permits remarkable insight into miRNAs that are dysregulated in carcinogenesis for consideration as novel treatment strategies.

The high-throughput technologies outlined above aid in the identification of novel disease-specific miRNA targets and miRNA signatures for diagnostic and therapeutic approaches. It is, however, necessary to validate these findings using additional techniques. The most commonly employed validation methods are northern blotting and quantitative real time polymerase chain reaction (qRT-PCR). Northern blotting requires large amounts of total RNA, is time consuming and requires handling of radioactive material (if conducted according to standard protocol). RT-PCR requires less RNA, is a simple technique to perform, and is more commonly employed in validation experiments.

5. miRNA AS THERAPEUTIC TARGETS IN CANCER

MiRNAs are rapidly becoming the next generation of disease-specific therapeutics. There is an increasing body of knowledge and experience in miRNA profiling and functional analyses, particularly in cancer. These tiny RNA molecules are critical to almost every aspect of the cell cycle, and appear to be key players in carcinogenesis in which they mediate specific gene silencing. The ability to safely manipulate specific miRNA expression levels as a treatment modality, and in doing so to change the natural history of a cancer, would revolutionize both the pharmacologic and medical fields.

Given the dual functionality of miRNAs in cancer, it stands to reason that a tumour's growth could potentially be halted or even reversed, by returning the disease-specific dysregulated miRNA to 'normal' expression levels. This could be achieved through replacement of down-regulated miRNAs, or knock-down/inactivation of miRNAs which are over-expressed. Several molecular and pharmacological approaches have been developed as potential treatment strategies.

5.1 Potential Treatment Strategies for Oncogenic miRNAs

Over-expressed 'oncogenic' miRNAs need to be knocked-down or deactivated in order to restore their normal expression level and reverse the mRNA inhibition that results from these aberrantly expressed miRNAs. Several strategies of RNA interference exist, some of which were originally developed for mRNA silencing but have been tailored for miRNA manipulation. The main approaches include anti-miRNA oligonucleotides (AMOs), miRNA sponges and miRNA masking.

5.1.1 Anti-miRNA oligonucleotides (AMOs)

Anti-miRNA oligonucleotides are synthetic oligonucleotides which represent a new application of the antisense concept which was originally applied to mRNA. AMOs for miRNA use are designed with perfect reverse

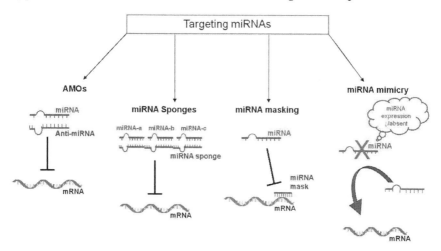

Figure 1. Potential miRNA-based therapeutic strategies. Oncogenic (over-expressed) miRNAs can be knocked-down or inactivated in an effort to restore 'normal' expression levels by anti-miRNA oligonucleotides (AMOs), miRNA sponges or miRNA masking. MiRNAs with reduced expression in the cancer state (tumour suppressor genes) can be replaced by miRNA mimicry. Other replacement techniques include induction of miRNA over-expression by viral vectors and reversal of epigenetic silencing using small molecules. *Reproduced with permission from Pharmaceutical Research.*

Color image of this figure appears in the color plate section at the end of the book.

base pair complementarity and competitively inhibit binding between the miRNA and its mRNA target by binding with the miRNA molecule. Unmodified DNA oligonucleotides were the first AMOs investigated. DNA AMOs were microinjected into *Drosophila* embryos, resulting in inactivation of *miR-2* and *miR-13* (Boutla et al. 2003). However, DNA AMOs are susceptible to degradation by nucleases and this experiment was likely to be successful due to direct microinjection of the AMO. RNA AMOs have distinct advantages over DNA AMOs; binding affinity is higher for duplex formation with miRNA target (Freier and Altmann 1997); modified RNA AMOs are less susceptible to degradation by nucleases and other enzymes. The most commonly used types of chemically modified AMO are two prime-O-methyl AMOs, two prime-methoxyethyl AMOs and locked nucleic acids (LNAs) (Weiler et al. 2006).

Chemical modifications of RNA AMOs permit enhancement of the above properties. For example, substituting a sulphur atom for a non-bridging oxygen atom in the phosphate backbone, called the phosphorothioate (PS) modification reduces the ability of enzymatic degradation. However, PS modifications also reduce binding affinity (Lennox et al. 2006) and these linkages have been deemed necessary for *in vivo* studies to permit cellular uptake and enhancement of pharmacokinetics. The safety of PS application has been reviewed and has been deemed acceptable (Levin 1999). However, a drawback of using AMOs with PS modifications lies in their activation of the immune system and non-specific interaction with proteins. The two-O-methyl- and two-O-methoxyethyl-AMOs are modified to provide enhanced binding affinity and enhanced resistance to nucleases, the latter in particular.

LNAs are a newer class of AMOs. They are chemically modified with the furanose ring in the sugar-phosphate backbone is locked in an RNA mimicking N-type configuration, as a consequence of a two prime-O, four prime-C methylene bridge (Petersen et al. 2000). LNAs have several advantages over the more traditional AMOs; they do not require a vector, have a lower toxicity profile and have superior thermal stability (Elmen et al. 2008). In addition, in comparison to other AMOs, LNAs display excellent mismatch discrimination as base pairing occurs in keeping with Watson-Crick complementarity and as such they provide potent antisense based gene silencing. LNAs are employed in most current *in vivo* studies in the miRNA field. LNAs are however, not the ideal therapeutic modality, they have several associated weaknesses, namely their inability to target more than one specific miRNA at a time and their short duration of action.

5.1.2 miRNA masking

MiRNA masking is another approach to anti-miRNA antisense oligonucleotides, with potentially important clinical utility. A miRNA mask is typically a single stranded two prime-O-methyl-modified oligonucleotide which exhibits perfect complementarity not to the miRNA itself, but rather to the miRNA binding site on the three prime UTR of a protein coding mRNA gene (Wang 2011). In this way, miRNA masking has the ability to provide gene-specific and miRNA-specific miRNA targeting, as it does not directly interact with the miRNA, but prevents binding of the miRNA. In addition, this approach avoids off-target effects and provides a better platform to study the effects of a specific gene on miRNA function. MiRNA masking technology has been successfully applied to *in vivo* studies. Choi et al. used a zebrafish model to investigate the role of *miR-430* on regulating expression transforming growth factor beta (TGF-ß) nodal agonist *squint* and antagonist *lft2* (lefty), which are important in zebrafish development namely of mesendoderm induction and left-right axis discrimination (Choi et al. 2007). They disrupted the specific miRNA-mRNA pairing by introducing miRNA masks which successfully inhibited the repressive action of *miR-430* on the target mRNAs *squint* and *lft2*.

5.1.3 miRNA sponges

MiRNA sponges provide an alternative to antimiR inhibition of miRNA function. First described in 2007, miRNA sponges can be expressed in cells as RNA molecules produced from transgenes and have the advantage of being able to bind multiple over-expressed miRNAs at the same time, a remarkable challenge to have overcome (Ebert et al. 2007). This is particularly attractive in the cancer setting, where multiple miRNAs are simultaneously deregulated and function together, particularly those within the same miRNA family. These competitive inhibitors display numerous and tandem binding sites complementary to the miRNAs of interest. When a sponge is present at high levels, it specifically deactivates the effects of the family of miRNAs sharing the common seed sequence. The miRNAs become bound to the miRNA sponge and are unable to exert their effects at the mRNA level. To ensure that the sponge transcript is present in high enough levels to maximize sponge to miRNA ratio, the sponges are frequently under the control of potent promoters, such as CMV. Typical sponge constructs contain between four and ten binding sites, each separated by a few nucleotides. Interestingly, increasing the number of binding sites is not beneficial, as it increases the likelihood of sponge degradation. The most effective structure for a miRNA sponge is that which contains 'bulged-sites' that are miss-paired opposite miRNA positions nine to twelve (Ebert et al. 2007, Gentner et al. 2009), this presumably forms a more stable sponge-miRNA interaction, one that is less susceptible to Ago2-mediated endonucleolytic cleavage. MiRNA sponges offer several advantages over AMOs. Firstly, the model of producing dominant negative transgenes is perhaps simpler than knock-outs such as AMOs. Secondly, all miRNA family members with the same seed sequence are targets, irrespective of their site of encoding or where they exert their functionality (Ebert and Sharp 2010). MiRNA sponges target only mature miRNAs and therefore do not exert any effect on other miRNA precursors within the same cluster. It is also plausible to suggest that sponges could be modified to include regulatory elements such that they may be induced by specific drugs or even by specific tissues. Delivery of miRNA sponges can be achieved by viral vectors. The efficacy of miRNA sponges has been demonstrated by Ebert et al. in an *in vitro* model. Cultured cells were transiently transfected with vectors encoding miRNA sponges resulting in a reduced level of miRNA to at least that achieved by AMOs (Ebert et al. 2007, Lu et al. 2005). *Drosphila* miR-SP is a novel miRNA sponge technology which aims to elucidate the functional outputs of miRNA-mRNA target interactions (Loya et al. 2009). It is designed to inform about the functional output of miRNAs with precise *in vivo* spatial resolution, while overcoming the lack of tissue specificity associated with more traditional miRNA sponges. This study by Loya et al. was the first to produce stable, germline sponge expression in an animal model organism. The sponge constructs consist of five Upstream Activation Sequence (UAS) elements, ten bulged miRNA binding sites in the three prime UTR and a fluorescent reporter system. These transgenic miRNA sponges are synthesized by placing modified miRNA complementary oligonucleotides downstream of recurring UAS. This remarkable system provides insight into the contribution of specific miRNAs to complex biological processes, and permits insight into their interactions with other genes.

5.2 Potential Treatment Strategies for Tumour Suppressor miRNAs

In the cancer state, miRNAs with decreased expression are regarded as tumour suppressor genes. The fundamental aim of the therapeutic approach in this case is to restore miRNA expression to a normal level. This can be achieved by miRNA mimicry, the induction of miRNA over-expression by viral vectors, or by reversing the epigenetic silencing of specific miRNAs using small molecules. These strategies are desirable as therapeutic approaches for cancer, as in general miRNA levels are found to be decreased in the cancer state (Lu et al. 2005).

5.2.1 miRNA mimicry

MiRNA oligonucleotide mimics are double stranded RNA molecules that contain a 'guide strand' that is identical to the mature miRNA, and is designed to mimic its function. The 'passenger strand', the second RNA strand, is usually perfectly complementary to the guide strand. The duplex formation provides better loading onto the mRISC and therefore improves the biological activity (Behlke 2008). MiRNA oligonucleotide mimics require chemical alteration to reduce degradation by nuclease. In addition, modifications can be made to enhance target organ specific uptake, for example cholesterol can be added to increase hepatic uptake. In theory, as miRNA mimics have the same functionality as the endogenous miRNA, this technique should have reduced off-target effects. MiRNA mimics have successfully been applied to both *in vitro* and *in vivo* studies, the details of which will be discussed later in this chapter.

5.2.2 Viral vector delivery of miRNAs

Increasing the level of tumour suppressor miRNAs can also be achieved by DNA plasmid delivery of the pri-miRNA, usually with the aid of a viral vector. This form of gene therapy typically exploits adenoviral or lentiviral vectors. The vector is based on the adeno-associated virus (AAV) and the specific miRNA is incorporated into this system to enable tissue delivery. Adeno-associated viral (AAV) vectors are thought to have reduced insertional mutagenesis compared to other viruses, and as such are particularly promising (Schnepp et al. 2003). These viruses are appealing as they do not stably integrate, unlike retro- or lenti-viruses which may randomly integrate and cause gene disruption, even perhaps resulting in *trans*-activation of oncogenes. MiRNAs are particularly suited to AAV delivery techniques as they are small. Indeed AAV miRNA replacement approaches have been studied *in vivo* and obtained encouraging results (Trang et al. 2010). MiRNA therapeutics companies such as Mirna Therapeutics and Asuragen have also shown interest in this strategy. AAV vectors have been reported to be utilized with minimal toxicity and transduction efficiency. However, this approach is not free from serious adverse effects. A study by Grimm et al. in 2006 highlighted the danger associated with this type of miRNA replacement. A delivery vector based on duplex-DNA-containing adeno-associated virus type eight (AAV8) was utilized to systemically administer short RNAs to adult mice. This resulted in down-regulation of crucial hepatic miRNAs, and the mice suffered morbidities and even fatalities (Grimm et al. 2006). It was hypothesized that the fatalities were due to over-saturating endogenous miRNA pathways, a consequence that can be avoided by monitoring and optimizing dose and sequence carefully.

6. OBSTACLES TO THE CLINICAL UTILITY OF miRNA THERAPEUTICS

Over the past two decades we have witnessed major scientific advances, from miRNA discovery to their application as novel treatment tools. Progress has been made but there are still several issues which must be overcome before we can safely move each of the above methods of miRNA suppression and miRNA replacement to the clinical setting.

Many of the studies to date regarding miRNAs and cancer have focused in the identification of disease-specific miRNA profiles. However, there is an extensive gap in our knowledge on the precise function and metabolic pathway involvement of these miRNAs, both in health and disease. It would not be safe to launch a miRNA treatment strategy without first characterizing the role of each therapeutic miRNA in other tissues and biological pathways. The complexity of the situation is augmented by the knowledge that miRNAs can have dual functionality, with the potential to act both as an oncogene and tumour suppressor gene depending on the tissue and cancer state in question, and in addition, a single miRNA can have multiple gene targets. Gene target prediction is currently conducted computationally in a sequence specific manner, with software programmes such as miRBase, TargetScan and Pictar capable of providing numerous potential targets. Yet it seems that sequence complementarity is not all that governs an efficient interaction between a miRNA and its mRNA target, as not all predicted targets are functional. Validation of such predicted targets and *in vivo* functional analysis need to be carried out in order to accurately predict the wide range of side effects and toxicities which may be expected from a miRNA treatment approach. It would be prudent to identify all existing miRNAs, and their functions, before miRNA therapeutics can safely be employed.

Standardization of miRNA analysis should also be enforced before these platforms are approved for clinical utility. Academic research and industrial laboratories employ different methods of identifying and validating the specified miRNA, or indeed its precursor. To ensure reproducible results across laboratories and continents it will also be necessary to share methods of miRNA analysis as well as detailed information on the type of specimen evaluated and the ideal storage conditions. If patients are to be treated on the basis of their individual tumour's miRNA profile, it is necessary to ensure an international standard of tumour miRNA expression is achieved.

Another challenge which must be overcome before miRNA-based therapeutics can become a clinical reality is delivery of these remarkable technologies. In order for this strategy to be safe, effective and tissue-specific, improvements in delivery modalities must be achieved. Delivery systems employed to date include direct, indirect and viral vector delivery. However, all of these methods have inherent weaknesses, as discussed earlier. The cell membrane is a natural barrier to exogenous miRNA-based drugs; it limits uptake of oligonucleotides into cells. To overcome this obstacle, methods of delivery focus on the specific target site, or on methods of enhancing uptake of the drug into the target cell. Oligonucleotides have been administered by several routes in *in vivo* studies including subcutaneous injections, intravenous injections, inhalation, oral intake, intraocular and topical applications. Enhanced cellular uptake can be achieved by either improving overall uptake or by permitting tissue-specific or disease-specific uptake by conjugation or coating. Conjugating the oligonucleotide to cholesterol, for example, improves hepatic uptake. Liposomes and lipoplexes (cationic lipid DNA complexes) have also been reported to improve cellular uptake (Akhtar et al. 2000, Jaaskelainen and Urtti 2002). Specific cells or tissues can be targeted by endocytosis techniques, which opportunistically exploit the cell membranes natural process of receptor mediated endocytosis as a consequence of protein (or antibody) cell surface binding. Coupling miRNA therapeutics with antibodies or other proteins that recognize tumour specific antigens, proteins which are uniquely expressed or over-expressed in the cancer state, would provide the ideal scenario of targeted treatment delivery. Potential targets, include the folate receptor and transferrin receptor (both over-expressed in some cancer cells), the ERBB2 receptor (breast and ovarian cancers) and GM-CSF (leukaemic blast cells) (Derycke and De Witte 2002, Frankel et al. 2002, Gabizon et al. 1999, Noonberg and Benz 2000). Such advances unveil ligand targeted therapeutics (LTT) as a promising method to create selective toxicity in neoplastic cells, while sparing already compromised patients from the immunogenic safety issues inherent to viral vector delivery systems. The main advantage of viral vector delivery is the potential to deliver more than one miRNA therapeutic agent at any one particular time. This would have remarkable potential in the cancer setting where multiple miRNAs are dysregulated. The ability to simultaneously replace deficient tumour suppressor genes while reduce oncogenes would theoretically result in a more powerful anti-tumour effect, particularly considering the complexity of most cancer states with multiple oncogenic pathways. However, viral vectors also have several associated weaknesses; they are likely to trigger an immune response, their effect is usually transient, and in the case of retro- or lenti-viruses, may even stably integrate into the patient's genome and cause disruption or even oncogenic activation. Advanced delivery technologies and increased knowledge on miRNA functionality will aid the reduction in off-target effects. Optimal dosing of miRNA is a further pharmacokinetic challenge. *In vivo* animal studies are likely to be most informative in this setting.

The future of miRNA therapeutics is exciting but these obstacles must be overcome so that disease- or tissue- specific uptake can be achieved. This is critical in order to minimize off-target effects and so reduce the exposure of healthy cells to potential side effects.

7. DISEASE SPECIFIC APPLICATION OF miRNA TREATMENT STRATEGIES

MiRNA-based treatment approaches have been successfully applied to a number of common cancers to date. We will now elaborate on the rationale and advances made in this setting, through both *in vitro* and *in vivo* work.

7.1 Hepatocellular Carcinoma

The major advances made in the realm of miRNA therapeutics have been in liver disorders and hepatocellular carcinoma (HCC). HCC is amongst the most commonly diagnosed cancers and is one of the leading causes of cancer-related death worldwide (Parkin et al. 2005). The majority of HCC arises in the setting of pre-existing chronic liver disease. On a worldwide scale this is largely related to infection with viral hepatitis B or C (Bosch et al. 2005). In this context, methods of preventing or treating viral hepatitis and liver fibrosis, will prevent the development of HCC and hence warrants discussion here. Our current understanding of the molecular mechanisms underpinning viral hepatitis, hepatic fibrosis and subsequent HCC remains fragmented, however it is clear that miRNAs play an intricate and complex role in this pathway. The introduction of a viral component affects miRNA-mediated RNA silencing pathways which impact on viral-host cell interactions (Gottwein and Cullen 2008). Host miRNAs are exploited and in addition, viruses even have the capacity to encode their own miRNAs (Lin and Flemington 2010, Nair and Zavolan 2006). These miRNAs evolve at a rapid pace and regulate both viral life-cycle and viral-host interactions. Some miRNAs were found capable of enhancing or inhibiting viral replication by directly acting on viral RNAs (Cullen 2009).

7.1.1 Viral hepatitis B

Hepatitis B Virus (HBV) is a small, enveloped DNA virus which belongs to the Hepadnaviridae family of viruses. HBV can cause acute or persistent infection, and is strongly associated with chronic cirrhosis which predisposes to HCC (Yang et al. 2008). There is robust evidence to support a role for miRNAs in the development of, and host response to, HBV infection (Bala et al. 2009). Jin et al. further examined the miRNA-encoding potential of HBV. By employing computational analysis, they identified that HBV encodes only one putative candidate pri-miRNA. Using a database of three prime UTR from the human genome, the authors also deduced that this viral miRNA could not target any host mRNA, but only viral mRNA. In this way, it appears that HBV has evolved to use viral miRNAs (vmiRNAs) to its advantage to regulate its own gene expression (Jin et al. 2007). This statistically derived conclusion was later examined *in vitro* by Gao et al. HepG2.2.15 cells were transfected with three artificial miRNA-HBV plasmids. HBV antigen secretion was detected using time resolved fluoroimmunoassays (TRFIA). Fluorescence quantitative-PCR was used to evaluate HBV DNA replication, while HBV mRNA was detected by real-time PCR (RT-PCR). In this study, vector-based artificial miRNA successfully inhibited viral replication and expression (Gao et al. 2008). The *in vivo* utility of this potential miRNA-centred treatment approach was confirmed by Ely et al. RNA polymerase II promoter cassettes that transcribe anti-HBV pri-miRNA-122 and pri-miR-31 shuttles were generated and studied in a murine hydrodynamic injection model resulting in reduced HBV expression (Ely et al. 2008). Zhang et al. aimed to identify whether host cellular miRNAs affect HBV replication by creating a loss of function approach. HepG2.2.15 cells were transfected with anti-miRNA oligonucleotides of 328 human miRNAs. Compared to controls, *miR-199a-3p* and *miR-201* transfection was associated with a reduction in HBV surface antigen without impacting on HepG2.2.15 cell proliferation, reflecting decreased viral replication (Zhang et al. 2010).

7.1.2 Viral hepatitis C

Hepatitis C Virus (HCV) is a member of the *Flaviviridae* family. Vaccination for this particular virus is not yet available. This highly dynamic virus replicates in human hepatocytes and is translated as an approximately 3,000 amino acid polyprotein which undergoes post-translational processing, by both viral and human enzymes, to result in ten viral structural and non-structural proteins (Moradpour et al. 2007). HCV is a major cause of chronic hepatitis, liver cirrhosis and HCC. The role of miRNAs in HCV replication and treatment is not as well documented as HBV, however further investigation is certainly warranted. Human host cells appear to modulate HCV replication through RNA interference (RNAi) (Randall et al. 2007). *MiR-122*, the first described liver-specific cellular miRNA which comprises over 70% of miRNAs expressed in the liver, is crucial for HCV replication and accumulation in cultured Huh-7 cells. In fact *miR-122* has two potential binding sites for HCV. Replication of HCV is enhanced by the presence of *miR-122* as it targets the five prime viral end of the non-coding region (Jopling et al. 2005). Jopling et al. highlighted a potential miRNA derived antiviral treatment strategy. This group inactivated *miR-122* by transfection with a two prime-O-methylated RNA oligonucleotide that displayed a complementary sequence to *miR-122* and reported an 80% reduction in HCV replication (Jopling et al. 2008). This *in vitro* model was later followed by an *in vivo* model using chimpanzees. Krutzfeld et al. elegantly showed that intravenously delivery of LNA-mediated oligonuclotides could successfully inhibit *miR-122* for 23 days (Jopling et al. 2008). The mechanism by which *miR-122* supports HCV infection may involve the Heme-Oxygenase-1 (HO-1) pathway. This pathway has the potential to inhibit HCV duplication and *miR-122* down-regulates this particular pathway. It therefore stands to reason that combination therapy with *miR-122* antagonism and HO-1 stimulation may act as a potent anti-HCV treatment approach (Shan et al. 2007). However, a study by Sarasin-Filipowicz et al. revealed that human *miR-122* expression and its role in natural HCV infection may not correlate with experimentally derived findings (Sarasin-Filipowicz et al. 2009). Further *in vivo* analysis is necessary before miRNA based treatment approaches are designed and approved for this disease.

7.1.3 Hepatic fibrosis

Liver fibrosis occurs as a response to chronic tissue injury. The most common insults are viral hepatitis (particularly HBV and HCV), autoimmune hepatitis, alcohol and metabolic diseases (resulting in iron or copper overload). Similar to other organs, the liver is comprised of numerous cell types. These include hepatocytes, an endothelial lining, Kuppfer cells (macrophages) and perivascular hepatic mesenchymal cells, known as Stellate cells (Friedman 1998). The Hepatic Stellate cells (HSCs) are critical to the process of fibrosis. Injury and inflammation causes these cells to undergo 'activation', a process which transforms these cells into highly proliferative, contractile and fibrogenic fibroblasts. Activated HSCs secrete large amounts of extracellular matrix (ECM), with a different composition which then undergoes repeated remodelling (Friedman 2000). Liver fibrosis is reversible, unlike cirrhosis which occurs as a consequence of longstanding fibrosis. However, once cirrhosis has occurred, there is an increased risk of developing HCC. There are currently no FDA-approved therapeutic strategies for the treatment of liver fibrosis, although a select cohort of patients with HCV may respond to interferon therapy, with regression in fibrosis and a viral response (Shiratori et al. 2000). Surgical intervention with resection or liver transplantation remains the only option for certain cases, however this is not ideal as the number of patients requiring treatment for hepatic fibrosis dramatically exceeds the number of organs available for transplantation. It is in this context that miRNA therapeutic approaches hold great promise. Several miRNAs have been studied in fibrosis and liver fibrosis in particular. Down-regulation of *miR-29* family members has been observed in fibrosis, including cardiac and renal fibrosis (Jiang et al. 2010). Roderburg et al. conducted a microarray on murine livers and concluded that *miR-29* family was also significantly down-regulated in liver fibrosis. This study also revealed that *miR-29* played a regulatory role in TGF-ß and NF-κß pathways. Moreover, over-expression of *miR-29b* caused down-regulation of collagen expression in murine HSCs (Roderburg et al. 2011). It is clear from this work that the *miR-29* family could have a future application as an anti-fibrotic miRNA-based agent. *MiR-27a* and *miR-27b* have also recently been evaluated in hepatic fibrosis. These miRNAs are typically over-expressed during inflammation. Down-regulation of these miRNAs in rat HSCs *in vitro* was associated with the HSCs returning to their normal state, with reduced proliferation (Ji et al. 2009).

7.1.3 Hepatocellular carcinoma

Seminal advances in the realm of miRNA therapeutics have been based on their application to HCC, a disease with limited successful management approaches thus far. *MiR-122*, a liver specific miRNA which was previously discussed with respect to HCV, also has a critical role to play in HCC (Gramantieri et al. 2007). This miRNA is known to act as a tumour suppressor in HCC; expression profiling has revealed that *miR-122* is down-regulated in approximately 50% of human HCC cases (Kutay et al. 2006). Restoring *miR-122 in vitro* in metastatic Mahlavu and SK-HEP-1 cells has been shown to inhibit migration and invasion. Furthermore, replenishing *miR-122* levels in an *in vivo* model reduces tumourigenesis, angiogenesis and metastases (Tsai et al. 2009). Additional studies have confirmed similar findings *in vitro*, with cells displaying reduced viability and increased apoptosis (Bai et al. 2009, Lin et al. 2008). Young et al. applied this knowledge, in an effort to test the feasibility of *miR-122* replacement as a therapeutic strategy for HCC. They were the first to attempt small molecule inhibition and activation of *miR-122*, by delivering miRNA modifiers (named 1–3) which act at the transcriptional level. Small molecule *miR-122* inhibitor two inhibited HCV replication and small molecule miR-122 inhibitor three resulted in an increased expression of *miR-122*, caspase activation and ultimately reduced cell viability in the HepG2 cell line for HCC (Young et al. 2010).

Other putative tumour suppressors which could have a therapeutic application in HCC include *miR-101* (Baffa et al. 2009, Li et al. 2009), *miR-26a* (Kota et al. 2009) and *miR-223*. *MiR-26a* is down-regulated in human liver cancers. Replacement of *miR-26a* using an AAV vector *in vivo* strongly prevents proliferation and induces tumour-cell-specific apoptosis. This is a remarkable finding as tumour-specific cell death could potentially provide a reduced toxicity and side effect profile. This phenomenon occurs as a consequence of HCC cells entering a state of G1 arrest due to cyclin D2 and E2 down-regulation (Kota et al. 2009). MiRNA-based therapeutic strategies for HCC appear to have potential in the clinical setting.

7.2 Cancers of the Lung

Lung cancer is the leading cause of cancer mortality worldwide and represents a disease in need of new, effective therapeutic strategies. Over 80% of lung cancers are of the non-small cell lung cancer (NSCLC) type (Jemal et al. 2009). The overall five year survival for NSCLC remains low (15%) and the recurrence rate is high, even amongst those with early stage disease (Miller 2005). As with other tumour types, several miRNAs have been found to be dysregulated in NSCLC and indeed various miRNA signatures have been devised to as diagnostic and prognostic tools (Boeri et al. 2011, Shen et al. 2010). MiRNAs have multiple potential therapeutic roles in lung cancer: as direct anti-cancer agents, as inhibitors of invasion and metastases, and in mediating the reversal of chemo- and radio-resistance.

The majority of the work to date in this setting has focused on the *let-7* family. This group of miRNAs are stably expressed in normal lung tissue, but some *let-7* members have been shown to be under-expressed in NSCLC, revealing a potential role as tumour suppressor genes (Johnson et al. 2007). Slack's group has evaluated the *in vitro* and *in vivo* application of *let-7* based miRNA treatment approaches. Esquela-Kerscher et al. conducted *in vitro* experiments on NSCLC cell lines in 2008 and concluded that *let-7* inhibited the growth of multiple human lung cancer cell lines. They also demonstrated the *in vivo* inhibitory effects of *let-7* on lung tumour formation in a murine model of intranasal *let-7* delivery, in which the mice displayed a G12D activating mutation for the *K-ras* oncogene (Esquela-Kerscher et al. 2008). Trang et al. supported the role of the *let-7* family of miRNAs as a potential treatment approach again in 2010 with another study confirming the *in vivo* application in lung cancer. It was confirmed that loss of *let-7* activity enhances lung tumour formation in a murine model, further evidence to support its tumour suppressor role. Additionally, this group studied the *in vivo* effects of *let-7* delivery on already established non-small cell lung tumours in a mouse model. Immuno-deficient NOD/SCID mice were inoculated with subcutaneous injections of human H460 cells and monitored until palpable tumours had formed. Then *let-7b* or a negative control miRNA (miR-NC) were administered by intra-tumoural injections every three days. Intracellular uptake of these synthetic miRNAs was enhanced by siPORTamine, a lipid based uptake reagent. Those tumors treated with *let-7b* displayed remarkably increased tumoural necrosis and cellular debris than the control group. Intranasal delivery of *let-7b* was then confirmed as a potential delivery approach as effective remission of a *K-ras* activated NSCLC murine model was achieved (Trang et al. 2010). Kumar et al. also explored the value of *let-7* as a therapeutic agent in cancer. A lentiviral miRNA delivery system was utilized in a mouse model to transfect a *let-7g* duplex into *K-ras* (GD12) expressing lung adenocarcinoma cells (LKR 13). This resulted in cell cycle arrest and cell death. Furthermore, these lentiviral vectors were then utilized to deliver *let-7g* in a mouse model, a reduction in tumour burden was observed (Kumar et al. 2008). These studies provide proof-of-concept supporting *let-7* based treatments for use in lung cancer.

Other miRNAs have been studied for therapeutic applications in lung cancer. Wiggins et al. replaced deficient *miR-34a*, both locally and systemically, in a mouse model of NSCLC using synthetic miRNA and a lipid based delivery vehicle. This study design resulted in inhibition of lung tumour growth and interestingly was not associated with any elevation of cytokines, liver or kidney enzymes, implying that it was safely tolerated (Wiggins et al. 2010).

The role of miRNA therapeutics is not confined to direct anti-tumour effects. Reducing the invasion and metastatic capabilities of lung cancer cells is equally as important. *MiR-221* and *miR-222* have been shown to enhance cellular migration by targeting PTEN and TIMP3, by inducing TRAIL resistance in aggressive NSCLC cells (Garofalo et al. 2009). MiRNA-silencing strategies for *miR-221* and *miR-222* could prove advantageous in this regard. Gibbons et al. demonstrated that forced expression of *miR-200* abrogated the ability of metastatic lung cancer cells to undergo EMT, epithelial-to-mesenchymal transition, invade and metastasize, identifying a potential valuable miRNA target for future studies (Gibbons et al. 2009). Finally, miRNAs have a role in targeting genes governing drug sensitivity, resulting in the ability to alter sensitivity of cancer cells to anti-cancer drugs and radiotherapy. The sensitivity to Cisplatin, Etoposide and Doxorubicin *in vitro* was greatly increased following transfection with *miR-134* in a small cell lung cancer cell line (Guo et al. 2010). Down-regulation of *miR-*

200b has been implicated in docetaxel-resistant lung adenocarcinoma cells, with replacement of this miRNA being associated with reversal of this chemo-resistance (Feng et al. 2011). Indeed *let-7* has also been implicated in radio-sensitivity via altered *K-ras* signalling. Two studies have reported that increasing *let-7* expression levels in lung cancer results in increased sensitivity to radiation therapy (Oh et al. 2010, Weidhaas et al. 2007).

7.3 Breast Cancer

Breast cancer is the most common female malignancy among women in almost all of Europe and North America. Each year, more than 1.3 million women worldwide will be diagnosed with this disease, and approximately 4,652,000 women will die as a result of the disease (Garcia 2007). The role of miRNAs in breast tumorigenesis, progression and metastases has been well studied and there is a large body of evidence supporting their role as key players in each of these events (Heneghan et al. 2009). Much of the breast cancer related research in this field to date focuses on miRNA profiling of breast cancer and perhaps it is surprising that relatively few studies have investigated miRNA replacement or knock-down strategies for use in breast cancer. As in other cancer states, miRNAs may be up- or down-regulated in breast cancer and therefore may act as oncogenes or tumour suppressor genes, respectively. As such, miRNA-based treatments have three main approaches in breast cancer: inhibition or loss of function of miRNAs which are over-expressed (oncomiRs), replacement of miRNAs which are under-expressed (tumour suppressor genes) and modulation of miRNAs which are implicated in augmenting response to chemotherapy, hormonal therapy or radiotherapy.

7.3.1 miRNA-based interference with the metastatic cascade

It is widely accepted that chemo-resistance and metastases are responsible for most breast cancer related deaths (Fossati et al. 1998). As such, targeting and interrupting the breast cancer metastatic cascade has formed the focus of miRNA based therapeutic strategies. Baffa et al. conducted a miRNA microarray on paired tumour tissues and metastatic lymph nodes in an effort to identify differentially expressed miRNAs, with a putative role in metastases. Several miRNAs were identified including *miR-10b, miR-21, miR-30a, miR-30e, miR-125b, miR-141, miR-200b, miR-200c* and *miR-105* (Baffa et al. 2009). Indeed *miR-10b* has been implicated in several cancers, particularly in the metastatic setting. *MiR-10b* expression is induced by Twist (a transcription factor) and exerts its pro-metastatic effects by inhibiting translation of the HOXD10 protein, leading to increased expression of RHOC, a pro-metastatic gene (Ma et al. 2007). Ma et al. confirmed that *miR-10b* positively regulates cell invasion and migration *in vitro*. Furthermore, this group showed that *in vivo* over-expression of *miR-10b* in a murine model of otherwise non-metastatic breast tumours resulted in increased invasion and metastases (Ma et al. 2007). The potential application of *miR-21* in breast cancer management has been evaluated *in vitro*. Yan et al. conducted *in vitro* LNA-mediated *miR-21* silencing in breast cancer cell lines, specifically MCF-7 and MDA-MB-231, which reduced proliferation and cell migration. This group also showed a similar effect in an *in vivo* model (Yan et al. 2011). Indeed, the mechanism by which *miR-21* promotes invasion may be via the regulation of TIMP3 (Song et al. 2010). *MiR-10b* and *miR-21* appear to have oncomiR properties in breast cancer, with a potential function for loss of function or knockdown based treatment approaches. *MiR-145* is under-expressed in breast cancer cells, therefore potentially acting as a tumour suppressor gene, and adenoviral vector deliver of *miR-145 in vitro* and *in vivo* suppresses tumour growth and cellular motility (Kim et al. 2011). Interestingly, a relatively newly described miRNA, *miR-1258*, is reported to have inhibitory properties in the development of breast cancer brain metastases by targeting heparanase (Zhang et al. 2011).

7.3.2 miRNAs to augment response to adjuvant therapies

Adjuvant therapies routinely used for the management of breast cancer include chemotherapeutic drugs, hormonal therapies and radiotherapy. However, a large proportion of women do not derive any benefit from these approaches, or develop resistance to these strategies over time (Gonzalez-Angulo et al. 2007). It has been postulated that miRNAs may have the ability to augment response to these adjuvant therapies and preliminary *in vitro* and *in vivo* studies appear to be promising. This is particularly promising for triple negative or basal breast tumours, which are characteristically hormone receptor negative and HER2/*neu* negative, for which no targeted therapy currently exists. Indeed miRNA signatures capable of predicting hormone receptor and HER2/*neu* status have been described (Lowery et al. 2009). This reveals the potential for inducing ER expression in ER negative disease and permitting targeted hormonal therapy for basal and HER2-overexpressing (ER negative) subtypes. *MiR-21* has already shown promise in augmenting the response to chemotherapy in breast cancer. Combining taxol chemotherapy with a *miR-21* inhibitor was successful in producing reduced cell viability and invasiveness, compared to taxol treatment alone. This study elegantly displays the increased chemotherapeutic effect of taxol in the presence of reduced *miR-21* levels (Mei et al. 2010). Slack's group demonstrated that *miR-34* is required for the DNA damage response, and is upregulated by p53 following radiation exposure. Moreover, loss of *miR-34 in vivo* in *C. elegans* results in increased radio-sensitivity in the soma of the animals, with radio-resistance of the germline. When applied to breast cancer cells *in vitro* it was observed that replacing *miR-34* post-radiation exposure altered cell survival (Kato et al. 2009). This work highlights the plausible role of anti-*miR-34* treatment strategies in radio-resistant breast cancer in a clinical setting. The manipulation of several other specific miRNAs have been described as single agents with a putative role in augmenting the response of breast tumours to adjuvant chemotherapy. *MiR-128a, miR-125b, miR-155* and *miR-342* have all been associated with the regulation of chemo-sensitivity (Cittelly et al. 2010, Kong et al. 2010, Masri et al. 2010, Zhou et al. 2010).

7.4 Haematological Malignancies

The initial work on miRNA expression profiling and functional analyses progressed from the studies on haematological malignancies. However, when compared to the solid tumours, very little research has been done on the application of miRNA-based therapeutics to this group of cancers.

7.4.1 Leukaemia

MiR-15a and *miR-16* were the first miRNAs identified as having a role in cancer. A translocation induced deletion at chromosome 13q14.3 in B-cell chronic lymphcytic leukemia (CLL) was identified (Calin et al. 2002). Loss of *miR-15a* and *miR-16* from the locus creates an increased expression of BCL2, an anti-apoptotic gene (Cimmino et al. 2005). A reasonable hypothesis would be to inhibit BCL2 by replacing *miR-15a* and *miR-16* for the treatment of B-cell CLL, but this has yet to be reported. Acute myeloid leukaemia (AML) has been the focus of more promising miRNA-centred treatment approaches. Transfection of synthetic *miR-193a in vitro* produces decreased AML cell proliferation in primary cell lines and primary AML blasts but not in normal bone marrow cells, which correlated with a reduction in the *c-kit* oncogene (Gao et al. 2011). Eyholzer et al. reported that *miR-29b* is under-expressed in AML patients with either loss of chromosome 7q or CEBPA deficiency (Eyholzer et al. 2010). *MiR-29b* mimicry was applied to an *in vitro* cell line and an *in vivo* mouse xenograft model of AML and resulted in reduced growth and increased cellular apoptosis (Garzon et al. 2009).

7.4.2 Lymphoma

The role for miRNAs as therapeutic agents in lymphoma is also relatively understudied. Most of the work in this realm focuses on the *miR-17-92* cluster. This group of miRNAs is comprised of *miR-17-5p, miR-17-3p, miR-18a, miR-19a, miR-20a* and *miR-92*. This cluster is located at chromosome 13q31-q32, an area that is often amplified in B-cell lymphoma (Ota et al. 2004). The *miR-17-92* cluster has been shown to have oncogenic effects in this disease process, with enforced expression associated with accelerated tumour growth and reduced Myc-induced apoptosis in a mouse model of B-cell lymphoma (He et al. 2005). Targeting this cluster and reducing its expression level could reduce the pro-tumour properties of these miRNAs. Additional functions of targeting the *miR-17-92* cluster have been described for increasing radio-sensitivity of Mantle Cell Lymphoma (MCL), as overexpression of this cluster is described in MCL cells with reduced cell death after radiotherapy. It would therefore be plausible that inhibition of this cluster could also sensitise MCL cells to radiation, thus improving the prognosis for these patients. An *in vivo* xenograft MCL murine model demonstrated how inhibition of the *miR-17-92* cluster was associated with reduced tumour growth (Rao et al. 2011).

8. CONCLUSIONS

The application of miRNAs as therapeutic agents and disease targets in cancer is a relatively new but rapidly evolving field, as evidenced by an increasing body of research in this setting. These tiny molecules play important roles of almost all aspects of the cell cycle, both in health and disease. Their individual roles in different cancer states make them ideal tissue-specific and disease-specific therapeutic agents. Moreover, our increasing level of understanding allows us to devise methods of replacing deficient miRNAs and reducing the activity of over-expressed miRNAs. Such technologies have been applied to liver disorders, lung cancers, breast cancer and haematological malignancies. However, several challenges must be overcome before miRNA-based therapeutic strategies can be safely translated to the clinical setting.

9. SUMMARY POINTS

- The application of miRNAs as therapeutic agents and disease targets in cancer is a relatively new and rapidly evolving field.
- MiRNAs are aberrantly expressed in the cancer state where they have dual functionality, acting as oncogenes or tumour suppressor genes.

- Specific miRNA profiles can be generated for almost every tumour type, unveiling the potential for cancer-specific diagnostic biomarkers and therapeutic strategies.
- There are several experimental techniques for miRNA expression profiling available. The most common methods are: Oligonucleotide based-microarray, bead-based flow cytometry technology, tag-based sequencing methods and deep sequencing technology.
- MiRNA-based therapeutics aim to restore the disease-specific dysregulated miRNA to 'normal' expression levels.
- Over-expressed oncogenic miRNAs may be knocked down by anti-miRNA oligonucleotides (AMOs), miRNA sponges and miRNA masking.
- MiRNAs with decreased expression in cancer (tumour suppressor genes) can be replaced by miRNA mimicry, induction of overexpression by viral vectors or by reversal of epigenetic silencing of specific miRNAs using small molecules.
- MiRNA-based treatment approaches have been applied to a number of common cancers both *in vitro* and *in vivo* including hepatocellular carcinoma, lung cancer breast cancer and some haematological malignancies.
- However, there are several challenges which must be overcome before miRNA-based treatment strategies can be safely transferred to the clinical setting.

ABBREVIATIONS

AAV	:	Adeno-associated virus
AAV8	:	Adeno-associated virus type 8
AGO	:	Argonaute
AML	:	Acute myeloid leukaemia
AMO	:	Anti-miRNA oligonucleotide
cDNA	:	Complementary DNA
CEBPA	:	CCAAT/enhancer-binding protein alpha
CLL	:	Chronic lymphocytic leukaemia
CMV	:	Cytomegalovirus
DNA	:	Deoxyribonucleic acid
dsDNA	:	Double-stranded DNA
ECM	:	Extracellular matrix
EMT	:	Epithelial mesenchymal transition
GM-CSF	:	Granulocyte-macrophage colony-stimulating factor
HBV	:	Hepatitis B virus

HCC	:	Hepatocellular carcinoma
HCV	:	Hepatitis C virus
HO-1	:	Heme-Oxygenase-1
HSC	:	Hepatic stellate cells
LNA	:	Locked nucleic acid
LTT	:	Ligand targeted therapeutics
MCL	:	Mantle cell lymphoma
miRAGE	:	MiRNA serial analysis of gene expression
miRNA	:	microRNA
mRNA	:	Messenger RNA
miR-NC	:	miRNA normal control
miRISC	:	miRNA-associated RNA-induced silencing complex
NOD	:	Non-obese diabetic
NSCLC	:	Non small cell lung cancer
PCR	:	Polymerase chain reaction
PS	:	Phosphorothioate
qRT-PCR	:	Quantitative real-time polymerase chain reaction
RAKE	:	RNA-primed array-based Klenow enzyme
RNA	:	Ribonucleic acid
RNAi	:	RNA interference
SAGE	:	Serial analysis of gene expression
SCID	:	Severe combined immunodeficiency
UAS	:	Upstream activation sequence
UTR	:	Untranslated region
vmiRNA	:	viral miRNAs

REFERENCES

Akhtar, S., M.D. Hughes, A. Khan, M. Bibby, M. Hussain, Q. Nawaz, J. Double, and P. Sayyed. 2000. The delivery of antisense therapeutics. Adv Drug Deliv Rev. 44: 3–21.

Baffa, R., M. Fassan, S. Volinia, B. O'Hara, C.G. Liu, J.P. Palazzo, M. Gardiman, M. Rugge, L.G. Gomella, C.M. Croce, and A. Rosenberg. 2009. MicroRNA expression profiling of human metastatic cancers identifies cancer gene targets. J Pathol. 219: 214–21.

Bai, S., M.W. Nasser, B. Wang, S.H. Hsu, J. Datta, H. Kutay, A. Yadav, G. Nuovo, P. Kumar, and K. Ghoshal. 2009. MicroRNA-122 inhibits tumorigenic properties of hepatocellular carcinoma cells and sensitizes these cells to sorafenib. J Biol Chem. 284: 32015–27.

Bala, S., M. Marcos, and G. Szabo. 2009. Emerging role of microRNAs in liver diseases. World J Gastroenterol. 15: 5633–40.

Bartel, D.P. 2004. MicroRNAs: genomics, biogenesis, mechanism, and function. Cell. 116: 281–97.

Behlke, M.A. 2008. Chemical modification of siRNAs for *in vivo* use. Oligonucleotides. 18: 305–19.

Boeri, M., C. Verri, D. Conte, L. Roz, P. Modena, F. Facchinetti, E. Calabro, C.M. Croce, U. Pastorino, and G. Sozzi. 2011. MicroRNA signatures in tissues and plasma predict development and prognosis of computed tomography detected lung cancer. Proc Natl Acad Sci USA. 108: 3713–8.

Bosch, F.X., J. Ribes, R. Cleries, and M. Diaz. 2005. Epidemiology of hepatocellular carcinoma. Clin Liver Dis. 9: 191–211, v.

Boutla, A., C. Delidakis, and M. Tabler. 2003. Developmental defects by antisense-mediated inactivation of micro-RNAs 2 and 13 in Drosophila and the identification of putative target genes. Nucleic Acids Res. 31: 4973–80.

Calin, G.A., C.D. Dumitru, M. Shimizu, R. Bichi, S. Zupo, E. Noch, H. Aldler, S. Rattan, M. Keating, K. Rai, L. Rassenti, T. Kipps, M. Negrini, F. Bullrich, and C.M. Croce. 2002. Frequent deletions and down-regulation of micro- RNA genes miR15 and miR16 at 13q14 in chronic lymphocytic leukemia. Proc Natl Acad Sci USA. 99: 15524–9.

Calin, G.A. and C.M. Croce. 2006. MicroRNA signatures in human cancers. Nat Rev Cancer. 6: 857–66.

Choi, W.Y., A.J. Giraldez, and A.F. Schier. 2007. Target protectors reveal dampening and balancing of Nodal agonist and antagonist by miR-430. Science. 318: 271–4.

Cimmino, A., G.A. Calin, M. Fabbri, M.V. Iorio, M. Ferracin, M. Shimizu, S.E. Wojcik, R.I. Aqeilan, S. Zupo, M. Dono, L. Rassenti, H. Alder, S. Volinia, C.G. Liu, T.J. Kipps, M. Negrini, and C.M. Croce. 2005. miR-15 and miR-16 induce apoptosis by targeting BCL2. Proc Natl Acad Sci USA. 102: 13944–9.

Cittelly, D.M., P.M. Das, N.S. Spoelstra, S.M. Edgerton, J.K. Richer, A.D. Thor, and F.E. Jones. 2010. Downregulation of miR-342 is associated with tamoxifen resistant breast tumors. Mol Cancer. 9: 317.

Cullen, B.R. 2009. Viral and cellular messenger RNA targets of viral microRNAs. Nature. 457: 421–5.

Cummins, J.M., Y. He, R.J. Leary, R. Pagliarini, L.A. Diaz, Jr., T. Sjoblom, O. Barad, Z. Bentwich, A.E. Szafranska, E. Labourier, C.K. Raymond, B.S. Roberts, H. Juhl, K.W. Kinzler, B. Vogelstein, and V.E. Velculescu. 2006. The colorectal microRNAome. Proc Natl Acad Sci USA. 103: 3687–92. Epub 2006 Feb 27.

Derycke, A.S. and P.A. De Witte. 2002. Transferrin-mediated targeting of hypericin embedded in sterically stabilized PEG-liposomes. Int J Oncol. 20: 181–7.

Ebert, M.S., J.R. Neilson, and P.A. Sharp. 2007. MicroRNA sponges: competitive inhibitors of small RNAs in mammalian cells. Nat Methods. 4: 721–6.

Ebert, M.S. and P.A. Sharp. 2010. MicroRNA sponges: progress and possibilities. Rna. 16: 2043–50.

Elmen, J., M. Lindow, S. Schutz, M. Lawrence, A. Petri, S. Obad, M. Lindholm, M. Hedtjarn, H.F. Hansen, U. Berger, S. Gullans, P. Kearney, P. Sarnow, E.M. Straarup, and S. Kauppinen. 2008. LNA-mediated microRNA silencing in non-human primates. Nature. 452: 896–9.

Ely, A., T. Naidoo, S. Mufamadi, C. Crowther, and P. Arbuthnot. 2008. Expressed anti-HBV primary microRNA shuttles inhibit viral replication efficiently *in vitro* and *in vivo*. Mol Ther. 16: 1105–12.

Esquela-Kerscher, A., P. Trang, J.F. Wiggins, L. Patrawala, A. Cheng, L. Ford, J.B. Weidhaas, D. Brown, A.G.Bader, and F.J. Slack. 2008. The let-7 microRNA reduces tumor growth in mouse models of lung cancer. Cell Cycle. 7: 759–64.

Eyholzer, M., S. Schmid, L. Wilkens, B.U. Mueller, and T. Pabst. 2010. The tumour-suppressive miR-29a/b1 cluster is regulated by CEBPA and blocked in human AML. Br J Cancer. 103: 275–84.

Feng, B., R. Wang, H.Z. Song, and L.B. Chen. 2012. MicroRNA-200b reverses chemoresistance of docetaxel-resistant human lung adenocarcinoma cells by targeting E2F3. Cancer. 118(13): 3365–76.

Filipowicz, W., S.N. Bhattacharyya, and N. Sonenberg. 2008. Mechanisms of post-transcriptional regulation by microRNAs: are the answers in sight? Nat Rev Genet. 9: 102–14.

Fossati, R., C. Confalonieri, V. Torri, E. Ghislandi, A. Penna, V. Pistotti, A. Tinazzi, and A. Liberati. 1998. Cytotoxic and hormonal treatment for metastatic breast cancer: a systematic review of published randomized trials involving 31,510 women. J Clin Oncol. 16: 3439–60.

Frankel, A.E., B.L. Powell, P.D.Hall, L.D. Case, and R.J. Kreitman. 2002. Phase I trial of a novel diphtheria toxin/granulocyte macrophage colony-stimulating factor fusion protein (DT388GMCSF) for refractory or relapsed acute myeloid leukemia. Clin Cancer Res. 8: 1004–13.

Freier, S.M. and K.H. Altmann. 1997. The ups and downs of nucleic acid duplex stability: structure-stability studies on chemically-modified DNA:RNA duplexes. Nucleic Acids Res. 25: 4429–43.

Friedlander, M.R., W. Chen, C. Adamidi, J. Maaskola, R. Einspanier, S. Knespel, and N. Rajewsky. 2008. Discovering microRNAs from deep sequencing data using miRDeep. Nat Biotechnol. 26: 407–15.

Friedman, S.L. 1998. Diseases of the Liver, pp. 371–386. *In:* E. Schiff, M. Sorrell, and W. Maddrey (eds.). Philadelphia: Lippincott-Raven.

Friedman, S.L. 2000. Molecular regulation of hepatic fibrosis, an integrated cellular response to tissue injury. J Biol Chem. 275: 2247–50.

Gabizon, A., A.T. Horowitz, D. Goren, D. Tzemach, F. Mandelbaum-Shavit, M.M. Qazen, and S. Zalipsky. 1999. Targeting folate receptor with folate linked to extremities of poly(ethylene glycol)-grafted liposomes: *in vitro* studies. Bioconjug Chem. 10: 289–98.

Gao, X.N., J. Lin, Y.H. Li, L. Gao, X.R. Wang, W. Wang, H.Y. Kang, G.T. Yan, L.L. Wang, and L. Yu. 2011. MicroRNA-193a represses c-kit expression and functions as a methylation-silenced tumor suppressor in acute myeloid leukemia. Oncogene. 30: 3416–28.

Gao, Y.F., L. Yu, W. Wei, J.B. Li, Q.L. Luo, and J.L. Shen. 2008. Inhibition of hepatitis B virus gene expression and replication by artificial microRNA. World J Gastroenterol. 14: 4684–9.

Garcia, M. 2007. Global Cancer Facts and Figures 2007. American Cancer Society, Atlanta, GA, USA.

Garofalo, M., G. DiLeva, G. Romano, G. Nuovo, S.S. Suh, A. Ngankeu, C. Taccioli, F. Pichiorri, H. Alder, P. Secchiero, P. Gasparini, A. Gonelli, S. Costinean, M. Acunzo, G. Condorelli, and C.M. Croce. 2009. miR-221&222 regulate TRAIL resistance and enhance tumorigenicity through PTEN and TIMP3 downregulation. Cancer Cell. 16: 498–509.

Garzon, R., C.E. Heaphy, V. Havelange, M. Fabbri, S. Volinia, T. Tsao, N. Zanesi, S.M. Kornblau, G. Marcucci, G.A. Calin, M. Andreeff, and C.M. Croce. 2009. MicroRNA 29b functions in acute myeloid leukemia. Blood. 114: 5331–41.

Gentner, B., G. Schira, A. Giustacchini, M. Amendola, B.D. Brown, M. Ponzoni, and L. Naldini. 2009. Stable knockdown of microRNA *in vivo* by lentiviral vectors. Nat Methods. 6: 63–6.

Gibbons, D.L., W. Lin, C.J. Creighton, Z.H. Rizvi, P.A. Gregory, G.J. Goodall, N. Thilaganathan, L. Du, Y. Zhang, A. Pertsemlidis, and J.M. Kurie. 2009. Contextual extracellular cues promote tumor cell EMT and metastasis by regulating miR-200 family expression. Genes Dev. 23: 2140–51.

Gonzalez-Angulo, A.M., F. Morales-Vasquez, and G.N. Hortobagyi. 2007. Overview of resistance to systemic therapy in patients with breast cancer. Adv Exp Med Biol. 608: 1–22.

Gottwein, E. and B.R. Cullen. 2008. Viral and cellular microRNAs as determinants of viral pathogenesis and immunity. Cell Host Microbe. 3: 375–87.

Gramantieri, L., M. Ferracin, F. Fornari, A. Veronese, S. Sabbioni, C.G. Liu, G.A. Calin, C. Giovannini, E. Ferrazzi, G.L. Grazi, C.M. Croce, L. Bolondi, and M. Negrini. 2007. Cyclin G1 is a target of miR-122a, a microRNA frequently down-regulated in human hepatocellular carcinoma. Cancer Res. 67: 6092–9.

Grimm, D., K.L. Streetz, C.L. Jopling, T.A. Storm, K. Pandey, C.R. Davis, P. Marion, F. Salazar, and M.A. Kay. 2006. Fatality in mice due to oversaturation of cellular microRNA/short hairpin RNA pathways. Nature. 441: 537–41.

Guo, L., Y. Liu, Y. Bai, Y. Sun, F. Xiao, and Y. Guo. 2010. Gene expression profiling of drug-resistant small cell lung cancer cells by combining microRNA and cDNA expression analysis. Eur J Cancer. 46: 1692–702.

Hackenberg, M., N. Rodriguez-Ezpeleta, and A.M. Aransay. 2011. miRanalyzer: an update on the detection and analysis of microRNAs in high-throughput sequencing experiments. Nucleic Acids Res. 39: W132-8.

He, L., J.M. Thomson, M.T. Hemann, E. Hernando-Monge, D. Mu, S. Goodson, S. Powers, C. Cordon-Cardo, S.W. Lowe, G.J. Hannon, and S.M. Hammond. 2005. A microRNA polycistron as a potential human oncogene. Nature. 435: 828–33.

Heneghan, H.M., N. Miller, A.J. Lowery, K.J. Sweeney, and M.J. Kerin. 2009. MicroRNAs as Novel Biomarkers for Breast Cancer. J Oncol. 2009: 950201.

Heneghan, H.M., N. Miller, and M.J. Kerin. 2010. MiRNAs as biomarkers and therapeutic targets in cancer. Curr Opin Pharmacol. 10: 543–50. Epub 2010 Jun 10.

Jaaskelainen, I. and A. Urtti. 2002. Cell membranes as barriers for the use of antisense therapeutic agents. Mini Rev Med Chem. 2: 307–18.

Jackson, R.J. and N. Standart. 2007. How do microRNAs regulate gene expression? Sci STKE. 2007: re1.

Jemal, A., R. Siegel, E. Ward, Y. Hao, J. Xu, and M.J. Thun. 2009. Cancer statistics, 2009. CA Cancer J Clin. 59: 225–49.

Ji, J., J. Zhang, G. Huang, J. Qian, X. Wang, and S. Mei. 2009. Over-expressed microRNA-27a and 27b influence fat accumulation and cell proliferation during rat hepatic stellate cell activation. FEBS Lett. 583: 759–66. Epub 2009 Jan 29.

Jiang, X., E. Tsitsiou, S.E. Herrick, and M.A. Lindsay. 2010. MicroRNAs and the regulation of fibrosis. FEBS J. 277: 2015–21.

Jin, W.B., F.L. Wu, D. Kong, and A.G. Guo. 2007. HBV-encoded microRNA candidate and its target. Comput Biol Chem. 31: 124–6.

Johnson, C.D., A. Esquela-Kerscher, G. Stefani, M. Byrom, K. Kelnar, D. Ovcharenko, M. Wilson, X. Wang, J. Shelton, J. Shingara, L. Chin, D. Brown, and F.J. Slack. 2007. The let-7 microRNA represses cell proliferation pathways in human cells. Cancer Res. 67: 7713–22.

Jopling, C.L., M. Yi, A.M. Lancaster, S.M. Lemon, and P. Sarnow. 2005. Modulation of hepatitis C virus RNA abundance by a liver-specific MicroRNA. Science. 309: 1577–81.

Jopling, C.L., S. Schutz, and P. Sarnow. 2008. Position-dependent function for a tandem microRNA miR-122-binding site located in the hepatitis C virus RNA genome. Cell Host Microbe. 4: 77–85.

Kato, M., T. Paranjape, R.U. Muller, S. Nallur, E. Gillespie, K. Keane, A. Esquela-Kerscher, J.B. Weidhaas, and F.J. Slack. 2009. The mir-34 microRNA is required for the DNA damage response *in vivo* in *C. elegans* and *in vitro* in human breast cancer cells. Oncogene. 28: 2419–24.

Kim, S.J., J.S. Oh, J.Y. Shin, K.D. Lee, K.W. Sung, S.J. Nam, and K.H. Chun. 2011. Development of microRNA-145 for therapeutic application in breast cancer. J Control Release. 155: 427–34.

Kong, W., L. He, M. Coppola, J. Guo, N.N. Esposito, D. Coppola, and J.Q. Cheng. 2010. MicroRNA-155 regulates cell survival, growth, and chemosensitivity by targeting FOXO3a in breast cancer. J Biol Chem. 285: 17869–79.

Kota, J., R.R. Chivukula, K.A. O'Donnell, E.A. Wentzel, C.L. Montgomery, H.W. Hwang, T.C. Chang, P. Vivekanandan, M. Torbenson, K.R. Clark, J.R. Mendell, and J.T. Mendell. 2009. Therapeutic microRNA delivery suppresses tumorigenesis in a murine liver cancer model. Cell. 137: 1005–17.

Kumar, M.S., S.J. Erkeland, R.E. Pester, C.Y. Chen, M.S. Ebert, P.A. Sharp, and T. Jacks. 2008. Suppression of non-small cell lung tumor development by the let-7 microRNA family. Proc Natl Acad Sci USA. 105: 3903–8.

Kutay, H., S. Bai, J. Datta, T. Motiwala, I. Pogribny, W. Frankel, S.T. Jacob, and K. Ghoshal. 2006. Downregulation of miR-122 in the rodent and human hepatocellular carcinomas. J Cell Biochem. 99: 671–8.

Lennox, K.A., J.L. Sabel, M.J. Johnson, B.G. Moreira, C.A. Fletcher, S.D. Rose, M.A. Behlke, A.L. Laikhter, J.A. Walder, and J.M. Dagle. 2006. Characterization of modified antisense oligonucleotides in Xenopus laevis embryos. Oligonucleotides. 16: 26–42.

Levin, A.A. 1999. A review of the issues in the pharmacokinetics and toxicology of phosphorothioate antisense oligonucleotides. Biochim Biophys Acta. 1489: 69–84.

Li, S., H. Fu, Y. Wang, Y. Tie, R. Xing, J. Zhu, Z. Sun, L. Wei, and X. Zheng. 2009. MicroRNA-101 regulates expression of the v-fos FBJ murine osteosarcoma viral oncogene homolog (FOS) oncogene in human hepatocellular carcinoma. Hepatology. 49: 1194–202.

Lin, C.J., H.Y. Gong, H.C. Tseng, W.L. Wang, and J.L. Wu. 2008. miR-122 targets an anti-apoptotic gene, Bcl-w, in human hepatocellular carcinoma cell lines. Biochem Biophys Res Commun. 375: 315–20. Epub 2008 Aug 8.

Lin, Z. and E.K. Flemington. 2010. miRNAs in the pathogenesis of oncogenic human viruses. Cancer. 305: 186–99.

Liu, C.G., G.A. Calin, B. Meloon, N. Gamliel, C. Sevignani, M. Ferracin, C.D. Dumitru, M. Shimizu, S. Zupo, M. Dono, H. Alder, F. Bullrich, M. Negrini, and C.M. Croce. 2004. An oligonucleotide microchip for genome-wide microRNA profiling in human and mouse tissues. Proc Natl Acad Sci USA. 101: 9740–4.

Lowery, A.J., N. Miller, R.E. McNeill, and M.J. Kerin. 2008. MicroRNAs as prognostic indicators and therapeutic targets: potential effect on breast cancer management. Clin Cancer Res. 14: 360–5.

Lowery, A.J., N. Miller, A. Devaney, R.E. McNeill, P.A. Davoren, C. Lemetre, V. Benens, S. Schmidt, J. Blake, G. Ball, and M.J. Kerin. 2009. MicroRNA signatures predict oestrogen receptor, progesterone receptor and HER2/neu receptor status in breast cancer. Breast Cancer Res. 11: R27.

Loya, C.M., C.S. Lu, D. Van Vactor, and T.A. Fulga. 2009. Transgenic microRNA inhibition with spatiotemporal specificity in intact organisms. Nat Methods. 6: 897–903.

Lu, J., G. Getz, E.A. Miska, E. Alvarez-Saavedra, J. Lamb, D. Peck, A. Sweet-Cordero, B.L. Ebert, R.H. Mak, A.A. Ferrando, J.R. Downing, T. Jacks, H.R. Horvitz, and T.R. Golub. 2005. MicroRNA expression profiles classify human cancers. Nature. 435: 834–8.

Ma, L., J. Teruya-Feldstein, and R.A. Weinberg. 2007. Tumour invasion and metastasis initiated by microRNA-10b in breast cancer. Nature. 449: 682–8.

Mann, M., O. Barad, R. Agami, B. Geiger, and E. Hornstein. 2010. miRNA-based mechanism for the commitment of multipotent progenitors to a single cellular fate. Proc Natl Acad Sci USA. 107: 15804–9.

Masri, S., Z. Liu, S. Phung, E. Wang, Y.C. Yuan, and S. Chen. 2010. The role of microRNA-128a in regulating TGFbeta signaling in letrozole-resistant breast cancer cells. Breast. 124: 89–99.

Mei, M., Y. Ren, X. Zhou, X.B. Yuan, L. Han, G.X. Wang, Z. Jia, P.Y. Pu, C.S. Kang, and Z. Yao. 2010. Downregulation of miR-21 enhances chemotherapeutic effect of taxol in breast carcinoma cells. Technol Cancer Res Treat. 9: 77–86.

Miller, Y.E. 2005. Pathogenesis of lung cancer: 100 year report. Am J Respir Cell Mol Biol. 33: 216–23.

Miranda, K.C., T. Huynh, Y. Tay, Y.S. Ang, W.L. Tam, A.M. Thomson, B. Lim, and I. Rigoutsos 2006. A pattern-based method for the identification of MicroRNA binding sites and their corresponding heteroduplexes. Cell. 126: 1203–17.

miRBase, Version 18, November 2011, University of Manchester [Online]. [Accessed December 2011].

Moradpour, D., F. Penin, and C.M. Rice. 2007. Replication of hepatitis C virus. Nat Rev Microbiol. 5: 453–63.

Nair, V. and M. Zavolan. 2006. Virus-encoded microRNAs: novel regulators of gene expression. Trends Microbiol. 14: 169–75.

Nam, E.J., H. Yoon, S.W. Kim, H. Kim, Y.T. Kim, J.H. Kim, J.W. Kim, and S. Kim. 2008. MicroRNA expression profiles in serous ovarian carcinoma. Clin Cancer Res. 14: 2690–5.

Nelson, P.T., D.A. Baldwin, L.M. Scearce, J.C. Oberholtzer, J.W. Tobias, and Z. Mourelatos. 2004. Microarray-based, high-throughput gene expression profiling of microRNAs. Nat Methods. 1: 155–61.

Noonberg, S.B. and C.C. Benz. 2000. Tyrosine kinase inhibitors targeted to the epidermal growth factor receptor subfamily: role as anticancer agents. Drugs. 59: 753–67.

Oh, J.S., J.J. Kim, J.Y. Byun, and I.A. Kim. 2010. Lin28-let7 modulates radiosensitivity of human cancer cells with activation of K-Ras. Int J Radiat Oncol Biol Phys. 76: 5–8.

Ota, A., H. Tagawa, S. Karnan, S. Tsuzuki, A. Karpas, S. Kira, Y. Yoshida, and M. Seto. 2004. Identification and characterization of a novel gene, C13orf25, as a target for 13q31-q32 amplification in malignant lymphoma. Cancer Res. 64: 3087–95.

Ozen, M., C.J. Creighton, M. Ozdemir, and M. Ittmann. 2008. Widespread deregulation of microRNA expression in human prostate cancer. Oncogene. 27: 1788–93.

Parkin, D.M., F. Bray, J. Ferlay, and P. Pisani. 2005. Global cancer statistics, 2002. CA Cancer J Clin. 55: 74–108.

Petersen, M., C.B. Nielsen, K.E. Nielsen, G.A. Jensen, K. Bondensgaard, S.K. Singh, V.K. Rajwanshi, A.A. Koshkin, B.M. Dahl, J. Wengel, and J.P. Jacobsen. 2000. The conformations of locked nucleic acids (LNA). J Mol Recognit. 13: 44–53.

PicTar, updated March 2007 [Online]. [Accessed December 2011].

Randall, G., M. Panis, J.D. Cooper, T.L. Tellinghuisen, K.E. Sukhodolets, S. Pfeffer, M. Landthaler, P. Landgraf, S. Kan, B.D. Lindenbach, M. Chien, D.B. Weir, J.J. Russo, J. Ju, M.J. Brownstein, R. Sheridan, C. Sander, M. Zavolan, T. Tuschl, and C.M. Rice. 2007. Cellular cofactors affecting hepatitis C virus infection and replication. Proc Natl Acad Sci USA. 104: 12884–9.

Rao, E., C. Jiang, M. Ji, X. Huang, J. Iqbal, G. Lenz, G. Wright, L.M. Staudt, Y. Zhao, T.W. Mckeithan, W.C. Chan, and K. Fu. 2011. The miRNA-17 approximately 92 cluster mediates chemoresistance and enhances tumor growth in mantle cell lymphoma via PI3K/AKT pathway activation. Leukemia. 305.

Roderburg, C., G.W. Urban, K. Bettermann, M. Vucur, H. Zimmermann, S. Schmidt, J. Janssen, C. Koppe, P. Knolle, M. Castoldi, F. Tacke, C. Trautwein, and T. Luedde. 2011. Micro-RNA profiling reveals a role for miR-29 in human and murine liver fibrosis. Hepatology. 53: 209–18.

Sarasin-Filipowicz, M., J. Krol, I. Markiewicz, M.H. Heim, and W. Filipowicz. 2009. Decreased levels of microRNA miR-122 in individuals with hepatitis C responding poorly to interferon therapy. Nat Med. 15: 31–3.

Schnepp, B.C., K.R. Clark, D.L. Klemanski, C.A. Pacak, and P.R. Johnson. 2003. Genetic fate of recombinant adeno-associated virus vector genomes in muscle. J Virol. 77: 3495–504.

Shan, Y., J. Zheng, R.W. Lambrecht, and H.L. Bonkovsky. 2007. Reciprocal effects of micro-RNA-122 on expression of heme oxygenase-1 and hepatitis C virus genes in human hepatocytes. Gastroenterology. 133: 1166–74.

Shen, J., N.W. Todd, H. Zhang, L. Yu, X. Lingxiao, Y. Mei, M. Guarnera, J. Liao, A. Chou, C.L. Lu, Z. Jiang, H. Fang, R.L. Katz, and F. Jiang. 2010. Plasma microRNAs as potential biomarkers for non-small-cell lung cancer. Lab. 91: 579–87.

Shiratori, Y., F. Imazeki, M. Moriyama, M. Yano, Y. Arakawa, O. Yokosuka, T. Kuroki, S. Nishiguchi, M. Sata, G. Yamada, S. Fujiyama, H. Yoshida, and M. Omata. 2000. Histologic improvement of fibrosis in patients with hepatitis C who have sustained response to interferon therapy. Ann Intern Med. 132: 517–24.

Song, B., C. Wang, J. Liu, X. Wang, L. Lv, L. Wei, L. Xie, Y. Zheng, and X. Song. 2010. MicroRNA-21 regulates breast cancer invasion partly by targeting tissue inhibitor of metalloproteinase 3 expression. J Exp Clin Cancer Res. 29: 29.

TargetScan Human, Release 6.0, November 2011 [Online]. [Accessed December 2011].

Trang, P., P.P. Medina, J.F. Wiggins, L. Ruffino, K. Kelnar, M. Omotola, R. Homer, D. Brown, A.G. Bader, J.B. Weidhaas, and F.J. Slack. 2010. Regression of murine lung tumors by the let-7 microRNA. Oncogene. 29: 1580–7.

Tsai, W.C., P.W. Hsu, T.C. Lai, G.Y. Chau, C.W. Lin, C.M. Chen, C.D. Lin, Y.L. Liao, J.L. Wang, Y.P. Chau, M.T. Hsu, M. Hsiao, H.D. Huang, and A.P. Tsou. 2009. MicroRNA-122, a tumor suppressor microRNA that regulates intrahepatic metastasis of hepatocellular carcinoma. Hepatology. 49: 1571–82.
Visone, R., P. Pallante, A. Vecchione, R. Cirombella, M. Ferracin, A. Ferraro, S. Volinia, S. Coluzzi, V. Leone, E. Borbone, C.G. Liu, F. Petrocca, G. Troncone, G.A. Calin, A. Scarpa, C. Colato, G. Tallini, M. Santoro, C.M. Croce, and A. Fusco. 2007. Specific microRNAs are downregulated in human thyroid anaplastic carcinomas. Oncogene. 26: 7590–5.
Volinia, S., G.A. Calin, C.G. Liu, S. Ambs, A. Cimmino, F. Petrocca, R. Visone, M. Iorio, C. Roldo, M. Ferracin, R.L. Prueitt, N. Yanaihara, G. Lanza, A. Scarpa, A. Vecchione, M. Negrini, C.C. Harris, and C.M. Croce. 2006. A microRNA expression signature of human solid tumors defines cancer gene targets. Proc Natl Acad Sci USA. 103: 2257–61. Epub 2006 Feb 3.
Wang, W.C., F.M. Lin, W.C. Chang, K.Y. Lin, H.D. Huang, and N.S. Lin. 2009. miRExpress: analyzing high-throughput sequencing data for profiling microRNA expression. BMC Bioinformatics. 10: 328.
Wang, Z. 2011. The principles of MiRNA-masking antisense oligonucleotides technology. Methods of Mol Biol. 676: 43–9.
Weidhaas, J.B., I. Babar, S.M. Nallur, P. Trang, S. Roush, M. Boehm, E. Gillespie, and F.J. Slack. 2007. MicroRNAs as potential agents to alter resistance to cytotoxic anticancer therapy. Cancer Res. 67: 11111–6.
Weiler, J., J. Hunziker, and J. Hall. 2006. Anti-miRNA oligonucleotides (AMOs): ammunition to target miRNAs implicated in human disease? Gene Ther. 13: 496–502.
Wiggins, J.F., L. Ruffino, K. Kelnar, M. Omotola, L. Patrawala, D. Brown, and A.G. Bader. 2010. Development of a lung cancer therapeutic based on the tumor suppressor microRNA-34. Cancer. 70: 5923–30.
Yan, L.X., Q.N. Wu, Y. Zhang, Y.Y. Li, D.Z. Liao, J.H. Hou, J. Fu, M.S. Zeng, J.P. Yun, Q.L. Wu, Y.X. Zeng, and J.Y. Shao. 2011. Knockdown of miR-21 in human breast cancer cell lines inhibits proliferation, *in vitro* migration and *in vivo* tumor growth. Breast. 13: R2.
Yang, H.I., S.H. Yeh, P.J. Chen, U.H. Iloeje, C.L. Jen, J. Su, L.Y. Wang, S.N. Lu, S.L. You, D.S. Chen, Y.F. Liaw, and C.J. Chen. 2008. Associations between hepatitis B virus genotype and mutants and the risk of hepatocellular carcinoma. J Natl Cancer Inst. 100: 1134–43.
Yang, J.H., P. Shao, H. Zhou, Y.Q. Chen, and L.H. QU. 2009. deepBase: a database for deeply annotating and mining deep sequencing data. Nucleic Acids Res. 38: 4.
Young, D.D., C.M. Connelly, C. Grohmann, and A. Deiters. 2010. Small molecule modifiers of microRNA miR-122 function for the treatment of hepatitis C virus infection and hepatocellular carcinoma. J Am Chem Soc. 132: 7976–81.
Zhang, G.L., Y.X. Li, S.Q. Zheng, M. Liu, X. Li, and H. Tang. 2010. Suppression of hepatitis B virus replication by microRNA-199a-3p and microRNA-210. Antiviral. 88: 169–75.
Zhang, L., P.S. Sullivan, J.C. Goodman, P.H. Gunaratne, and D. Marchetti. 2011. MicroRNA-1258 suppresses breast cancer brain metastasis by targeting heparanase. Cancer. 71: 645–54.
Zhou, M., Z. Liu, Y. Zhao, Y. Ding, H. Liu, Y. Xi, W. Xiong, G. Li, J. Lu, O. Fodstad, A.I. Riker, and M. Tan. 2010. MicroRNA-125b confers the resistance of breast cancer cells to paclitaxel through suppression of pro-apoptotic Bcl-2 antagonist killer 1 (Bak1) expression. J Biol Chem. 285: 21496–507.

CHAPTER 14

REGULATION OF MicroRNAs FOR POTENTIAL CANCER THERAPEUTICS: THE PARADIGM SHIFT FROM PATHWAYS TO PERTURBATION OF GENE REGULATORY NETWORKS

Wei Wu,[1,a,]* Wei Cao[2] and Jennifer A. Chan[1,b,]*

ABSTRACT

MicroRNAs (miRNAs) are evolutionarily conserved small non-protein coding RNA transcripts that, typically, negatively regulate gene expression at a post-transcriptional level. miRNAs exert these effects on diverse physiological functions in development and homeostasis in a variety of organisms including *Homo sapiens*. Deregulation of endogenous miRNAs has been found in the pathogenesis of diseases, particularly in cancer. Extensive miRNAs-association studies in cancer

[1]Departments of Pathology & Laboratory Medicine, Clinical Neurosciences and Oncology, University of Calgary, Calgary, Canada.
[a]Email: wuwei@ucalgary.ca
[b]Email: jawchan@ucalgary.ca
[2]Clinical Research Center, ZhenZhou Hospital, South University, China.
Email: caoweiyu@hotmail.com
*Corresponding authors

have characterized unique 'oncomiRs' (e.g. miR-21, miR-155, miR-17-92 and C19MC miRNAs), 'TS-miRNAs' (tumor suppressor miRNAs e.g. miR-15a/16-1, let-7), 'metasta-miRNAs' (metastasis regulatory miRNAs e.g. miR-10a, miR-34) and 'epi-miRNAs' (epigenetically regulated miRNAs e.g. miR-29a/b/c) in cancer development, progression and metastasis. These findings have inspired the cancer research community to utilize "fingerprint-miRNAs" to aid in cancer diagnosis and prognosis and it has emerged as an exciting new approaches for potential alternative cancer therapy. Restoration of miRNAs in cancer cells with either miRNA mimics or antagomiRs has demonstrated an effective approach to change the cancer cell fate in numerous cultured systems and a few animal models. Due to the nature of miRNAs targeting multiple genes and miRNA-transcription factor co-regulation of gene expression networks, modulation of one or a subgroup of miRNAs could reprogram whole regulatory networks as reflected in multiple core pathways, such as proliferation and apoptosis, differentiation, cell adhesion and invasion as well as epigenomic and genomic stability. In this chapter, we will discuss the miRNA regulation in the genome, highlight the progress of miRNA-based exploratory cancer treatment, and underscore the challenges of developing "miRNA-drugs" with network perturbation.

Keywords: microRNA, gene regulatory networks, experimental therapeutics, cancer, C19MC, premitive neuroectodermal brain tumor, PNET, miR-17-92

1. INTRODUCTION

"Genomic dark matter" is beginning to be uncovered with the completion of human genome project. We have learned that the vast majority of RNA transcripts are not translated into proteins; these non-coding RNAs include small nucleolar RNAs (snoRNAs), Piwi interacting RNAs (piRNAs), large intergenic non-coding RNAs (lincRNAs), small non-coding RNAs (microRNAs and short interfering RNAs) and transcribed ultraconserved regions (T-UCRs) (Esteller 2011) (Fig. 1). The function of non-coding RNAs is not completely understood, but they are likely to play critical roles in normal cellular processes, RNA splicing, genome defence and chromosome structure (Michalak 2006). Among them, microRNAs (miRNAs) have been relatively well studied and have become rising stars in cancer research in the past decade. MiRNAs are evolutionarily conserved small non-coding RNAs of ~22 nucleotides in length, which negatively regulate the gene expression at a posttranscriptional level mainly through binding to 3'-untranslated region (UTR) of mRNAs (Lee et al. 1993). Increasingly, studies show that miRNAs participate in a wide spectrum of physiological functions such as

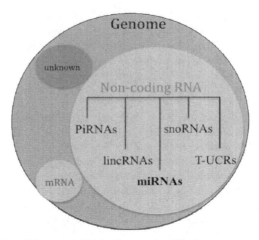

Figure 1. Non-coding RNA transcripts in the genome. The human genome (largest circle) with 3 billion DNA codes transcribes only ~2% of protein-coding mRNAs (left low small circle), and a majority of non-coding RNA transcripts including piRNAs, lincRNAs, miRNAs snoRNAs and T-UCRs. These non-coding RNA transcripts are speculated to act as regulatory elements in the genome. There are still some regions (upper left circle) that remain unknown. This diagram is simplified and not drawn in scale.

stem cell self-renewal, cellular development, differentiation, proliferation and apoptosis (Bartel 2004, Wightman et al. 1993).

Small miRNAs have big impacts on cancer development. A subset of miRNAs was identified as a regulator of neoplastic transformation, tumour progression, invasion and metastasis. The widespread deregulation of miRNomes (genome-wide miRNA expression profiling) in diverse types of cancers compared with normal tissues have been unveiled (Lu et al. 2005). The oncomirs (oncogenic miRNAs) (He et al. 2005), TS-miRNAs (tumor suppressive miRNAs) (Johnson et al. 2005) and Metasta-miRNAs (miRNAs associated with cancer metastasis) (Hurst et al. 2009) and Epi-miRNAs (epigenetically regulated miRNAs) (Iorio et al. 2010) comprise important elements of cancer genomics. Cancer associated miRNAs are providing unique biomarkers for cancer diagnosis and prognosis (Calin and Croce 2006). Moreover, the advantage of miRNAs modulating two-thirds of the whole transcriptome shift the paradigm from developing "one to one" pathway targeting to "one to all" network targeting for putative cancer therapeutics. For examples, miR-21 expression is increased in many solid tumors, such as glioblastoma (Chan et al. 2005), lung, breast, stomach, colon and prostate cancer (Volinia et al. 2006). The underlying mechanism of miR-21 oncogenic effect has been shown to inhibit multiple target genes including programmed cell death 4 (PDCD4) (Frankel et al. 2008), phosphatase and tensin homologue (PTEN) (Meng et al. 2007), Sprouty2 (Sayed et al. 2008), MMP (Gabriely et al. 2008) and GRHL3 (Darido et al.

2011). miR-15a suppresses the expression of BCL-2 and MCL1 (myeloid cell leukaemia sequence 1) (Cimmino et al. 2005), which have been observed in certain experimental conditions. These findings are reminiscence of miRNA-mediated gene expression that fine-tunes genetic circuits. The complexity of miRNA::mRNA interactome (Shirdel et al. 2011) and miRNA-transcription factor (TF) co-regulation of the transcriptome (Shalgi et al. 2007) turn our attention to regulatory network thinking. Here we will highlight the progress of miRNA biology, particularly the miRNA-mediated regulatory networks, describe the applications of miRNA-based experimental cancer treatments, and finally, discuss some challenges of developing "miRNA-drugs" for alternative cancer therapeutics in network perturbation.

2. MicroRNAS IN THE GENOME, NOT JUNK ANY MORE

With the completion of the human genome project, we now learned that only ~2% human genomic DNA is transcribed into protein-coding RNA, and the vast majority of RNA transcripts, which were previously considered noise or junk, are emerging as important elements in controlling biological pathways and processes (Michalak 2006). Since the first miRNA lin-4 was identified in 1993 with the genetic screen in *Caenorhabditis elegans* for defects in the temporal control of post-embryonic development (Lee et al. 1993), 1424 human miRNAs have been identified and deposited in the miRBase registry (http://www.mirbase.org/, release 18, November 2011). Non-canonical miRNAs were reported recently with the whole genome transcriptome sequencing using massively parallel sequencing technology (Friedlander et al. 2008, Ryu et al. 2011).

MiRNAs are located universally in the human genome. A total of 53% miRNAs are encoded in intergenic regions, followed by 31% in the intron regions and 13% in the UTR (Kim and Nam 2006). The fact that 44% are located in the intron or UTR of known genes suggests that there is sharing of regulation elements between protein-coding genes and miRNAs-coding genes (Hinske et al. 2010). In addition, miRNAs in the human genome appear to organize in clusters. Any two microRNAs within 1kb distance are considered to be in the same cluster, and with this criterion, 30.5% of miRNAs in the human genome form clusters (Megraw et al. 2007). The clustered miRNAs most likely share cis-acting regulatory regions or higher-order, chromatin-mediated regulatory coordination. Among these clusters, the miR-17-92 gene cluster, which contains six miRNAs (mir-17, mir-18a, mir-19a, mir-19b-1, mir-20a and mir-92-1), was mapped into the chromosome 13q22 of non-coding region C13orf25 (He et al. 2005). A small miR-93 cluster (miR-106b, miR-93 and miR-25) lies in close proximity inside an intron of the MCM7 gene, which is located in Chromosome 7q22.1 (Petrocca et al. 2008). The miR-15a-miR-16-1 cluster is located in the chromosome 13q14 region (between exon 2 and exon 5 of the non-coding gene LEU2) (Calin et al. 2002). The largest human miRNA gene cluster, also called chromosome 19 microRNA cluster (C19MC), is located in chromosome 19q13.41 and extends over a ~100 kb-long region. It consists of 46 tandemly repeated, primate-specific pre-miRNA genes that are flanked by Alu elements. Due to Alu elements co-localizing with C19MC, RNA polymerase III is used to transcribe C19MC transcript (Borchert et al. 2006). However, C19MC expression is also shown to be transcribed by RNA polymase II in choriocarcinoma cell lines (Bortolin-Cavaille et al. 2009). Overall, these miRNA clusters are likely to have functional significance.

In general, a miRNA gene is transcribed from intragenic or intergenic regions by RNA polymerase II or III into the primary miRNA transcripts with 1~3 kb in a nuclei, followed by a multi-step process of biogenesis (details see other chapters) to form intermediate precursor microRNA and, finally, the mature miRNA (Kim and Nam 2006). The RNA-induced silencing complex (RISC) directs the regulation of mRNA by recognizing a complementary sequence in a mRNA, which is generally located in the 3'UTR, and it exerts its inhibitory effect by either triggering mRNA degradation or suppression of the translation initiation (Bartel 2004). In addition to post-transcriptional regulation, a recent study has reported that miRNAs can regulate gene expression at the transcriptional level by binding directly to promoter regions for epigenetic modification (Khraiwesh et al. 2010).

Although individual miRNAs can function as biological switches, and numerous papers have focused on single targets or single pathways targeted by individual miRNAs, the promiscuity of miRNAs is well recognized, and it is likely that these biological functions are the aggregate of many targeting events. Furthermore, mature miRNAs fine tune the gene expression level in concert with other regulators in order to influence the diverse physiological and biological functions in homeostasis, development (Reinhart et al. 2000), metabolism (Poy et al. 2004) and immune responses (Rodriguez et al. 2007). Taken together, these data showed miRNAs are an important component of the genome and they regulate transcriptome at a stable state to ensure normal function. In the next few decades, we expect that virtually all non-coding RNA transcripts in the genome will be defined and many of their biological functions will be delineated.

3. DEREGULATION OF MicroRNAs IN CANCER, NEW PLAYERS IN THE DRAMA

The central dogma of the mono-directional flow of DNA-RNA-protein has dominated cancer research for more than fifty years. The identification and characterization of cancer-causing genes (oncogenes and tumor suppressors) undoubtedly deepened our fundamental understanding of the molecular mechanisms of cancer initiation, progression and metastasis (Hanahan and Weinberg 2011). Nevertheless, while the brunt of cancer research was focusing on digging more deeply into the cancer genome to find more cancer-related coding genes, a new layer of complexity was added. Calin and his colleagues unexpectedly found that a non-coding region containing miR-15-a/miR-16-1 were down-regulated or deleted in chronic lymphocytic leukemia (Calin et al. 2002). This was the first study linking miRNA to cancer. Three papers, published in 2005 in the journal *Nature*, shed light on potential causal functions of microRNAs in cancer and changed the landscape of cancer genetics by identifying the miRNAs expressed in most common cancers, and investigating the effects of miRNAs on cancer development and cancer genes. The first milestone publication by Lu and colleagues (Lu et al. 2005) defined the pattern of miRNA expression by measuring the expression of 217 human miRNAs in cancer samples and they found miRNA-expression profiling can distinguish cancer type and origin. Remarkably, they found that miRNA signatures are more accurate/robust in classifying cancers than mRNA signatures. The other two papers both investigated different aspects of miRNA-17-92 cluster. He and his colleagues (He et al. 2005) were the first to identify microRNAs as potential oncogenes by observing the over-expression of miR-17-92 polycistron within the C13orf25 genomic region amplification in human B-cell lymphoma. Moreover, ectopic expression of a subset of miR-17-92 encoded miRNAs promoted Myc-driven tumor development in a mouse model, thus implicating that miR-17-92 cluster functions as a non-coding oncogene. At the same time, O'Donnell and his colleagues (O'Donnell et al. 2005) demonstrated the regulatory circuit between mir-17-92 cluster (particularly mir-17-5p and mir-20a) and a transcription factor, the oncoprotein Myc. They showed that Myc could directly bind the promoter of these two miRNAs to increase mir-17-5p and mir-20a expression. They further found that the cell cycle regulator E2F is the target of the cluster, and the feedback loop between Myc-miRNAs-E2F constitutes an important regulatory mechanism controlling proliferation.

Subsequently, microRNA profiling using miRNA microarray technology was intensively carried out in common solid tumours (lung cancer, brain tumor, breast cancer, colorectal cancer, ovarian cancer, melanoma) (Volinia et al. 2006) and hematologic malignancies (Calin et al. 2004a), identifying numerous differentially expressed miRNAs in cancer compared to corresponding normal tissues. Functionally, the up-regulated miRNAs may act as oncogenes (Esquela-Kerscher and Slack 2006, He et al. 2005) and down-regulated miRNAs may act as tumor suppressors (Cimmino et al. 2005, Esquela-Kerscher et al. 2008) expanding the possibilities for potential targets for therapies (Calin and Croce 2006). Aside from basic cancer biology, the potential of miRNAs signatures as biomarkers for diagnosis and prognosis also has not gone unnoticed. Although microRNA biomarkers have not been yet integrated into the routine clinical diagnostic setting (Mattie et al. 2006), it is worth mentioning that specific miRNAs can be detected in patients' body fluids with regular polymerase chain reaction technology and other techniques due to the stability and specificity of small RNAs (Bianchi et al. 2011, Bryant et al. 2012). Such advances have important implications for early diagnosis and disease monitoring. All these findings have motivated targeting miRNAs for putative cancer treatment, which will be discussed in greater detail in another section.

miRNAs are master regulators of gene expression, but who regulates the regulators? Genetic studies for miRNAs have provided hints to four different aspects of mechanistic alteration of miRNAs expression in cancer.

A) Genomic miRNA copy number changes correlate with miRNA expression level. The genes encoding miRNAs are indeed frequently located inside or close to fragile sites and in minimal regions of loss of heterozygosity, in minimal regions of amplification and in common breakpoints associated with cancer (Calin et al. 2004b). The gain of miRNA function through gene amplification is exemplified by several miRNAs such as miR-17-92 gene cluster and the recently identified C19MC cluster miRNAs in brain tumors (Li et al. 2009). The miR-17-92 gene cluster, which contains six miRNAs (mir-17, mir-18a, mir-19a, mir-19b-1, mir-20a and mir-92-1) is frequently amplified in a subset of human B cell lymphoma (Ota et al. 2004) and over-expressed in a variety of other human cancers (Dews et al. 2006, He et al. 2005, Takakura et al. 2008). The C19MC miRNA cluster is found to be amplified in primitive neuroectodermal brain tumors with remarkable increase of mir-520c and mir-517a expression (Li et al. 2009), and other members of C19MC, such as mir-516a, mir-518s, mir-519s, mir-522, mir-524, mir-525 and mir-526 (Wu and Chan, unpublished data). Conversely, loss of miRNA function is often due to the deletion of miRNA genes in chromosomal sites. For example, mir-15a and mir-16-1 cluster is deleted in the majority of chronic lymphocytic leukemias and in a subset of mantle cell lymphoma and prostate cancer (Calin et al. 2002). With the high-resolution array-based comparative genomic hybridization of 227 ovarian cancer, breast cancer and melanoma specimens, the authors found that, in 41 of them, miRNA genes had altered the DNA copy numbers across these three cancer types (e.g. mir-9-1 gene gains and mir-320 gene loss of copy number in genomic loci). Interestingly, Dicer 1, Argonaute 2, and other miRNA-biogenesis associated genes were also found to harbour copy number changes in cancer cells (Zhang et al. 2006). The copy number changes of these miRNA genes are correlated to miRNA transcript expression levels.

B) Epigenetic regulation of miRNAs. Genes are subject to epigenetic modifications during normal development and physiopathological conditions such as tumorigenesis. miRNAs are directly connected to the epigenetic machinery through a regulatory loop. The interplay between miRNA and epigenetic machinery forms an intricate network: miRNA expression is regulated by epigenetic mechanisms, in turn, miRNA can affect the epigenetic machinery (reviewed in Iorio et al. 2010). Altered methylations status can be responsible for the deregulated expression of miRNAs in cancer. The changes of DNA methylations of miRNA promoters or chromatin histone deacetylases (HDAC) of several miRNAs in various cancer cell lines or cancer samples from patients were reported (Lujambio and Esteller 2007, Scott et al. 2006), either silencing the putative tumor suppressor miRNAs by hypermethylation of CpG island in promoters or enhancing the oncogenic miRNAs by hypomethylation of promoter regions. For an example, promoter hypermethylation of miR-9-1, miR-34b,c and miR-148 were reversed by the treatment with DNA methylation inhibitors (5-Aza-2'-deoxycytidine) or histone deacetylase inhibitors and their expression level significantly increased (Saito and Jones 2006). The epigenetic regulation of miRNAs is likely to be tissue specific and dynamic.

C) Transcriptional regulation of miRNAs. Accumulating evidence demonstrates that a subset of miRNA genes are regulated by known transcription factors. It is clear that c-MYC regulates the expression of the miR-17-92 cluster through direct binding to E-boxes of the promoter of miR-17-92 gene (O'Donnell et al. 2005). P53 directly binds to miR-34 gene promoter to activate its transcription (Chang et al. 2007) and that of miR-519d, a member of C19MC cluster (Fornari et al. 2012). C19MC miRNA cluster (miR-524-5p, miR-520d-5p) is postulated to be regulated by transcription factor OCT4 and/or NANOG by deep sequencing of ChIp-chip assay (Bar et al. 2008). Large proportions of miRNAs may be co-regulated with protein-coding genes due to their localization in the host genes as previously mentioned.

D) Polymorphisms and mutations of microRNAs. The mutation of mature RNAs change the miRNA-mRNA interaction and specificity and, therefore, abrogate the miRNA regulator effects. Particularly, the drug-related miRNA polymorphisms have drawn great attention and will be discussed in another section. We expect that a larger scale cancer genome re-sequencing will provide more information of miRNA mutations in different types of cancers (Ramsingh et al. 2010). Overall, miRNAs themselves are subject to genomic regulation to ensure normal cellular functions. Any deregulation of miRNAs from different layers results in rewiring of the regulatory circuits that can contribute to the malignant phenotype of cancer.

4. miRNA-MEDIATED REGULATORY NETWORKS, LESS IS MORE

The discovery of miRNAs has revolutionized the regulation of gene expression. These tiny RNA molecules likely act as molecular switches in extensive regulatory networks that involve thousands of transcripts. The roles of miRNAs as molecular switches in the integrated circuits of cancer cells are emerging. The initial theoretical analysis indicated that as many as 30% of the genes in the human genome may be the targets of miRNA (Lewis et al. 2003). On average, each miRNA has about 500 putative targets, conversely, >60% of mRNAs have one or more evolutionarily conserved sequences that are predicted to interact with miRNAs (Friedman et al. 2009). The diversity of abundance of miRNA targets offers an enormous level of combinatorial possibilities and suggests that miRNAs and their targets (the targetome) appear to form a complex regulatory network. The non-liner effects of miRNA regulatory networks tend to control or modulate many "hubs" or "nodes" in the biological information flow. Using multiple miRNA prediction databases, the miRNA-associated networks can be identified with the multiple miRNA prediction databases, the miRNA-associated networks can be identified with the signalling pathways linked to protein-protein interaction (Shirdel et al. 2011). Intriguingly, using evolutionarily conserved potential miRNA binding sites in mRNAs and conserved binding sites of transcription factors in promoters of the genes encoding these mRNAs, the miRNA target hubs (genes that are regulated by dozens of miRNAs) are also transcriptionally regulated, and these genes are involved in a diversity of developmental processes. In particular, the network of miR-TF co-regulation reveals recurring local architectures of network motifs (Shalgi et al. 2007). The TFs appear to regulate the miRNAs or to be regulated by the miRNAs, forming several feed-forward loops (Fig. 2A). One such an example is miR-17-92 cluster and E2F/Myc network motif in cancer gene network (Aguda et al. 2008, He et al. 2005). Myc and its targets, E2Fs (1,2,3), are both TFs, which directly up-regulate the expression of the miRNAs encoded in mir-17-92 cluster, while these miRNAs in turn act in a feedback loop to inhibit E2Fs mRNA (Fig. 2B). The consequences of this are to switch the state of the 'on' and 'off' in E2F/Myc protein levels and control the cell fate: either proliferation or apoptosis. From more global analysis with large scale samples, Volinia et al. (2010) built up a comprehensive miRNA network in normal tissues and cancers including solid tumors, leukemias and mouse models, and the miRNA networks are reprogrammed during cancer development and inevitably change the global "target hubs" in the interactive networks (Watanabe and Kanai 2011). All these network analyses provide insight to rewiring the topology of network architecture in order to achieve therapeutic effects.

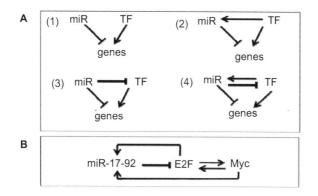

Figure 2. miRNA-transcription factor co-regulation network: miRNAs (miRs) and transcription factor (TFs) have been computationally predicted and experimentally validated to co-regulate gene expression at transcriptional and posttranscriptional level. (A) This network motifs are predicted with TF binding and miRNA target databases. (1) Common function of miRNAs to inhibit gene expression regulated by TFs; (2–4) feed forward loop (FFL) for various interactions between miRNAs and TFs to maintain gene expression at a precise level at spatiotemporal states. (B) Oncogenic miR-17-92 cluster has been shown to be induced by TFs, Myc and E2F and, in turn, to suppress E2F function.

5. miRNA-BASED CANCER TREATMENT, A LONG WAY TO GO

In cancer, step-wise genetic and epigenetic alternation changes the genomic landscape, rewires the interactome and leads to malignant transformation. miRNAs function as regulatory molecules, act as oncogenes or tumor suppressors (Wu et al. 2006) and reprogram cancer networks (Volinia et al. 2010). Therefore, regulation of miRNAs aims to reprogram the interactive networks and to drive cancerous cells to a normal state in the multidimensional state space. Technically, either miRNA mimics or miRNA inhibitors (antagomirs) could be introduced into cells to restore the miRNA physiological function (Wang and Wu 2009, Wu 2011). The effort has been made to target every step of miRNA regulation from the endogenous induction of miRNA gene expression with small molecules to the enzymatic modification (i.e. Drosha or Dicer, Argonautes) involved in miRNA biogenesis (Winter et al. 2009). Some new approaches have come out with creative ideas such as chemical inhibitor (miR-21 inhibitor, diazobenzene) (Gumireddy et al. 2008) to increase the specificity and decrease the off-target effects and other side effects. We summarize partial interactive roadmap illustrating potential applications of regulation miRNAs for experimental cancer treatment (Fig. 3). As we do this, we keep in mind that miRNA-mediated phenotype changes reflected not only alterations of intrinsic

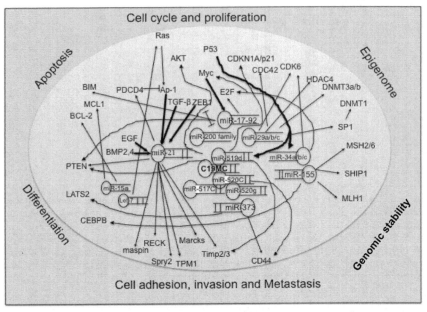

Figure 3. Diagram of miRNA::mRNA interaction contributing to cancer development. We summarize predominantly changed and well-studied miRNAs in cancer and their interaction with protein coding genes to form a complicated regulatory circuit, which results in diverse biological outcomes including cell cycle and proliferation, apoptosis, differentiation, cell adhesion, invasion and metastasis as well as epigenetic modification and genome stability. One miRNA can interact with multiple target genes, whereas, one gene could be regulated by multiple miRNAs. Therefore, the regulation of miRNAs is to reprogram the whole network. [Dark lines with arrows indicate proteins or molecules regulate miRNAs; light lines with arrows represent miRNAs interacting with protein coding genes.]

or cell-autonomous biological pathways, but likely also alterations of cellular networks at the tissue level such rewiring by miRNAs in reality is undoubtedly much more complicated.

5.1 Targeting miRNAs Mediates Apoptosis

Cancer cells gain uncontrolled growth property in part due to the loss of inhibition of the apoptotic hub in the regulatory networks. Multiple miRNAs have been found to be linked with this hub which became a rationale for therapeutic targets. miR-21is over-expressed in various tumors and functionally considered as an oncomiR (Lu et al. 2008, Si et al. 2006, Zhu et al. 2007) and multiple targets of mir-21 have been identified and mapped to the different signalling pathways, including the anti-apoptotic signalling pathways (reviewed in Selcuklu et al. 2009). In fact, the application of 2'-O-methyl-and/or DNA/LNA-mixed oligonucleotides to specifically inhibit miR-21 in glioblastoma and breast cancer cells results in significant tumor cells growth inhibition and enhanced apoptotic cascases activation *in vitro* (Chan et al. 2005, Si et al. 2006). On the other hand, reduced expression of miR-15, miR-16 and let -7 has been observed in different types of cancers and as one consequence, the anti-apoptotic genes are activated in cancer cells (Bottoni et al. 2005, Takamizawa et al. 2004). Restoration of these miRNAs exerts the activation of apoptotic signalling pathways (Cimmino et al. 2005, Johnson et al. 2007). With similar strategy, supplement of miR-491 induces apoptosis by targeting Bcl-X(L) in colorectal cancer cells (Nakano et al. 2010). In parallel, we could design specific miRNAs to down-regulate survival gene BCL-2 to induce cancer cell death (Singh and Saini 2012). Collectively, apoptotic genes and anti-apoptotic genes involved signalling pathways are critical to cancer cells regulated by miRNAs. This well-studied signalling pathway hub is one of the mechanisms of action for manipulation of miRNAs (see Fig. 3).

5.2 Steering miRNAs Sensitizes Chemo or Radio-therapy

Therapeutic resistance of cancer cell subpopulations is the main cause of recurrence or relapse. Therefore, discovering agents or molecules to enhance cancer cell sensitivity to therapy is a long-term goal for improving the efficacy of treatment. By definition of sensitivity enhancer, the agents (e.g. miRNAs) themselves (at certain expression levels) may not have any effect or they may only have minor effects on cancer cell proliferation, apoptosis or cell cycle related networks, but in combination with other treatment, additive or synergistic effects can be achieved. This sensitivity enhancer effect of miRNAs provides a novel platform to achieve the aim of targeting the resistant cancer cells. For instance, knockdown of miR-221 and/or miR-222 sensitized MDA-MB-468 cells to tamoxifen-induced cell growth arrest and apoptosis (Zhao et al. 2008). Restoring miR-15b and miR-16 expression in gastric cancer cells demonstrated the enhancement of the sensitivity of cancer cells to chemotherapy through inhibition of BCL-1 (Xia et al. 2008). miR-326 is inversely related to multidrug resistant associated protein (MRP-1) expression in VP-16-resistant MDR cell line MCF-7/VP (Liang et al. 2010). It has been speculated to regulate miR-326 for preventing and reversing multi-drug resistant effect in tumour cells for increasing efficacy of treatment. Restoration of miR-34 in pancreatic cancer stem cells could increase the sensitivity of cells to docetaxel, cisplatin and gemcitabine treatment and irradiation exposure (Ji et al. 2009). The difference of miRNAs-inducing apoptosis or increasing the sensitivity is likely to be determined by the doses of miRNAs and the topology in the regulatory network. When a given miRNA expression level crosses over the critical threshold, the cell death process will take place; in other words, a resulting change in the network drives cells to a death state through a defined trajectory. Taken together, increasing the cancer cell sensitivity to radio/chemo-therapy by regulating specific miRNAs provides an additional layer to treat cancer cells, at least *in vitro*.

5.3 Modulating miRNAs Induces Cancer Cell Differentiation

Emerging evidence suggests that miRNAs can regulate cell-fate decisions. A subgroup of miRNAs expression is markedly reduced in stem cell state and increased their expression during differentiation. A rare tumor initiating cell population has been documented in leukemia and solid tumors, induction of the differentiation of these tumor initiating cells becomes a future direction for developing new anticancer agents (Gupta et al. 2009). miRNA expression profiling revealed that miRNAs are differentially expressed in breast cancer stem cells and in differentiated breast cancer cells (Wang and Wu 2009). let-7 was found to be a gatekeeper for cancer cell differentiation. let-7 expression was blocked in breast cancer stem cells derived from either cultured mammospheres or clinical cancer specimens, whereas, the oncoprotein Ras or HMGA2 proteins level were very high. The Ras and HMGA2 are negatively regulated by let-7. In contrast, when changing the cell fate from "stemness" state to differentiation state, let-7 expression increased and in turn silencing target gene expression. Moreover, increasing let-7 expression inhibits tumorigenesis and metastasis in NOD/SCID mice model (Yu et al. 2007). More recently, miR-200c was linked to breast cancer stem cell differentiation through inhibition of stem cell self-renewal factor BMI1 gene (Shimono et al. 2009). The muscle-specific miR-206 suppresses human rhabdomyosarcoma growth and promotes myogenic differentiation in xenotransplanted mice mainly via inhibiting its target gene MET expression (Taulli et al. 2009). Taken together, the increasing evidence supports that the re-expression of specific miRNAs could induce cancer cell differentiation, which could be utilized for cancer therapy.

5.4 Monitoring miRNA Mutations for Individualized Therapy

Cancer cells are heterogeneous, even within the same type of tumours. Subclonal cell populations give rise to different treatment responses, and "one size does not fit all" increases the demand for individualized treatment. Growing evidence indicates that miRNAs may be reliable biomarkers for diagnosis, prognosis and treatment evaluation in different types of cancer

(Heneghan et al. 2010, Hu et al. 2010). MiRNAs' expression profiles could be used to define the cell types (Calin and Croce 2006, Lu et al. 2005); therefore, miRNA biomarkers can be developed in particular tumors or an individual patient, and then an individualized therapeutic plan could be designed to manage cancer treatment. Furthermore, the understanding of how an individual's genetic inheritance of miRNA polymorphisms affects the body's response to certain drugs will be key to select chemotherapy agents with maximum efficacy and minimum unwanted side effects. miRNA polymorphisms can occur in the miRNA itself or in its binding site of the 3'UTR of mRNA, resulting in loss of miRNA regulatory functions. The drug-target related miRNA polymorphisms have been reported such as the existence of a miR-24 binding site SNP 829C—T in 3' UTR of dihydrofolate reductase gene that contributes to dihydrofolate reductase overexpression and methotrexate resistance (Mishra et al. 2007). Thus, although miRNA biomarkers have yet to be routinely applied to the clinical setting, they have potential for informing the selection of therapy, for monitoring disease burden, and for evaluating pharmacologic responses. The monitoring or detection of miRNA polymorphisms could also itself lead to development of miRNA-based therapeutics.

5.5 MiRNA Modulation Toward Clinical Trial

The development of miRNA-based therapies is still in its infancy. Numerous experiments demonstrate that modulating miRNAs could serve as treatment in cultured cells and exnograft animal model (Park et al. 2007, Si et al. 2006, Takeshita et al. 2010) (Table 1). The small interfering RNA delivery techniques have been adapted to miRNA-based experimental therapy. No acute or sub-acute toxicity was observed in the locked nucleotide acid (LNA) anti-miRNA treated mice (Elmen et al. 2008) and non-human primates (Elmen et al. 2008). Due to the advance of the delivery technology and the relative safety of miRNAs based treatment in the preclinical trial (Chen et al. 2009), LNA-antimiRTM-122 (SPC3649, developed by Santaris Pharma) clinical trial in human subjects has been started in 2008. This first miRNA based therapy in clinical trial for HCV will pave the way for treating other diseases including cancer. However, the hurdles of miRNA-based cancer treatment *in vivo* prevent the rapid translation from benchsides to bedsides. These issues such as tissue-specific delivery and efficiency of targeted cells taking up the miRNA molecules could hopefully be solved with the nanoparticle technology and chemical inhibitor in the future.

Table 1. Systematic delivery of therapeutic microRNAs *in vivo* for experimental cancer treatment.

microRNA	Animal model	Mode of delivery	Refs.
Let-7	LSL-K-ras G12D	Intranasal administration of let-7	(Esquela-Kerscher et al. 2008)
Let-7 g	Autochthonous model of NSCLC in the mouse	Lentiviruses infection	(Kumar et al. 2008)
Let-7a, let-7g	Lung cancer xenograft NOD/SCID mice Exografts lung cancer LSL-K-ras G12D	Intratumoral delivery Intranasal	(Trang et al. 2010)
miR-16	Prostate cancer xenograft model with bone metastasis	Atelocollegen via tail vein injection	(Takeshita et al. 2010)

6. CHALLENGE OF REGULATION OF miRNA NETWORKS, FROM REDUCTIONISM TO HOLISM

Current miRNA-based therapy *in vitro* and *in vivo* only targets single miRNA, which is relatively easy to manipulate in laboratories. As we discussed above, 27% of known miRNAs tend to present in clusters and act in orchestrated fashion for gene-expression regulation. As a snapshot, the primate-specific C19MC consists of 46 pre-miRNA genes close to downstream small miR-371-3 cluster in Chr19q13.41 region (Borchert et al. 2006) (Fig. 4). In human embryonic stem cells, transcription factors, such as OCT4 or NANOG, SOX2, KLF and P53 (Fornari et al. 2012) could bind to a subset of miRNAs promoters of C19MC genes and activate their expression. The promoter hypermethylation-mediated silencing of C19MC was observed in several tumor cell lines (Fornari et al. 2012, Tsai et al. 2009). The genomic amplification of C19MC was identified in a subset of primitive neuroectodermal brain tumors with predominant miR-520 and 517a over-expression (Li et al. 2009, Nobusawa et al. 2012). We recently found that expression of 18 of 46 miRNAs in C19MC cluster was detected with high-throughput miRNA screening in brain tumors, there are almost 1000 potential targets predicted with TargetScan. These targets tend to be enriched in several core pathways such as Wnt, MAPK, TGF-beta and cell cycle (Wu and Chan, unpublished data). To understand the function of C19MC in brain tumorigenesis, systematic regulation of C19MC miRNAs may be considered from the transcriptional level (e.g. inhibition of TF binding to turn off its expression) through the combinatorial knockdown of key miRNAs to the pathway inhibitors (e.g. Wnt signalling). Ideally, blocking transcription factors binding to the promoter of C19MC will be the most efficient strategy, but the specific TFs have not been clearly categorized. Therefore, this leaves the option to define the miRNAs and their targets network topology and to find the core hub to reprogram the network leading

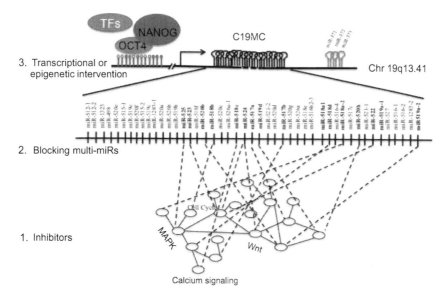

Figure 4. Multidimensional regulation of C19MC cluster. Chromosome 19 microRNA cluster (C19MC) is a primate-specific microRNA gene cluster located in Chr19q13.41 regions. This largest miRNA cluster contains 46 miRNAs across 100 kb span. Specific transcription factors or modification of the promoter or amplification of this region could regulate expression of C19MC region. A subset of miRNAs expression in this cluster has been linked to gene regulatory networks. To restore aberrant function of C19MC systemic strategies may be employed from indicated three layers (1, 2, 3) instead of blocking single miRNA.

to the change in the cell fate. This is a unique challenge because it requires the combination of extensive experiments such as high-throughput miRNA library screening with computational prediction. Altogether, it is time to rethink the function of miRNAs from a systems biology point view and to take the miRNA network into account in the design of a new therapeutic medicine.

7. CONCLUSIONS

MiRNAs are the central components of gene regulatory networks, and the aberrant expression of miRNAs contributes to a wide variety of diseases including cancer. Genome-wide miRNA expression profiling in cancers demonstrated that miRNA expression signatures can discriminate type and origin of cancer with high accuracy. A subgroup of miRNAs contribute to cancer stem-cell maintenance, cancer development, metastasis and drug-resistance. Therefore, great efforts have been made to develop technology and strategies to restore miRNA expression levels and, in turn, reverse the cancerous phenotype. With the understanding of miRNA biology and miRNA-mediated regulatory networks, the paradigm shift from targeting pathways to whole networks is emerging. The multi-dimensional regulation of miRNA clusters for therapeutics is challenging; it needs to combine functional genomic approaches with computational power to dissect the "hubs" in the regulatory networks.

8. SUMMARY POINTS

- miRNAs play crucial functions in the regulation of important physiological processes.
- Deregulation of miRNAs contributes to the malignant phenotype of cancer.
- Novel miRNAs have been identified through whole genome sequencing, and the complex circuitry of miRNAs-mediated gene expression has been revealed in the last few years.
- We are witnessing advances in miRNA-based therapeutics in cultured cells, animal models and into clinical trials.
- Remaining challenges are to manage the complexity of miRNA regulatory networks and to harness this complexity to reprogram the gene regulatory networks for potential cancer therapeutics.
- Rethinking miRNA regulatory networks from systems biology view, and targeting the hubs of the networks may represent principles for the future development of miRNA-oriented cancer therapies.

ACKNOWLEDGEMENTS

JAC is an Alberta Innovates Health Solutions (AIHS) Clinical Investigator. JAC and WW are supported by funds from the Kids Cancer Care Foundation of Alberta, Genome Canada, the Clark H. Smith Brain Tumor Centre, and the Ross Family Fund and Family of Kathleen Lorette. We are grateful to Dr. Fred Biddle for the stimulating discussion about the gene regulatory networks and for critical reading of the manuscript.

ABBREVIATIONS

C19MC	:	chromosome 19 miRNA cluster
Epi-miRNA	:	epigenetic machinery associated miRNA
LincRNA	:	large intergenic non-coding RNA
miR	:	microRNA
miRNA	:	microRNA
Metasta-miRNA	:	metastasis associated miRNA

nt	:	nucleotide
OncomiR	:	oncogenic miRNA
PiRNA	:	Piwi interacting RNA
SNORNA	:	small nucleolar RNA
TF	:	transcription factor
TSmiRNA	:	Tumour Suppressor miRNA
T-UCRs	:	transcribed ultraconserved regions
UTR	:	untranslated region

REFERENCES

Aguda, B.D., Y. Kim, M.G. Piper-Hunter, A. Friedman, and C.B. Marsh. 2008. MicroRNA regulation of a cancer network: consequences of the feedback loops involving miR-17-92, E2F, and Myc. Proc Natl Acad Sci USA. 105: 19678–19683.

Bar, M., S.K. Wyman, B.R. Fritz, J. Qi, K.S. Garg, R.K. Parkin, E.M. Kroh, A. Bendoraite, P.S. Mitchell, A.M. Nelson et al. 2008. MicroRNA discovery and profiling in human embryonic stem cells by deep sequencing of small RNA libraries. Stem Cells. 26: 2496–2505.

Bartel, D.P. 2004. MicroRNAs: genomics, biogenesis, mechanism, and function. Cell. 116: 281–297.

Bianchi, F., F. Nicassio, M. Marzi, E. Belloni, V. Dall'olio, L. Bernard, G. Pelosi, P. Maisonneuve, G. Veronesi, and P.P. Di Fiore. 2011. A serum circulating miRNA diagnostic test to identify asymptomatic high-risk individuals with early stage lung cancer. EMBO Mol Med. 3: 495–503.

Borchert, G.M., W. Lanier, and B.L. Davidson. 2006. RNA polymerase III transcribes human microRNAs. Nat Struct Mol Biol. 13: 1097–1101.

Bortolin-Cavaille, M.L., M. Dance, M. Weber, and J. Cavaille. 2009. C19MC microRNAs are processed from introns of large Pol-II, non-protein-coding transcripts. Nucleic Acids Res. 37: 3464–3473.

Bottoni, A., D. Piccin, F. Tagliati, A. Luchin, M.C. Zatelli, and E.C. degli Uberti. 2005. miR-15a and miR-16-1 down-regulation in pituitary adenomas. J Cell Physiol. 204: 280–285.

Bryant, R.J., T. Pawlowski, J.W. Catto, G. Marsden, R.L. Vessella, B. Rhees, C. Kuslich, T. Visakorpi, and F.C. Hamdy. 2012. Changes in circulating microRNA levels associated with prostate cancer. Br J Cancer.

Calin, G.A., C.D. Dumitru, M. Shimizu, R. Bichi, S. Zupo, E. Noch, H. Aldler, S. Rattan, M. Keating, K. Rai et al. 2002. Frequent deletions and down-regulation of micro- RNA genes miR15 and miR16 at 13q14 in chronic lymphocytic leukemia. Proc Natl Acad Sci USA. 99: 15524–15529.

Calin, G.A., C.G. Liu, C. Sevignani, M. Ferracin, N. Felli, C.D. Dumitru, M. Shimizu, A. Cimmino, S. Zupo, M. Dono et al. 2004a. MicroRNA profiling reveals distinct signatures in B cell chronic lymphocytic leukemias. Proc Natl Acad Sci USA. 101: 11755–11760.

Calin, G.A., C. Sevignani, C.D. Dumitru, T. Hyslop, E. Noch, S. Yendamuri, M. Shimizu, S. Rattan, F. Bullrich, M. Negrini et al. 2004b. Human microRNA genes are frequently located at fragile sites and genomic regions involved in cancers. Proc Natl Acad Sci USA. 101: 2999–3004.

Calin, G.A. and C.M. Croce. 2006. MicroRNA signatures in human cancers. Nat Rev Cancer. 6: 857–866.

Chan, J.A., A.M. Krichevsky, and K.S. Kosik. 2005. MicroRNA-21 is an antiapoptotic factor in human glioblastoma cells. Cancer Res. 65: 6029–6033.

Chang, T.C., E.A. Wentzel, O.A. Kent, K. Ramachandran, M. Mullendore, K.H. Lee, G. Feldmann, M. Yamakuchi, M. Ferlito, C.J. Lowenstein et al. 2007. Transactivation of miR-34a by p53 broadly influences gene expression and promotes apoptosis. Mol Cell. 26: 745–752.

Chen, X., X. Guo, H. Zhang, Y. Xiang, J. Chen, Y. Yin, X. Cai, K. Wang, G. Wang, Y. Ba et al. 2009. Role of miR-143 targeting KRAS in colorectal tumorigenesis. Oncogene. 28(10): 1385–1392.

Cimmino, A., G.A. Calin, M. Fabbri, M.V. Iorio, M. Ferracin, M. Shimizu, S.E. Wojcik, R.I. Aqeilan, S. Zupo, M. Dono et al. 2005. miR-15 and miR-16 induce apoptosis by targeting BCL2. Proc Natl Acad Sci USA. 102: 13944–13949.

Darido, C., S.R. Georgy, T. Wilanowski, S. Dworkin, A. Auden, Q. Zhao, G. Rank, S. Srivastava, M.J. Finlay, A.T. Papenfuss et al. 2011. Targeting of the tumor suppressor GRHL3 by a miR-21-dependent proto-oncogenic network results in PTEN loss and tumorigenesis. Cancer Cell. 20: 635–648.

Dews, M., A. Homayouni, D. Yu, D. Murphy, C. Sevignani, E. Wentzel, E.E. Furth, W.M. Lee, G.H. Enders, J.T. Mendell et al. 2006. Augmentation of tumor angiogenesis by a Myc-activated microRNA cluster. Nat Genet. 38: 1060–1065.

Elmen, J., M. Lindow, A. Silahtaroglu, M. Bak, M. Christensen, A. Lind-Thomsen, M. Hedtjarn, J.B. Hansen, H.F. Hansen, E.M. Straarup et al. 2008. Antagonism of microRNA-122 in mice by systemically administered LNA-antimiR leads to up-regulation of a large set of predicted target mRNAs in the liver. Nucleic Acids Res. 36: 1153–1162.

Esquela-Kerscher, A. and F.J. Slack. 2006. Oncomirs - microRNAs with a role in cancer. Nat Rev Cancer. 6: 259–269.

Esquela-Kerscher, A., P. Trang, J.F. Wiggins, L. Patrawala, A. Cheng, L. Ford, J.B. Weidhaas, D. Brown, A.G. Bader, and F.J. Slack. 2008. The let-7 microRNA reduces tumor growth in mouse models of lung cancer. Cell Cycle. 7: 759–764.

Esteller, M. 2011. Non-coding RNAs in human disease. Nat Rev Genet. 12: 861–874.

Fornari, F., M. Milazzo, P. Chieco, M. Negrini, E. Marasco, G. Capranico, V. Mantovani, J. Marinello, S. Sabbioni, E. Callegari et al. 2012. In hepatocellular carcinoma miR-519d is upregulated by p53 and DNA hypomethylation and targets CDKN1A/p21, PTEN, AKT3 and TIMP2. J Pathol. 227(3): 275–285.

Frankel, L.B., N.R. Christoffersen, A. Jacobsen, M. Lindow, A. Krogh, and A.H. Lund. 2008. Programmed cell death 4 (PDCD4) is an important functional target of the microRNA miR-21 in breast cancer cells. J Biol Chem. 283: 1026–1033.

Friedlander, M.R., W. Chen, C. Adamidi, J. Maaskola, R. Einspanier, S. Knespel, and N. Rajewsky. 2008. Discovering microRNAs from deep sequencing data using miRDeep. Nat Biotechnol. 26: 407–415.

Friedman, R.C., K.K. Farh, C.B. Burge, and D.P. Bartel. 2009. Most mammalian mRNAs are conserved targets of microRNAs. Genome Res. 19: 92–105.

Gabriely, G., T. Wurdinger, S. Kesari, C.C. Esau, J. Burchard, P.S. Linsley, and A.M. Krichevsky. 2008. MiR-21 promotes glioma invasion by targeting MMP regulators. Mol Cell Biol.

Gumireddy, K., D.D. Young, X. Xiong, J.B. Hogenesch, Q. Huang, and A. Deiters. 2008. Small-molecule inhibitors of microrna miR-21 function. Angew Chem Int Ed Engl. 47: 7482–7484.

Gupta, P.B., T.T. Onder, G. Jiang, K. Tao, C. Kuperwasser, R.A. Weinberg, and E.S. Lander. 2009. Identification of selective inhibitors of cancer stem cells by high-throughput screening. Cell. 138: 645–659.

Hanahan, D. and R.A. Weinberg. 2011. Hallmarks of cancer: the next generation. Cell. 144: 646–674.

He, L., J.M. Thomson, M.T. Hemann, E. Hernando-Monge, D. Mu, S. Goodson, S. Powers, C. Cordon-Cardo, S.W. Lowe, G.J. Hannon et al. 2005. A microRNA polycistron as a potential human oncogene. Nature. 435: 828–833.

Heneghan, H.M., N. Miller, A.J. Lowery, K.J. Sweeney, J. Newell, and M.J. Kerin. 2010. Circulating microRNAs as novel minimally invasive biomarkers for breast cancer. Ann Surg. 251: 499–505.

Hinske, L.C., P.A. Galante, W.P. Kuo, and L. Ohno-Machado. 2010. A potential role for intragenic miRNAs on their hosts' interactome. BMC Genomics. 11: 533.

Hu, Z., X. Chen, Y. Zhao, T. Tian, G. Jin, Y. Shu, Y. Chen, L. Xu, K. Zen, C. Zhang et al. 2010. Serum microRNA signatures identified in a genome-wide serum microRNA expression profiling predict survival of non-small-cell lung cancer. J Clin Oncol. 28: 1721–1726.

Hurst, D.R., M.D. Edmonds, and D.R. Welch. 2009. Metastamir: the field of metastasis-regulatory microRNA is spreading. Cancer Res. 69: 7495–7498.

Iorio, M.V., C. Piovan, and C.M. Croce. 2010. Interplay between microRNAs and the epigenetic machinery: an intricate network. Biochim Biophys Acta. 1799: 694–701.

Ji, Q., X. Hao, M. Zhang, W. Tang, M. Yang, L. Li, D. Xiang, J.T. Desano, G.T. Bommer, D. Fan et al. 2009. MicroRNA miR-34 inhibits human pancreatic cancer tumor-initiating cells. PLoS One. 4: e6816.

Johnson, C.D., A. Esquela-Kerscher, G. Stefani, M. Byrom, K. Kelnar, D. Ovcharenko, M. Wilson, X. Wang, J. Shelton, J. Shingara et al. 2007. The let-7 microRNA represses cell proliferation pathways in human cells. Cancer Res. 67: 7713–7722.

Johnson, S.M., H. Grosshans, J. Shingara, M. Byrom, R. Jarvis, A. Cheng, E. Labourier, K.L. Reinert, D. Brown, and F.J. Slack. 2005. RAS is regulated by the let-7 microRNA family. Cell. 120: 635–647.

Khraiwesh, B., M.A. Arif, G.I. Seumel, S. Ossowski, D. Weigel, R. Reski, and W. Frank. 2010. Transcriptional control of gene expression by microRNAs. Cell. 140: 111–122.

Kim, V.N. and J.W. Nam. 2006. Genomics of microRNA. Trends Genet. 22(3): 165–173.

Kota, J., R.R. Chivukula, K.A. O'Donnell, E.A. Wentzel, C.L. Montgomery, H.W. Hwang, T.C. Chang, P. Vivekanandan, M. Torbenson, K.R. Clark et al. 2009. Therapeutic microRNA delivery suppresses tumorigenesis in a murine liver cancer model. Cell. 137: 1005–1017.

Kumar, M.S., S.J. Erkeland, R.E. Pester, C.Y. Chen, M.S. Ebert, P.A. Sharp, and T. Jacks. 2008. Suppression of non-small cell lung tumor development by the let-7 microRNA family. Proc Natl Acad Sci USA. 105: 3903–3908.

Lee, R.C., R.L. Feinbaum, and V. Ambros. 1993. The *C. elegans* heterochronic gene lin-4 encodes small RNAs with antisense complementarity to lin-14. Cell. 75: 843–854.

Lewis, B.P., I.H. Shih, M.W. Jones-Rhoades, D.P. Bartel, and C.B. Burge. 2003. Prediction of mammalian microRNA targets. Cell. 115: 787–798.

Li, M., K.F. Lee, Y. Lu, I. Clarke, D. Shih, C. Eberhart, V.P. Collins, T. Van Meter, D. Picard, L. Zhou et al. 2009. Frequent amplification of a chr19q13.41 microRNA polycistron in aggressive primitive neuroectodermal brain tumors. Cancer Cell. 16: 533–546.

Liang, Z., H. Wu, J. Xia, Y. Li, Y. Zhang, K. Huang, N. Wagar, Y. Yoon, H.T. Cho, S. Scala et al. 2010. Involvement of miR-326 in chemotherapy resistance of breast cancer through modulating expression of multidrug resistance-associated protein 1. Biochem Pharmacol. 79: 817–824.

Lu, J., G. Getz, E.A. Miska, E. Alvarez-Saavedra, J. Lamb, D. Peck, A. Sweet-Cordero, B.L. Ebert, R.H. Mak, A.A. Ferrando et al. 2005. MicroRNA expression profiles classify human cancers. Nature. 435: 834–838.

Lu, Z., M. Liu, V. Stribinskis, C.M. Klinge, K.S. Ramos, N.H. Colburn, and Y. Li. 2008. MicroRNA-21 promotes cell transformation by targeting the programmed cell death 4 gene. Oncogene. 27: 4373–4379.

Lujambio, A. and M. Esteller. 2007. CpG island hypermethylation of tumor suppressor microRNAs in human cancer. Cell Cycle. 6: 1455–1459.

Mattie, M.D., C.C. Benz, J. Bowers, K. Sensinger, L. Wong, G.K. Scott, V. Fedele, D. Ginzinger, R. Getts, and C. Haqq. 2006. Optimized high-throughput microRNA expression profiling provides novel biomarker assessment of clinical prostate and breast cancer biopsies. Mol Cancer. 5: 24.

Megraw, M., P. Sethupathy, B. Corda, and A.G. Hatzigeorgiou. 2007. miRGen: a database for the study of animal microRNA genomic organization and function. Nucleic Acids Res. 35: D149–155.

Meng, F., R. Henson, H. Wehbe-Janek, K. Ghoshal, S.T. Jacob, and T. Patel. 2007. MicroRNA-21 regulates expression of the PTEN tumor suppressor gene in human hepatocellular cancer. Gastroenterology. 133: 647–658.

Michalak, P. 2006. RNA world—the dark matter of evolutionary genomics. J Evol Biol. 19: 1768–1774.

Mishra, P.J., R. Humeniuk, G.S. Longo-Sorbello, D. Banerjee, and J.R. Bertino. 2007. A miR-24 microRNA binding-site polymorphism in dihydrofolate reductase gene leads to methotrexate resistance. Proc Natl Acad Sci USA. 104: 13513–13518.

Nakano, H., T. Miyazawa, K. Kinoshita, Y. Yamada, and T. Yoshida. 2010. Functional screening identifies a microRNA, miR-491 that induces apoptosis by targeting Bcl-X(L) in colorectal cancer cells. Int J Cancer. 127: 1072–1080.

Nobusawa, S., H. Yokoo, J. Hirato, A. Kakita, H. Takahashi, T. Sugino, K. Tasaki, H. Itoh, T. Hatori, Y. Shimoyama et al. 2012. Analysis of Chromosome 19q13.42 Amplification in Embryonal Brain Tumors with Ependymoblastic Multilayered Rosettes. Brain Pathol.

O'Donnell, K.A., E.A. Wentzel, K.I. Zeller, C.V. Dang, and J.T. Mendell. 2005. c-Myc-regulated microRNAs modulate E2F1 expression. Nature. 435: 839–843.

Ota, A., H. Tagawa, S. Karnan, S. Tsuzuki, A. Karpas, S. Kira, Y. Yoshida, and M. Seto. 2004. Identification and characterization of a novel gene, C13orf25, as a target for 13q31-q32 amplification in malignant lymphoma. Cancer Res. 64: 3087–3095.

Park, S.M., S. Shell, A.R. Radjabi, R. Schickel, C. Feig, B. Boyerinas, D.M. Dinulescu, E. Lengyel, and M.E. Peter. 2007. Let-7 prevents early cancer progression by suppressing expression of the embryonic gene HMGA2. Cell Cycle. 6: 2585–2590.

Petrocca, F., R. Visone, M.R. Onelli, M.H. Shah, M.S. Nicoloso, I. de Martino, D. Iliopoulos, E. Pilozzi, C.G. Liu, M. Negrini et al. 2008. E2F1-regulated microRNAs impair TGFbeta-dependent cell-cycle arrest and apoptosis in gastric cancer. Cancer Cell. 13: 272–286.

Poy, M.N., L. Eliasson, J. Krutzfeldt, S. Kuwajima, X. Ma, P.E. Macdonald, S. Pfeffer, T. Tuschl, N. Rajewsky, P. Rorsman et al. 2004. A pancreatic islet-specific microRNA regulates insulin secretion. Nature. 432: 226–230.

Ramsingh, G., D.C. Koboldt, M. Trissal, K.B. Chiappinelli, T. Wylie, S. Koul, L.W. Chang, R. Nagarajan, T.A. Fehniger, P. Goodfellow et al. 2010. Complete characterization of the microRNAome in a patient with acute myeloid leukemia. Blood. 116: 5316–5326.

Reinhart, B.J., F.J. Slack, M. Basson, A.E. Pasquinelli, J.C. Bettinger, A.E. Rougvie, H.R. Horvitz, and G. Ruvkun. 2000. The 21-nucleotide let-7 RNA regulates developmental timing in Caenorhabditis elegans. Nature. 403: 901–906.

Rodriguez, A., E. Vigorito, S. Clare, M.V. Warren, P. Couttet, D.R. Soond, S. van Dongen, R.J. Grocock, P.P. Das, E.A. Miska et al. 2007. Requirement of bic/microRNA-155 for normal immune function. Science. 316: 608–611.

Ryu, S., N. Joshi, K. McDonnell, J. Woo, H. Choi, D. Gao, W.R. McCombie, and V. Mittal. 2011. Discovery of novel human breast cancer microRNAs from deep sequencing data by analysis of pri-microRNA secondary structures. PLoS One. 6: e16403.

Saito, Y. and P.A. Jones. 2006. Epigenetic activation of tumor suppressor microRNAs in human cancer cells. Cell Cycle. 5: 2220–2222.

Sayed, D., S. Rane, J. Lypowy, M. He, I.Y. Chen, H. Vashistha, L. Yan, A. Malhotra, D. Vatner, and M. Abdellatif. 2008. MicroRNA-21 targets sprouty2 and promotes cellular outgrowths. Mol Biol Cell. 19(8): 3272–3282.

Scott, G.K., M.D. Mattie, C.E. Berger, S.C. Benz, and C.C. Benz. 2006. Rapid alteration of microRNA levels by histone deacetylase inhibition. Cancer Res. 66: 1277–1281.

Selcuklu, S.D., M.T. Donoghue, and C. Spillane. 2009. miR-21 as a key regulator of oncogenic processes. Biochem Soc Trans. 37: 918–925.

Shalgi, R., D. Lieber, M. Oren, and Y. Pilpel. 2007. Global and local architecture of the mammalian microRNA-transcription factor regulatory network. PLoS Comput Biol. 3: e131.

Shimono, Y., M. Zabala, R.W. Cho, N. Lobo, P. Dalerba, D. Qian, M. Diehn, H. Liu, S.P. Panula, E. Chiao et al. 2009. Downregulation of miRNA-200c links breast cancer stem cells with normal stem cells. Cell. 138: 592–603.

Shirdel, E.A., W. Xie, T.W. Mak, and I. Jurisica. 2011. NAViGaTing the micronome--using multiple microRNA prediction databases to identify signalling pathway-associated microRNAs. PLoS One. 6: e17429.

Si, M.L., S. Zhu, H. Wu, Z. Lu, F. Wu, and Y.Y. Mo. 2007. miR-21-mediated tumor growth. Oncogene. 26(19): 2799–2803.

Singh, R. and N. Saini. 2012. Downregulation of BCL2 by miRNAs augments drug induced apoptosis: Combined computational and experimental approach. J Cell Sci. 125(Pt 6): 1568–1578.

Takakura, S., N. Mitsutake, M. Nakashima, H. Namba, V.A. Saenko, T.I. Rogounovitch, Y. Nakazawa, T. Hayashi, A. Ohtsuru, and S. Yamashita. 2008. Oncogenic role of miR-17-92 cluster in anaplastic thyroid cancer cells. Cancer Sci. 99: 1147–1154.

Takamizawa, J., H. Konishi, K. Yanagisawa, S. Tomida, H. Osada, H. Endoh, T. Harano, Y. Yatabe, M. Nagino, Y. Nimura et al. 2004. Reduced expression of the let-7 microRNAs in human lung cancers in association with shortened postoperative survival. Cancer Res. 64: 3753–3756.

Takeshita, F., L. Patrawala, M. Osaki, R.U. Takahashi, Y. Yamamoto, N. Kosaka, M. Kawamata, K. Kelnar, A.G. Bader, D. Brown et al. 2010. Systemic delivery of synthetic microRNA-16 inhibits the growth of metastatic prostate tumors via downregulation of multiple cell-cycle genes. Mol Ther. 18: 181–187.

Taulli, R., F. Bersani, V. Foglizzo, A. Linari, E. Vigna, M. Ladanyi, T. Tuschl, and C. Ponzetto. 2009. The muscle-specific microRNA miR-206 blocks human rhabdomyosarcoma growth in xenotransplanted mice by promoting myogenic differentiation. J Clin Invest. 119: 2366–2378.

Trang, P., P.P. Medina, J.F. Wiggins, L. Ruffino, K. Kelnar, M. Omotola, R. Homer, D. Brown, A.G. Bader, J.B. Weidhaas et al. 2010. Regression of murine lung tumors by the let-7 microRNA. Oncogene. 29: 1580–1587.

Tsai, K.W., H.W. Kao, H.C. Chen, S.J. Chen, and W.C. Lin. 2009. Epigenetic control of the expression of a primate-specific microRNA cluster in human cancer cells. Epigenetics. 4: 587–592.

Volinia, S., G.A. Calin, C.G. Liu, S. Ambs, A. Cimmino, F. Petrocca, R. Visone, M. Iorio, C. Roldo, M. Ferracin et al. 2006. A microRNA expression signature of human solid tumors defines cancer gene targets. Proc Natl Acad Sci USA. 103: 2257–2261.

Volinia, S., M. Galasso, S. Costinean, L. Tagliavini, G. Gamberoni, A. Drusco, J. Marchesini, N. Mascellani, M.E. Sana, R. Abu Jarour et al. 2010. Reprogramming of miRNA networks in cancer and leukemia. Genome Res. 20: 589–599.

Wang, V. and W. Wu. 2009. MicroRNA-based therapeutics for cancer. Bio Drugs. 23: 15–23.

Watanabe, Y. and A. Kanai. 2011. Systems Biology Reveals MicroRNA-Mediated Gene Regulation. Front Genet. 2: 29.

Wightman, B., I. Ha, and G. Ruvkun. 1993. Posttranscriptional regulation of the heterochronic gene lin-14 by lin-4 mediates temporal pattern formation in C. elegans. Cell. 75: 855–862.

Winter, J., S. Jung, S. Keller, R.I. Gregory, and S. Diederichs. 2009. Many roads to maturity: microRNA biogenesis pathways and their regulation. Nat Cell Biol. 11: 228–234.

Wu, W. 2011. Modulation of microRNAs for potential cancer therapeutics. Methods Mol Biol. 676: 59–70.

Wu, W., M. Sun, G.M. Zou, and J. Chen. 2007. MicroRNA and cancer: Current status and prospective. Int J Cancer. 120(5): 953–960.

Xia, L., D. Zhang, R. Du, Y. Pan, L. Zhao, S. Sun, L. Hong, J. Liu, and D. Fan. 2008. miR-15b and miR-16 modulate multidrug resistance by targeting BCL2 in human gastric cancer cells. Int J Cancer. 123: 372–379.

Yu, F., H. Yao, P. Zhu, X. Zhang, Q. Pan, C. Gong, Y. Huang, X. Hu, F. Su, J. Lieberman et al. 2007. let-7 regulates self renewal and tumorigenicity of breast cancer cells. Cell. 131: 1109–1123.

Zhang, L., J. Huang, N. Yang, J. Greshock, M.S. Megraw, A. Giannakakis, S. Liang, T.L. Naylor, A. Barchetti, M.R. Ward et al. 2006. microRNAs exhibit high frequency genomic alterations in human cancer. Proc Natl Acad Sci USA. 103: 9136–9141.

Zhao, J.J., J. Lin, H. Yang, W. Kong, L. He, X. Ma, D. Coppola, and J.Q. Cheng. 2008. MicroRNA-221/222 negatively regulates estrogen receptor alpha and is associated with tamoxifen resistance in breast cancer. J Biol Chem. 283: 31079–31086.

Zhu, S., M.L. Si, H. Wu, and Y.Y. Mo. 2007. MicroRNA-21 targets the tumor suppressor gene tropomyosin 1 (TPM1). J Biol Chem. 282: 14328–14336.

CHAPTER 15

SYSTEMS AND NETWORK BIOLOGY IN MicroRNA BASED PERSONALIZED MEDICINE

Asfar S. Azmi,[1] Bin Bao,[1] Philip A. Philip,[2]
Ramzi M. Mohammad[2] and Fazlul H. Sarkar[1,2,*]

ABSTRACT

MicroRNAs (miRNAs), a class of endogenous small noncoding RNAs, mediate post-transcriptional regulation of protein-coding genes by binding to the 3' untranslated region of target mRNAs, leading to translational inhibition, mRNA destabilization and/or degradation. A single miRNA could concurrently down-regulate hundreds of target mRNAs, and as it may be context dependent for specificity, it fine-tunes gene expression involved in diverse cellular functions, such as development, differentiation, proliferation, apoptosis and metabolism. Emerging research on this topic has revealed that miRNA alterations are involved in the initiation and progression of human cancer. The differential expression of miRNA genes in malignant compared with normal cells could in part be explained by the location of these genes in cancer-associated genomic regions, and by epigenetic mechanisms

[1]Department of Pathology and Karmanos Cancer Institute, Wayne State University School of Medicine, 740 HWCRC, 4100 John R Street, Detroit, MI 48201.
[2]Oncology, Karmanos Cancer Institute, Wayne State University School of Medicine, 740 HWCRC, 4100 John R Street, Detroit, MI 48201.
*Corresponding author: Department of Pathology, Wayne State University School of Medicine, Karmanos Cancer Institute, 4100 John R. HWCRC Room 732, Detroit MI, 48201 USA.
Email: fsarkar@med.wayne.edu

due to deregulations in the miRNA processing machinery. Additionally, miRNA-expression profiling of human tumors has identified signatures associated with diagnosis, staging, progression, prognosis and response to treatment. The expression of miRNAs appears to serve as hubs of regulatory networks underlying complex diseases. Recent evidence has suggested that miRNAs might be viable therapeutic targets for a wide range of diseases, including cancer. However, it remains unknown whether the set of miRNA target genes designated "targetome" regulated by an individual miRNA constitutes the biological network of functionally-associated molecules or reflects a random set of functionally-independent genes. Prior to clinical application of miRNA as therapeutics for cancer, experimental evidence for coordinated regulation of a large number of genes by miRNAs that includes the entire set of on and off target pathways is absolutely necessary. Nevertheless, such type of detailed analysis cannot be achieved by traditional molecular biology techniques that have their own limitations. The complexity of the action of miRNAs calls for comprehensive, integrative system approaches to examine the effect of these miRNAs in their entirety. In order to reveal such complexity, "systems biology" is the science that combines integrated methodologies with molecular analysis and intervention and can be helpful in addressing clinically relevant questions. Notably, advanced proteomic techniques have begun to be utilized in the analysis of widespread effects of miRNAs. Other approaches that have been used include large scale sequencing of miRNA and potential miRNA targets, mRNA expression profiling, and bioinformatic modeling. Systems technologies can help in mining large miRNA data-sets, help rapidly translate complex miRNA target interactomes that can aid in the identification of target patient population which may be most effectively respond to any newly designed miRNA based therapy. We are of the view that when combined together, systems biology can accelerate miRNA therapeutic development and the two can become very powerful tools in personalized cancer medicine.

1. INTRODUCTION

MicroRNAs (miRNA) comprise of a class of short noncoding RNAs that are 18–25 nucleotides in length found in all animal and plant cells. In 1993, the first miRNAs were recognized in *Caenorhabditis elegans* by Lau et al (Lau et al. 2001). Later on, various small regulatory RNAs were discovered in plants and mammals and designated as 'microRNA' (Lagos-Quintana et al. 2003, Lagos-Quintana et al. 2001). Currently, more than 1200 human miRNAs are registered in the miRBase database and have been extensively studied for their involvement in RNA interference (RNAi) i.e. how they regulate gene expression post-transcriptionally, and how they contribute to diverse physiological and pathological functions (Bartel 2004). The biogenesis and RNAi functions of miRNA (i.e. how miRNAs are generated and processed into a mature form, and how they regulate gene expression) have been intensively investigated and well-described. Furthermore, developments in miRNA-related technologies, such as miRNA expression profiling and synthetic oligoRNA, have contributed to the identification of miRNAs involved in a number of physiological and pathological phenotypes. A Pubmed search for microRNAs returns 15,549 hits (on July 2, 2012), similarly key words microRNA and cancer return 5,711 research articles. Most striking is the observation that research publications from 2000 to 2011 show an exponential increase in research indicating that the field is advancing rapidly. These studies have led to a deeper understanding of microRNA biogenesis, their regulatory control on different genes and strategies to target them for anti-cancer therapeutic benefits. There has been a drive at the pharmaceutical front to identify and target miRNAs for therapeutic benefit. A number of strategies have been proposed as potential miRNA targeted therapy and these include, anti-sense to miRNA target sequence, pre miRNA synthetic oligonucleotide, peptide nucleic acids (PNAs) and many more (Garzon et al. 2010). Nevertheless, the design of pre- and anti-miRNA agents is not as straightforward as originally thought because key questions remain largely unanswered, such as how miRNA expression is controlled and which genes are regulated by each miRNA. Additionally it is not known whether individual miRNAs regulate certain set of genes in isolation or they are part of a complex network and work in tandem with other miRNA. Additional degree of complexity is added by the recent finding that a majority of miRNAs are regulated epigenetically as well (Sato et al. 2011). Here, we will introduce the concept that novel integrated technologies such as systems biology and network modeling can be utilized for better understanding of the complex interaction network between miRNAs and that will, in turn, aid in the design of personalized miRNA based therapy.

A number of key observations early in the history of miRNA discovery suggested their potential role in human cancer. Firstly, the earliest miRNAs discovered in the roundworm *C. elegans* and the fruit fly *Drosophila* were shown to control cell proliferation and apoptosis (Lu et al. 2005b). These observations led to the hypothesis that miRNA deregulation may contribute to proliferative diseases that loose proliferation control such as cancer. Secondly, human miRNAs genes were located at fragile sites in the genome or regions that are commonly amplified or deleted in human cancers (Calin et al. 2004). Last but not the least, malignant tumors and tumor cell lines were found to have widespread deregulated miRNA expression compared to normal tissues. The questions remained whether the altered miRNA expression observed in cancer is a cause or consequence of malignant transformation. A number of very recent and comprehensive reviews deeply evaluated all the currently known miRNAs and their functions in different cancers (Lovat et al. 2011). Since the discovery of the causal link between miRNAs and cancer, investigators have shifted their focus towards development of therapeutic strategies involving these small RNAs. Nevertheless, as discussed below, miRNAs form a highly complex network that needs to be thoroughly investigated using integrated science prior to designing therapies for cancer.

2. COMPLEX MicroRNA INTERACTION NETWORK

Even though miRNAs account for <1% of predicted genes in higher eukaryotic genomes yet they have been predicted to regulate between 10–30% of genes (Axtell 2008). Furthermore, miRNA targets include diverse set of macromolecules ranging from signaling proteins, enzymes and transcription factors (TFs) (Hieronymus and Silver 2004). The diversity and abundance of miRNA targets offer an exponential level of combinatorial possibilities suggesting that miRNAs and their targets appear to form a complex regulatory network intertwined with other cellular networks such as signal transduction networks. However, it is unclear how miRNAs might orchestrate the regulation of cellular signaling networks and how regulation of these networks might contribute to the biological functions of miRNAs. Additionally, recent evidence has shown that miRNAs either regulate or are themselves under epigenetic control thereby forming a regulatory circuit that maintains normal physiological functions (Azmi et al. 2011). Alteration in this multi-tier regulatory control results in disruption of the regulatory circuit that manifests itself in different diseases including cancer. The most important question is whether in cancer, a set of miRNA target genes regulated by an individual miRNA generally constitute the biological network of functionally-associated molecules or simply reflect a random set of functionally-independent genes. If former is the case, then what kind of biological networks does the human microRNAome most actively regulate? These are critical issues that need to be addressed if clinical progress is to be made in the direction of incorporating miRNA mediated regulation as a therapy for cancer.

2.1 Utilizing Systems and Network Biology to Unwind the Complexity of miRNA Network

Binding of miRNA to 3'-UTR does not require perfect complementarities and this promiscuity makes it possible for a miRNA to regulate several genes in a pathway or even multiple pathways (Orom and Lund 2010). The effects could simply be parallel but could also be coordinated, additive, synergistic, or antagonistic. Any coordinated actions on multiple target genes would provide a powerful mechanism for a single miRNA to have significant impacts on a complex regulatory network and ultimately the physiological process or disease. Nevertheless, experimental evidence for coordinated regulation of a large number of genes by miRNAs is still rare. This is mainly due to the lack of proper tools in traditional molecular biology that cannot analyze multi-pathway changes at the same time (Qavi et al. 2010). It leaves open several fundamental questions that are important for determining the value of these small non-coding RNAs in complex regulatory networks. If a miRNA often regulates multiple genes that do not have close functional relationships, how does miRNA achieve any specificity for its effect on the cellular and organ systems function? It is possible that target genes of a miRNA may have functional connections among them that are not yet recognized. It is also imperative that the specificity of the effect of a miRNA may be partially determined by which target mRNAs are present in a given biological setting, which would suggest interaction between the miRNA mechanism and other mechanisms that could regulate gene expression. It is anticipated that answering these complex questions can pave the way to future design of personalized therapies. Consistent with the notion of miRNAs working with other mechanisms to fine-tune gene expression, many studies have shown that the effect of a miRNA on the abundance of a target is often modest. That also leads to the question of the regulatory relationship between different miRNAs and between miRNAs and their host or adjacent protein-encoding genes.

The complexity of the action of miRNAs calls for comprehensive, integrative systems level approaches to examine the effect of miRNAs (Wilbert and Yeo 2011). Notably, advanced genomics, proteomic, epigenomic techniques have begun to be utilized in the analysis of widespread effects of miRNAs (Hawkins et al. 2010). Other approaches that have been used include large-scale sequencing of miRNA and potential miRNA targets, mRNA expression profiling, and bioinformatics modeling (Lu et al. 2005a). Sequencing of cleaved fragments of mRNAs has been used to identify miRNA targets (Henderson and Jacobsen 2008), the applicability of which would depend on the extent defining which miRNAs would induce which mRNA cleavage in a given species. Advanced genomic and genome-related approaches that are coupled with systems biology methodologies in the study of miRNA have been appreciated recently. Integration of genome-related approaches with physiological and clinical approaches has been shown to be valuable for further elucidating the role of miRNAs in systems and personalized medicine.

Due to the large number of putative microRNA gene targets predicted by sequence-alignment databases and the relative low accuracy of such predictions which are conducted independently of biological context by design, systematic experimental identification and validation of every functional microRNA target is currently a challenging area. Consequently, biological studies have yet to identify, on a genome scale, key regulatory networks perturbed by altered microRNA functions in the context of cancer. However, some in-roads have been made in understanding these deregulatory miRNA networks in cancer. Aided by high-throughput technologies such as systems biology and network modeling, a number of investigations have been performed that have led to a greater understanding of miRNA targets and genome-wide studies to computationally predict miRNA target genes have been performed in a number of different laboratories (Krek et al. 2005). These investigations have further been utilized to map all the overlapped miRNA targets in the human signaling network proteins to conduct a network-structural analysis. Such analyses have demonstrated strategies as to how miRNA regulates signaling networks and have provided clues on weaker miRNA nodes targeting which can have maximal impact. Through these investigations, it was found that miRNAs more frequently target signaling proteins than others e.g. ~30% of the network proteins and the total genes in human genome, respectively, are miRNA targets. This discovery implies that miRNAs might play a relatively more important role in regulating signaling networks than in other cellular processes. Normally, in signaling networks, cellular signal information flow initiates from extra-cellular space: a ligand binds to a cellular membrane receptor to start the signal, which is then transmitted by intracellular signaling components in cytosol and finally reaches the signaling components in the nucleus. Systems biology shows that the fraction of miRNA targets increases with the signal information flow from the top to bottom e.g. from ligands, cell surface receptors, and intracellular signaling proteins to nuclear proteins (Cui et al. 2007, Cui et al. 2006). For example, less than 10% of the ligands are miRNA targets, whereas ~50% of the nuclear proteins, most of which are transcription factors, are miRNA targets. In other words, the miRNA targets are enriched more than five times in the most downstream proteins compared to the most upstream proteins. This indicates that in order to achieve therapeutic benefit, targeting ligand or receptor modulating miRNA would not be the ideal strategy. On the other hand, a strategy designed against nuclear protein, or transcription factor will have a superior effect that could translate into clinical efficacy. As will be discussed in the following paragraphs, our laboratory has shown some proof of principal strategies that target transcription factors such as NF-κB, p53 and their miRNA regulatory networks for efficient killing of cancer cells [Fig. 1 showing complex miRNA network developed using Ingenuity Pathway Analysis (IPA) intertwined with critical master regulatory genes]. In signaling networks, adaptor proteins recruit downstream signaling components to the vicinity of receptors. They activate, inhibit or relocalize downstream components

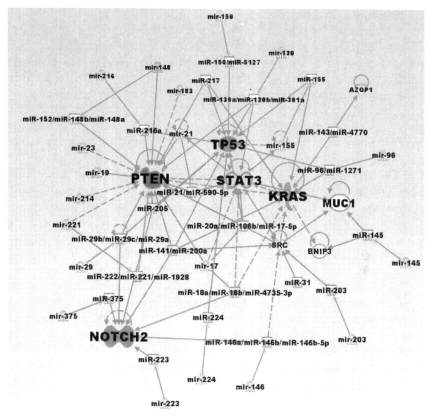

Figure 1. Systems Biology in Understanding Disease Specific miRNA Networks. Ingenuity Network analysis for miRNA interaction networks targeting major genes in pancreatic ductal adenocarcinoma from existing knowledgebase. Pathway was designed using Ingenuity Pathway Analysis (IPA) under defined limits (key entries miRNA and pancreatic ductal adenocarcinoma (PDAC)) incorporating direct (regular lines) and in-direct interactions (dashed lines). The entire miRNA network was found to center around six key pathways (KRAS, TP53, PTEN, STAT3, MUC and NOTCH). Critical miRNAs that formed multi-tier interaction with the six genes of interest were verified in a transgenic PDAC model as shown in Fig. 2.

Color image of this figure appears in the color plate section at the end of the book.

through direct protein–protein interactions. Adaptors do not have enzyme activity, but physically interact with upstream and downstream signaling proteins. One adaptor is able to recruit distinct downstream components in different cellular conditions. Studies have suggested that miRNAs preferentially target the downstream components of adaptors, which have potential to recruit additional downstream components.

As demonstrated by Cui and group (Cui et al. 2006), for example, the adaptor Grb2 (growth factor receptor-bound protein 2) directly interacts with 14 downstream signaling proteins, half of which are miRNA targets. These downstream components are functionally involved in different signaling pathways that lead to different cellular outputs. In the same study these authors showed that SHC (Src homology two domain containing) regulates cell growth and apoptosis through activation of small GTPases of the Ras family, while N-WASP (Wiskott–Aldrich syndrome protein) is involved in the regulation of actin-based cytoskeleton through activation of small GTPases of the Rho family and is targeted by different miRNAs. In yet another study, *cancer-miRNA* network was developed by mining the literature of experimentally verified cancer-miRNA relationships (Bandyopadhyay et al. 2010). This network revealed several new and interesting biological insights which were not evident in individual experiments, but become evident when studied in the global perspective. From the network a number of *cancer-miRNA* modules have been identified based on a computational approach to mine associations between cancer types and miRNAs. The modules that are generated based on these associations are found to have a number of common predicted target onco/tumor suppressor genes. This suggests a combinatorial effect of the module associated miRNAs on target gene regulation in selective cancer tissues or cell lines.

Moreover, neighboring miRNAs (group of miRNAs that are located within 50 kb of genomic location) of these modules show similar deregulation patterns, suggesting common regulatory pathway. Besides this, neighboring miRNAs may also show a similar deregulation patterns (differentially co-expressed) in the cancer tissues. Among the main findings included the observation that in 67% of the cancer types, at least, two neighboring miRNAs showing statistically significant down-regulation. A similar result was obtained for the neighboring miRNAs showing up-regulation in specific cancer type. These results elucidate the fact that the neighboring miRNAs might be differentially co-expressed in cancer tissues as that of the normal tissue types. Additionally, *cancer-miRNA* network efficiently detects hub miRNAs deregulated in many cancer types and identifies cancer specific miRNAs. Depending on the expression patterns, it is possible to identify those hubs that have strong oncogenic or tumor suppressor characteristics. Limited work has been done towards revealing the fact that a number of miRNAs can control commonly altered regulatory pathways. However, this becomes immediately evident by accompanying the analysis of cancer-miRNA relationships in the proposed network model. These raise many unanswered questions in miRNA research that have never been reported previously because such knowledge will have a significant implication in miRNA targeted therapeutics. Recently Bandyopadhye and group developed a cancer miRNA network map using bipartite graph theoretic approach. Their obtained network showed a number of new and interesting biological insights that were not visible during individual experiments, and were evident only when investigated utilizing a global perspective i.e. through systems level analysis.

From the network, a number of *cancer-miRNA* modules were identified based on a computational approach to mine associations between cancer types and miRNAs. The modules that are generated based on these associations were found to have a number of common predicted target onco/tumor suppressor genes. This suggests a combinatorial effect of the module associated miRNAs on target gene regulation in selective cancer tissues or cell lines. Moreover, neighboring miRNAs (group of miRNAs that are located within 50 kb of genomic location) of these modules show similar deregulation patterns, suggesting common regulatory pathway. Besides this, neighboring miRNAs may also show a similar deregulation patterns (differentially co-expressed) in the cancer tissues. In our study, we found that in 67% of the cancer types have at least two neighboring miRNAs showing down-regulation, which is statistically significant ($P < 10-7$, Randomization test). A similar result is obtained for the neighboring miRNAs showing up-regulation in specific cancer type.

These results elucidate the fact that the neighboring miRNAs might be differentially co-expressed in cancer tissues as that of the normal tissue types. Additionally, *cancer-miRNA* network efficiently detect hub miRNAs deregulated in many cancer types, and identify cancer specific miRNAs. Depending on the expression patterns, it is possible to identify those hubs that have strong oncogenic or tumor suppressor characteristics. Limited work has been done towards revealing the fact that a number of miRNAs can control commonly altered regulatory pathways. However, this becomes immediately evident by accompanying the analysis of cancer miRNA relationships in the proposed network model. These raise many unaddressed issues in miRNA research that have never been reported previously. These observations are expected to have an intense implication in cancer and may be useful for further research. Many more examples exist in literature that indicate to the usefulness of utilizing systems biology and other integrated technologies to unwind the exact role of miRNA in a complex network of yet to be identified disease networks.

3. SYSTEMS BIOLOGY CAN ADDRESS CHALLENGES IN THE DESIGN OF miRNA TARGETED THERAPY

The miRNA have enormous but as of yet unrealized potential as therapeutics for cancer. In recent years tumor suppressor miRNA mimics and oligonucleotide inhibitors complementary to oncogenic miRNA have been shown to restore normal cell programming (Medina and Slack 2009). However, the history of nucleic acid based drug development strategies proved that there are many important challenges in this area (Rupaimoole et al. 2011). As one miRNA regulates many genes, it is still not clear whether targeted inhibition or up-regulation of single miRNA would result in restricted set of gene modulations with minimal desired effect or it may cause multi-gene changes leading to undesirable side effect. Therefore, true understanding of miRNA regulatory network is integral to the successful design of miRNA based targeted therapy. Despite several technical advances, there are still many challenges such as determining the ideal design of anti-miRNA sequence, bypassing the activation of the immune system, off-target effects, and competition with endogenous microRNAs for cellular miRNA-processing machinery. Additionally, one must first identify the correct and relevant target miRNA that is critical in the disease of choice prior to designing anti- or pre-miRNA therapy.

Therefore, the translation of miRNA targeted technology into the clinic depends on resolving these challenges. In the following paragraph we will discuss some of the promising approaches that are currently being explored in identifying the most potent miRNA that could have maximal efficacy without unwanted effects. It is important to understand the specificity of any reagent employed in miRNA-knockdown studies. As a general rule, modifications that increase binding affinity also make an oligonucleotide more likely to hybridize to a mismatched target. It may be necessary to adjust the length and modification content of an oligonucleotide (that is, to adjust T_m) to achieve the desired balance between potency and specificity. From a practical standpoint, sometimes cross reactivity problems seen in an *in vitro* setting where high oligonucleotide concentrations are easily achieved are reduced when used *in vivo*, where high intracellular oligonucleotide concentrations are seldom seen. Thus, miRNA specificity should be studied in the actual system of interest and results cannot always be generalized between model systems.

In an elegant review from Lieberman's group the challenges associated with the design of miRNA based therapeutic strategies were evaluated (Petrocca and Lieberman 2011). These authors discussed how miRNA based drugs need to survive in extracellular fluids rich in RNA-degrading ribonucleases, and to be taken up into the cytoplasm of cells where they can interact with the RISC in the cytoplasm. Although endocytosis via cell surface receptors might be a good first step for entry into cells, which has the potential and added advantage of cell-specific targeting, a workable delivery strategy needs to release the RNA efficiently from the endosome. This step appears to be a stumbling block for many delivery approaches. The half-life of small RNAs in extracellular fluids and serum can be extended to several days without significantly affecting gene silencing activity by the same sorts of chemical modifications that reduce off-target effects and by changing the nucleotide linkages at the ends to phosphorothioates that are resistant to exonuclease cleavage.

In addition, naked small RNAs are rapidly excreted by the kidney; therefore, they need to be modified or complexed into a larger particle to circumvent rapid elimination via glomerular filtration (Whitehead et al. 2009). Moreover, it is well known that miRNAs do not get across the cell's plasma membrane and are not taken up even by cells, like microphages, that are constantly sampling their environment. The major obstacle to RNAi-based therapeutics is intracellular delivery. This is especially true for cancer therapy, given that delivery is more problematic for cells disseminated throughout the body, as is the case for most cancers that are not cured by surgery. Several hundred biotechnology and pharmaceutical companies are working to solve the small RNA delivery problem.

Systems biology and associated mathematical modeling have helped in developing robust models for understanding miRNA turn-over/decay. For example, Bryan Williams group have utilized mathematical modeling to investigate in-depth the turn-over of miRNAs and concluded that the average miRNA half-life is independent of cell proliferation and provides direct evidence that miRNA turn-over can vary widely among miRNAs (Gantier et al. 2011). Honing on three miRNAs mir-155, mir-21 and mir-146a, these authors demonstrated that the decay or turn-over profile was unique to each miRNA. The authors showed that mir-155 was less stable than mir-21 or mir-146a. Additionally, their modeling results revealed that some miRNAs such as miR-125b are more persistent than others. Increased stability of select miRNAs is unlikely to be related to a 3' modification of these miRNAs, as indicated by a recent genome-wide study that failed to demonstrate an impact of 3' adenylation on miRNA stability (Burroughs et al. 2010). Even though performed on a small cohort of cells, these systems level investigations do provide insights as to which particular miRNA can be more stable and also be an ideal target for oligonucleotide drug design.

4. SYSTEMS AND NETWORK BIOLOGY AND ITS APPLICATION IN miRNA BASED PERSONALIZED MEDICINE

Multi-dimensional data-sets permit the identification of multimodality aberrations that occur when a critical signaling protein is targeted in multiple ways in different patients. Interspecies comparative integrated analysis may provide a novel approach to identify driver miRNAs and can also aid in linking evolutionary conservation of miRNA aberrations across species points to their critical in tumorigenesis. High-throughput microRNA screening assays have been developed to facilitate a comprehensive evaluation of the role of different miRNAs in cellular functions and potential synthetic lethality. Our laboratory has used systems biology to first identify deregulated miRNA data-sets in complex cancers followed by anti-miR targeted therapy against the identified targets (Li et al. 2010).

In the first study of its kind, using miRNA microarrays, we reported lower expression of miR-146a in pancreatic cancer cells compared with normal human pancreatic duct epithelial cells. Re-expression of miR-146a inhibited the invasive capacity of pancreatic cancer cells with concomitant down-regulation of EGFR and the NF-κB regulatory kinase interleukin-1 receptor-associated kinase 1 (IRAK-1). Cellular mechanism studies revealed crosstalk between EGFR, IRAK-1, IκBα, NF-κB, and MTA-2, a transcription factor that regulates metastasis. Most significantly, treatment of pancreatic cancer cells with the natural products 3,3'-diinodolylmethane (DIM) or isoflavone, which increased miR-146a expression, caused a down-regulation of EGFR, MTA-2, IRAK-1, and NF-κB, resulting in an inhibition of pancreatic cancer cell invasion. These findings revealed that DIM and isoflavone as nontoxic activators of a miRNA that can block pancreatic cancer cell invasion and metastasis, offering starting points to design novel miRNA targeted anticancer strategies (Table 1 showing some of the Systems Biology based miRNA studies). In another proof of concept study, we investigated the expression of miRNAs in pancreas tissues obtained from transgenic mouse models of K-Ras (K), Pdx1-Cre (C), K-Ras; Pdx1-Cre (KC) and K-Ras; Pdx1-Cre; INK4a/Arf (KCI), initially from pooled RNA samples using miRNA profiling, and further confirmed in individual specimens by quantitative RT-PCR (Fig. 2) (Ali et al. 2011).

Expression profiles of miRNAs could be highly informative in discriminating malignant from the normal pancreas because of the unique patterns of expression of miRNAs in tumors. We found over-expression of miR-21, miR-221, miR-27a, miR-27b and miR-155, and down-regulation of miR-216a, miR-216b, miR-217 and miR-146a expression in tumors derived from KC and KCI mouse model, which was consistent with data from KCI-derived RInk-1 cells. Mechanistic investigations revealed a significant induction of EGFR, K-Ras, and MT1-MMP protein expression in tissues from both KC and KCI mouse compared to tissues from K or C, and these results were consistent with similar findings in RInk-1 cells compared to human MIAPaCa-2 cells. We performed network modeling on the miRNAs microarray data-sets and have identified a total of 187 deregulated miRNAs. These miRNAs provided key miRNAs that were targeted using different strategies to induce cell killing in transgenic derived Rink-1 cells. For example, instead of targeting all deregulated miRNAs, we focused on few key miR's (miR-21, miR-155, miR-220, miR-143 and miR-217 and miR-217) that were central nodes in the entire miRNA network. Knock-down of

Table 1. Some examples of miRNA systems level investigations.

Method	Type of Mirna Expression Analysis	Advantage	Representative Reference
High Throughput sequencing for identification of miRNAs	Identification of miRNAs	Ability to detect evidence of expressed miRNAs in a cell and context-specific manner	(Lu et al. 2005a)
miRNA target prediction algorithms	Identification of global miRNAs	Genome-wide miRNA predictions for evolutionarily-important miRNAs	(Brennecke et al. 2005)
Transcriptome and proteome profiling for target finding	experimental identification of miRNA targets	detects mRNA and proteome changes in response to miRNA treatment	(Baek et al. 2008)
IP-based approaches for miRNA target finding	identification of miRNA targets	immunoprecipitation of RNA-binding proteins, followed by microarray analysis, to study associated RNA transcripts, cell specific posttranscriptional regulation	(Karginov et al. 2007)
CLIP and HT-sequencing based miRNA target finding	miRNA targets	crosslinking immunoprecipitation (CLIP) coupled with high-throughput sequencing (HITS-CLIP) to identification of RNA-binding protein target sites at nucleotide-level resolution	(Ule et al. 2003, Ule et al. 2005)

miR-155 in RInk-1 cells resulted in the inhibition of cell growth and colony formation, which was consistent with down-regulation of EGFR, MT1-MMP and K-Ras expression. While forced re-expression of Ras targeting miR-216b showed inhibition of cell proliferation and colony formation, which was correlated with reduced expression of Ras, EGFR and MT1-MMP. These findings suggest that systems biology would be useful for preclinical evaluation of novel miRNA-targeted agents for designing personalized therapy for pancreatic cancer (PC).

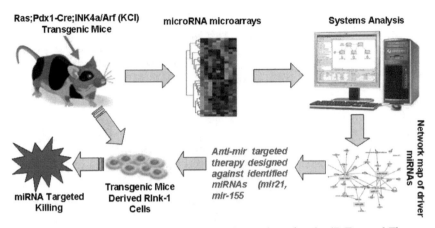

Figure 2. Ingenuity Pathway Analysis Guides the Design of anti-miR Targeted Therapy against Disease miRNA Networks in PDAC Transgenic Model. The miRNA microarrays were run on pooled RNA from Ras; Pdx1-Cre; INK4a/Arf (KCI) transgenic PDAC model followed by IPA analysis for miRNA networks. The transgenic network centers around few key miR's (miR-21, mir-155, miR-220, miR-143 and miR-217 and miR-217) that correlates with those observed from IPA knowledgebase as obtained in Fig. 1 above.

Color image of this figure appears in the color plate section at the end of the book.

5. CONCLUSIONS

As cancer biology has provided direction for investigations of miRNA biology, the discovery of miRNAs has also changed our thinking in cancer research. Profiling of human tumors has identified signatures associated with risk, diagnosis, prognosis and therapeutic response. Reports of differential expression of highly conserved non-coding genes in malignant cells relative to normal cells are providing better understanding of disease progression. Correlation between genomic locations of miRNAs and cancer-associated genomic regions has coupled aberrant miRNA expression to the genetic basis of cancer. The expanded study of pathways relevant to cancer processes is leading to new insights on the cellular and epigenetic regulation of miRNA genes and miRNA pathway components. Taken together, these advances comprise revisiting of molecular oncology dogma, and re-opening of Pandora's box, from which miRNAs and highly conserved non-coding genes emerge as principal candidates for cancer-predisposing genes. Genetics and biochemistry have been of crucial importance in developing the models and rules that define where the field is today. The next steps are clearly in the direction of systems biology, which will take us from investigations of single miRNA and single targets to more comprehensive explorations of the regulatory networks through which they interact and function (Fig. 3

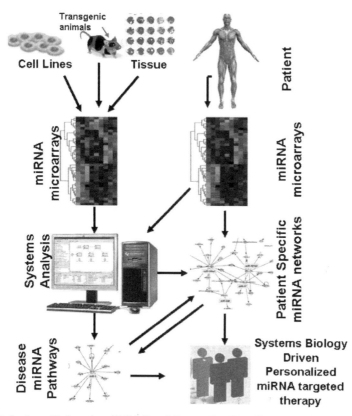

Figure 3. Systems Biology in miRNA Based Personalized Medicine. Advanced integrated technologies can be applied in different ways to derive solid evidence for development of anti- or pre-miR therapy against identified miRNAs. Expression analysis identifying miRNA maps can be performed on RNA isolated from different model systems such as cell lines, corresponding patient derived samples (both active diseased patients or prospective samples such as paraffin imbedded tissue samples, and transgenic animal model systems). The miRNA expression data-sets then can be mapped using miRNA ontology tools such as IPA. Once the critical disease specific of driver miRNAs are identified, they can be correlated with miRNA network maps on RNA isolated from patients undergoing treatment. Depending on the presence or absence of certain miRNAs in the patient anti- or pre-miRNA therapy can be designed for beneficial clinical outcome.

Color image of this figure appears in the color plate section at the end of the book.

summarizing how systems and network biology can aid in miRNA directed personalized medicine). For miRNA personalized therapeutic aspirations to be realized, researchers can expect that the synergy between molecular techniques and systems level approaches, which has been such a critical element of the miRNA field, will be more essential than ever.

6. SUMMARY POINTS

- miRNAs have complex biology that cannot be interrogated using traditional molecular biology techniques alone.
- Their complexity of interactions call for computational analysis that takes advantage of newer computational tools such as systems and network biology.
- Using systems and network biology we and others have been able to prioritize cancer related miRNAs.
- Network modeling and miRNA-mRNA pathway interaction analysis has revealed the role of certain key miRNAs that could be targeted for beneficial outcome in cancer.

ACKNOWLEDGEMENTS

National Cancer Institute, NIH grant R01CA109389 (R.M. Mohammad) and NIH grant 5R01CA101870, 5R01CA131151, 1R01CA132794 and 1R01CA154321 awarded to FHS, are acknowledged. We also sincerely acknowledge the Guido foundation for their support.

ABBREVIATIONS

miRNAs	:	microRNAs
NF-κB	:	Nuclear Factor Kappa Beta
EGFR	:	Epidermal Growth Factor Receptor
PDAC	:	Pancreatic Ductal Adenocarcinoma
IPA	:	Ingenuity Pathway Analysis
TFs	:	Transcription Factors
RNAi	:	RNA interference
MMP	:	Matrix Metalloproteinase
KC	:	K-Ras; Pdx1-Cre
KCI	:	K-Ras; Pdx1-Cre; INK4a/Arf

REFERENCES

Ali, S., S. Banerjee, F. Logna, B. Bao, P.A. Philip, M. Korc, and F.H. Sarkar. 2011. Inactivation of Ink4a/Arf leads to deregulated expression of miRNAs in K-Ras transgenic mouse model of pancreatic cancer. J Cell Physiol. 227(10): 3373–80.

Axtell, M.J. 2008. Evolution of microRNAs and their targets: are all microRNAs biologically relevant? Biochim Biophys Acta. 1779: 725–734.

Azmi, A.S., F.W. Beck, B. Bao, R.M. Mohammad, and F.H. Sarkar. 2011. Aberrant epigenetic grooming of miRNAs in pancreatic cancer: a systems biology perspective. Epigenomics. 3: 747–759.

Baek, D., J. Villen, C. Shin, F.D. Camargo, S.P. Gygi, and D.P. Bartel. 2008. The impact of microRNAs on protein output. Nature. 455: 64–71.

Bandyopadhyay, S., R. Mitra, U. Maulik, and M.Q. Zhang. 2010. Development of the human cancer microRNA network. Silence. 1: 6.

Bartel, DP. 2004. MicroRNAs: genomics, biogenesis, mechanism, and function. Cell. 116: 281–297.

Brennecke, J., A. Stark, R.B. Russell, and S.M. Cohen. 2005. Principles of microRNA-target recognition. PLoS Biol. 3: e85.

Burroughs, A.M., Y. Ando, M.J. de Hoon, Y. Tomaru, T. Nishibu, R. Ukekawa, T. Funakoshi, T. Kurokawa, H. Suzuki, Y. Hayashizaki, and C.O. Daub. 2010. A comprehensive survey of 3' animal miRNA modification events and a possible role for 3' adenylation in modulating miRNA targeting effectiveness. Genome Res. 20: 1398–1410.

Calin, G.A., C. Sevignani, C.D. Dumitru, T. Hyslop, E. Noch, S. Yendamuri, M. Shimizu, S. Rattan, F. Bullrich, M. Negrini, and C.M. Croce. 2004. Human microRNA genes are frequently located at fragile sites and genomic regions involved in cancers. Proc Natl Acad Sci USA. 101: 2999–3004.

Cui, Q., Z. Yu, E.O. Purisima, and E. Wang. 2006. Principles of microRNA regulation of a human cellular signaling network. Mol Syst Biol. 2: 46.

Cui, Q., Z. Yu, Y. Pan, E.O. Purisima, and E. Wang. 2007. MicroRNAs preferentially target the genes with high transcriptional regulation complexity. Biochem Biophys Res Commun. 352: 733–738.

Gantier, M.P., C.E. McCoy, I. Rusinova, D. Saulep, D. Wang, D. Xu, A.T. Irving, M.A. Behlke, P.J. Hertzog, F. Mackay, and B.R. Williams. 2011. Analysis of microRNA turnover in mammalian cells following Dicer1 ablation. Nucleic Acids Res. 39: 5692–5703.

Garzon, R., G. Marcucci, and C.M. Croce. 2010. Targeting microRNAs in cancer: rationale, strategies and challenges. Nat Rev Drug Discov. 9: 775–789.

Hawkins, R.D., G.C. Hon, and B. Ren. 2010. Next-generation genomics: an integrative approach. Nat Rev Genet. 11: 476–486.

Henderson, I.R. and S.E. Jacobsen. 2008. Sequencing sliced ends reveals microRNA targets. Nat Biotechnol. 26: 881–882.

Hieronymus, H. and P.A. Silver. 2004. A systems view of mRNP biology. Genes Dev. 18: 2845–2860.

Karginov, F.V., C. Conaco, Z. Xuan, B.H. Schmidt, J.S. Parker, G. Mandel, and G.J. Hannon. 2007. A biochemical approach to identifying microRNA targets. Proc Natl Acad Sci USA. 104: 19291–19296.

Krek, A., D. Grun, M.N. Poy, R. Wolf, L. Rosenberg, E.J. Epstein, P. Macmenamin, da P, I, K.C. Gunsalus, M. Stoffel, and N. Rajewsky. 2005. Combinatorial microRNA target predictions. Nat Genet. 37: 495–500.

Lagos-Quintana, M., R. Rauhut, W. Lendeckel, and T. Tuschl. 2001. Identification of novel genes coding for small expressed RNAs. Science. 294: 853–858.

Lagos-Quintana, M., R. Rauhut, J. Meyer, A. Borkhardt, and T. Tuschl. 2003. New microRNAs from mouse and human, RNA. 9: 175–179.

Lau, N.C., L.P. Lim, E.G. Weinstein, and D.P. Bartel. 2001. An abundant class of tiny RNAs with probable regulatory roles in Caenorhabditis elegans. Identification of novel genes coding for small expressed RNAs. Science. 294: 858–862.

Li, Y., T.G. Vandenboom, Z. Wang, D. Kong, S. Ali, P.A. Philip, and F.H. Sarkar. 2010. iR-146a suppresses invasion of pancreatic cancer cells. Cancer Res. 70: 1486–1495.

Lovat, F., N. Valeri, and C.M. Croce. 2011. MicroRNAs in the pathogenesis of cancer. Semin Oncol. 38: 724–733.

Lu, C., S.S. Tej, S. Luo, C.D. Haudenschild, B.C. Meyers, and P.J. Green. 2005a. Elucidation of the small RNA component of the transcriptome. Science. 309: 1567–1569.

Lu, J., G. Getz, E.A. Miska, E. varez-Saavedra, J. Lamb, D. Peck, A. Sweet-Cordero, B.L. Ebert, R.H. Mak, A.A. Ferrando, J.R. Downing, T. Jacks, H.R. Horvitz, and T.R. Golub. 2005b. MicroRNA expression profiles classify human cancers. Nature. 435: 834–838.

Medina, P.P. and F.J. Slack. 2009. Inhibiting microRNA function *in vivo*. Nat Methods. 6: 37–38.

Orom, U.A. and A.H. Lund. 2010. Experimental identification of microRNA targets. Gene. 451: 1–5.

Petrocca, F. and J. Lieberman. 2011. Promise and challenge of RNA interference-based therapy for cancer. J Clin Oncol. 29: 747–754.

Qavi, A.J., J.T. Kindt, and R.C. Bailey. 2010. Sizing up the future of microRNA analysis. Anal Bioanal Chem. 398: 2535–2549.

Rupaimoole, R., H.D. Han, G. Lopez-Berestein, and A.K. Sood. 2011. MicroRNA therapeutics: principles, expectations, and challenges. Chin J Cancer. 30: 368–370.

Sato, F., S. Tsuchiya, S.J. Meltzer, and K. Shimizu. 2011. MicroRNAs and epigenetics. FEBS J. 278: 1598–1609.

Ule, J., K.B. Jensen, M. Ruggiu, A. Mele, A. Ule, and R.B. Darnell. 2003. CLIP identifies Nova-regulated RNA networks in the brain. Science. 302: 1212–1215.

Ule, J., K.B. Jensen, A. Mele, and R.B. Darnell. 2005. CLIP: a method for identifying protein-RNA interaction sites in living cells. Methods. 37: 376–386.

Whitehead, K.A., R. Langer, and D.G. Anderson. 2009. Knocking down barriers: advances in siRNA delivery. Nat Rev Drug Discov. 8: 129–138.

Wilbert, M.L. and G.W. Yeo. 2011. Genome-wide approaches in the study of microRNA biology. Wiley Interdiscip Rev Syst Biol Med. 3: 491–512.

INDEX

Color Plate Section

Chapter 1

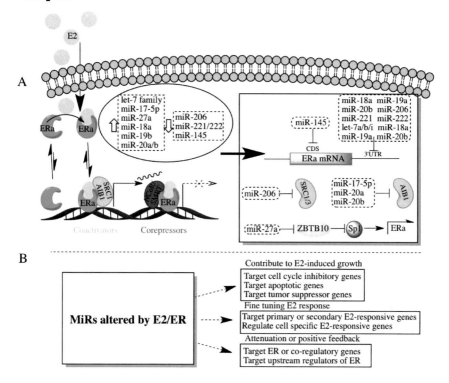

Figure 1. Estrogen-regulated miRs participate in autoregulatory feedback mechanisms. (A) E2 signaling modulates expression of numerous miRs that can directly or indirectly regulate ERα or co-regulatory proteins. E2 binding to ERα leads to ERα dimerization and localization to gene regulatory regions on DNA, followed by recruitment or dissociation of co-regulatory proteins. In response to E2, many miRs are up- or down-regulated that are involved in regulatory feedback loops through targeting ERα or SRC-1 and AIB1 co-activators. (B) In response to E2 signaling, miRs may regulate pathways that promote growth, fine tune the cellular response to E2, or provide positive or negative feedback.

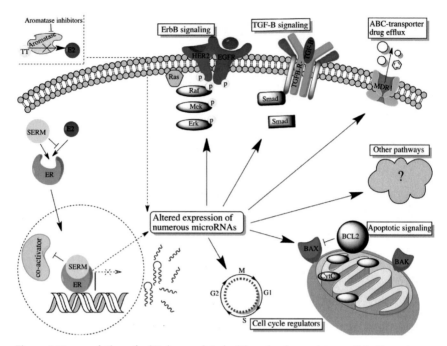

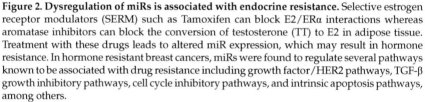

Figure 2. Dysregulation of miRs is associated with endocrine resistance. Selective estrogen receptor modulators (SERM) such as Tamoxifen can block E2/ERα interactions whereas aromatase inhibitors can block the conversion of testosterone (TT) to E2 in adipose tissue. Treatment with these drugs leads to altered miR expression, which may result in hormone resistance. In hormone resistant breast cancers, miRs were found to regulate several pathways known to be associated with drug resistance including growth factor/HER2 pathways, TGF-β growth inhibitory pathways, cell cycle inhibitory pathways, and intrinsic apoptosis pathways, among others.

Chapter 2

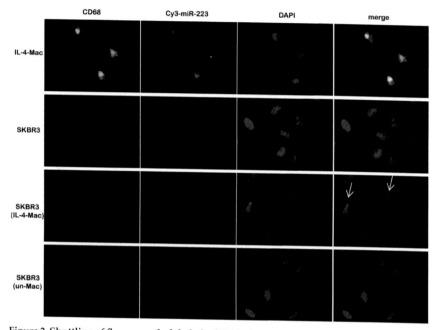

Figure 2. Shuttling of fluorescently-labeled miRNAs from macrophages to breast cancer cells.
SKBR3 cells were cultured alone or co-cultured with IL-4-activated (IL-4-Mac) or unactivated
macrophages (un-Mac) that were pre-transfected with Cy3-miR-223. Both macrophages and
SKBR3 cells were then stained using the macrophage marker CD68 followed by an Alexa
Fluor 488-conjugated secondary antibody. DAPI was used to visualize nuclei. Fluorescence
was observed by fluorescence microscopy. Arrows indicate the Cy3 signal in SKBR3 cells, and
images are shown at 1,000X magnification.

Chapter 5

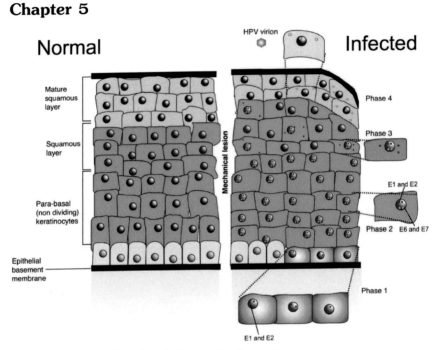

Figure 1. HPV viral cycle and cervical cancer development.

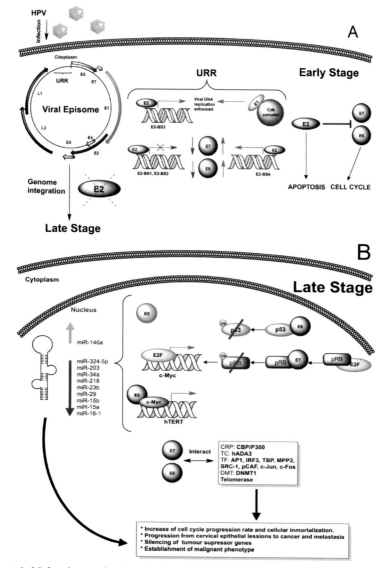

Figure 2. Molecular mechanisms induced by HPV early-expressed proteins. (A) While HPV replicates as an episome, E1 and E2 are actively expressed, as they are essential for viral genome replication and viral cycle completion. E2 is a sequence-specific transcription factor which regulates E6 and E7 expression rate; depending on occupancy of E2-BSs sites. (B) Human papilloma virus E6 and E7 interact with nuclear proteins such as transcription factors (TF), chromatin remodelers (CRP), transcriptional co-activators (TC) and DNA methyl-transferases (DMT) to influence gene expression and miRNA profile, which ultimately affects cellular processes leading to malignant phenotype and tumoral progression.

Chapter 6

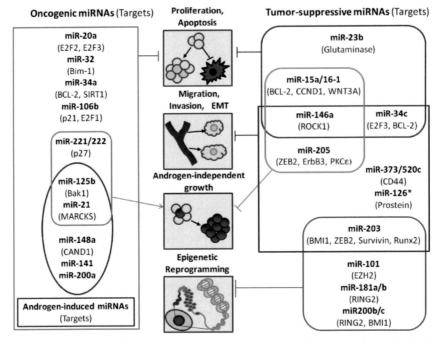

Figure 1. The targets and functions of oncogenic, tumor-suppressive, and androgen-induced miRNAs in prostate cancer. Oncogenic miRNAs enhance the cancer cell phenotype (cell proliferation and apoptosis resistance). Among the oncogenic miRNAs, miR-125b, miR-148a, miR-141, and miR-21 are induced by androgens, whereas miR-221/222, miR-125b, and miR-21 are involved in androgen-independent growth. miR-23b, miR-15/16-1, miR-34c, and miR-146a are known as tumor-suppressive miRNAs that decrease cellular proliferation and increase apoptosis. miR-34c, miR-146a, miR-373/520c, miR-126*, miR-205, and miR-203 suppress migration and invasion. miR-15a/16-1, miR-146a, and miR-205 are involved in androgen-independent growth. miR-203, miR-101, miR-181a/b, and miR-200b/c attenuate cancer cell phenotypes through epigenetic mechanisms.

Chapter 8

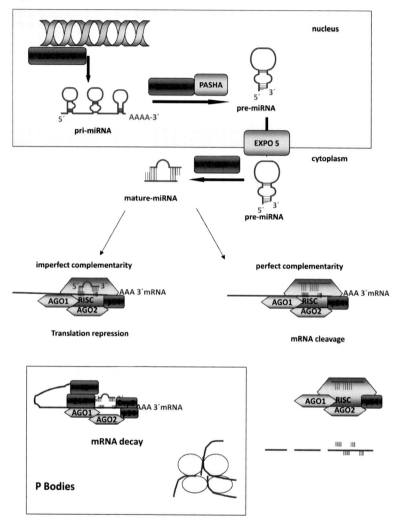

Figure 1. miRNAs biogenesis pathway. miRNAs genes are transcribed by RNA polymerase II in the nucleus to form large pri-miRNAs transcripts. These pri-miRNAs transcripts are processed by the RNase III enzyme Drosha and its co-factor, Pasha, to release the 70-nucleotide pre-miRNAs precursor product. RAN–GTP and exportin 5 transport the pre-miRNAs into the cytoplasm. Subsequently, Dicer processes the pre-miRNAs to generate a transient 22-nucleotide miRNA:miRNA* duplex. This duplex is then loaded into the miRNAs-associated multiprotein RNA-induced silencing complex (miRISC). The mature miRNAs then binds to complementary sites in the mRNA target to negatively regulate gene expression in one of the two ways that depend on the degree of complementarity between the miRNAs and its target. miRNAs that bind to mRNA targets with imperfect complementarity block target gene expression at the level of protein translation. miRNAs that bind to their mRNA targets with perfect complementarity induce target-mRNA cleavage.

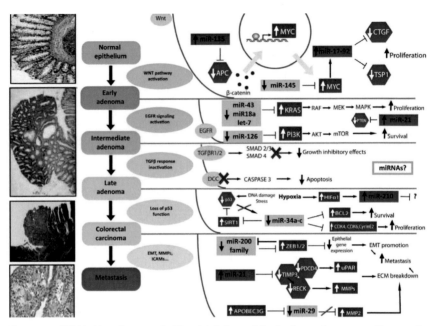

Figure 2. miRNAs' involvement in Vogelstein's model of colorectal cancer pathogenesis. According to the model proposed by Vogelstein different genes are involved in each step of the progression from normal mucosa to metastatic CRC. The knowledge of the mechanism of action of miRNAs in CRC adapts the model with the addition of specific miRNAs that modulate selective targets at each step of this model.

Chapter 13

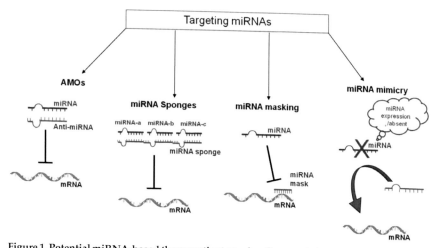

Figure 1. Potential miRNA-based therapeutic strategies. Oncogenic (over-expressed) miRNAs can be knocked-down or inactivated in an effort to restore 'normal' expression levels by anti-miRNA oligonucleotides (AMOs), miRNA sponges or miRNA masking. MiRNAs with reduced expression in the cancer state (tumour suppressor genes) can be replaced by miRNA mimicry. Other replacement techniques include induction of miRNA over-expression by viral vectors and reversal of epigenetic silencing using small molecules. *Reproduced with permission from Pharmaceutical Research.*

Chapter 15

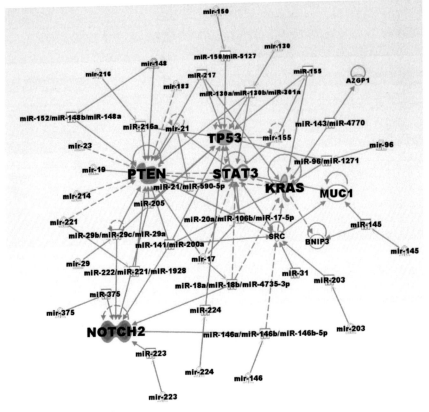

Figure 1. Systems Biology in Understanding Disease Specific miRNA Networks. Ingenuity Network analysis for miRNA interaction networks targeting major genes in pancreatic ductal adenocarcinoma from existing knowledgebase. Pathway was designed using Ingenuity Pathway Analysis (IPA) under defined limits (key entries miRNA and pancreatic ductal adenocarcinoma (PDAC)) incorporating direct (regular lines) and in-direct interactions (dashed lines). The entire miRNA network was found to center around six key pathways (KRAS, TP53, PTEN, STAT3, MUC and NOTCH). Critical miRNAs that formed multi-tier interaction with the six genes of interest were verified in a transgenic PDAC model as shown in Fig. 2.

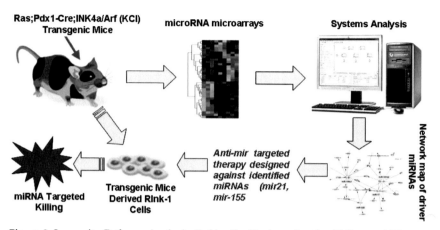

Figure 2. Ingenuity Pathway Analysis Guides the Design of anti-miR Targeted Therapy against Disease miRNA Networks in PDAC Transgenic Model. The miRNA microarrays were run on pooled RNA from Ras;Pdx1-Cre;INK4a/Arf (KCI) transgenic PDAC model followed by IPA analysis for miRNA networks. The transgenic network centers around few key miR's (miR-21, mir-155, miR-220, miR-143 and miR-217 and miR-217) that correlates with those observed from IPA knowledgebase as obtained in Fig. 1 above.

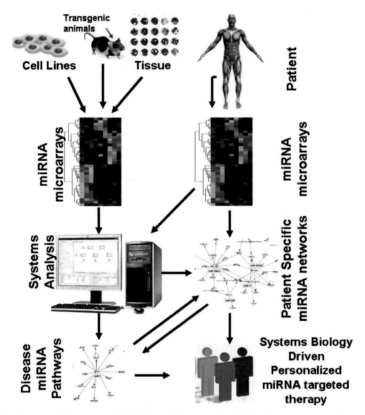

Figure 3. Systems Biology in miRNA Based Personalized Medicine. Advanced integrated technologies can be applied in different ways to derive solid evidence for development of anti- or pre-miR therapy against identified miRNAs. Expression analysis identifying miRNA maps can be performed on RNA isolated from different model systems such as cell lines, corresponding patient derived samples (both active diseased patients or prospective samples such as paraffin imbedded tissue samples, and transgenic animal model systems). The miRNA expression data-sets then can be mapped using miRNA ontology tools such as IPA. Once the critical disease specific of driver miRNAs are identified, they can be correlated with miRNA network maps on RNA isolated from patients undergoing treatment. Depending on the presence or absence of certain miRNAs in the patient anti- or pre-miRNA therapy can be designed for beneficial clinical outcome.